GeoWorld NSW 7

Foreword by
Tim Costello, World Vision

Susan Bliss, Greg Reid, Lorraine Chaffer

Series editor: Susan Bliss

This edition published in 2021 by

Matilda Education Australia, an imprint
of Meanwhile Education Pty Ltd
Melbourne, Australia
T: 1300 277 235
E: customersupport@matildaed.com.au
www.matildaeducation.com.au

First edition published in 2016 by Macmillan Science and Education Australia Pty Ltd

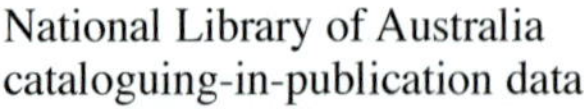

National Library of Australia
cataloguing-in-publication data

Creator: Bliss, Susan, author.
Title: Geoworld NSW 7 / Susan Bliss, Greg Reid, Lorraine Chaffer.
ISBN: 978 1 4586 6277 4 (paperback)
Target Audience: For secondary school age.
Subjects: Geography – Study and teaching – New South Wales.
New South Wales – Geography.
Other Creators/Contributors: Reid, Greg, 1955-, author; Chaffer, Lorraine, author

Dewey number: 910.71

Editor: Emma de Smit
Illustrator: Guy Holt
Cartographer: Laurie Whiddon
Cover designer: Dim Frangoulis
Text designer: Norma van Rees
Production control: Katherine Fullagar
Photo research and permissions clearance: Vanessa Roberts
Typeset in ITC Stone Serif Medium 10/13.5 pt by Norma van Rees
Cover image: Corbis/ALEX VIDAL BRECAS
Indexer: Fay Donlevy

Printed in Australia by Pegasus Media & Logistics

December 2023

Internet addresses
At the time of printing, the internet addresses appearing in this book were correct. Owing to the dynamic nature of the internet, however, we cannot guarantee that all these addresses will remain correct.

Warning: It is recommended that Aboriginal and Torres Strait Islander Peoples exercise caution when viewing this publication as it may contain images of deceased persons.

Contents

ISBN 978 1 4586 6277 4

ISBN 978 1 4586 6277 4

ISBN 978 1 4586 6277 4

ISBN 978 1 4586 6277 4

ISBN 978 1 4586 6277 4

Foreword

The new Australian Geography Curriculum presents a dynamic, contemporary and engaging approach to learning about the world of today and the world of the future. It offers teachers and students an exciting opportunity to understand the forces shaping the interconnected, globalised world of the 21st century.

Geography explores how and why places have their particular environmental and human characteristics, and how and why our world is changing. Importantly, the new curriculum uses an inquiry approach to equip our young people to analyse and evaluate to what extent this change is just and sustainable.

Geography has a unique role to play. No other learning area in the school curriculum draws on both natural and social sciences in such an integrated way. Major issues like climate change, migration, trade, urbanisation and global inequality are all addressed. The skills and insights students gain, and the global perspective it inspires, are invaluable.

Good global learning—learning about the world as it really is, in all its complexity and with all the challenges it faces—helps young people to develop resilience in the face of complex events. It can also nourish both students' emotional and psychological wellbeing and their development as informed and active citizens.

Geography really matters. Some will remember the story of Tilly Smith, an English primary school student who saved a hundred fellow tourists from the 2004 Indian Ocean tsunami when she recalled her geography lesson on earthquakes and tsunamis. She urged her family to get off Maikhao Beach in Thailand after seeing the tide rush out and boats on the horizon begin to bob violently. Tilly's parents alerted others and they cleared the beach just in time. It was one of the few beaches where no-one was killed or seriously injured. Tilly gave the credit to her geography teacher.

Susan Bliss has been an outstanding geography teacher and textbook writer for many years. I am confident this book will stimulate student curiosity and wonder at this incredible world.

Tim Costello
Chief Executive, World Vision Australia
Patron, Australian Geography Teachers Association

ISBN 978 1 4586 6277 4

Introduction

GeoWorld 7, 8, 9 and 10 have been written for Stages 4 and 5 in the New South Wales Syllabus (Board of Studies Teaching and Educational Standards NSW 2015). The four textbooks integrate geographical knowledge, understanding and skills, and values and attitudes. The three stages of the geographical investigation provide students with skills to acquire, process and communicate geographical information.

The *GeoWorld* series explores the characteristics of different places by integrating seven concepts—place, space, environment, interconnection, sustainability, scale and change. Inquiry-based learning is the focus of the books, which incorporate examples of critical literacy, empathy activities and group work.

Key features of the *GeoWorld* series:

- **Key inquiry** questions at the beginning of each unit provide a framework for developing students' geographical knowledge and understanding, and inquiry skills.
- A **quote** on the chapter opener provokes thinking and classroom discussion.
- **Think, puzzle, explore** on each chapter opener encourages deeper thinking within the seven key geographical concepts.
- **Geoskills in focus** highlights essential geographical skills relevant to each chapter.
- **Geovocab** is a glossary of terms relating to relevant concepts.
- **Geoactivities** are divided into Knowledge and Understanding, and Inquiry and Skills. They cater for students with diverse learning abilities, including students with disabilities and gifted and talented students.
- **Geolinks** provide links to relevant ICT resources, including websites, YouTube clips, web 2.0 tools, infographics, interactive skills exercises, photographs, topographic maps, weather maps, graphs, tables, cartoons, Geographic Information Systems (GIS) and satellite imagery.
- **Geoinfo** offers students interesting bite-sized snippets of geographical information.
- **Geothink** at the end of each chapter aims to expand and revise the concepts and skills developed in the chapter.
- **Cross Curriculum Priorities** and **General Capabilities** are integrated within each double-page spread, with icons to assist teachers in planning lessons.

Cross Curriculum priorities

 Asia and Australia's engagement with Asia

 Sustainability

 Aboriginal and Torres Strait Islander histories and cultures

General Capabilities

 Literacy

 Ethical understanding

 Numeracy

 Intercultural understanding

 Information and communications technology (ICT)

 Civics and citizenship

 Critical and creative thinking

 Difference and diversity

 Personal and social capability

 Work and enterprise

- **Studies** from across Australia and around the world are incorporated throughout the books, with a focus on countries in the Asia region.
- **Actual** and **virtual fieldwork** suggestions are integrated within the text, with links to ICT.
- **Geographical tools** such as maps, graphs, statistics, spatial technologies and visual representations are incorporated within the books.
- **Objectives** and **Outcomes**, **Stage Statements** and **Assessment Tasks** ('for' learning, 'as' learning, 'of' learning) assists students to meet the syllabus requirements.
- A curriculum map is available at www.macmillan.com.au/secondary. Go to Teacher Support.

The desirable outcome of the *GeoWorld* series is for students to develop curiosity and wonder about the diversity of the world's places, peoples, cultures and environments, and explore them from a variety of perspectives.

Dr Susan Bliss
Series Editor, author

ISBN 978 1 4586 6277 4

unit 1 Landscapes and landforms

Trolltunga in Hordaland Fylke, Norway

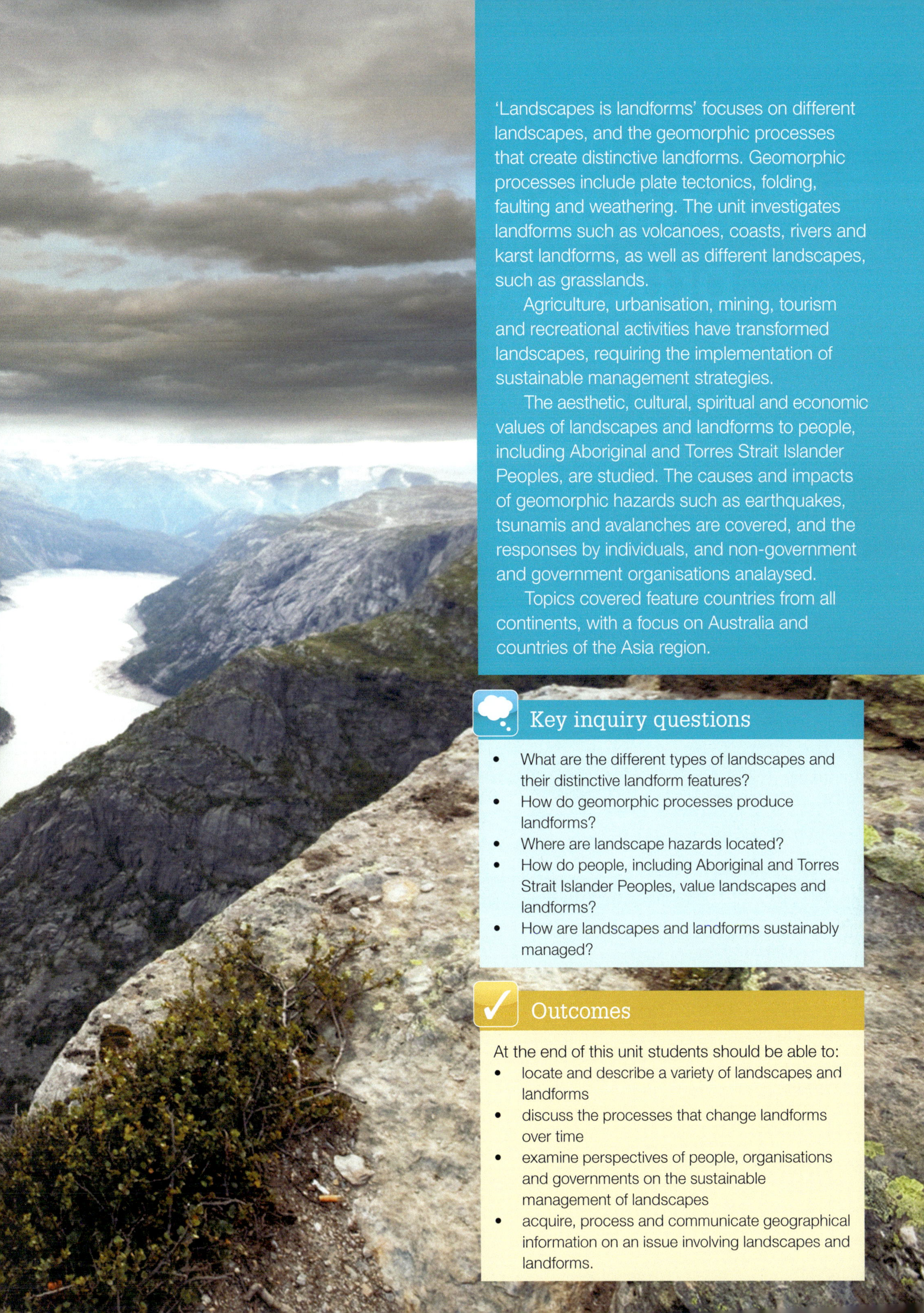

'Landscapes is landforms' focuses on different landscapes, and the geomorphic processes that create distinctive landforms. Geomorphic processes include plate tectonics, folding, faulting and weathering. The unit investigates landforms such as volcanoes, coasts, rivers and karst landforms, as well as different landscapes, such as grasslands.

Agriculture, urbanisation, mining, tourism and recreational activities have transformed landscapes, requiring the implementation of sustainable management strategies.

The aesthetic, cultural, spiritual and economic values of landscapes and landforms to people, including Aboriginal and Torres Strait Islander Peoples, are studied. The causes and impacts of geomorphic hazards such as earthquakes, tsunamis and avalanches are covered, and the responses by individuals, and non-government and government organisations analaysed.

Topics covered feature countries from all continents, with a focus on Australia and countries of the Asia region.

Key inquiry questions

- What are the different types of landscapes and their distinctive landform features?
- How do geomorphic processes produce landforms?
- Where are landscape hazards located?
- How do people, including Aboriginal and Torres Strait Islander Peoples, value landscapes and landforms?
- How are landscapes and landforms sustainably managed?

Outcomes

At the end of this unit students should be able to:

- locate and describe a variety of landscapes and landforms
- discuss the processes that change landforms over time
- examine perspectives of people, organisations and governments on the sustainable management of landscapes
- acquire, process and communicate geographical information on an issue involving landscapes and landforms.

chapter 1

Distinctive landforms and geomorphic processes

'… seen from above, landscapes are made up of mountains and watercourses. Just as a transparent model of the human body consists of a framework of bone and a network of arteries, the Earth's crust is structured in mountain ridges, rivers, creeks, and gullies.'

Stefan Dech and Reinhold Messner

Geovocab

aquifers: rocks and sediments that contain groundwater

area references (AR): four-figure coordinates consisting of a two-figure eastings reading and a two-figure northings reading

catchment area: area of land where precipitation drains into river drainage basin

coastal landforms: landforms that are part of the coastal landscape, such as beachs and cliffs, formed by waves and wind

contour interval: difference in height between successive contour lines on a map

contours: isolines on a map that join all places with the same height above sea level

coral reefs: biological landforms created by coral polyps

cross-section: side-on view

distributaries: small rivers that break off from the main river in a delta

geomorphic processes: processes including plate tectonics and physical and chemical weathering, such as erosion, transportation and deposition, that form landforms

grid references (GR): six-figure coordinates consisting of a three-figure eastings reading and a three-figure northings reading

karst: distinctive topography formed when rocks are dissolved by surface water or ground water

mountains: landforms greater than 600 m that were created by tectonic activity

plains: level, gently undulating surfaces, on the coast, inland or sea floor

plateaus: large, relatively level areas rising more than 450 m above the surroundings

sinkholes: closed surface depressions draining underground in a karst landscape

speleothems: deposits of calcium carbonate, formed in caves by chemical precipitation from drips or thin films of water

ISBN 978 1 4586 6277 4

Horseshoe Bend meander on the Colorado River, Arizona, USA

spot heights: drawn on maps to show the exact height in metres above sea level of a certain place (e.g. mountain peak)
stalactite: speleothem that hangs downwards from a roof or wall of a cave
stalagmite: speleothem that projects vertically upwards from a cave floor
tectonic activity: movement in Earth's crust
terrestrial: related to land
transect: line drawn from one point to another on a topographic map
trig station (or trig point): fixed surveying station established using trigonometrical measurement
vertical exaggeration (VE): relationship between the horizontal scale and the vertical scale on a cross-section
wetlands: mangroves, swamps, billabongs or marshes permanently or temporarily covered with fresh, salt or brackish water
World Heritage site: special area of universal value internationally protected by UNESCO

ISBN 978 1 4586 6277 4

A diversity of landforms is located at different places across Earth. These landforms have distinctive features, and include karst and submarine landforms, coasts, coral reefs, rivers and wetlands. As these landforms are dynamic and vary in scale, an investigation into their geomorphic processes is important for sustainable management.

Think, puzzle, explore

- **Place** Where are different landforms located?
- **Space** What is the spatial distribution of karst landforms?
- **Environment** What are the impacts of humans on coastal landforms?
- **Interconnection** How has technology improved our knowledge about submarine landforms?
- **Sustainability** How can wetlands be managed sustainably?
- **Scale** How does the Snowy Mountains in Australia differ to the Himalayan Mountains in Nepal?
- **Change** How have geomorphic processes changed landforms?

Geo**skills** in focus

- **Identifying** the problems of living on eroded headlands or areas vulnerable to karst subsidence.
- **Collecting and analysing** geographical data on landforms and their changes over time.
- **Communicating** results using a variety of strategies.
- **Reflecting** on the complexity of landforms and the need for their protection.

1.1 Weird and wonderful landforms

Flying in a hot air balloon over Earth offers a bird's-eye view of the wonderful array of different **terrestrial** landforms. Have you every wondered how these landforms were formed? How long have the landforms looked like this? What will they look like in a million years? Will they be under the sea?

Fairy chimneys, Turkey

Cappadocia is located in central Turkey. The present landform began with eruptions of three **volcanoes** about 10 million years ago. The eruptions spread hot volcanic ash over the region, which hardened into a soft stone called tufa.

Over millions of years, wind erosion and water erosion wore away portions of the soft tufa, leaving tall columns of soft tufa with a hard boulder perched on top. These are called 'fairy chimneys'.

As tufa is soft, simple tools could be used to cut homes into rocks. Elaborate multi-level cave cities were built. This spectacular landscape was declared a **World Heritage site**.

Colourful landforms, China

The Danxia landform in Gansu, north-west China, consists of red coloured sandstones and mineral deposits laid down over 24 million years ago in lakes and rivers. The resulting layer cake landform was then buckled into folds by the movement of tectonic plates. Wind and rain then carved towers, valleys and waterfalls. In 2010, several Danxia landscapes were inscribed as a World Heritage site.

Travertine Formations in Mono Lake, California, USA: formed at least 760 000 years ago, the lake is located in a geologically active area with volcanoes located nearby. The lake contains around 255 million tonnes of dissolved salts

Coloured Danxia landforms in Gansu, China

The Wave, Coyote Buttes, Colorado Plateau in Arizona, USA

Laguna Colorada in the Bolivian Andes at 4300 masl: a reddish orange colour is given by micro-organisms that live in its waters beside a salt crust

1.1.1 Flying over Earth in a hot air balloon to view amazing landforms

ISBN 978 1 4586 6277 4

Rock formation at Tassili Ahaggar, Algeria: an artichoke-shaped yardang (feature caused by wind erosion). The pedestal effect is created by wind-blown sand whose cutting force is strongest near ground level

Fairy chimneys in Cappadocia, Turkey

Bungle Bungle in Australia. The Beehives are made up of sandstones and conglomerates deposited in the Ord Basin 375–350 million years ago

Limestone Island in Palua: erosion due to the tide has caused mushroom-shaped formations

The world is filled with countless amazing landforms. Unit 1 investigates their geomorphic processes and resulting geomorphic hazards.

Geomorphology studies the original development of landforms, such as the fairy chimneys in Cappadocia and the towers in Gansu, and how those landforms combine to form landscapes. Geomorphology investigations include reconstructing past landforms and anticipating changes to future landforms.

Geo**info**

More than 160 meteorite craters are on Earth's surface. The energy of these impacts caused some rocks to melt.

Geo**activities 1.1**

Knowledge and understanding

1 *Landforms you see today will differ from those you would see in a million years.* Explain this statement.

2 How were layer cake formations produced in Gansu, China?

Inquiry and skills

3 Refer to 1.1.1.
   a On a world map locate these landforms.
   b Which landform would you like to visit? Give reasons for your answer.
   c Draw a sketch of one of these landforms and label its features after researching information on the internet.

4 Refer to the photo of ballooning over Cappadocia.
   a Describe the different processes required to produce fairy chimneys in Turkey.
   b Why was it easy to build underground cities in this type of landform?

5 Design a collage of weird and wonderful landforms using pictures from the internet. Include the name and location of each place.

6 Research meteor craters in Australia and how they change landforms. Include answers to these questions: What are the causes? Where are they located? How do they change landforms? Why is it difficult to determine the exact number of meteor craters? Evaluate your data and present as a photostory.

7 Investigate the main landforms in New South Wales. Draw a map locating these features.

ISBN 978 1 4586 6277 4

1.2 Landforms: fun and deadly

Holidays are often filled with fun activities that involve landforms, such as skiing in the Swiss Alps, cliff jumping in Hawaii, snorkelling in the Great Barrier Reef, climbing Mt Everest and cruising along the Danube River. However, landforms can be deadly, too. Recent media reports on landforms that have experienced disasters include:

- earthquakes in Nepal that killed over 8000 people
- volcanic ash from Mount Ruang in Indonesia, which affected air flights to and from Sydney
- low-lying islands such as Tuvalu at risk of disappearing under the ocean.

People change landforms for their own use. Hillsides are terraced to grow grapes in Italy and rice in China, and in the Appalachian Mountains in USA mountain tops are demolished for minerals. Chang es to landforms impact on animals. For example, polar bears require a platform of sea ice to reach the seals that they eat. Climate change is causing the sea ice to melt. With less access to food, it is anticipated that two-thirds of polar bears will disappear by 2050.

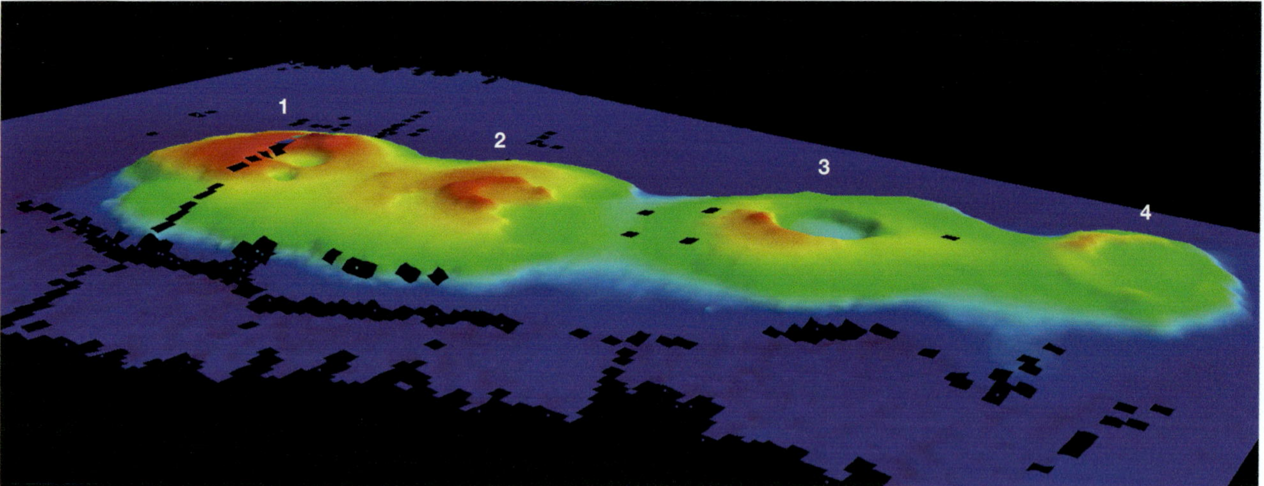

1.2.1 Old landforms: in 2015, four volcanoes about 50 million years old were found 250 km off the coast of Sydney. These old landforms are believed to have been created when geological plates moved and caused Australia to split from New Zealand

1.2.2 Vanishing landforms: Tuvalu is one of many low-lying Pacific islands vulnerable to rising sea levels resulting from climate change. If the global temperature increases by 2°C, Tuvalu will disappear under water. Already, salt water is flooding coastal houses, and killing coconuts—an important source of food, fuel and building material

ISBN 978 1 4586 6277 4

1.2.3 New landforms: Chinese-built islands in the South China Sea. Originally there was a small reef in the South China Sea. Today, satellite images show the Chinese have built an island by dredging the seabed and piling the sand onto the reef. The island now contains airstrips and harbours to support China's military forces. It is a source of contention between China and the USA.

1.2.5 Deadly landforms: in 2014, landslides caused two-thirds of the disaster-related deaths in Indonesia. Approximately 41 million people in Indonesia are exposed to moderate and high landslide danger

1.2.4 Fun landforms: cliff diving

Geo**activities 1.2**

Knowledge and understanding

1 What is meant by old, disappearing, deadly and new landforms?
2 Landforms are always changing. What does this mean?

Inquiry and skills

3 In groups collect two media articles on changing landforms in the last year. Summarise the articles as a TV report.
4 Design a collage linking sport and recreation to landforms.

ISBN 978 1 4586 6277 4

1.3 Different shapes and sizes

A landform is a feature on the Earth's surface. It can be above the ground or below the sea. The highest landform is Mt Everest, in Nepal, measuring 8848 metres above sea level, and the deepest landform is the Mariana Trench in the Pacific Ocean, measuring 10994 metres below see level.

Landforms are constantly changing: some are created by movements under the ground (**tectonic activity**), while others are shaped by water and wind above the ground. After the devastating earthquake in Nepal in 2015, the Himalayan Mountains dropped around one metre.

The four main landforms are **mountains**, hills, **plateaus** and **plains**. Minor landforms include valleys, rivers and waterfalls.

If you look at the first page of this chapter, the photograph shows Horseshoe Bend, which was carved by the Colorado River in Arizona, USA. It took 6 million years for the Colorado River to carve the Grand Canyon.

1.3.1 Different major landforms

Landforms

Mountains

Mt Everest (8848 m) is the world's highest continental mountain

Features

- Higher than 600 m
- Jagged profile
- Small summit
- Cover about 25% of Earth's land surface

Hills have similar features to mountains but are under 600m.

Examples

- Ranges: Himalayas, Alps, Rockies, Andes, Urals, Caucasus, Atlas Mountains, Transantarctic, Australian Alps, Southern Alps (NZ), Drakensberg (South Africa)
- Mountains: Mount Fuji (Japan), Ben Nevis (UK)

Plateaus

Blue Mountains, New South Wales

Features

- Higher than 450 m
- Flat summit
- Steep sides
- Also called tablelands
- Cover about 45% of Earth's land surface

The Blue Mountains in Australia are dissected plateaus, eroded by rivers. These landforms have been incorrectly called mountains. If you look at the skyline it is fairly even and the tops of surrounding landforms are almost the same height.

Examples

- Western Plateau (Australia)—covers two-thirds of the continent and is more than 500 million years old
- Tibetan Plateau (south-west China)—world's largest and most extensive
- Deccan Plateau (India)

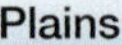

Plains

West Siberian plain

Features

- Flat or gently undulating
- Formed by erosion of surrounding higher areas, and material is then deposited by rivers, wind or glaciers
- Cover more than 50% of the interior of most continents
- Coastal plains cover less than 1% of Earth's land surface
- Abyssal plains or underwater plains cover more than 50% of the ocean's land surface. They are the flattest places on Earth

Examples

- West Siberian Plain (Asia)
- Great Plains (North America)

ISBN 978 1 4586 6277 4

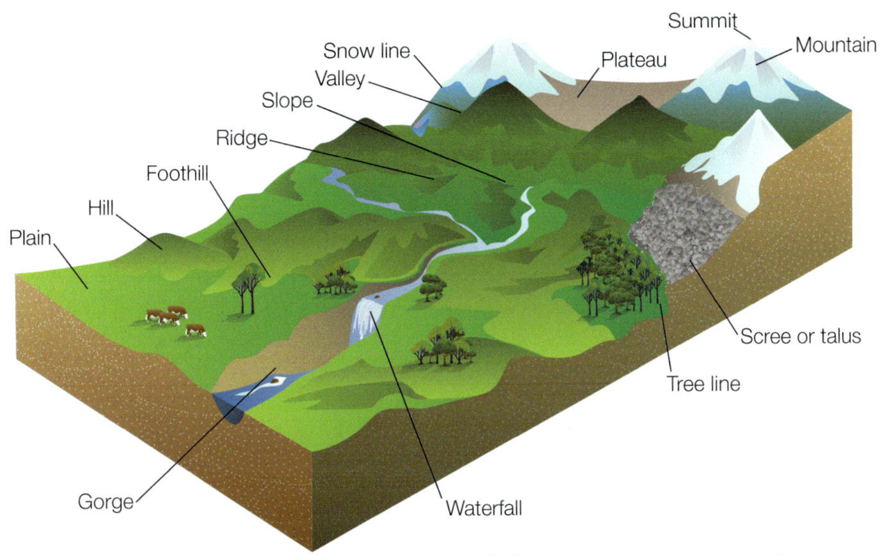

1.3.2 Features of mountain landforms

Scree: broken rock fragments at the bottom of cliffs or steep slopes

Snow line: the place above which snow and ice cover the ground all year

Tree line: the place above which trees stop growing

Gorge: steep valley

Summit: top

Valleys

Between mountains and plateaus there are valleys. A deep river valley is sometimes called a canyon or gorge. River valleys tend to be V-shaped, compared to glacial valleys, which are U-shaped. One famous valley is Death Valley, in the Mojave Desert, California. It is the lowest and driest area in North America, with average summer temperatures exceeding 37 °C. The valley is home to the Native American Timbisha tribe, who have inhabited the valley for over 1000 years. This landform has been designated a National Park.

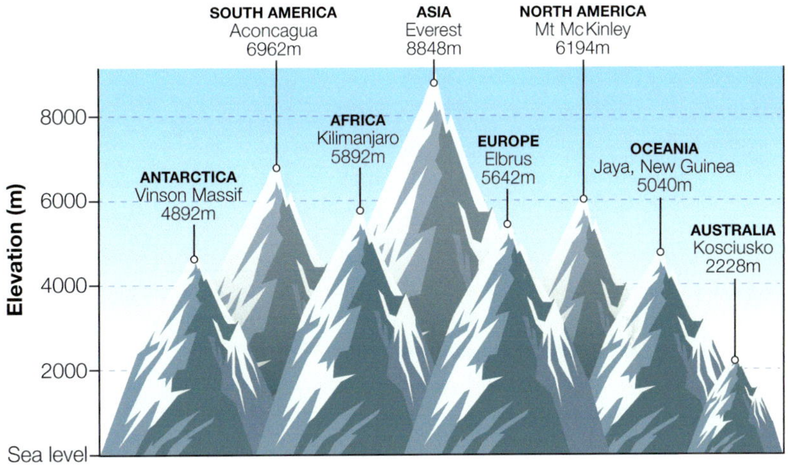

1.3.3 World's largest mountains by continent/region

As a geographer there are many questions concerning landforms to be answered, such as:

- Why are some parts of the Earth mountainous while other parts contain low plains?
- Why are the Himalayan Mountains so high?
- What landform produces the sand in the Sahara Desert?
- Are landform hazards, like landslides, increasing?

Geo**activities 1.3**

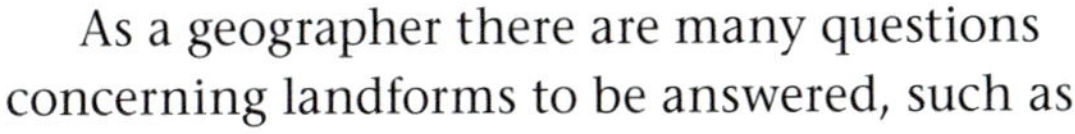

Knowledge and understanding

1 List the four main landforms.

Inquiry and skills

2 Refer to 1.3.1.
   a Compare a mountain with a plateau.
   b What is the flattest place on Earth?
   c The Blue Mountains are wrongly named. Explain this statement.
   d Name two mountain ranges.
   e Describe the Western Plateau in Australia.
   f What is the difference between a hill and a mountain?

3 Refer to 1.3.2.
   a Name two major landforms.
   b Explain two minor landform features such as scree and gorge.

4 Refer to 1.3.3. Calculate the difference in height between the highest mountain in Asia and the highest in Australia.

5 Research the main landforms in Australia. Present as a photo report.

ISBN 978 1 4586 6277 4

1.4 Geomorphic processes and geographical inquiry

Geomorphic processes create different landscapes containing distinctive landforms, such as volcanoes, rivers, glaciers, coasts, deserts, sinkholes and caves. These processes are studied in chapter 2, and are integrated throughout chapters 1, 3, 4 and 5.

Figure 1.4.1 provides an overview of geomorphic processes. The numbers in brackets refer to sections that provide more detailed information on the topics.

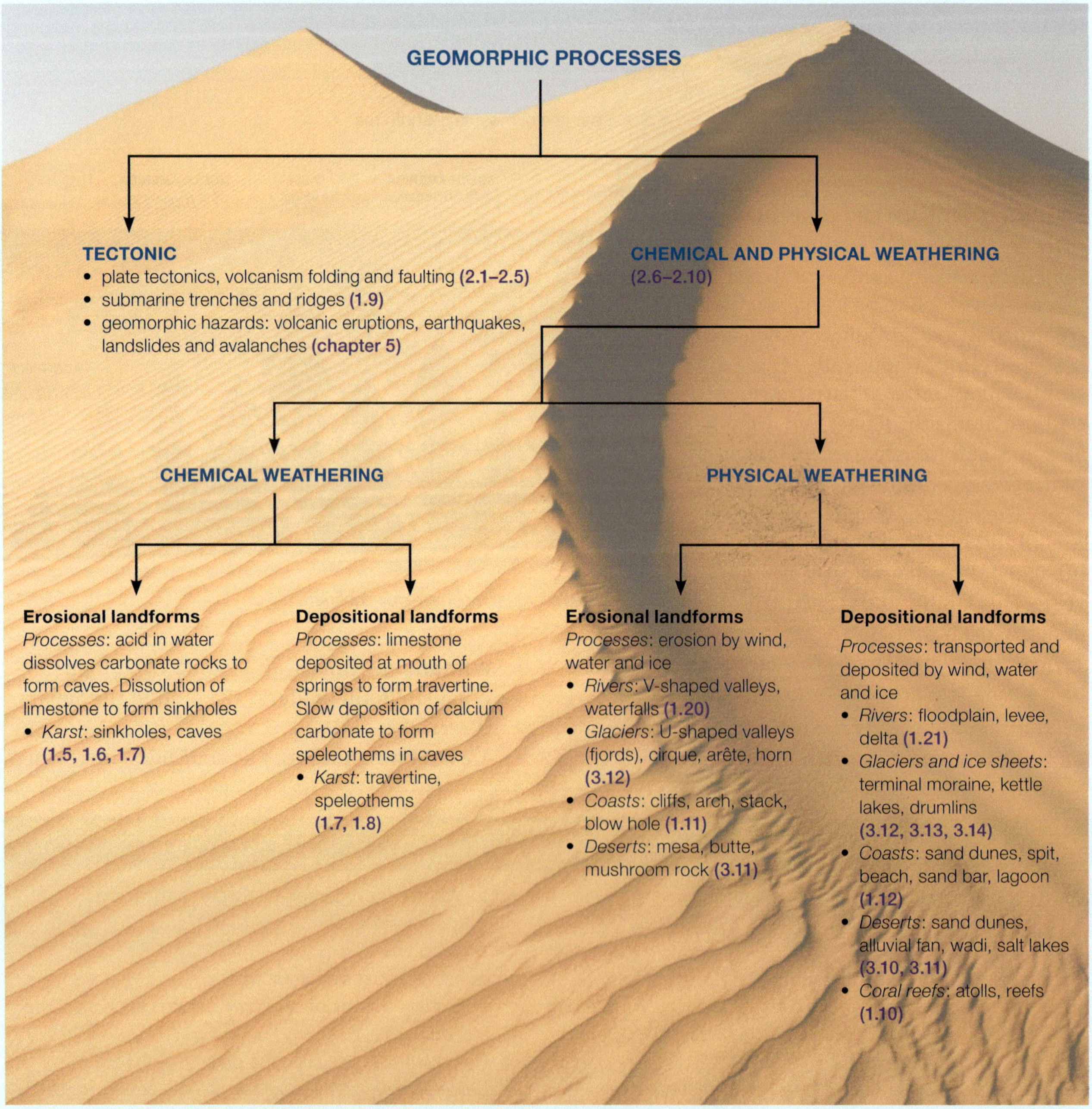

1.4.1 An overview of different geomorphic processes. The photo shows Oman's sandy desert, which was formed by the physical weathering processes of water and wind. The desert also contains high mountains that were once part of a seabed, while limestone caves were caused by chemical weathering

ISBN 978 1 4586 6277 4

Geographical investigation and tools

At the end of Unit 1, students will be able to complete a geographical investigation on different landscapes, including the geomorphic processes that create distinctive landforms.

Geographers would start the inquiry by asking questions such as: Why is there a diversity of landforms on Earth? Why do landscapes change? What will be the impacts of weathering on coastal landforms? How are waterfalls formed? Why are most large volcanoes located around the Pacific Ring of Fire?

Finding answers to these questions involves working through a geographical investigation by acquiring, processing and communicating geographical information. Additionally, the geographical investigation involves using and interpreting geographical tools such as maps and geographic information systems (GIS).

1. Acquire information
- identify an issue
- develop geographical questions to investigate the issue
- collect primary geographical data (e.g. fieldwork, interviews, questionnaires)
- gather geographical information from secondary sources (e.g. internet, journals, newspapers)
- record information

2. Process information
- evaluate data and information for bias and reliability
- represent information in appropriate forms such as maps, graphs, statistics, spatial technologies and visual representation
- interpret data and information gathered
- analyse findings and results
- draw conclusions

3. Communicate information
- communicate results using a variety of strategies
- reflect on the investigation findings
- propose individual or collective actions
- predict expected outcomes
- where appropriate, take action

1.4.2 Steps in a geographical investigation. The photo shows the Arenal volcano in Costa Rica, which was formed by tectonic forces. This is an active stratovolcano

1.4.3 Geographical tools

Maps	Graphs and statistics	Spatial technologies	Visual representations	Fieldwork
• pictorial maps • relief maps • choropleth maps • flowline maps • isoline maps • précis maps • thematic maps • topographic maps	• pictographs • line graphs • column graphs • pie graphs • climate graphs • weather charts • population profiles	• virtual maps • satellite images • global positioning systems (GPS) • geographic information systems (GIS) • augmented reality	• diagrams • photographs • paintings • models • collages • cartoons • infographics • mind maps	• integral and mandatory part of the study of Geography

ISBN 978 1 4586 6277 4

1.5 Karst landforms and processes

The awe-inspiring, beautiful colours and distinctive formations in the Jenolan Caves—nestled in the Blue Mountains World Heritage Area—make the caves a popular site for geographers and tourists alike. 'Legends, Mysteries and Ghosts' is a favoured tour.

Aboriginal Peoples tell of the Jenolan Caves being created in the Dreamtime when Gurangatch, part eel and fish, engaged in a deadly struggle with Mirragan, a native quoll. Gurangatch rested and lay regaining his strength where Blue Lake is now located near the caves.

1.5.2 The Jenolan Caves in the Blue Mountains, New South Wales, are a popular tourist attraction

Unique karst landforms

Hundreds of magnificent **karst** formations are scattered across the world, for instance the Nullarbor Region of Australia, Shan Plateau of China, Atlas Mountains of northern Africa and Belo Horizonte of Brazil. The Phong Nha-Ke Bang caves in Vietnam and Laos comprise 300 caves extending over a vast area of 70 km^2. The area is a national park and a World Heritage site.

Approximately 10% of the Earth's surface is occupied by karst landscapes, and 24% of the world's population depends on karst springs for water supplies. Karst landforms are shaped by surface or ground water, which chemically dissolves the rocks. This geological process occurs over millions of years, resulting in unusual landforms that vary from rolling hills dotted with sinkholes (New Zealand) or jagged hills (Guilin, China), to islets in Ha Long Bay (Vietnam).

Karst process

The process of karst formation is referred to as the 'carbon dioxide cascade'. Precipitation—rain or snow—soaks into the soil, and the water becomes slightly acidic after chemically reacting with carbon

1.5.1 Spatial distribution showing karst regions of the world

ISBN 978 1 4586 6277 4

dioxide in the atmosphere and soil. Precipitation infiltrates downward through the soil and through cracks in the rocks. The acid dissolves carbonate rocks along the cracks, forming underground caves.

The karst process results in formations above and beneath the Earth's surface. Above-ground characteristics include:

- *sinkholes*, also known as dolines or cenotes
- *disappearing streams* that reappear as springs
- *isolated hills* composed of limestone, marble or dolomite, surrounded by alluvial plains (e.g. Vinales Valley in Cuba)
- *limited surface water*, as rivers or lakes drain underground rather than flow along the surface of the ground.

Below the ground, underground drainage systems form, including karst **aquifers** and caves. In caves, a variety of features called **speleothems** appear, such as **stalactites** and **stalagmites**, which are formed by the slow deposition of calcium carbonate and other dissolved minerals.

Karst regions with caves, natural bridges and travertine pools (made from limestone deposited by mineral springs) attract tourists, who enjoy exploring caverns and diving in flooded caves. However, karst landforms support unique ecosystems—karst ecosystems in Croatia host 3500 species of flora plus over 300 species of resident birds, mammals, reptiles, amphibians and freshwater fish. These fragile ecosystems are vulnerable to anthropogenic (human-caused) and climatic stresses.

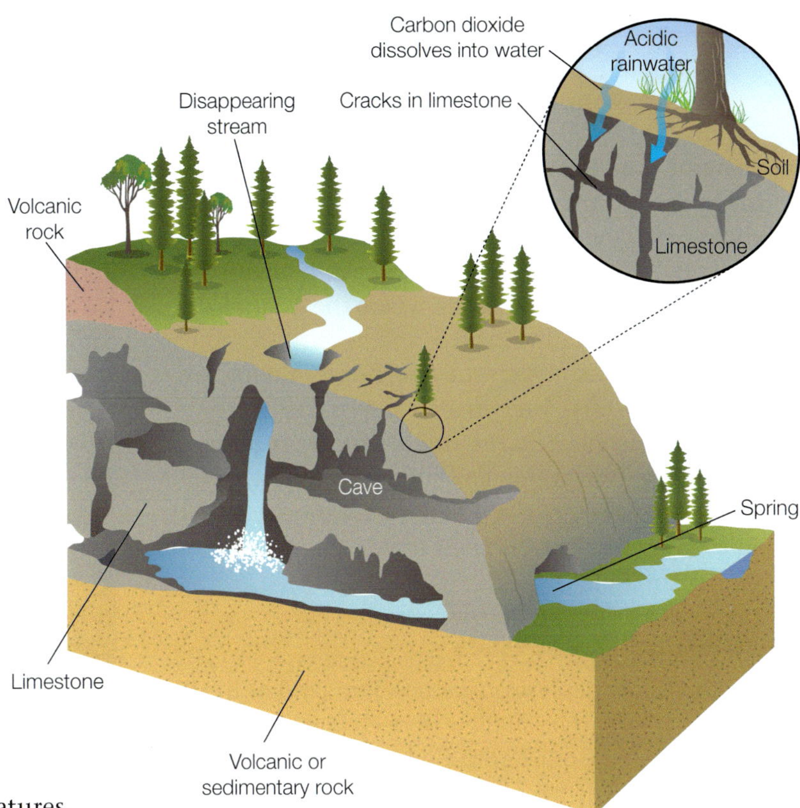

1.5.4 Karst formations: caves, disappearing streams, springs and underground rivers

Geo**info**

The world's largest karst-protected area is in Honduras, covering 2400 km^2 within the Patuca National Park and Tawahka Anthropological Reserve.

Geo**activities 1.5**

Knowledge and understanding

1. Explain two different views of how karst caves were created.
2. Why are karst formations important to humans?
3. List four karst formations.
4. List three karst landforms observed above the ground.

Inquiry and skills

5. Refer to 1.5.1. Explain why karst formations are located in most countries. Provide examples.
6. Refer to 1.5.2 and 1.5.3. Compare the different karst landforms above and below ground.
7. Refer to 1.5.4.
   a. Describe the karst process from precipitation to an underground spring.
   b. What features on land indicate karst processes are operating underground?

1.5.3 Jagged hills in Guilin, China

ISBN 978 1 4586 6277 4

1.6 Sinking world: sinkholes

Imagine that one day you are playing soccer in the backyard and the next day there is a gaping hole there with your house precariously hanging over the edge. This alarming scenario has been known to happen in places like Florida where the ground contains a lot of limestone.

For centuries, depressions in the ground formed by karst processes were used as disposal sites. The Maya on the Yucatan Peninsula, Mexico, regarded them as sacrificial sites and storage areas. Cave diving in sinkholes is a popular recreational activity. The Blue Hole—off the coast of Belize—is the world's largest sinkhole.

1.6.2 The Blue Hole off Belize. Sinkholes that form in coral reefs and islands collapse to enormous depths

Sinkholes: processes

Sinkholes are depressions or holes on land or in the ocean. They can be shallow or deep and small or large, ranging from 1 m to a massive 600 m in both diameter and depth.

Sinkholes are caused by the dissolution of limestone, dolomite, marble or other carbonate rocks. There are two main types:

- *solution sinkholes*—formed slowly when bedrock dissolves, creating a bowl-shaped depression
- *collapse sinkholes*—shallow caves formed by dissolution of the bedrock suddenly collapse.

Sinkholes can cause death and loss of property. Shallow sinkholes frequently fill with water to form lakes or ponds, such as the Winter Park Florida Sinkhole. The Sinkhole Plain in central Kentucky has 5.4 sinkholes per square kilometre over a 153 km^2 area, while Florida has 8 sinkholes per square kilometre over 427 km^2. When several sinkholes join, they form a larger hole known as a poljen.

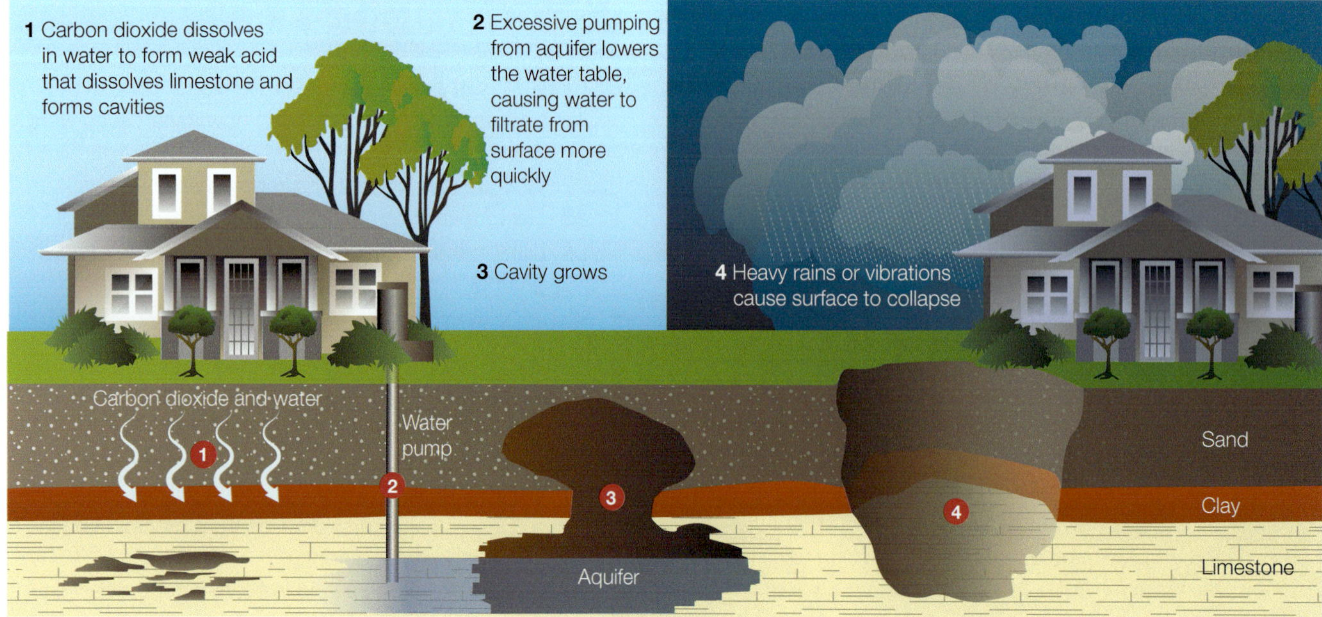

1.6.1 How sinkholes form in karst landscapes

ISBN 978 1 4586 6277 4

Geoinfo

- The name 'cenote' derives from a Mayan word meaning abysmal and deep.
- Winter Park Florida Sinkhole, which formed in 1981, is now a park.

Human activities

Sinkholes are formed by karst processes. Geologist do not consider large holes caused by humans to be sinkholes. However, humans have contributed to the number and size of karst-formed sinkholes in urban areas with broken water pipes and sewer systems plus the over-extraction of groundwater. In Berezniki, Russia, a huge 200 m deep, 80 m long and 40 m wide hole was created by underground mining.

In parts of the world, such as China and south-east USA, people live on top of karst landforms. Some areas have experienced subsiding ground, contaminated groundwater and destruction of cave ecosystems. Signs of sinkholes include cracks on buildings and footpaths, sediment in tap water, and depressions in the ground. To repair sinkholes, building foundations are anchored to bedrock and cement pumped into caverns to reduce the risk of collapsing.

Karst groundwater is a natural resource that feeds waterways and is a source of drinking water. The quality of the groundwater depends on how people use the land and how they protect the quality of groundwater recharge (which can occur naturally via precipitation and runoff, or artificially). An old saying is 'whatever goes up must come down', but in karst areas, whatever goes down comes up—through a cave, a spring or a well.

1.6.3 Aerial view of sinkholes pock-marking the limestone rock in a karst landscape west of Timaru, Pareora, New Zealand

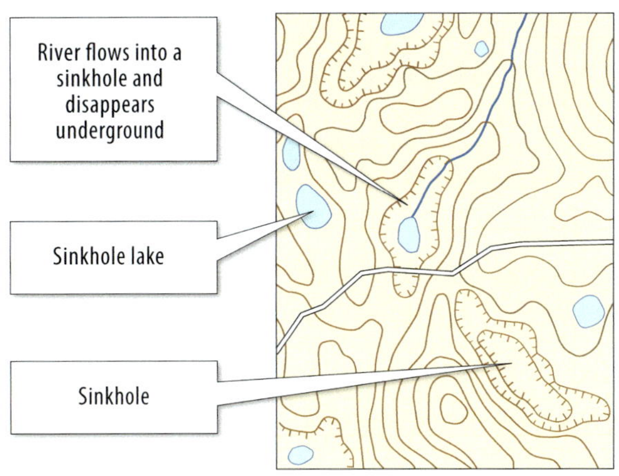

1.6.4 Topographic map section—the contour lines with hatch marks are depressions, indicating the land is lower than the surrounding area and has been subject to karst processes. It is a way of distinguishing a depression from a hill. (See 1.15–1.18 for topographic maps)

Geoactivities 1.6

Knowledge and understanding

1. How have humans used sinkholes?
2. The media called the Berezniki hole a 'sinkhole', but geologists would say that was incorrect. Explain this statement.
3. Discuss how humans have contributed to the increase in the number of sinkholes.

Inquiry and skills

4. Refer to 1.6.1.
   a. What is a sinkhole?
   b. Describe the formation of a collapse sinkhole.
   c. How can sinkholes be repaired and reduced in number?
5. Refer to 1.6.2. Describe the process of the formation of this sinkhole, and its problems and advantages.
6. Refer to 1.6.3 and imagine this was your property. Suggest strategies to rectify the problem.
7. Refer to 1.6.4.
   a. How can you distinguish between a depression in the ground and a hill?
   b. What happens when a river meets a sinkhole?
8. Design a tour called 'See the world's sinkholes'. Include an annotated travel map with latitude and longitude as well as the causes and impacts of the sinkhole on the landscape and surrounding communities. Present as a PowerPoint.
9. Explain why an understanding of karst development and the location of karst features is important for town planners.

ISBN 978 1 4586 6277 4

1.7 Going underground: karst caves

Caves form a tiny part of most karst landscapes. Caves made from karst processes are valued as places for recreation and shelter. The Blue Grotto on the island of Capri, Italy, is a limestone sea cave hollowed out by wave action. It attracts more than 250 000 visitors a year. Mammoth Cave National Park in Kentucky, USA, contains 640 km of passageways and is the largest cave system in the world. It is a World Heritage site and an international Biosphere Reserve.

Karst caves support unique ecosystems that have adapted to the dark environments. They are home to glow worms (e.g. Mole Creek Karst National Park, Tasmania) and to hundreds of species of bats around the world.

Speleothems: cave deposits

The breathtaking beauty of karst caves makes recreational caving a popular activity. Speleological Societies visit caves to view the mineral deposits called speleothems.

Speleothems are formed by the slow deposition of calcium carbonate and other dissolved minerals when water seeps through cracks in a cave's bedrock. Over thousands of years the accumulation of these deposits form dripstones (such as soda straws, stalactites, stalagmites and columns), shawls and flowstones, as seen in the Jenolan Caves.

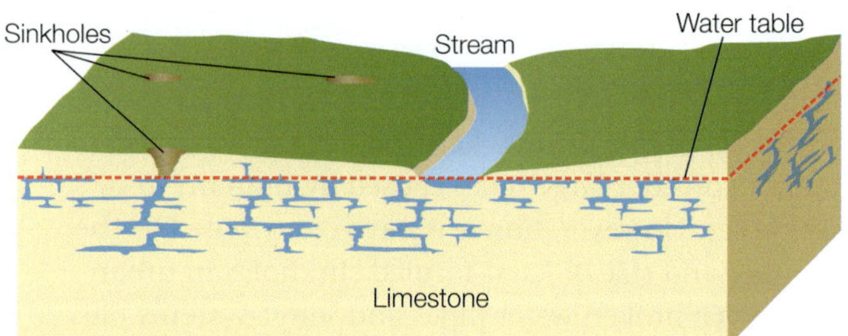

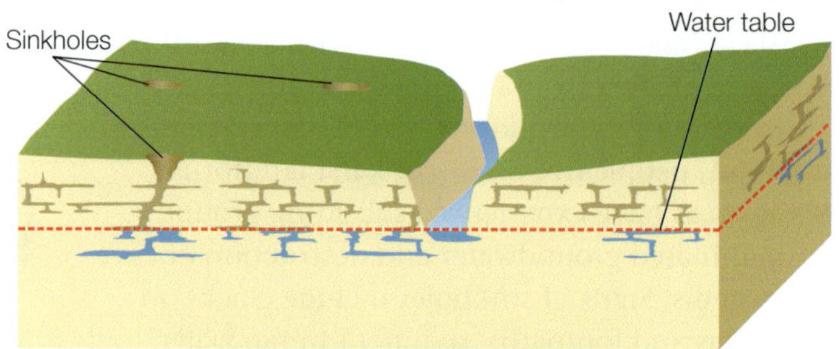

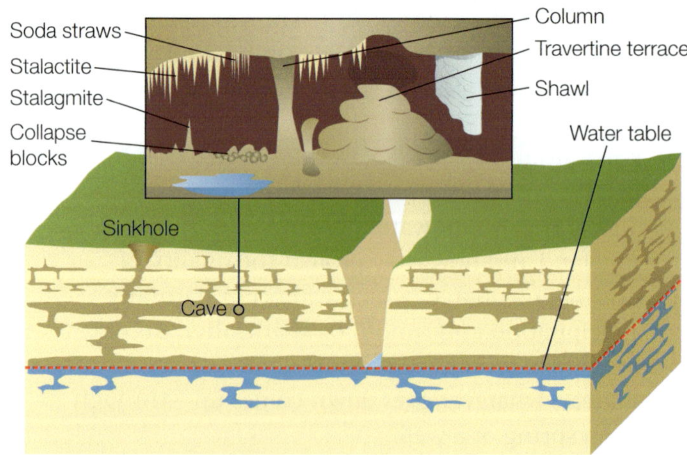

1.7.2 Formation and features of caves

Soda straws are thin-walled hollow formations resembling drinking straws

Stalactites grow downwards from the cave roof. Nearly all stalactites start their life as straws

Stalagmites grow upwards from the cave floor, formed by drops of water from the roof or stalactites overhead

Columns or **pillars** develop when stalactites meet stalagmites

1.7.1 Cave formations: dripstones

ISBN 978 1 4586 6277 4

Shawls are formed when water trickles down a rockface, depositing a narrow strip of calcite. It eventually results in a thin sheet, growing at an angle from the wall. Shawls often contain interesting folds and have richly coloured banding caused by other minerals in the solution, e.g. iron oxide

Flowstones are attractive formations occurring when flowing water leaves a film of calcite. They cover the original rock or mud floor. Sometimes the lower portions hang free, making a fringe or shawl of stalactites

1.7.3 Shawls and flowstones

Karst in Australia: Nullarbor Plain

Karst landforms are thought to cover 4% of Australia, but this could be underestimated due to incomplete geological mapping. The South Australian Limestone Coast contains the Naracoorte Caves—Australia's largest source of megafauna fossils, a World Heritage site and home to significant Aboriginal cultural heritage. Notable tourist locations include:

- *Nullarbor National Park*—the vast Nullarbor Plain is the world's largest limestone karst landscape. It covers an area of 270 000 km^2, extending 2000 km between Norseman and Ceduna. The Nullarbor National Park and Regional Reserve aims to protect the world's largest semi-arid cave karst system. Over 250 caves are recorded with the ocean flowing into some caves and creating blowholes, e.g. the Murrawjini Caves.
- *Nuytsland Nature Reserve*—located in south-east Western Australia, the reserve extends along the coast for 500 km. Within the reserve is Twilight Cove—famous for its 70 m limestone cliffs that overlook the Great Australian Bight.

Increasing tourism has resulted in environmental damage to caves. Project Underground aims to build responsible attitudes towards karst and cave resources and their management. WildCountry—an initiative of the Wilderness Society—aims to involve indigenous custodians in managing karst landforms and their unique habitats.

Geo**activities 1.7**

Knowledge and understanding

1 Name three underground caves.
2 Why are karst caves important?
3 What are speleothems?

Inquiry and skills

4 Refer to 1.7.1.
   a What is the difference between stalactites and stalagmites?
   b Describe how a column is formed.
5 Refer to 1.7.2 and describe how caves form over time.
6 Refer to 1.7.3 and explain the formation of a shawl.
7 Describe the importance of the Nullarbor Plain karst system.
8 Prepare a report on how the Nullarbor karst formations sustained Indigenous Australians.
9 Imagine you are a tour director for a speleological society that wants to visit famous caves around the world. On a map draw the location of 10 caves formed by karst processes and state what is significant about each cave.
10 A plan to commercialise Capri's famous Blue Grotto sea cave by installing floating booms on which businesses could advertise has sparked indignation in Italy. Where is the Blue Grotto located? How was it formed? How is it managed? Describe the advantages and disadvantages of advertising.
11 Caves support a unique community of bacteria, fungi and animals not seen on the surface of the Earth. Research the topic. List unique species and explain how they adapted to their environment.

ISBN 978 1 4586 6277 4

1.8 Above ground: towers, travertine and stone forests

Guilin in China and Halong Bay in Vietnam feature numerous karst towers, and Plitvice National Park in Croatia is the location of travertine lakes and waterfalls. These different types of karst landforms are World Heritage sites.

Karst towers: processes

Some karst landscapes feature steep limestone towers. These spectacular 30–300 m high towers with vertical sides were formed by a combination of tectonic uplift and erosion in tropical wet climates. Submerged karst towers on the coast of Thailand and in the South China Sea form steep limestone islands. These spectacular landforms have been used as locations for movies, including *The Beach* with Leonardo DiCaprio.

Halong Bay in Vietnam boasts seascape karst towers. The landscape consists of 1600 islands and islets, most of which are limestone. The formations have developed over the last 20 million years during a warm, wet climate. The landscape is shared with tourists and 1600 people who live on floating houses and survive by fishing and aquaculture.

Karst landforms in Guilin, China

Guilin is one of China's leading tourist destinations, with more than 5180 km^2 of karst landforms. Limestone peaks resemble giant teeth rising vertically 30–80 m above the green plain.

1.8.2 Above and below ground views of the karst formations in Guilin, China

1 Surface streams lose water to cave systems developing in the limestone plain. Surface drainage is diverted down sinkholes to below the water table.

2 Peaks develop from the land left after erosion by the streams. The cave system gets larger as fast-moving subsurface streams bore through the limestone and the water table drops.

3 Much of the liimestone has eroded past the caves down to a layer of shale. Limestone peaks remain, many fractured with small, waterless caves.

1.8.1 Formation of peaked landscapes in Guilin, China

ISBN 978 1 4586 6277 4

The topography is characterised by karst caves, sinkholes, isolated towers and stone forests.

Karst collapse is a major geological hazard in the area, with more than 200 cases in China. Furthermore, karst aquifers have been polluted from the discharge of industrial waste into caves.

Travertine: processes

Travertine is a form of limestone deposited at the mouth of springs. In Plitvice Lakes National Park in Croatia, travertine has built up over several millennia to form 16 huge dams and waterfalls.

Cascades of natural lakes formed behind travertine dams can be found in Band-e Amir in Afghanistan and Pamukkale, Turkey. Romans mined deposits of travertine for aqueducts, monuments and amphitheatres, and today it is used as paving for patios and garden paths.

Stone forests: processes

Karst stone forests contain a range of pinnacle shapes and colours. A Stone Forest Geopark has been established in Yunnan Province, China, covering an area of 400 km^2. The World Heritage listed Grand Tsingy landscape in Madagascar is the world's largest limestone forest, where high spiked towers of eroded limestone tower over the greenery. Formed 200 million years ago in a lagoon, a thick limestone bed was created. Later tectonic activity elevated the limestone and the sea level fell. Since then the limestone has been eroded by monsoonal rains, creating the dramatic landscape there today.

1.8.3 Travertine hot springs in Pamukkale, Turkey

1.8.4 Grand Tsingy, Madagascar

Geo**info**

Michaelangelo chose travertine as the material for the external ribs of the dome of St Peter's Basilica.

Geo**activities 1.8**

Knowledge and understanding

1 What are karst towers?
2 Name two countries where karst towers are located.
3 Describe the four conditions that allowed the formation of karst towers in Guilin.
4 What are the problems facing karst formations in China?

Inquiry and skills

5 Refer to 1.8.1 and explain how a limestone plain develops from caves to peaks.
6 Refer to 1.8.2 and the internet.
   a If you were visiting Guilin in China what five types of karst landforms might you observe?
   b What is the main river running through the area?
7 Refer to 1.8.3 and describe travertine formations.
8 Design a pamphlet advertising the importance of travelling to karst landforms—above and below the ground—for improved geographical understanding of landforms and landscapes.
9 Reflect on your learning and suggest actions to conserve karst landforms.

ISBN 978 1 4586 6277 4

1.9 Submarine: landforms and processes

Submarine landscapes occupy approximately 71% of Earth's surface. Until recently, they were largely unexplored or unknown. Submarine landscapes mirror similar features found on land, but on a much larger scale. The following are major landforms located in oceans:

- *Continental shelves* are the submerged outer edges of continents. The Australian continent lies on a continental shelf overlain by shallow seas such as the Arafura Sea, Torres Strait and Bass Strait.
- *Continental slopes* occur where continental shelves merge with the ocean's crust.
- *Abyssal plains* are underwater plains on the deep ocean floor. They cover 33% of the ocean floor and are usually located at depths of 4000–5000 m. The South Australian Plain is located in the Indian Ocean at 37°S and 130°E.
- *Mid-ocean ridges* are underwater mountain ranges formed by plate tectonics (see 2.2). They rise into the water when the ocean crust pulls apart, and magma (liquid rock) seeps out. As all mid-ocean ridges are connected around the world, they form a continuous mountain range stretching 65 000 km. This is longer than the Andes Mountains, the longest mountain range on land.

Geoinfo

The vast majority of volcanic rocks ejected onto the surface of the Earth is erupted at mid-ocean ridges.

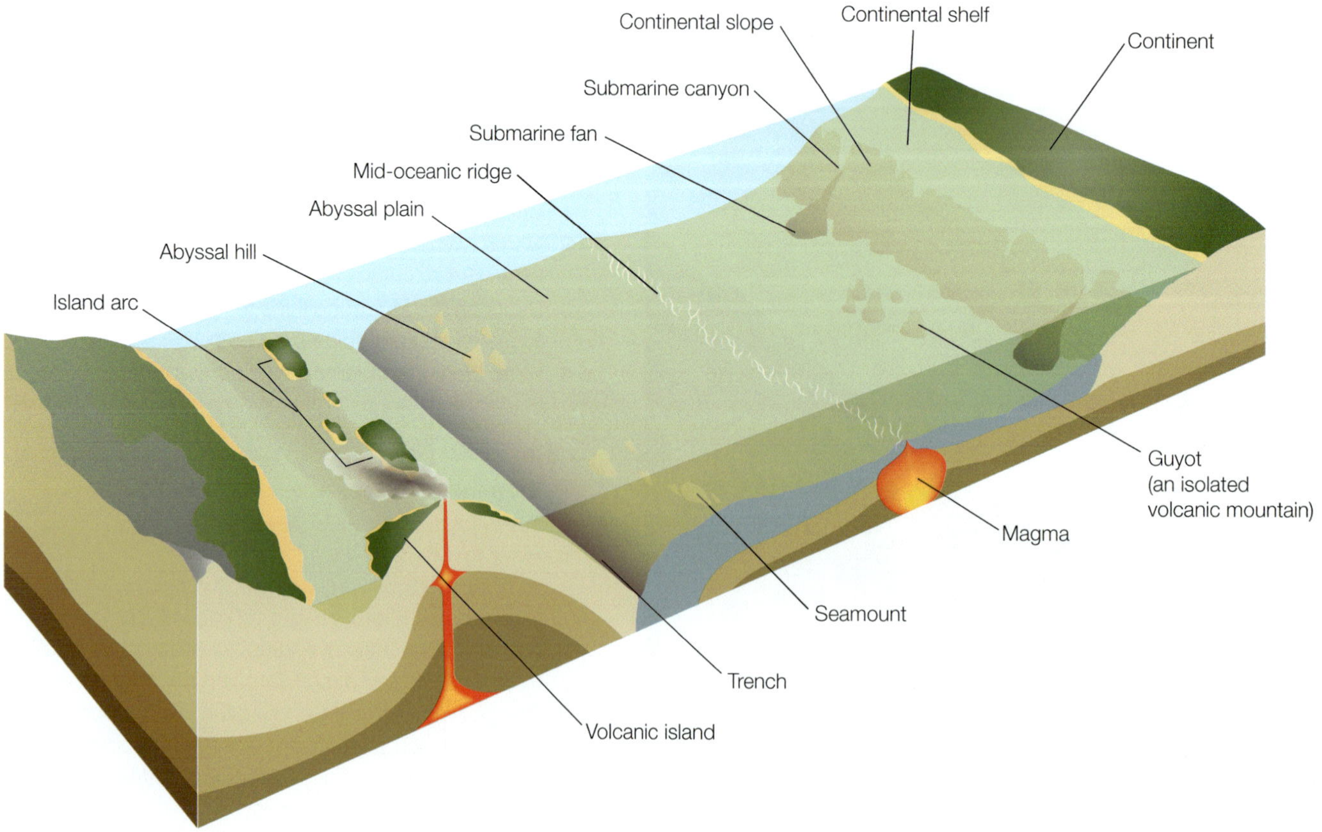

1.9.1 Submarine landforms

ISBN 978 1 4586 6277 4

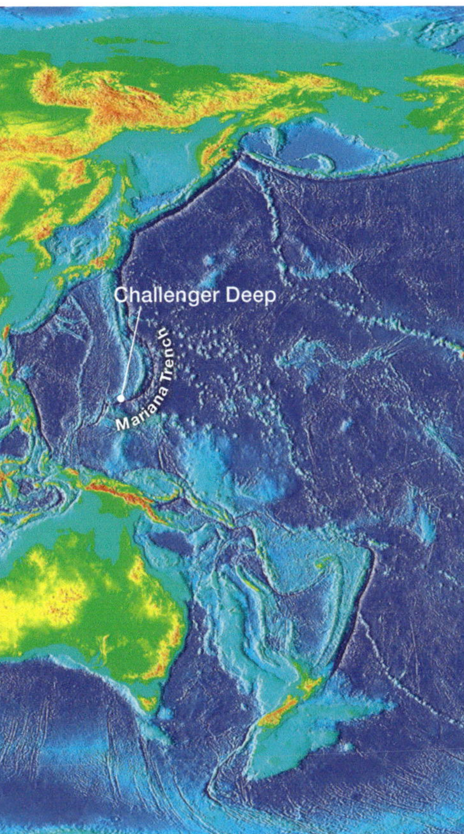

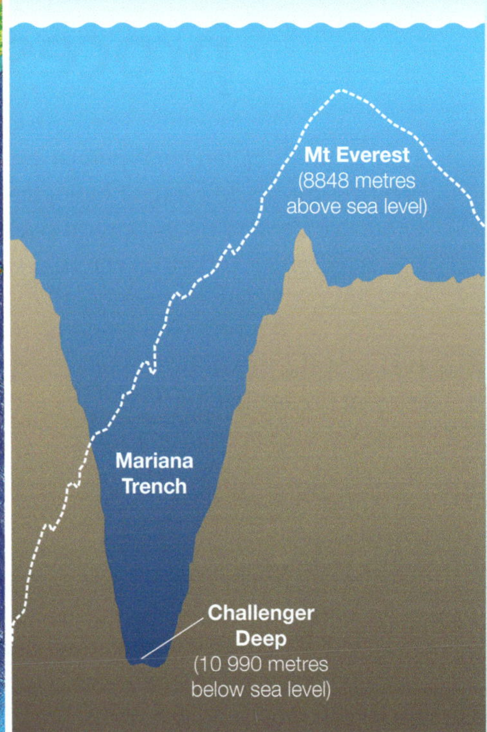

1.9.2 The Deepsea Challenger's dive to the bottom of the Mariana Trench

- *Trenches* are formed in the deepest parts of oceans. The vast majority of trenches (18 of 22) are located in the Pacific Ocean. The Mariana Trench, located in the Pacific Ocean, is the deepest place on Earth at 10 994 m.
- *Seamounts* are submarine volcanoes that rise from the ocean floor. Geologists estimate that there are 100 000 seamounts. They are an important spawning ground for marine life such as sharks and tuna.

Geo**activities 1.9**

Knowledge and understanding

1. How significant are submarine landforms?
2. List one way submarine landforms differ from equivalent continental landscapes.
3. What is the difference between a continental shelf and a continental slope?
4. Describe how mid-oceanic ridges are formed.
5. What are seamounts and how are they formed?

Inquiry and skills

6. Refer to 1.9.1.
   a. List five landforms in the ocean.
   b. Describe the formation of an abyssal plain.
7. Refer to 1.9.2 and secondary sources.
   a. How long is the world's deepest ocean trench?
   b. Undertake further research to describe the type of technology used to investigate the Mariana Trench.
   c. Using Geocontext-Profiler on the internet, draw a topographic profile of Challenger Deep.
8. Investigate the main threats to submarine landscapes (e.g. sea-floor mining, oil and gas drilling and commercial fishing). What can be done to protect these landscapes from overexploitation?
9. Use MapMaker Interactive to explore the landform features of the oceans (see Geolinks).
10. Using digital and spatial technologies, identify the spatial distribuion of marine landscapes. Present your findings as a photo story.

ISBN 978 1 4586 6277 4

1.10 Coral reefs: landforms and processes

Tiny animals called coral polyps are responsible for creating the world's largest biological landscapes—extensive **coral reefs** in tropical oceans. Coral polyps live in colonies and build external skeletons of calcium carbonate (limestone) that over time become coral landforms. Coral reefs occupy only 0.17% of the ocean floor, but they provide a habitat for more than 25% of all marine species.

Corals have specific environmental requirements for growth. They need sunlight and thrive best in shallow waters (2–30m deep). Corals also need clear water and do not survive in muddy waters near rivers. Coral polyps need to be protected from the sediment and nutrients in agricultural and urban runoff. Worldwide, coral reefs are under threat from mining, tourism and climate change.

There are four main types of coral reefs.

Fringing reefs and platform reefs

Fringing reefs form a narrow band between 0.5 and 1 km wide surrounding islands and continents. Ningaloo Reef and continental islands, such as the Whitsundays, have Australia's best fringing reefs.

Coral reefs can be categorised into several types according to their shape and origin.

Fringing reefs

Coral reefs occur adjacent to the coast. Most of the reefs in Indonesia are fringing reefs.

Platform reefs

Reefs are isolated from the land. Small examples can be seen in Japan and the Coral Sea.

Atolls

Reefs occur in circular shape where land is absent in the middle of the circle. This type of reef can be commonly seen in the South Pacific islands.

Barrier reefs

These reefs are similar to fringing reefs, but are much larger at distant offshore, forming a deepening (called a lagoon) between the land and reefs.

1.10.2 Cross section of four types of reefs

1.10.1 Aerial image of Bora Bora, French Polynesia

ISBN 978 1 4586 6277 4

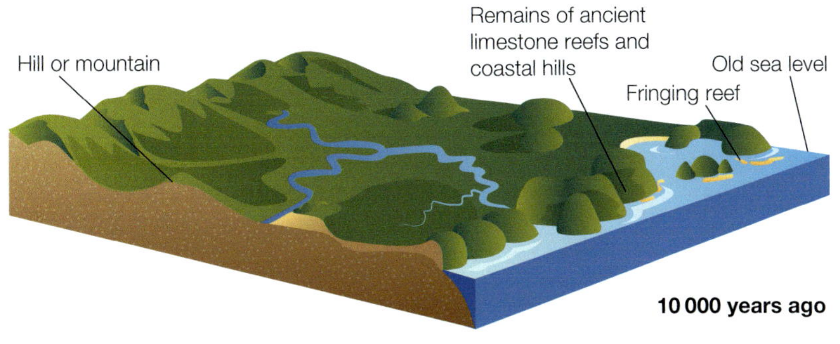

Geoinfo

The Great Barrier Reef is the world's largest living landform, covering about 270 000 km^2 and extending some 2300 km along Queensland's eastern coast.

1.10.3 How the Great Barrier Reef evolved with sea level change

Platform reefs or patch reefs are isolated flat topped reefs with small lagoons. They are common in calm water off Australia's east coast, such as the Great Barrier Reef.

Atolls

Atolls are circular reefs that rise from deep seas and enclose shallow lagoons. Most are located in the tropical Indo-Pacific region. Core drilling has confirmed that atolls form on subsiding volcanic islands. Coral on some atolls has grown up to 1400 m thick over a 40–50 million year period to keep pace with the sinking land. As the reef erodes, calcium-rich sand and reef and shell fragments are deposited in the lagoon.

Barrier reefs

Barrier reefs form long landforms separated from the mainland by a lagoon that can be wider than 20 km. Australia's Great Barrier Reef (GBR) is the world's most famous example. It comprises 2900 individual reefs, which vary greatly in shape and size. The GBR is about 15 million years old in the north (where it is 1–2 km thick) and two million years old in the south (where it is about 120 m thick).

The GBR has not always been in its present form. During the last ice age, the sea level dropped and exposed the coral landforms to weathering and erosion. Rivers cut channels through the reef limestone and the limestone dissolved in some places, forming solution holes. When the sea levels rose, corals recolonised the eroded surfaces to form a thin veneer over the relic surface.

The GBR was listed as a World Heritage site in 1981. The GBR Marine Park Authority protects and manages the marine park and the World Heritage site.

Geoactivities 1.10

Knowledge and understanding

1. How do coral polyps build landforms?
2. How are atolls formed?

Inquiry and skills

3. Refer to 1.10.2 and describe the different types of coral landforms.
4. Refer to 1.10.3 and explain the changes to the Great Barrier Reef during and after the last ice age (around 10 000 years ago).
5. Identify the problems facing coral reefs. Create a PowerPoint slide to explain the solutions to these problems.
6. Research what is being done globally to protect coral reefs. Outline the management strategies used by the GBR Marine Park Authority.
7. Discuss this statement: *Coral reefs are like the modern canary in the coal mine of global warming.*
8. Research the problems of developing coal export harbour facilities inside the GBR. Evaluate your sources for reliability and bias. Present your findings as an oral report (see Geolinks).

ISBN 978 1 4586 6277 4

1.11 Coastal landforms: erosion

In Australia, nine out of ten inhabitants live within 50 km of the coast, so **coastal landforms** are familiar features to most Australians. The majority of Australia's World Heritage sites are situated in coastal zones, such as the Great Barrier Reef, Lord Howe Island, Fraser Island and Shark Bay.

Coastal processes

The coast is a dynamic, narrow contact zone where the land and the sea interact. Coastal landforms are formed by two main processes:

- *erosion* (e.g. cliffs, headlands and blowholes)
- *deposition* (e.g. beaches, dunes and lagoons)

Most of the erosion and deposition in coastal landscapes is caused by wind-generated waves. In addition, coastal landforms are sculptured by tides, currents and sea-level changes over time.

Australian coastal processes

Australia's coastal length is 35 876 km with an additional 23 859 km of island coastlines. Most of the coastal features we see today were formed over the last 6000 years, after the sea level reached its present level. However, there are variations in coastal landforms between places depending on the type of rocks, climate, weather and natural disasters:

- *Rocks*—softer rocks erode faster than harder rocks. Easily eroded rocks, such as clay and shale, tend to form beaches, whereas more resistant rocks, such as limestone and chalk, tend to form steep cliffs and headlands.

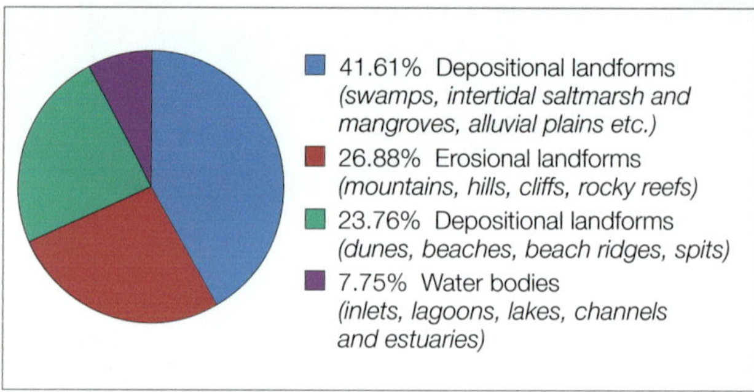

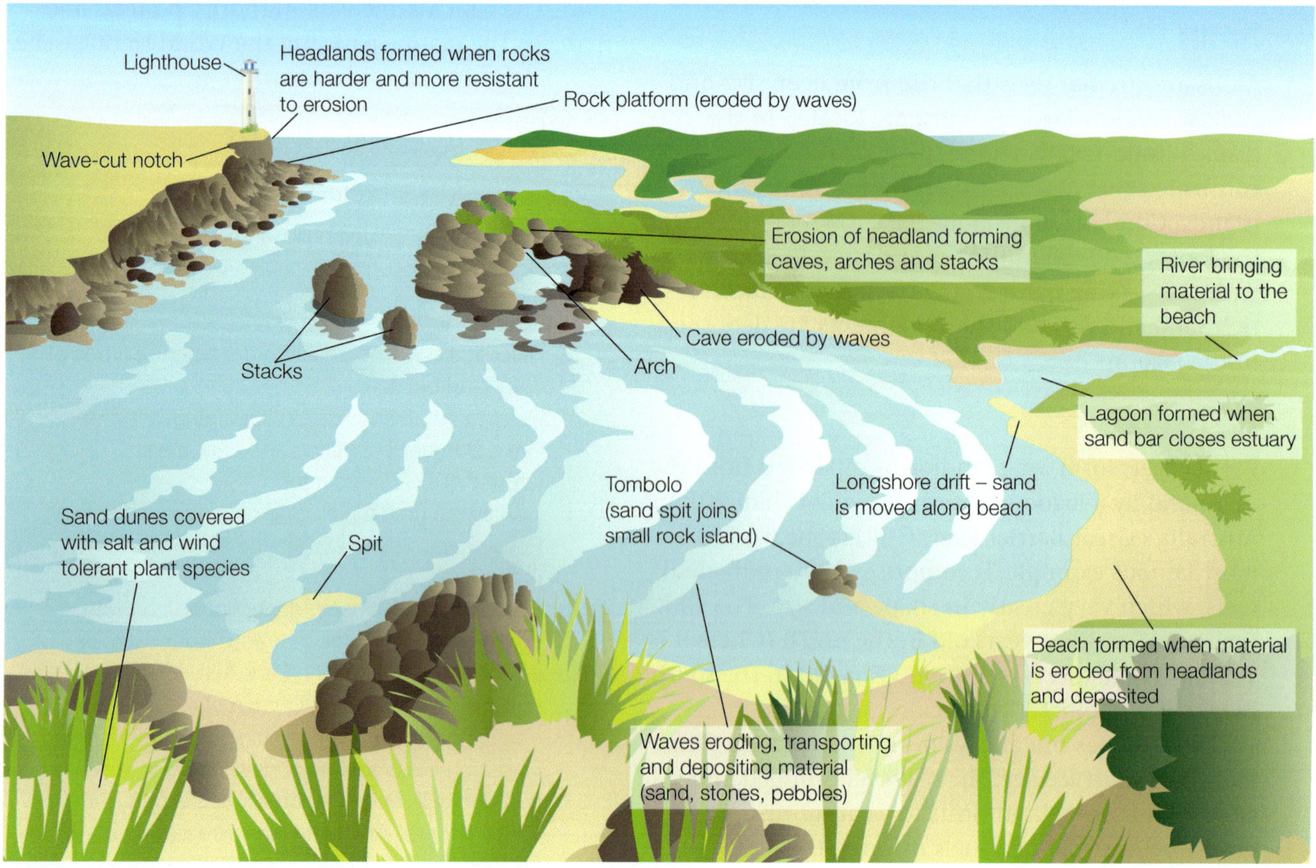

1.11.2 Types of coastal landforms and water bodies in Australia

1.11.1 Coastal landforms

ISBN 978 1 4586 6277 4

- *Climate*—in tropical Australia (north of the Tropic of Capricorn), coastal landscapes are dominated by finer sediments (mud and silt), while in temperate Australia coarser sediments (rocks and sands) dominate.
- *Weather*—during storms or windy weather, high-energy destructive waves erode landforms. Low-energy waves tend to construct landforms by adding sand and material to the beach.
- *Tsunamis and storm surges*—erode landforms.

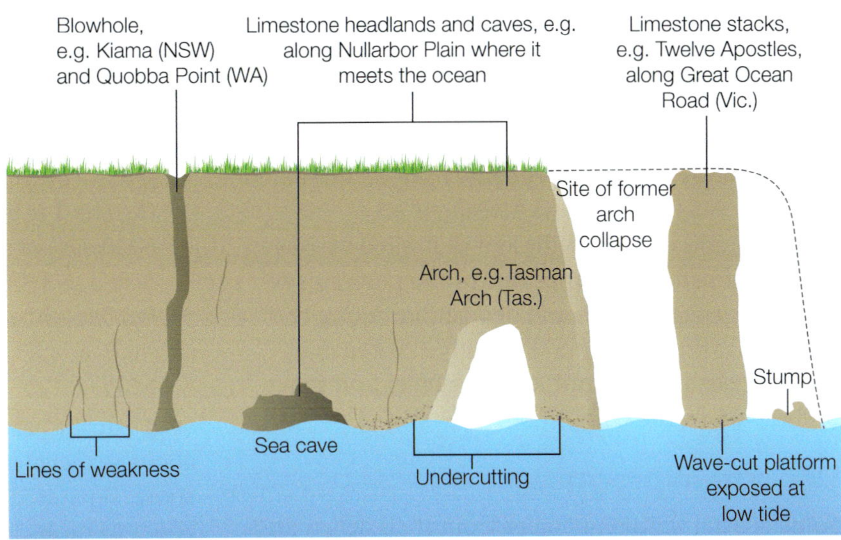

1.11.4 Erosional landform features

Erosional landforms

Coastal erosion occurs when waves, tidal currents and winds wear down and remove sediments. Ultimately, erosion causes the coastline to retreat. The rate of erosion depends on wave and wind energy, which is related to climate. Wave energy and erosion are increased during storms. Erosion produces distinctive coastal landforms, such as sea cliffs, rock platforms, arches, sea stacks, sea caves and blowholes.

Geo**info**

Mainland Australia's coastline, including Tasmania, is almost 37 000 km. The total area covered by Australian coastal landscapes is about 125 000 km^2.

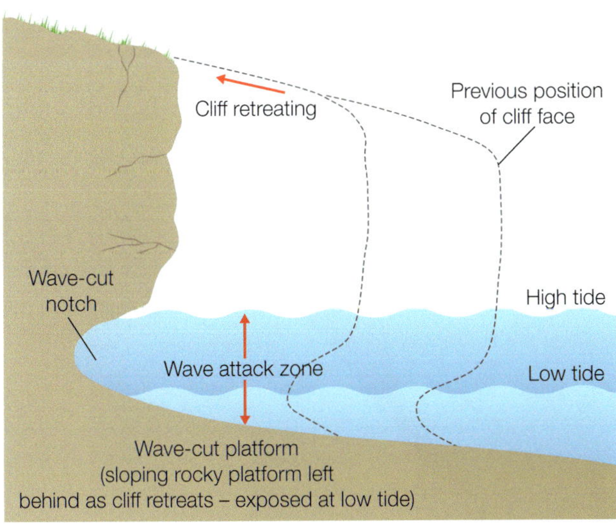

1.11.3 A wave-cut notch and retreating rock platform

Geo**activities 1.11**

Knowledge and understanding

1. Describe the main natural influences on coastal landforms.
2. How are coastal sediments in tropical and temperate Australia different?
3. What is coastal erosion and what distinctive landforms does it produce?

Inquiry and skills

4. Refer to 1.11.1.
   a. List five coastal landforms.
   b. What is the role of waves in forming coastal landforms?
5. Refer to 1.11.2 and state whether the following are erosional or depositional landforms: rocky reefs, cliffs, dunes and spits.
6. Refer to 1.11.3.
   a. Explain how waves erode cliffs and form wave-cut notches.
   b. Describe the processes that form a wave-cut platform.
7. Refer to 1.11.4.
   a. How can a line of weakness in rocks eventually form blowholes?
   b. Describe how the Twelve Apostles were formed.
   c. What are the processes that will cause the Twelve Apostles to disappear in the future?
8. Using Google Earth, choose a section of the Australian coastline and describe the erosional landforms, using geographical concepts.
9. Imagine you were one of the students stranded when the London Bridge Arch near Port Campbell, Victoria, collapsed. Recount your experience and the rescue as a short story.

ISBN 978 1 4586 6277 4

1.12 Coastal landforms: deposition

Over millions of years, sediments have accumulated along the Australian coast. A beach is an accumulation of loose sediment deposited by waves. Beaches constantly change due to the processes of erosion, transportation and deposition. Along the NSW coast there are approximately 720 beaches. Most consist of sand, unlike the pebbles and rocks on beaches in the UK.

Deposition processes

Some coastal sediment comes from shell debris and the erosion of headlands, but most comes from rivers transporting eroded sediments from the land. Sediment ranges from fine clays and silts to sand, gravel and boulders. Waves, tides and longshore currents transport sediment. Deposition occurs when sediment supply is greater than its removal. Sediments move along coasts in two ways:

- *landwards transport*—wind moves sediments to form sand dunes
- *seawards transport*—ocean currents move sediments to form sand bars.

Barrier beaches and islands

There are other depositional landforms that fringe low-lying coastal plains, for example:

- *bars* and *barriers*—off-shore deposits of sediment generally parallel to the coast (e.g. Caloundra bar, Queensland)
- *spits*—continuations of beaches that diverge from the coast (e.g. Noosa, Queensland)
- *tombolos*—where a spit joins an island (e.g. Palm Beach, NSW)
- *lagoons*—formed behind depositional features, such as spits and beaches (e.g. Dee Why Lagoon, NSW).

A barrier beach is a sand ridge that rises slightly above sea level and runs roughly parallel to the shore, separated by a lagoon. They serve as buffer zones to protect the coast from severe storms.

In the USA, barrier systems are under increasing pressure from human development, with more than 70% already developed. Cities such as Miami and Atlantic City are located on barrier islands, and summer beach houses occupy many barrier beaches.

Wave processes

Waves are the main agent of change in coastal landforms. The waves are generally formed by wind blowing over the ocean. The size of the wave depends on wind velocity, the distance the wind travels over the ocean, and the duration of time it blows. Tropical cyclones contribute to destructive or erosive waves that batter the Australian shoreline.

When waves approach the shore, they break in different ways depending on the ocean topography:

- *spilling*—rise and break gently over a distance of several metres

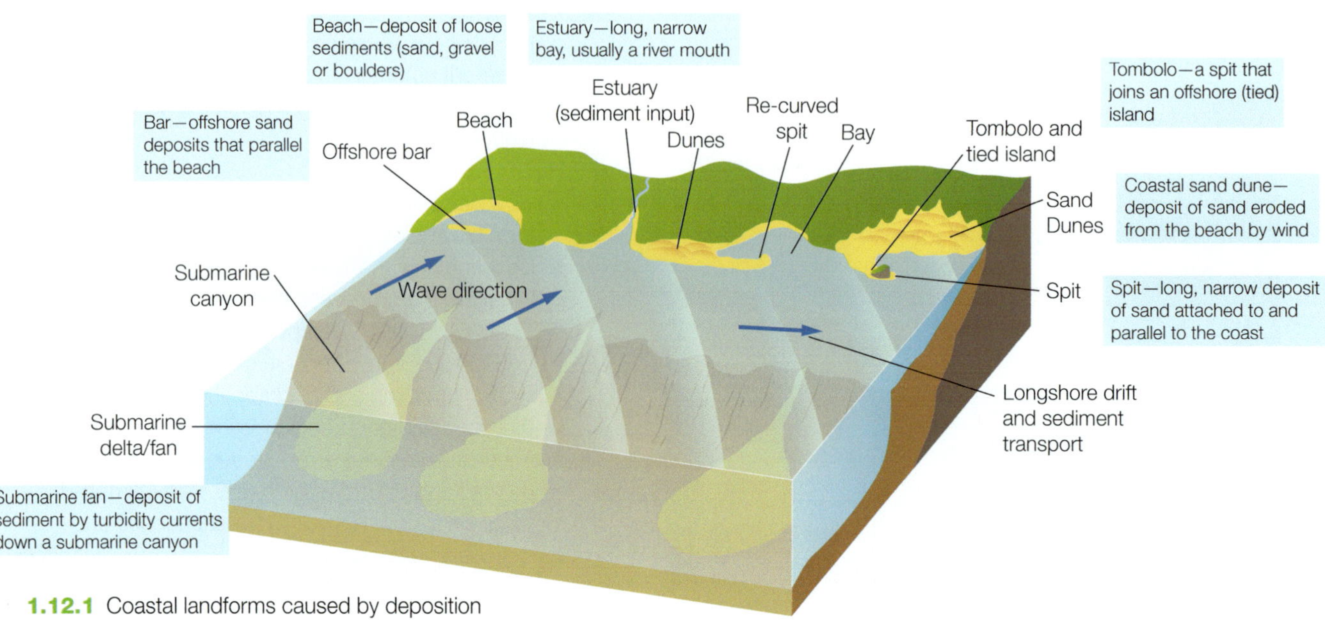

1.12.1 Coastal landforms caused by deposition

ISBN 978 1 4586 6277 4

- *plunging*—waves curl over the crest (popular with surfers)
- *surging*—waves run up beach slopes.

As waves move from the sea to the beach it is called 'swash', and when waves run backwards out to sea it is called 'backwash'. A constructive wave builds the beach when swash is more powerful than backwash, and a destructive wave erodes the beach when backwash is more powerful than swash.

Sediment is also moved by rips and tides. Rips are fast-flowing currents that are dangerous to swimmers, while tides move sediments in and out of bays and beaches twice a day. Waves move sediment along the beach by longshore drift—a process you might be familiar with if you have ever gone for a swim and come out of the water quite a distance from where you went in.

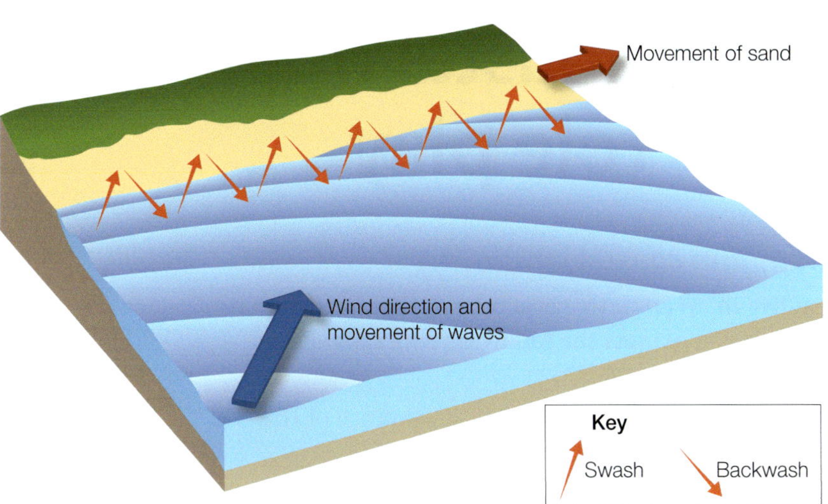

1.12.3 Longshore drfit process

When beaches lose sand from longshore drift, groynes are sometimes built along the shoreline or new sand is used to replenish the area, in a process called beach nourishment.

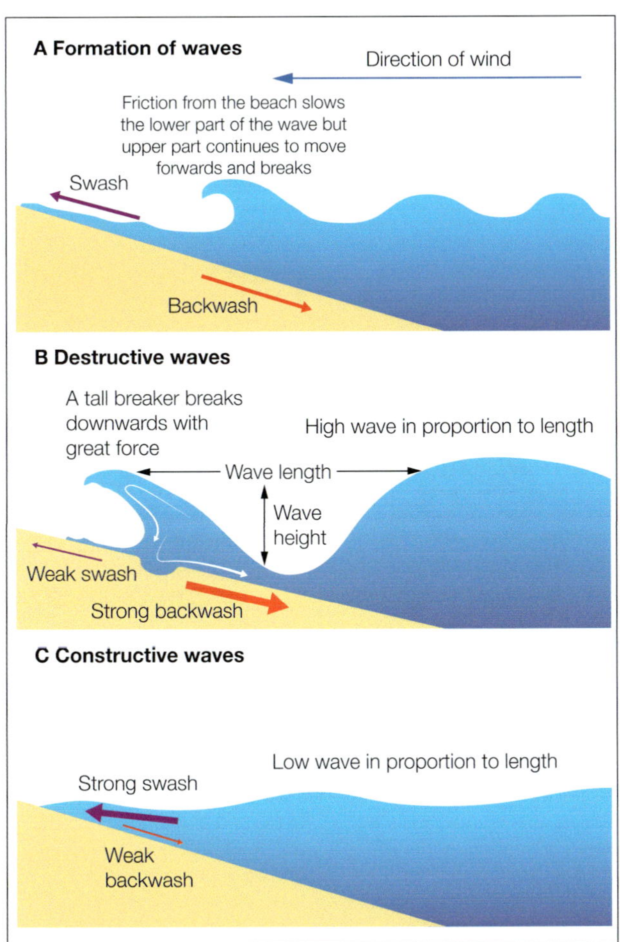

1.12.2 Waves are important in the formation of landforms

Geoinfo

The world's largest and most continuous barrier system lies along the USA's Atlantic Ocean and Gulf of Mexico coasts, extending over 4000 km.

Geoactivities 1.12

Knowledge and understanding

1. Why are coastal landforms dynamic?
2. Where do coasts obtain their sediment? How are sediments lost?
3. What are three different types of waves that transport material?
4. How do rips and tides affect coastal landforms?

Inquiry and skills

5. Refer to 1.12.1. Describe six coastal depositional landforms.
6. Refer to 1.12.2. Distinguish between swash and backwash on constructive and destructive waves.
7. Refer to 1.12.3.
   - a Explain how longshore drift moves sediment along beaches.
   - b How can coastal communities reduce the loss of sand?
8. In groups, research the advantages of deposition to the tourist and recreation industries. Present your solutions as a photo story using digital images or an annotated diagram.

ISBN 978 1 4586 6277 4

1.13 Coastal management: role play

Changing coastal landforms

You live in an attractive beachside town. At a council meeting you express the following concerns about the local beach:

- *erosion* of the cliff and the disappearance of the beach sand from destructive waves
- *cyclones* and *tsunamis* (see 5.8) could result in a storm surge that will flood low-lying areas
- *climate change* is predicted to increase sea levels.

Climate change is the long-term change in temperature, such as getting colder during the ice ages or getting warmer in the 21st century. As temperatures increase, the ice caps melt and the sea level rises. As the sea level rises, it floods low-lying coastlines. Climate change is also contributing to more severe cyclones affecting Australia's coastal areas.

1.13.1 Perspectives on future management of coastal landforms

Ethan Elect—Minister for the Environment

Ethan wants to please local residents and reduce loss of lives and property to imporve his chances of re-election. He also wants to attract tourists and businesses to the community. As the government is in debt his budget is tight. He suggests further study into the type of structures and procedures to be implemented.

Audrey Action

As a naturalist, Audrey is concerned the proposed scheme for an offshore break-water, seawall or groyne will be an eyesore. She says they will interfere with the natural movement of water and beach sediment and cause a decline in marine species. Audrey proposes the council buys properties on the cliff and foredune and revegetates the dunes.

Sandy Dune

Sandy lives in a luxury home on the sand dune and fears the loss of property due to sand erosion, especially during storms. The home has been in the family for generations and he plans to leave it to his grandchildren. Sandy feels the council should look after tax-paying residents. He suggests the building of seawalls and offshore breakwaters.

Tom Traveller

Tom has his holiday here each year as he enjoys the surfing, snorkelling, fishing and whale-watching. The town also offers activities such as movies, skating rinks, parks and shops. He would like the council to protect the beach or he will find a new holiday location. This will mean loss of business to the community.

Ben Builder

Ben is an engineer and has suggested both a seawall to stop erosion and a groyne in front of the swimming area. He wants to bid for the construction contract and intends to focus on negative publicity the council will receive from loss of lives, damaged homes and reduced number of tourists if erosion and future coastal disasters occur.

Sally Surfer

Surfing is Sally's main recreational activity and is unhappy about seawalls, groynes and an off shore breakwater as they would destroy the waves. She and her friends would be forced to travel to beaches further away with poorer surfing conditions if waves changed. Less money would be spent in local shops.

Cheri Cliff-hanger

Cheri owns a cliff-top motel and coffee shop, which bring money and jobs into the community. Her business attracts tourists who watch whales. She fears erosion by waves will undercut the cliff and it will collapse. Cheri blames the council for allowing construction on top of an eroding cliff and favours an off-shore breakwater.

Charlie Chairperson

Charlie is the local councillor and listens to all the different views and asks further questions when necessary. He provides a brief summary at the end of the presentations and then conducts a vote for and against the proposal to take action.

If a proposal is passed consultation with experts will be required.

ISBN 978 1 4586 6277 4

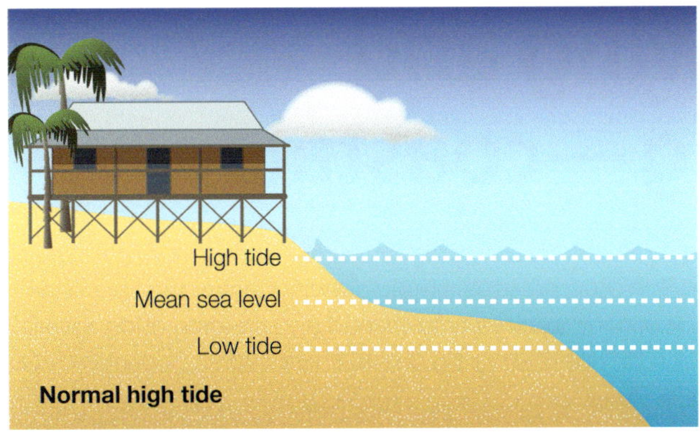

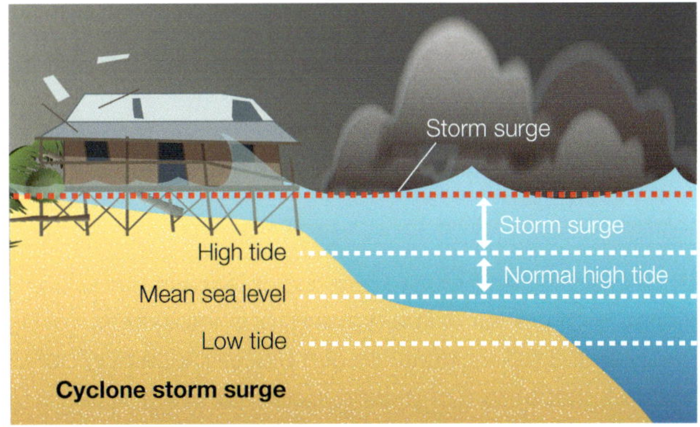

1.13.2 A storm surge is a rise in the water level, often caused by cyclones that bring rough seas, wild winds and heavy rain

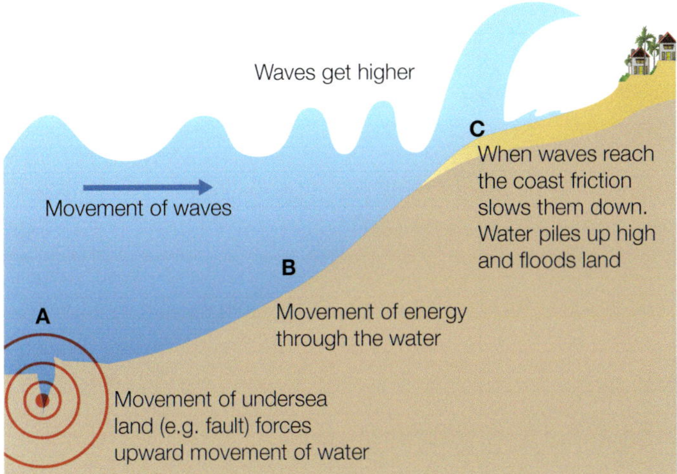

1.13.3 A tsunami is a series of huge waves, generally caused by underground earthquakes and volcanic eruptions. Over the last 130 years Australia has experienced 145 tsunamis

1.13.4 Management options

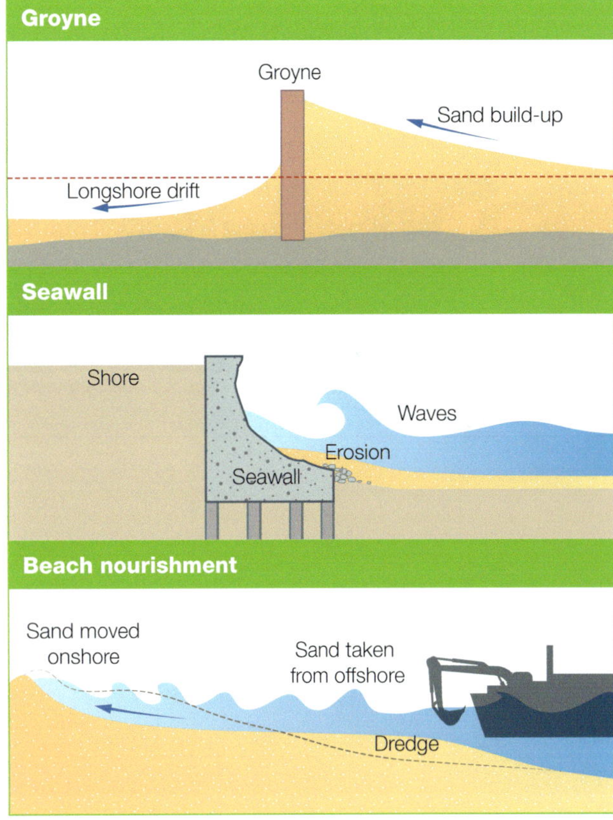

Management proposals

Assign roles from 1.13.1 and present a roleplay of a meeting discussing the different views for and against the proposal to 'take action to control beach erosion'. If action is to be taken, suggest the structures and procedures most suitable, for example:

- *seawall*—a structure placed close and parallel to the beach preventing erosion and protecting buildings. It is expensive and an eyesore
- *offshore breakwater*—a structure placed offshore and parallel to the beach. Waves break in deeper water reducing erosion. It destroys surfing and is expensive
- *groyne*—a structure to stop longshore drift and trap sand. It maintains the beach but is an eyesore
- *beach nourishment*—artificial placement of sand dredged from other places. It is a continuous process and is costly but has a low environmental impact on the beach
- *property purchase*—the removal of buildings threatened by erosion. It means loss of revenue for the council but it allows natural beach processes to continue. Increases public access to the beach.

At the end of the meeting, the group votes on whether to take action and if so, which proposal to proceed with. Present a short report to be published in the local newspaper.

ISBN 978 1 4586 6277 4

1.14 Skill: line drawing and field sketching—coastal

Photographs can be used as a tool to assess the liveability of places. A line drawing, or photosketch, summarises the key features of a place. These may be features that attract or deter people from living there.

A line drawing uses simple drawing techniques in pencil. Dominant features are outlined and shading adds detail. You do not have to be an artist to create an effective line drawing.

1 Draw a frame and divide it into a grid.

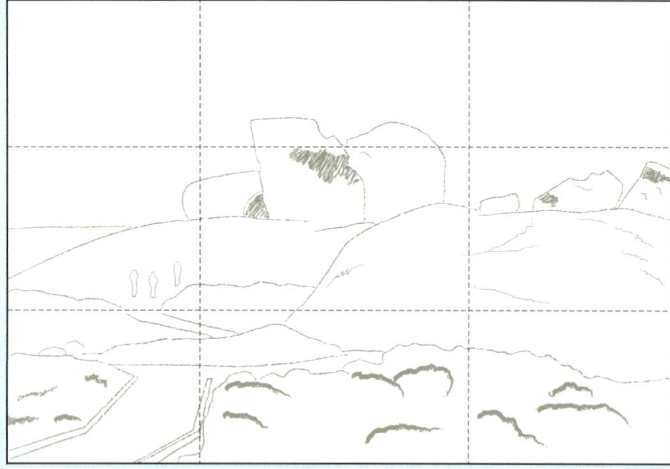

2 Create the outlines and add shading.

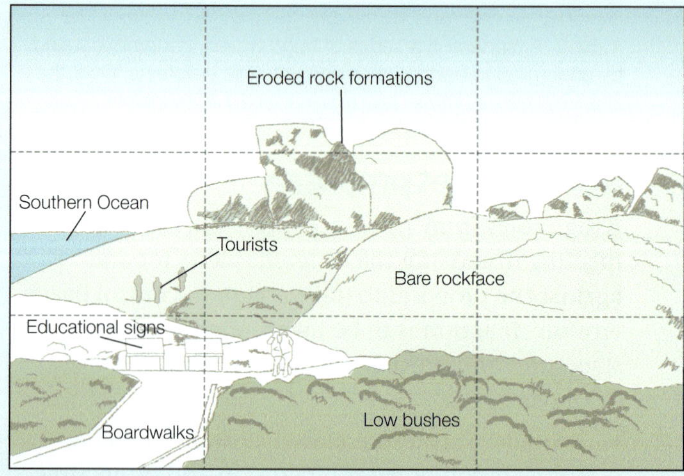

3 Finalise your drawing with labels, a heading and the date.

1.14.1 Constructing a line drawing of this scene on Kangaroo Island, South Australia

Constructing a line drawing

Step 1

- Using the photograph create a frame the same size as the photo.
- Use a pencil to divide the frame into a grid of three areas—foreground, middle ground and background.
- Add three vertical grid lines to divide your frame into middle, left and right sections.
- Use pencil so the grid can be erased when your drawing is completed.

ISBN 978 1 4586 6277 4

Step 2

- Use the grid as a guideline and outline the main features in the photograph.
- If there is a horizon it can be useful to put this in first.
- When you are satisfied you have the main outlines correct, make them darker using a permanent marker.
- Use lines and shading to highlight features (e.g. plants).

Step 3

- Label the features you have drawn.
- Erase the grid lines.
- Create a title.
- Add the date.

Field sketching

During fieldwork, students may be required to construct a field sketch of coastal landforms and landscapes, and indicate the geomorphic processes that are operating. This is the same as a line drawing but the source of the sketch is the real world. Start by deciding on a frame for your sketch. Decide what you will include in your frame. Then follow the same steps used for a line drawing.

Although many students now take photographs during fieldwork, the advantage of a quick field sketch is that key features can be labelled. On return to the classroom this information can be added digitally to photographs.

Geo**activities 1.14**

Inquiry and skills

1 Refer to 1.14.2.

a Construct a line drawing of the Gold Coast in Queensland. Indicate coastal depositional and erosional landforms.

b Explain how geomorphic processes could change coastal landscapes and their landforms. For example, an underwater earthquake and its tsunami, rising sea levels with climate change, and a cyclone with a large storm surge.

c Suggest strategies that could be implemented to protect this coastal resort.

1.14.2 Gold Coast, Queensland

ISBN 978 1 4586 6277 4

1.15 Skill: topographic maps—landforms

A topographic map is a detailed and accurate representation of landform features developed through measurements from satellite imagery, vertical aerial photography and fieldwork. Such maps are an essential component of managing water resources.

Topographic map features

Topographic maps show *natural* (e.g. water, landforms and vegetation), *managed* (e.g. parks) and *constructed* (e.g. roads and bridges) features. These features are generally divided into:

- *relief*—elevation or landforms; e.g. valleys, plains and mountains
- *hydrography*—e.g. rivers, lakes, wetlands and coasts
- *vegetation*—e.g. wooded and cleared areas, plantations and orchards
- *built features*—e.g. buildings, urban settlements, roads, airports, place names and borders.

The real world surface features are represented on a topographic map by the use of lines, symbols and colours. They are listed in the map key or legend.

Topographic maps are geographical tools that help people find locations and directions, measure distance and area, describe the distribution of natural and human features, identify the relationships between these features, and make management decisions. They have a wide variety of uses, including planning road trips or recreational activities such as hiking. Topographic maps are also used by governments and businesses as planning tools for roads, urban areas, and emergency responses. It is necessary to regularly update topographic maps to reflect changes to Earth's surface, particularly in regard to built features.

Spot heights and contour lines

Maps are usually two-dimensional (2D), flat pieces of paper, but the land surface they represent is three-dimensional (3D) as in 1.15.1.

To represent what exists in reality, topographic maps use lines, symbols and colour gradients.

Spot heights are drawn on maps to show the exact height in metres above sea level of certain places (e.g. mountain peaks). They are shown as small black dots on topographic maps.

Trig stations are accurately surveyed points with markers on the ground. They are generally shown as small black triangles on topographic maps.

Contours are imaginary lines (called isolines) that join all places with equal height above sea level. To avoid cluttering a map, contour lines are drawn at a set interval called the **contour interval**. Each topographic map has only one fixed contour interval (e.g. 50 m) to avoid confusion and make it easier to read.

1.15.1 Surface display of topographic maps using colour gradient and contour lines

ISBN 978 1 4586 6277 4

Different topographic maps may use different contour intervals depending on their purpose. Sometimes, key contour lines (called index contours) at set intervals are shaded darker than the others to help make map reading easier. Contour lines that are widely spaced represent gentle slopes. Contour lines that are closely packed represent steep slopes. A vertical cliff is represented by several contour lines joining into a single line.

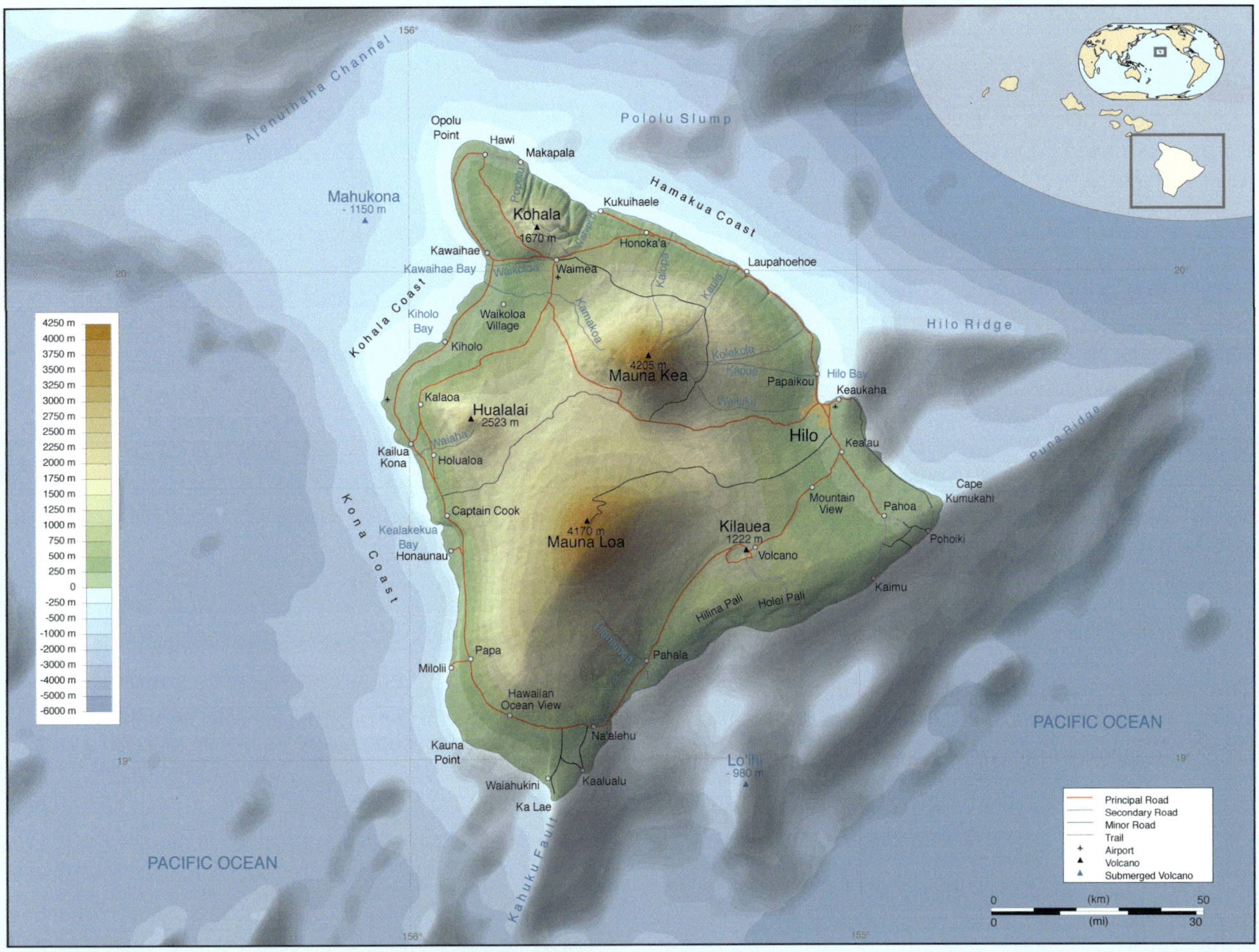

1.15.2 Two maps of Hawaii: topographic map (top) and Landsat mosaic (bottom)

Geo**activities 1.15**

Knowledge and understanding

1. What is a topographic map?
2. List the four features of Earth's surface represented by topographic maps.
3. Outline some practical uses of topographic maps.
4. Explain the difference between contours and contour interval.
5. Describe how contour spacing reflects different slopes.

Inquiry and skills

6. Refer to 1.15.2.
   a. What is the highest point in Hawaii?
   b. How is height depicted on the topographic map?
   c. Calculate the width of the island.
   d. How does the Landsat mosaic present a different perspective on topography of the island?

ISBN 978 1 4586 6277 4

1.16 Skill: topographic maps—relief and gradient

A topographic map is a two-dimensional representation of the three-dimensional features of the surface of the Earth. Two important features of topographic maps are relief and gradient.

Relief

Relief refers to the shape of the landforms in an area. Relief is represented on topographic maps using contour lines, colours, spot heights and symbols. Reading topographic maps enables you to visualise the shape of the land. Various landforms have distinctive patterns of relief. Relative relief is the height difference between the highest and lowest points in an area. The general topography of an area is known as the terrain. Mountainous areas have rugged terrain and plains have gentle terrain.

Gradient

The gradient of an area is the angle of the slope between two points. It is calculated by dividing the vertical difference in height (rise) by the horizontal distance (run).

$$\text{Gradient} = \frac{\text{Vertical distance in height (rise)}}{\text{Horizontal distance (run)}}$$

Gradient can be represented in four ways. The following example has a distance between two points of 1000 m and a rise of 100 m.

1 Fraction: $\frac{1}{5}$
2 Ratio: 1:5
3 Words: For every 5 m there is a rise of 1 m. Or there is a 1 m rise for every 5 m.
4 Percentage: 20%

rise
$a°$
run

Contour lines give a good indication of the gradient of an area. The closer the lines on a topographic map the steeper the gradient and vice versa. Surveyors and engineers need to accurately measure gradients when building dams, roads, railways and construction sites. Gradients are also important in activities like rock climbing, bushwalking and skiing.

Steeper gradients have greater runoff and potential erosion (removal of sediment). They also pose potential hazards in terms of landslides, rock falls, mudslides and avalanches. Authorities use topographic maps to identify and manage these areas of risk.

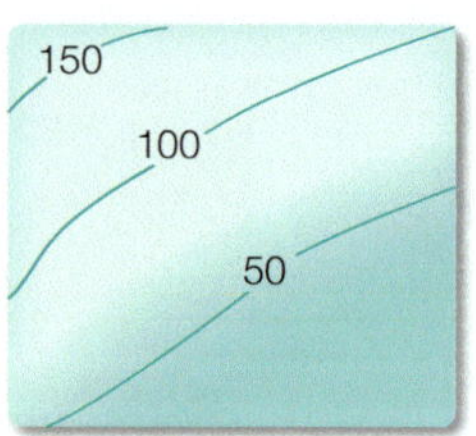

Flat land with gentle slopes has widely spaced contours.

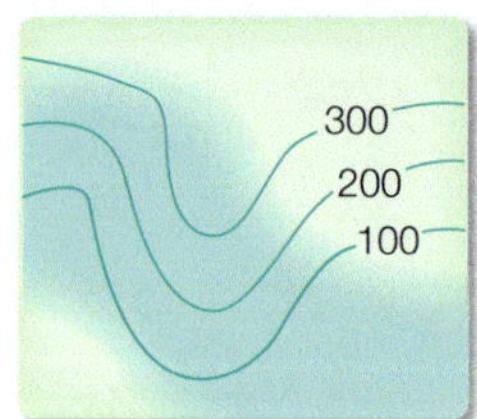

Spurs are extensions of hills or mountains with contours that point towards lowest land.

Knolls or round hills have evenly spaced circular contours.

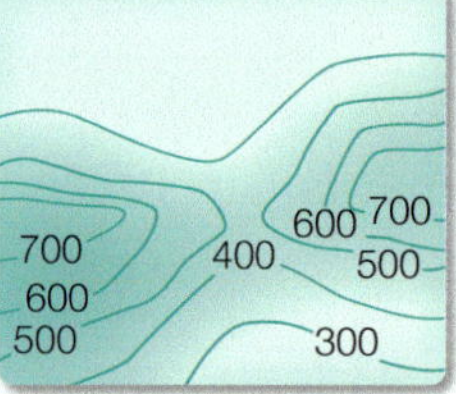

Saddles have two areas of higher land with a dip between them.

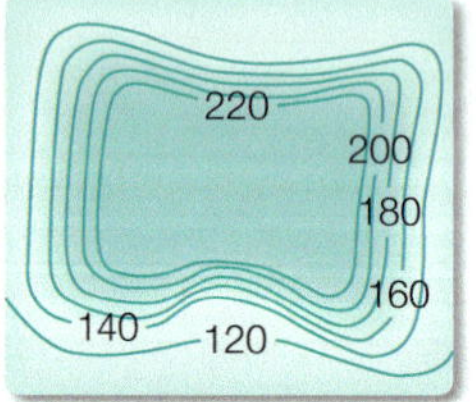

Plateaus are large areas of flat land with closely spaced contours at the steep edges.

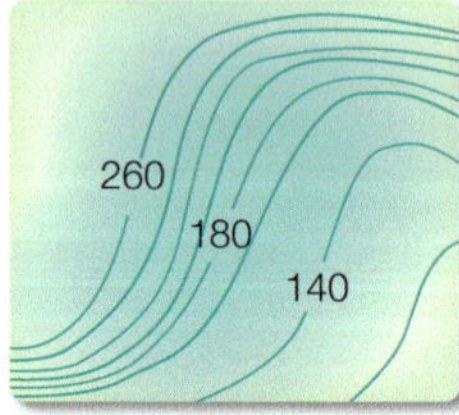

Steep slopes or cliffs have very closely spaced contours.

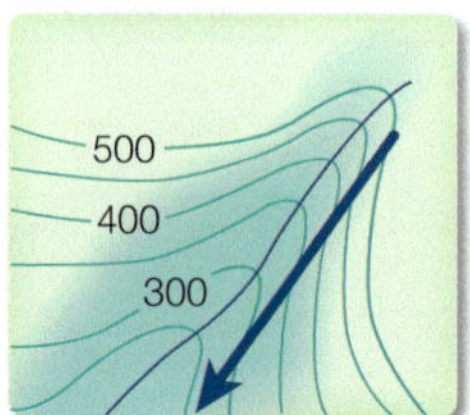

River Valleys flow from the highest to lowest point.

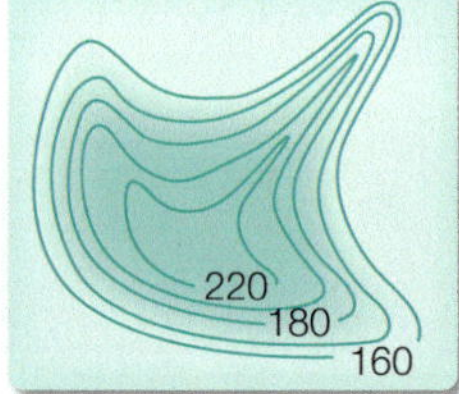

Ridges are the long, narrow tops of hills or mountains with contours that point away from the high land.

1.16.1 Contour patterns of some common landforms

ISBN 978 1 4586 6277 4

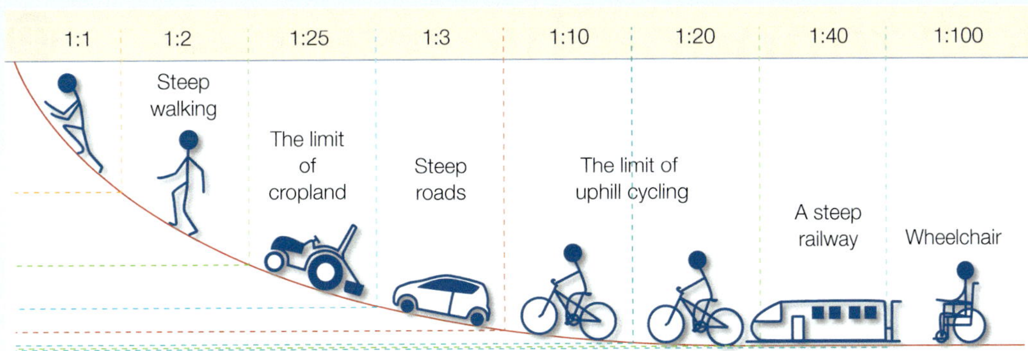

1.16.2 Slopes determine the limits of human activities

5.16.3 Rock climbing Point Perpendicular—one of Australia's premier rock climbing venues (90° vertical cliff)

Geoinfo

Contours with small internal marks at right angles show depressions. Sometimes these extend below sea level; for example, Australia's lowest point, Lake Eyre (–16 m), and the world's lowest point, the Dead Sea in the Middle East (–423 m).

Geoactivities 1.16

Knowledge and understanding

1. How is relief represented on topographic maps?
2. What is the formula for calculating the gradient between two points?
3. List the four ways of calculating gradient.
4. Explain how contour lines indicate the gradient of an area.
5. Why is an understanding of gradient important for work and play?

Inquiry and skills

6. Refer to 1.16.1.
   a. Distinguish beween a plateau and a hill.
   b. Explain the direction of a river on a topographic map.
7. Refer to 1.16.2. Describe the slope limits for human activities such as cycling, walking and building a road.
8. Calculate the following gradients.
   a. If I walked 2 km from sea level (0 masl, or metres above sea level) to 2000 masl.
   b. If my car travelled 10 km from 100 masl to 1500 masl.
   c. If I was cycling in the Tour de France and travelled 2 km from 500 masl to 1000 masl.
9. Access GEFS—a free, online flight simulator based on Google Earth. Notice the changing topography as you 'fly' over the areas (see Geolinks).

ISBN 978 1 4586 6277 4

1.17 Skill: topographic maps—locating features and places

Map coordinates are used to accurately locate features on topographic maps. There are two ways these are shown: geographical coordinates (latitude and longitude) and grid coordinates (eastings and northings).

Geographical coordinates

Latitude and longitude for places are listed in the index of atlases and are shown on Google Earth. Latitude is always given first. Lines of latitude are imaginary horizontal lines that run parallel to each other north and south of the Equator (0° latitude). Lines of longitude are imaginary vertical lines that run from the North Pole (90°N) to the South Pole (90°S) east or west of the Prime Meridian at Greenwich, England (0° longitude). They are measured in degrees (°), minutes (') and seconds ("). Each degree is divided into 60' (minutes), and each minute is divided into 60" (seconds). Latitude and longitude coordinates are shown on a topographic map.

Grid coordinates

The second way of locating places on topographic maps is by using eastings and northings. These make it possible to inform another person where on a map a particular house or river, for example, is located.

Eastings	Northings
Vertical lines on maps	Horizontal lines on maps
Numbers get larger going west to east	Numbers get larger going south to north
Numbers get larger from the left to the right of the map	Numbers get larger from the bottom to the top of the map

When locating a place, an area reference or a grid reference is used:

- An **area reference (AR)** is a four-figure map coordinate (e.g. AR 1943). It is formed by the intersection of eastings (i.e. 19) and northings (i.e. 43). The first two numbers are the eastings and the second two numbers are the northings. Eastings are always written before northings—in alphabetical order E is before N. (e.g. 1943). ARs are useful for locating large features such as towns or lakes.

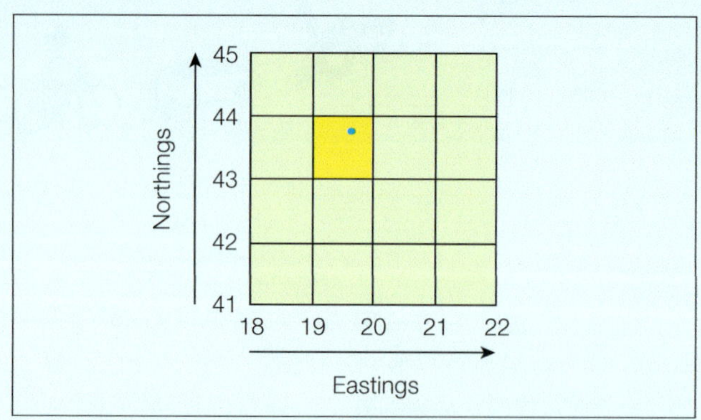

1.17.1 AR locating a lake. The grid square containing the lake has easting 19, northing 43, giving the four-digit AR 1943

- A **grid reference (GR)** is a six-figure map coordinate (e.g. GR 196437) used by geographers to locate smaller features such as a building.

  GRs include extra eastings (i.e. 19**6**) and extra northings (i.e. 43**7**). To find these extra numbers, each easting and northing is subdivided into 10 equal parts. The extra numbers in GRs are the sixth (i.e. 6) and seventh numbers (i.e. 7).

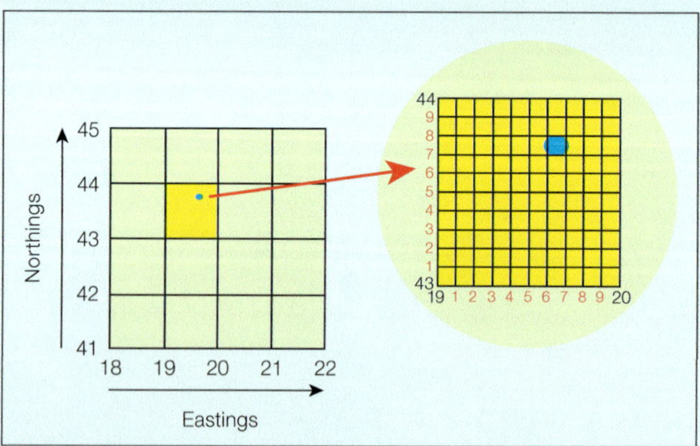

1.17.2 GR locating a lake. We can give a more accurate reference by imagining that the 1 km square is divided up into a further 10 by 10 grid. This allows us to find the six-digit grid reference for the lake: GR 196437

Geo**info**

On 1:100 000 scale topographic maps, each area reference represents an area of 100 ha or 1 km^2.

ISBN 978 1 4586 6277 4

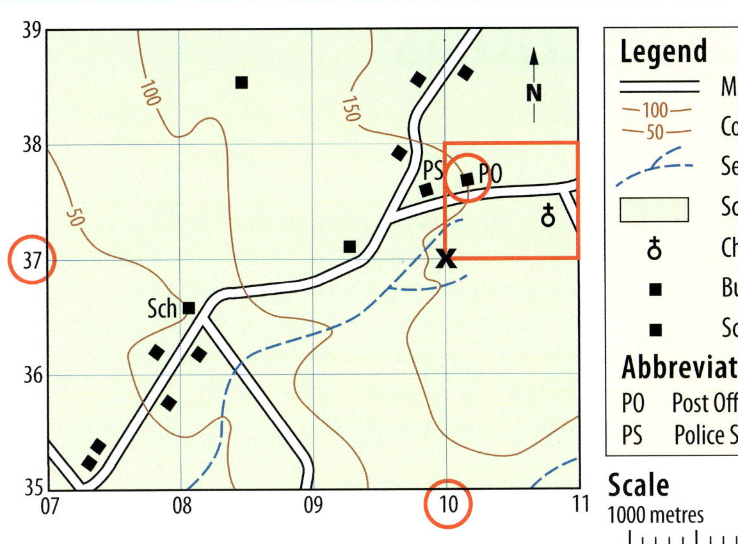

1.17.3 AR on a topographic map

- Find the AR of point X.
- Read the easting (two figures: 10) and the northing (two figures: 37) that intersect at the bottom left corner of the grid square. The four-figure AR is 1037.

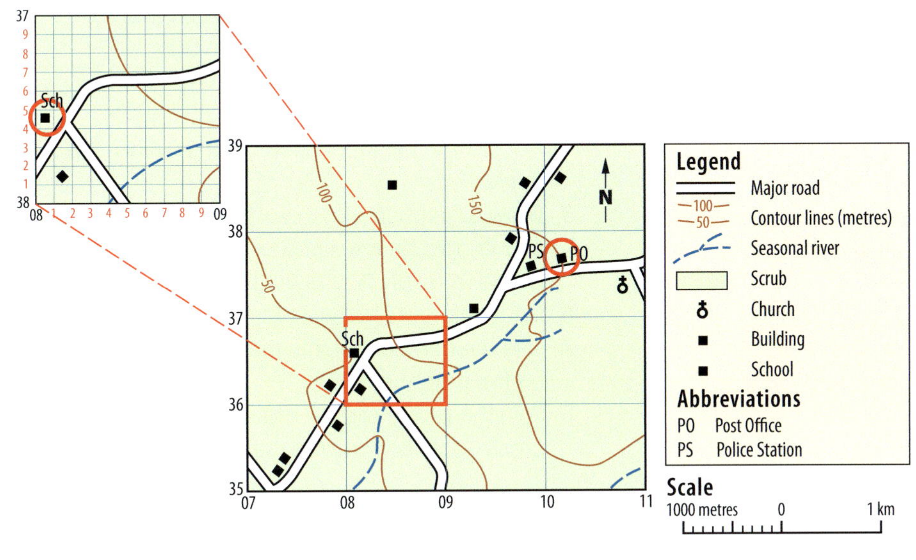

1.17.4 Find the GR of the school

- Repeat the above steps for the AR. The school is AR 0836. Find the missing numbers GR 08_36_.
- Divide the space between the two eastings (08 and 09) into 10 equal parts. Number the parts 1 to 9.
- Divide the space between the two northings (36 and 37) into 10 equal parts. Number the parts 1 to 9.
- Locate the small square where the school is found.
- Read the easting (three figures: 081) and the northing (three figures: 365)
- The six-figure GR of the school is GR 081365.

Geoactivities 1.17

Knowledge and understanding

1. What are the two types of map coordinates?
2. Explain eastings and northings.
3. Explain the differences between ARs and GRs.

Inquiry and skills

4. Refer to 1.17.3 and 1.17.4.
   a. List the natural and built features.
   b. What is the contour intervals (difference between contour lines)?
   c. What is the scale of each map?
   d. What are the numbers from 07 to 11 called?
   e. What is the AR of point X?
   f. What is the AR of the post office, police station and school?
   g. What is the GR of the post office, police station and school?
   h. What is the straight-line distance between the church and point X?

1.18 Skill: cross-sections

Contour lines provide an imaginary view of the earth from directly above—you imagine yourself looking down at a landscape and trying to recognise hills and valleys on a flat map.

The next step is to try to visualise what you see if you take a vertical slice out of the landscape. This is done by drawing a **cross-section** (or topographic profile) that shows elevations along a **transect**.

Vertical exaggeration

For a cross-section to fit on a page and be visually useful it has to be distorted vertically. Each cross-section has its own **vertical exaggeration (VE)** that is a measure of this distortion of shape. The VE is written on the cross-section, along with the horizontal and vertical scales, to ensure that users do not make false interpretations of the relative relief.

The formula for calculating VE is:

$$\text{VE} = \frac{\text{vertical scale}}{\text{horizontal scale}}$$

Calculating VE—an example

If a cross-section is constructed with a:

- horizontal scale of 1 cm to 1 km (a ratio of 1:100000)
- vertical scale of 1 cm to 200 m (a ratio of 1:20000)

then

$$\text{VE} = \frac{1/100000}{1/20000} = 5$$

This means the vertical scale is five times that of the horizontal scale.

This might seem a considerable step away from reality. However, on a cross-section with a VE of 1 (i.e. the horizontal and vertical scales are the same), the cross-section (or profile) of the landscape appears very low or flat. We need to exaggerate or distort the view of the landscape in order to highlight the differences (main features).

Steps in constructing a cross-section

1.18.2 shows these steps in constructing a cross-section:

1. Choose two points on a map for your transect. Label them A and B.
2. Place an edge of a sheet of paper along the transect between points A and B. Write the heights underneath these two marks.
3. Work carefully from left to right; mark where each contour line meets the edge of the paper.
4. Record the height of each contour under the mark.
5. Draw a graph to plot the contours. The horizontal axis should be the same length as the distance between points A to B on the map.
6. Choose a suitable scale for the vertical axis.
7. Use the contour heights marked to plot each point on the graph.
8. Join the points with a smooth line to show the cross-section (or profile) between points A and B.

1.18.1 A cross-section is a side-on view of the topography

ISBN 978 1 4586 6277 4

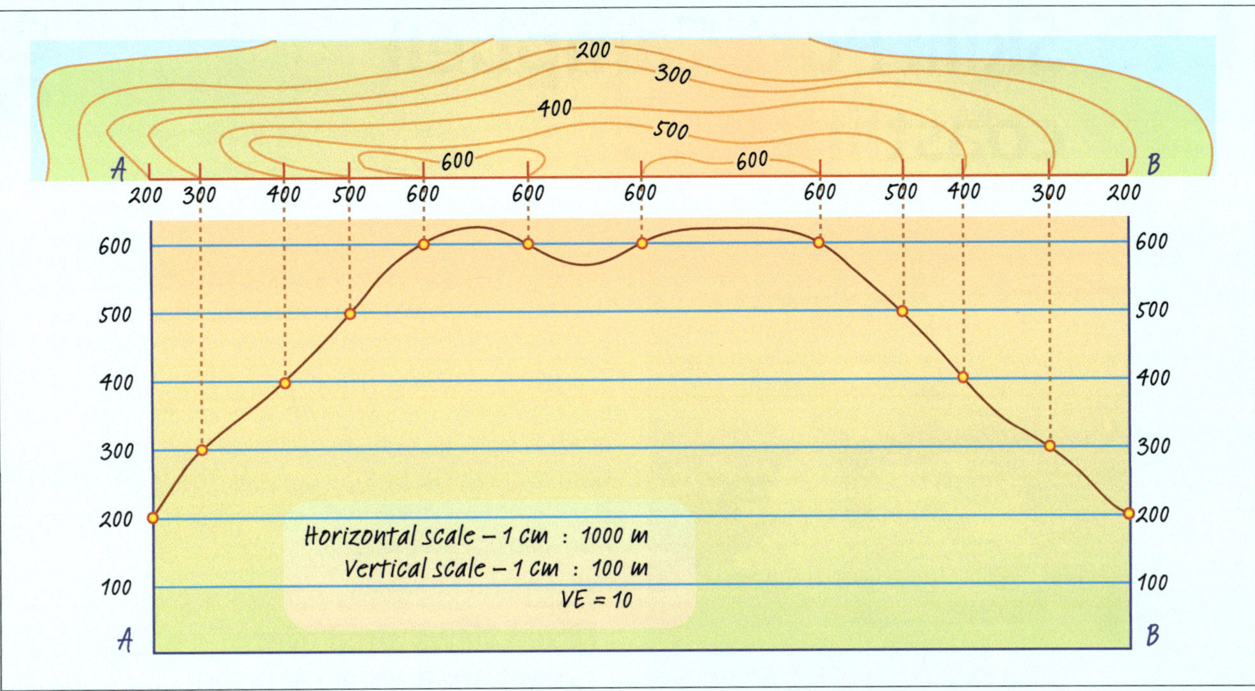

1.18.2 Constructing a cross-section from a topographic map

9 Give the cross-section a border and a title then label the important features.
10 Show the horizontal and vertical scales and the VE.

Sight lines

It is easy to find out from a cross-section whether places are visible from another location by using sight lines. It is much more difficult to do this by looking at a map alone. Where places can be seen from one to another it is said that they are inter-visible. If one place cannot be seen from the other it is referred to as dead ground (1.18.3).

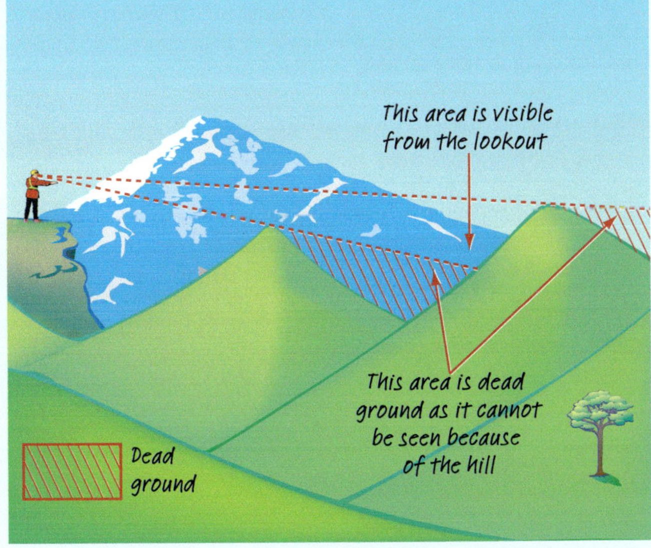

1.18.3 Areas of inter-visibility or dead ground

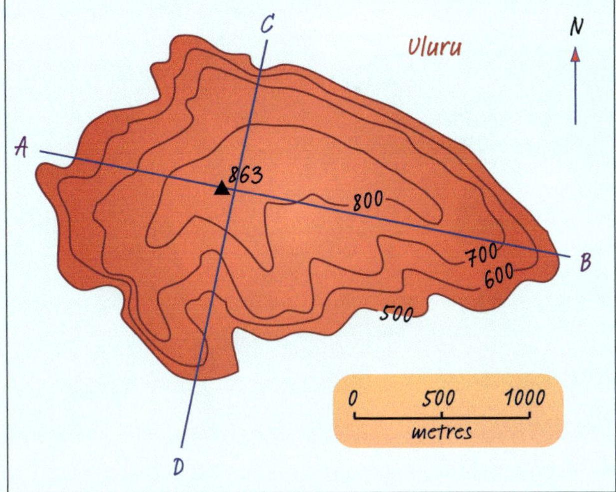

1.18.4 Contour map of Uluru

Geoactivities 1.18

Knowledge and understanding

1 Explain what is meant by VE.
2 What is meant by 'inter-visable'?

Inquiry and skills

3 Refer to 1.18.4 for this activity.
  a Trace the line from point A to point B or C to D and complete a cross-section for this transect.
  b Calculate the length and width of Uluru.

1.19 Skill: Port Campbell coastline

1.19.1 Oblique aerial photograph of the Port Campbell coastline

One of the most spectacular coastlines in Australia is located near Port Campbell in western Victoria. The soft limestone cliffs have been battered by strong winds and waves from the Southern Ocean for millions of years. This constant erosion has left rock stacks, caves, arches and blowholes. The Twelve Apostles (or stacks) are now down to eight, as high-energy waves have eroded the base of the stacks by 2 cm per year. This fragile coastline is protected within the Port Campbell National Park and managed by Parks Victoria.

Mixed views aired over Great Ocean Road rezoning

Residents along the Great Ocean Road are divided over the Victorian Government's decision to rezone tracts of land for tourism development. Mayor Chris O'Connor says the creation of new tourist facilities will be a boon for the economy.

'We only have about 200 000 overnight stays but we have 2.6 million people come through and we just don't think that is good enough,' he said.

However, Marion Manifold, from the Port Campbell Community Group, says the [planning] Minister has ignored residents' concerns. She says development will destroy the untouched landscape which draws people to the region. 'You just don't do this because it is going to destroy the product that the people come here for,' she said.

13 November, 2012, www.abc.net.au

London Bridge with its double arches in 1959

London Arch after the collapse in 1990

1.19.2 London Bridge was renamed London Arch after its collapse

ISBN 978 1 4586 6277 4

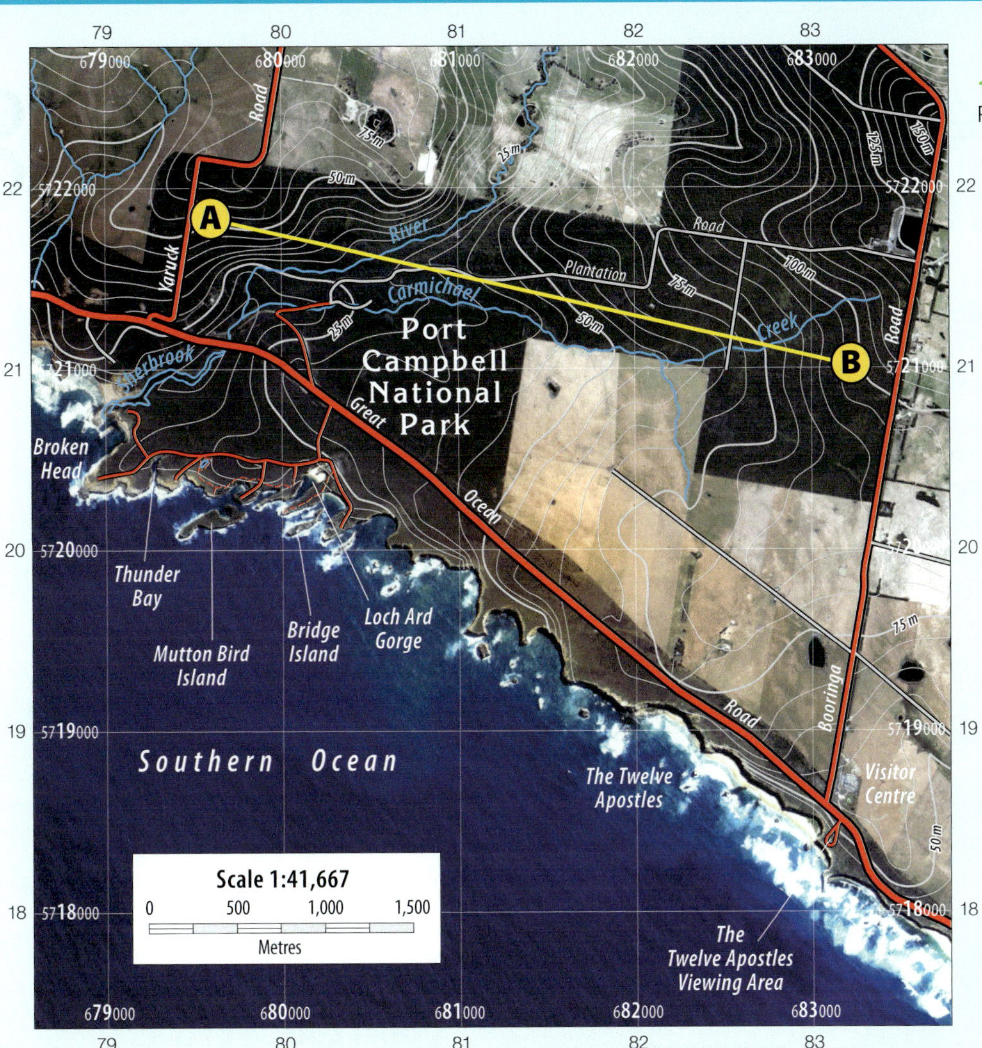

1.19.3 Topographic map of Port Campbell National Park

Geoactivities 1.19

Knowledge and understanding

1. Where is Port Campbell located?
2. What coastal landforms are found in this location?

Inquiry and skills

3. Refer to 1.19.1.
   a. Label the coastal features A to E.
   b. Draw and label a line drawing of the coastline showing erosional, transportation and depositional processes.
4. Refer to 1.19.2.
   a. Describe the processes that led to the collapse of one of the arches in 1990.
   b. Draw and label a sketch of what could occur to these landforms in the next 1000 years.
5. Refer to 1.19.3.
   a. What are the advantages and disadvantages of this map?
   b. Using Google Earth, find the latitude and longitude of the Twelve Apostles.
   c. What is the grid reference (GR) of the Twelve Apostles Viewing Area?
   d. What are the area references (AR) for Bridge Island and Mutton Bird Island?
   e. How was sand formed in AR 8218?
   f. Explain why this area has more erosional features than depositional features.
   g. What is the name of the ocean where the Twelve Apostles are located?
   h. Draw a cross-section from A to B.
6. Refer to the web article.
   a. What are the different views on developing Port Campbell?
   b. Discuss the impacts of tourists on the environment.
   c. Suggest five management strategies to reduce the impacts of tourism.
7. Imagine you were to visit the Port Campbell coastline. Research the attractions you would visit and tourist facilities in the area. Present as a photo story.
8. Using digital and spatial technologies, design a tour along the Great Ocean Road. Draw an annotated map of the tour including sketches of the landforms along the route. Describe how two of the landforms have been formed.

ISBN 978 1 4586 6277 4

1.20 Rivers: landforms and processes

A river is a natural watercourse, flowing from its source in upland areas to its mouth in lower areas. As the river flows it becomes deeper and wider until it reaches the sea. The area that supplies water to the river is referred to as the drainage basin or the **catchment area**. A watershed is located at the highest point of the catchment area and is the source of the river.

As rivers journey from their source to their mouth, they erode, transport and deposit material. As a result, rivers can be classified by either of the following:

- *processes*—features indicating erosion or deposition of the river
- *location*—where they are found, such as the upper, middle or lower course of a river.

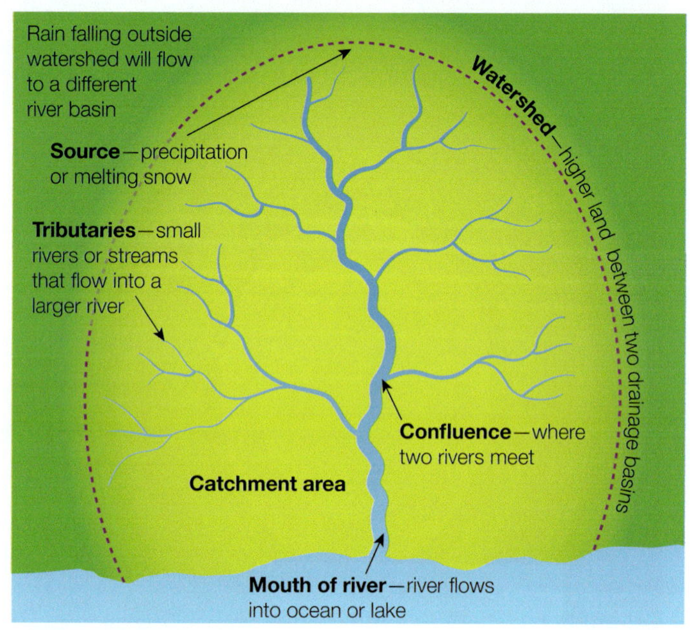

1.20.2 Drainage basin: an area drained by a river and its tributaries

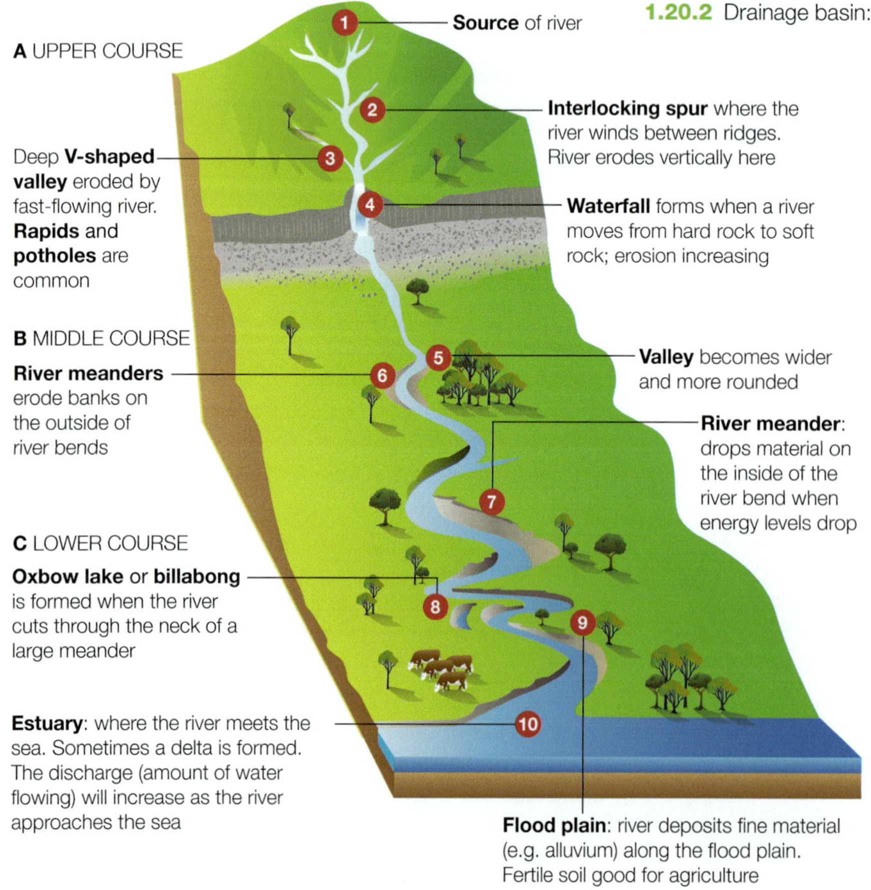

1.20.1 River journey from source to estuary

Upper course of rivers

Rivers change over time and space as they move from source to mouth. In the upper course of the river, fast-flowing water erodes the landforms resulting in a range of features:

- *Waterfalls* occur when water descends vertically (e.g. over a cliff). The formation of a waterfall occurs when the river flows over different types of rocks that are soft and hard. The softer rocks erode faster. As the water falls over the hard rock it erodes the riverbed at the bottom, forming a plunge pool. The continual undercutting of the soft rock creates overhanging rocks, which eventually become too heavy and collapse into the plunge pool. Waterfalls are popular tourist sites and include Niagara Falls (USA) and Iguazu Falls (Brazil and Argentina) which are both World Heritage listed. However, they

ISBN 978 1 4586 6277 4

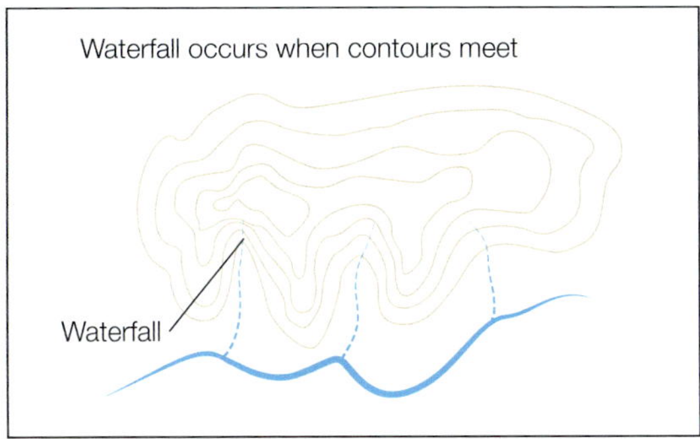

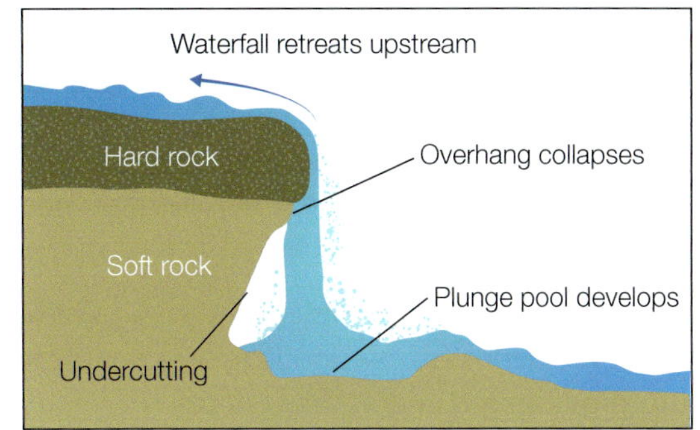

1.20.3 Topographic map and a cross-section of a waterfall

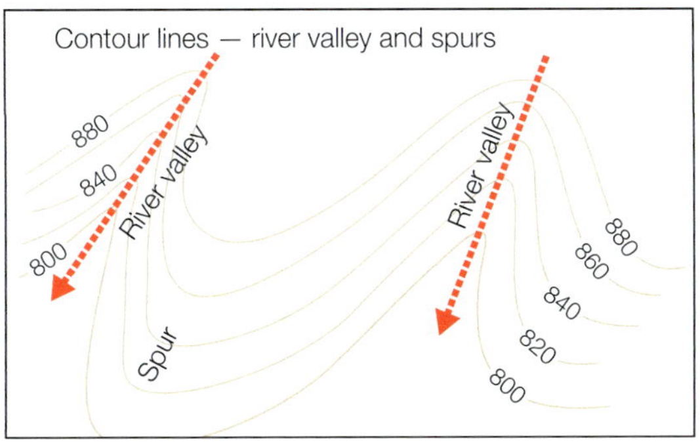

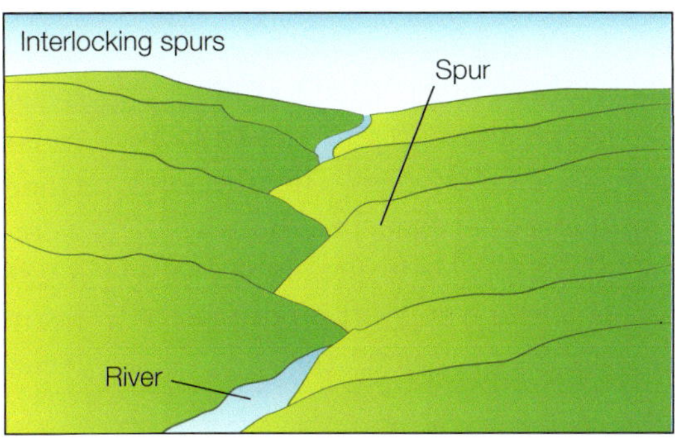

1.20.4 Topographic map and diagram shows river valleys and interlocking spurs

present difficulties for boat transport and can become polluted from unsustainable tourism.

- *Rapids* are rough, turbulent water formed when water flows over layers of hard and soft rock. Rivers with rapids make popular rafting rivers, Examples are the Zambezi River (Zimbabwe) and the Tully River (Queensland). Rapids also create problems for boat transport.
- *Potholes* are holes found in a riverbed. Rivers carry rocks, which are washed around in a circular motion causing erosion.
- *V-shaped valleys* occur when rivers erode downwards while the sides are weathered by rain and wind. The weathered material falls into the river and is transported downstream. V-shaped valleys are suitable for the construction of reservoirs and dams that produce hydroelectricity.
- *Interlocking spurs* occur when hills jut out into the river valleys resulting in a formation resembling the teeth of a zip.

Geo**activities 1.20**

Knowledge and understanding

1 Define the words riverine, river and catchment.

2 Which part of a river would be the best location for a dam and which would be best for a farm?

Inquiry and skills

3 Refer to 1.20.1.
  a What landforms are located in the upper course of the river?
  b Explain the formation of rapids and potholes.
  c Where are V-shaped valleys located and how are they formed?
  d What occurs when the river starts to meander?
  e *Valley shapes change from the source to the mouth of a river.* Explain this statement.

4 Refer to 1.20.2.
  a What are watersheds, tributaries and confluences?
  b Where are the source and mouth of a river located?

5 Refer to 1.20.3.
  a Explain the formation of a waterfall.
  b How could you identify a waterfall on a topographic map?

6 Refer to 1.20.4.
  a How can you identify a spur on a topographic map?
  b What is meant by interlocking spurs?

7 From their source to mouth rivers erode, transport and deposit material. Imagine you were a rock located at the source of the river. Describe your long journey over time and the different landforms you would pass. Present your story as an oral report.

ISBN 978 1 4586 6277 4

1.21 River depositional landforms and processes

Billions of people live on floodplains, where sediments called alluvium are deposited by flooding rivers. The alluvium provides fertile soil for agriculture along the Nile River in Egypt, the Ganges–Brahmaputra Rivers in Bangladesh and the Mekong River in Vietnam.

In Australia, the World Heritage listed Kakadu National Park is located on a floodplain. The Kakadu landscape undergoes seasonal changes. The summer monsoonal rains produce water that spreads over hundreds of square kilometres. As it spreads, it drops nutrient-rich sediments, which support prolific plant and animal life. During the dry season the floodplain dries up.

Traditional Aboriginal Peoples relied on the floodplain landscape as a source of food, and today the area is both culturally significant and a biodiversity wonderland.

The Kakadu floodplain is subject to unsustainable human actions, such as tourism, pastoralism and uranium mining.

There are fears that salt water from the ocean will intrude into the wetlands as a result of climate change.

Hazardous floodplains

Most Australian cities, towns and farms are located on floodplains. Floods have caused tragic loss of lives and properties, as experienced in Brisbane in January 2011. At the time, three-quarters of Queensland was declared a disaster zone, 38 people were confirmed dead and the damage bill reached $2.38 billion. In response to the floods, dams have been built to reduce future flooding and artificial levees built to control the spread of floodwaters.

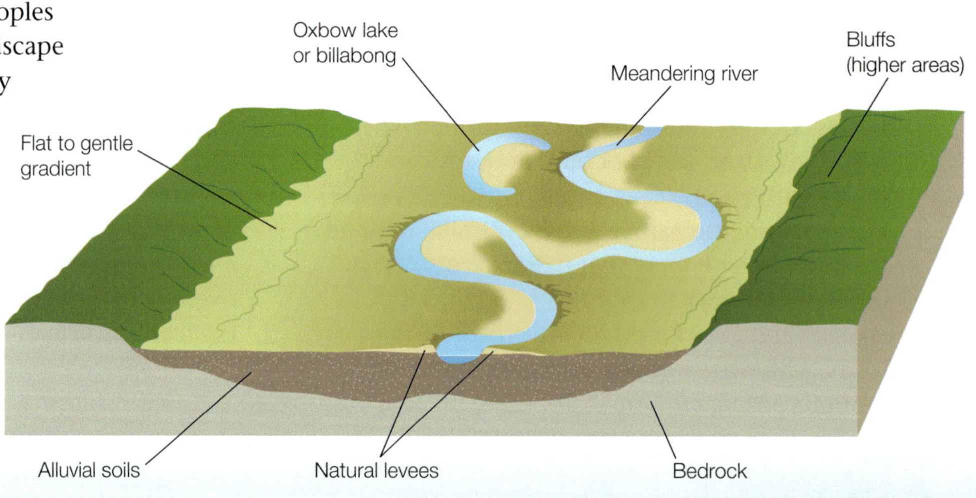

1.21.2 Landform features on a floodplain

1.21.1 Kakadu National Park is on a floodplain with abundant wetlands

ISBN 978 1 4586 6277 4

However, one downside of flood mitigation is that floodplains are deprived of fertile alluvium in the long term. The Brisbane City Council has initiated a property buy-back and land-swap program available to properties experiencing an average flood recurrence every two years.

Landforms

Floodplains include a variety of landforms, including oxbow lakes, meandering rivers, levees, alluvial soils and braided streams:

- *Floodplains*—this is an area of land bordering a river that stretches from its banks to the base of the enclosing valley walls. When water in a river increases in volume due to melting snow upstream or heavy rain from a cyclone, the river overflows its banks and floods the plain.
- *Meandering rivers*—rivers meander across flood plains as they erode laterally from side to side. The river erodes on the outside of the bend, which creates a steep-sided riverbank. On the inside of the bend the river deposits sediment, which forms a gently sloping bank and small beach.
- *Natural levees*—when the river floods, more silt is deposited close to the river at the riverbank. Less silt is deposited on the wider floodplain. Frequent floods build up the levee. Artificial levees are constructed parallel to rivers—they are made of sandbags in a flood emergency or built as permanent earth works. Over 3000 years ago, levees were constructed along the left bank of the River Nile for 970 km. To protect levees from erosion, vegetation is planted on their banks.

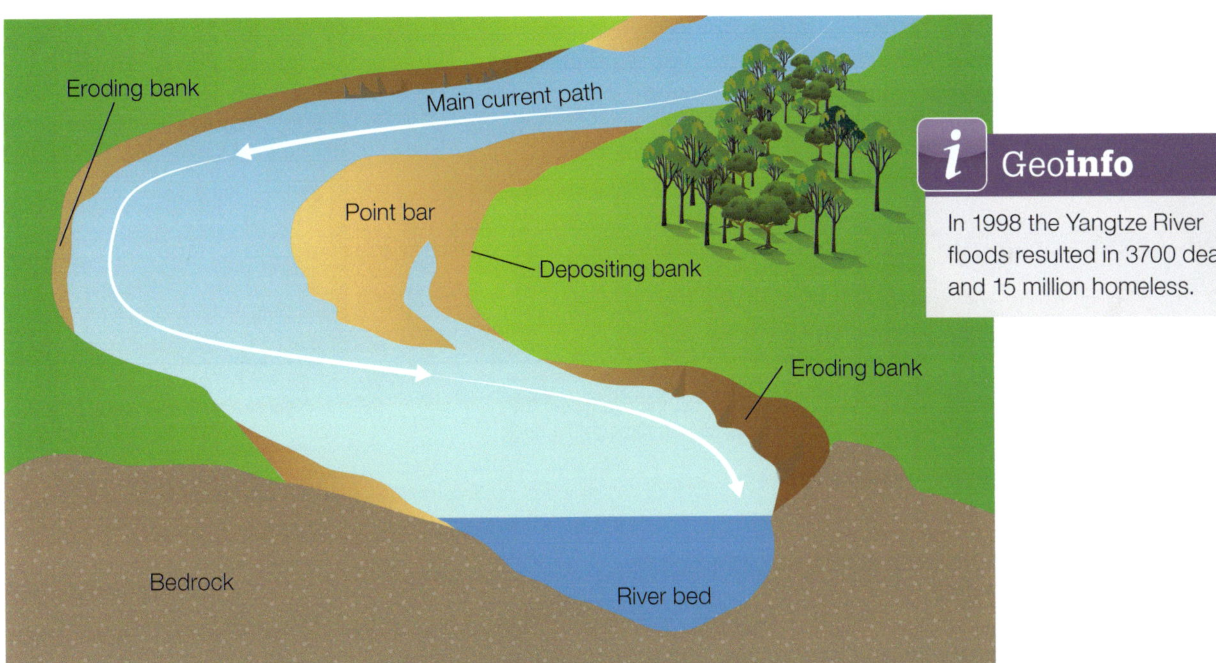

Geoinfo

In 1998 the Yangtze River floods resulted in 3700 dead and 15 million homeless.

1.21.3 Erosion and deposition on a meandering river (process)

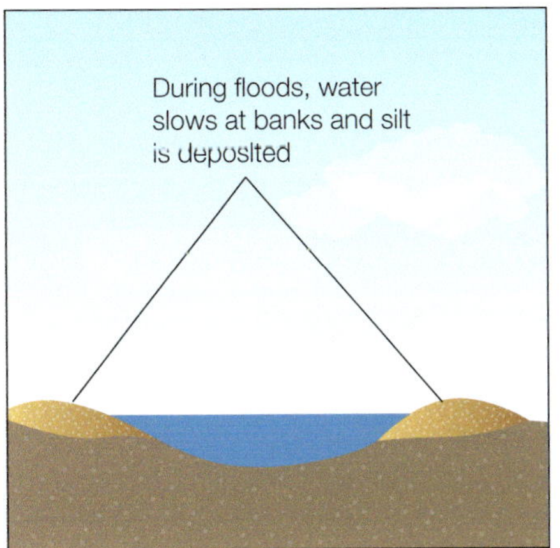

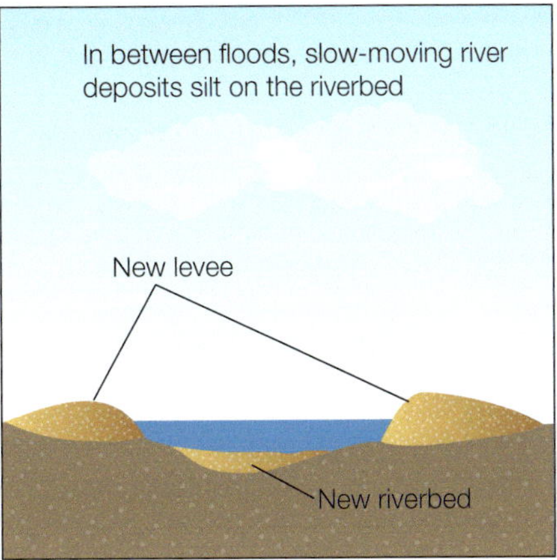

1.21.4 Formation of a levee and higher river bed (process)

ISBN 978 1 4586 6277 4

- *Oxbow lakes or billabongs*—over time, the erosion and deposition of a meandering river is responsible for changing the river's course. Oxbow lakes (or billabongs) are created when two meanders connect, isolating part of the loop. The river continues in the new bed and the old meander path is abandoned.
- *Braided river*—this consists of a network of river channels separated by small temporary islands. Extensive braided river systems are found in Alaska, New Zealand's South Island and the Himalayas.
- *Deltas*—when a river reaches a lake or the sea, it slows down and drops its sediment. The dropped material builds up in layers to form a delta. Sometimes deposited sediment blocks river channels, forcing the water to find alternative routes to the sea by forming **distributaries**. Deltas produce a variety of landforms (such as arcuate, cuspate and bird's foot) and contain fertile soil for agriculture.

1.21.5 Formation of an oxbow lake or billabong

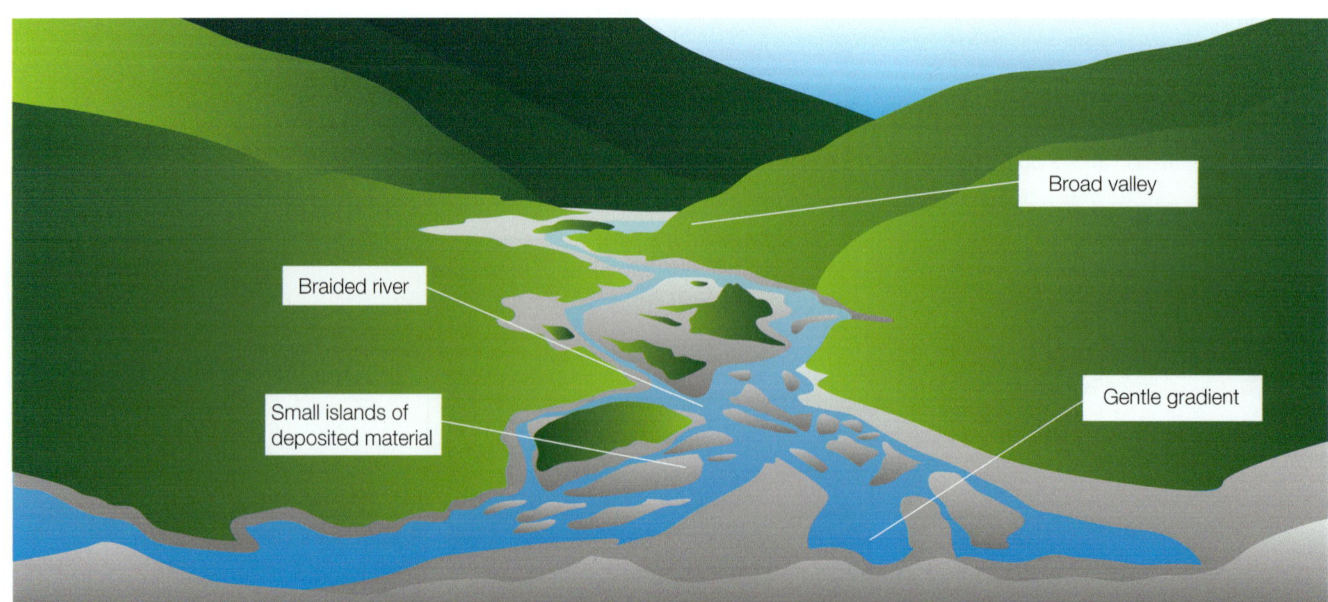

1.21.6 Braided rivers and depositional landforms on a flood plain

ISBN 978 1 4586 6277 4

Arcuate delta
(e.g. Niger River in Nigeria, Africa). They are called arcuate deltas because the outside edge is a regular shape, like the arc of a circle

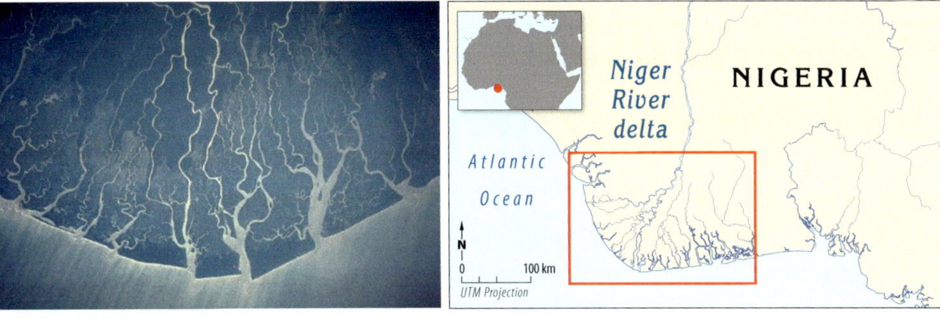

Bird's foot delta
(e.g. Mississippi River in the USA). They are called bird's foot deltas because they look like a bird's foot.

Cuspate delta
(e.g. River Ebro in Spain). Cuspate deltas have been shaped by currents of fairly equal force.

1.21.7 Types of deltas: satellite images and diagrams

Geo**activities 1.21**

Knowledge and understanding

1 Name three floodplains and three deltas.
2 The Kakadu landscape undergoes seasonal changes. Explain what this means.
3 What are the problems when towns and cities are located on floodplains?
4 Suggest strategies to control the damage caused by floods on floodplains.

Inquiry and skills

5 Refer to 1.21.1. Explain how this Kakadu landscape will change in the dry season.
6 Refer to 1.21.2.
  a Describe a floodplain.
  b List the features located on a floodplain.
7 Refer to 1.21.3 and distinguish between the processes occurring on the outside and inside of a meandering river.
8 Refer to 1.21.4.
  a Explain how a levee is formed.
  b Describe how levees and riverbeds build up over time.
  c What is the difference between a natural levee and one constructed by humans?
9 Refer to 1.21.5 and describe the formation of an oxbow lake or billabong.
10 Refer to 1.21.6 and explain how braided rivers are formed.
11 Refer to 1.21.7.
  a What is a delta?
  b List the three types of deltas shown and where they are located.
12 The gradient of a river generally gets less steep as it travels from source to mouth. What does this mean?
13 Investigate one large river delta, such as that of the Nile, Ganges, Amazon or Mekong. Include a map locating the delta, how it was formed and how it has been used by humans. Present your findings as a Prezi.
14 It is hard to believe that a slow-flowing river could cause serious damage to people and properties. Investigate two riverine disasters and their impacts. Select one example in Australia and one in another country.

1.22 Wetlands: threatened landforms

Wetlands are a combination of landforms and bodies of water. They are found on flood plains, coastal lagoons and high mountains. Their landforms extend across all climate zones, from the hot tropics to the cold polar. Wetlands range in type of vegetation from mangroves (trees) to grass and moss. They can contain fresh, brackish or saline water.

The common feature of all wetlands is that the water table (water under the ground) is near the surface of the ground, or shallow water covers the ground for part of the year.

Wetlands are of value to humans as they provide clean water, raw materials and food. Despite their importance, more than 50% of wetlands have been lost in the past 100 years, mostly for coastal and river development.

1.22.2 Wetlands in Thailand cleared for shrimp farming

Wetlands in China

China's wetlands, which comprise 10% of the global total, are suffering as a result of the country's rapid economic development. Coastal wetlands are under pressure from pollution, drainage, and urban and agricultural development. A 2013 report by the China Council for International Cooperation on Environment and Development (CCICED) says 57% of coastal wetlands have disappeared since the 1950s due to land reclamation. The CCICED report also estimates that, based on government-approved development projects, another 5800 km^2 of coastal wetlands will be lost by 2020.

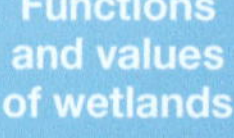

1.22.1 Important functions and values of wetlands

ISBN 978 1 4586 6277 4

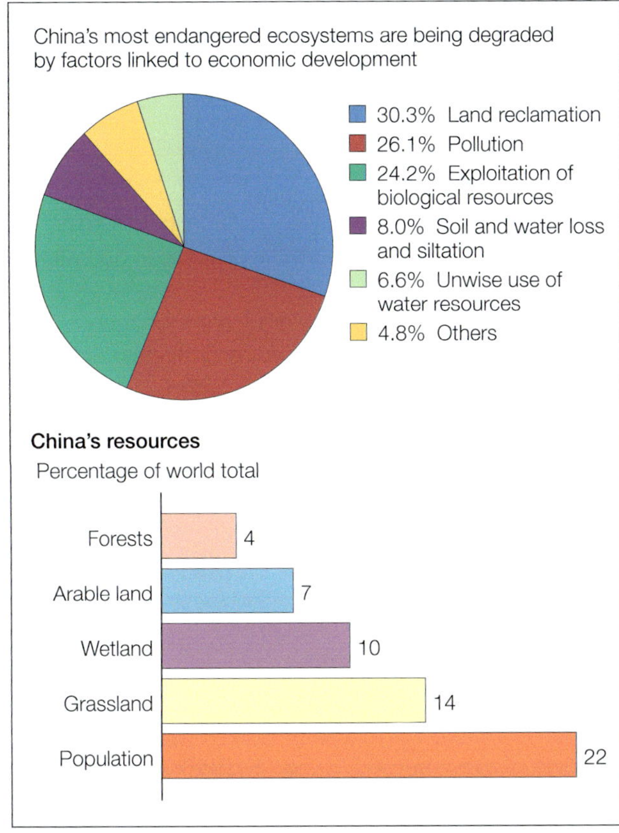

1.22.3 Wetland threats in China

Geoinfo

Recent scientific studies found it takes around 100 years for restored wetlands to function at similar levels to natural wetlands.

Sundarbans World Heritage site

The Sundarbans (Bengali for 'beautiful forest') is an extensive area of mangroves found at the mouth of the Ganges River in the Bay of Bengal. The Sundarbans is protected as a transnational park and World Heritage Area by India and Bangladesh, despite pressure from their rapidly increasing populations. The wetlands are home to species such as the endangered Bengal tiger and freshwater dolphins. It acts as a buffer zone for tropical cyclones and storm surges. The satellite image in 1.22.4 shows the dark green of the Sundarbans backed by the lighter green agricultural land and the tan coloured towns. Many wetlands that lie at the edge of the protected area have been cleared for shrimp aquaculture, which may pose a potential threat to water quality. However, the greatest threat to the Sundarbans is rising sea levels as a result of climate change.

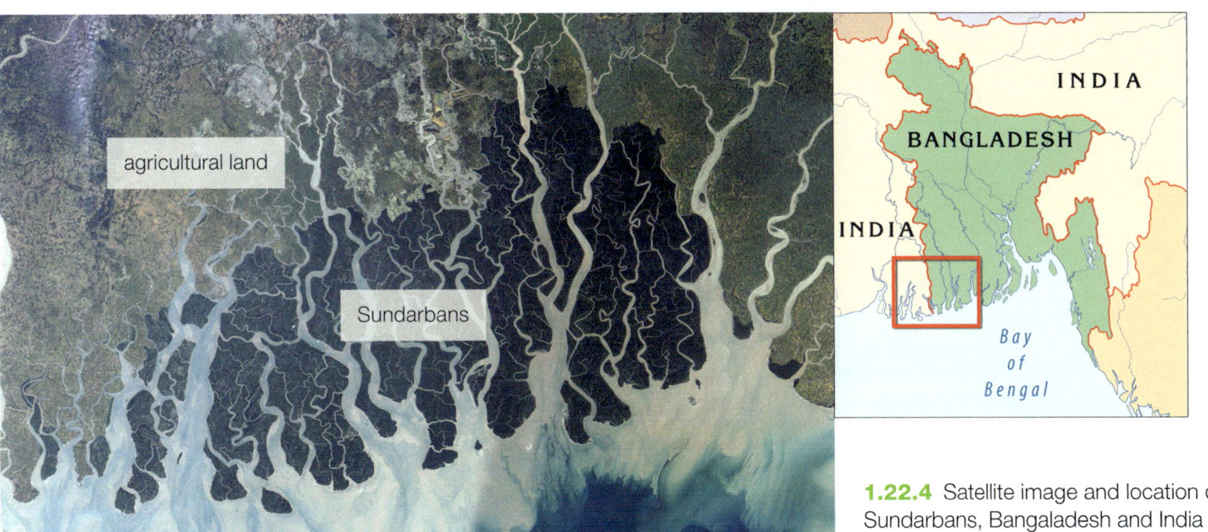

1.22.4 Satellite image and location of Sundarbans, Bangaladesh and India

Geoactivities 1.22

Knowledge and understanding

1 Give statistics to support the fact that wetlands are a threatened landscape.
2 What does the TEEB report state about wetlands and people's attitudes towards them?
3 Describe how China's wetlands are suffering.
4 What is the Sundarbans and what role does it play?

Inquiry and skills

5 Refer to 1.22.1. Outline the functions and values of wetlands.
6 Refer to 1.22.3. Describe the threats to China's wetlands.
7 Investigate the concept of environmental flow to rivers and wetlands in Australia. What is it and how does it work?

ISBN 978 1 4586 6277 4

Geo**think**

Coastal fieldwork

Fieldwork on coastal landforms and their geomorphic processes enables people to better understand how beach environments work so they can propose sustainable management strategies.

Fieldwork process

There are five steps in the fieldwork process.

Step 1

Develop inquiry questions such as:

- How have coasts changed over time?
- What are the geomorphic processes that change coasts?
- How have humans impacted on coastal processes?
- How should coasts be sustainably managed?

Step 2

Conduct fieldwork by identifying, collecting, recording and interpreting information on the processes operating at a local beach.

- Before going to the beach, draw a sketch map locating the beach, using an atlas or the internet.
- When at the beach, draw an annotated diagram of coastal landforms including:
  - natural features, such as cliffs, rock platforms, caves, wave-cut notch, beach, foredune, spit, lagoon and vegetation
  - human features, such as a car park, surf club, homes, roads and litter.
- Take photographs and label landforms and processes.

1.23.1 Annotated photograph showing natural and human features at Narrabeen, located on the northern beaches in Sydney

ISBN 978 1 4586 6277 4

1.23.2 Label coastal landforms such as Bilgola Beach, located on Sydney's northern beaches. The beach is 500m long. To the north is Bilgola headland, where homes have been built, and to the south is a rock platform. What are the natural and human impacts on Bilgola? What are the geomorphic processes operating at this beach?

1.23.3 Impacts of humans on Bilgola Beach, Sydney. How have humans changed the coastal landscape?

1.23.4 Natural impacts on Bilgola Beach, Sydney. Over time, waves eroded the soft rock on the headland leaving behind a rock platform. What are the processes operating over time at Bilgola Beach? What will be the effect of future geomorphic processes on the beach?

Step 3

Observe waves at the beach and answer the following questions:

- Are the waves breaking at right angles or obliquely?
- What is the approximate height of the waves?
- Are the waves spilling, plunging or surging?
- Can you see a rip? Where is it located? What are the problems with swimming in rips?
- Can you see surfboard riders? At what part of the beach are the riders located, and why?
- What is the clarity of the water? (clean, dirty)
- What evidence can you see that sand is moved in the surf?
- In which direction are the waves moving the sand?
- How far does the water run up the beach at high tide? (swash)

1.23.5 Waves at Bilgola Beach. Bilgola is protected from south-east waves by Newport reef, which extends 1 km out to sea. Waves average at 1.5 m.

Step 4

Carry out the following tasks on the sand:

- Collect a handful of sand close to shore and at the back of the beach near the foredune. What are the differences in size of the grain and particles such as shell content? Give reasons for the differences.
- Is the beach patrolled? Where are the flags located? Why are they located at this spot?
- What evidence can you find that the sand is eroding and/or the cliff is eroding?
- Imagine you came back to the beach in a million years. Describe the scene.

Step 5

Present your work.

- Present fieldwork as a poster, PowerPoint or multi-media display. Include a map of the beach, sketches, diagrams and photographs.
- Where appropriate, propose individual and group action.

ISBN 978 1 4586 6277 4

chapter

2 Restless Earth: geomorphic processes

'Each geomorphologic landscape tells a story and unravels pages from the history of the Earth.'

Piotr Migoń

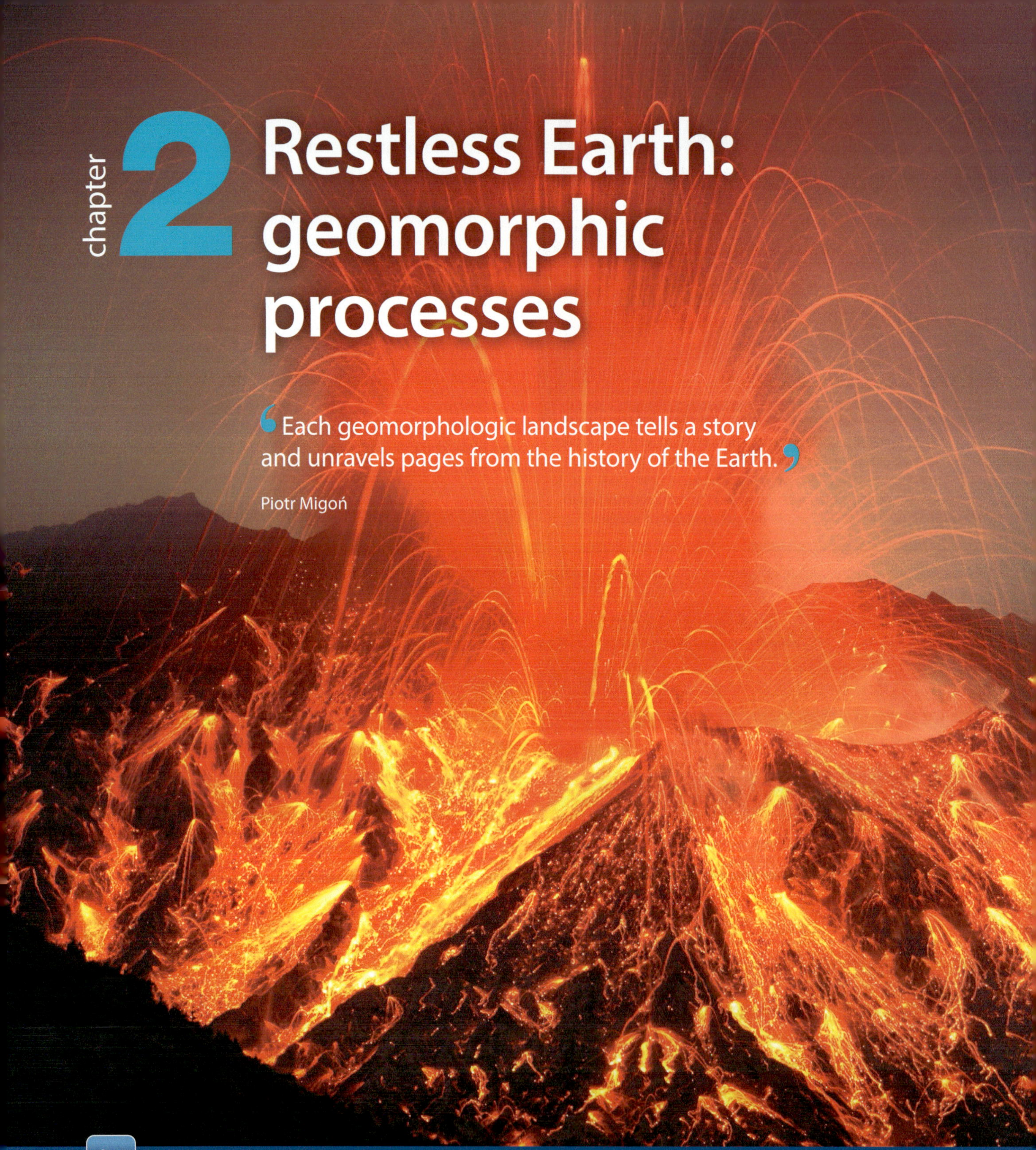

Aa Geovocab

continental drift: theory describing how continents broke away and drifted from their original landmass

dyke: magma cutting through adjacent rock

faulting: cracking in Earth's crust resulting from pressure exerted, ranging from a few centimetres to hundreds of kilometres in length

folding: bending of layered rock when pressure in Earth is exerted forming synclines and anticlines

geomorphic processes: physical and chemical interactions that produce landforms

inselberg: isolated hill or mountain rising abruptly from a plain

laccolith: landform created when magma between two layers of sedimentary rock forces the overlying strata upwards, giving a mushroom-like form at the top with a flat base

plate tectonics: the theory of slowly moving plates that make up Earth's crust

ISBN 978 1 4586 6277 4

Eruption at Sakurajima Volcano, Japan

Ring of Fire: zone of volcanic activity surrounding the Pacific Ocean
rock cycle: cycle of changes to rocks; for example igneous rocks can change into sedimentary rocks or into metamorphic rocks
sill: magma running parallel to rock strata
tectonic plates: slowly moving plates that make up Earth's crust
volcano: opening in Earth's crust through which molten lava, ash and gases are ejected
weathering: breakdown of rock by physical and/or chemical processes

ISBN 978 1 4586 6277 4

This chapter investigates the geomorphic processes that produce Earth's landforms such as plate tectonics, and chemical and physical weathering. These natural forces change rivers, coasts and hot and cold desert landforms, through the physical process of erosion, transportation and deposition.

Movement of tectonic plates and folding and faulting contributes to geomorphic hazards such as volcanic activity, and earthquakes that cause tsunamis and landslides (see Chapter 5). As the tectonic plates constantly move, they contribute to changes in landscapes and their landforms.

Think, puzzle, explore

- **Place:** How do geomorphic processes produce landforms in different places?
- **Space:** What is the spatial distribution of volcanoes across Earth?
- **Environment:** How do different environments cause physical and chemical weathering?
- **Interconnection:** What is the interconnection between rock type and formation of a landform?
- **Sustainability:** How can humans use volcanic soils for sustainable agriculture?
- **Scale:** What are the scales of geomorphic processes from local to global?
- **Change:** How have geomorphic processes changed landforms over time?

Geo**skills** in focus

- **Identifying** the variety of geomorphic processes and landforms and their changes over time using the inquiry process
- **Collecting** and **recording** relevant geographical data on landforms and their changes over time due to geomorphic processes from primary and secondary sources, such as satellite and digital images
- **Evaluating** sources for reliability and usefulness
- **Representing** spatial distribution of volcanoes and changes over time by constructing maps using data from NASA and Geoscience Australia
- **Interpreting** topographic maps, digital terrain models, cross-sections and block diagrams to investigate geomorphic processes
- **Concluding** by presenting an oral report supported by an audio-visual display to communicate that geomorphic processes are combinations of physical and chemical interactions between Earth's surface and the natural forces that operate at a variety of scales
- **Reflecting** on the complexity of geomorphic processes and landforms and the need for their protection

2.1 Geomorphic processes

Plate tectonics, chemical weathering and physical weathering (such as erosion, transportation and deposition) are the major **geomorphic processes**. Geomorphic processes vary according to rock type and climate, and operate at speeds ranging from slow to fast.

Geomorphic processes are part of the **rock cycle**, which continually recycles Earth's materials. As new crust is created, older crust is subducted back into the mantle to be melted once again. The slow-motion movement of Earth's crustal plates is called **plate tectonics**.

Landforms undergo several stages of development. The Sydney Basin was under the sea 250 million years ago, and then volcanoes exploded 200 million years ago. About 14 to 20 million years ago, a huge lava flow, 50–60 m thick, covered the Blue Mountains. The mountains have since been eroded away by rivers, leaving high areas such as Mt Tomah.

Geomorphic processes at Uluru

Uluru is a famous Australian **inselberg**, or isolated hill. The landform began hundreds of millions of years ago. Its current shape was formed about 60 million years ago by geomorphic processes, such as folding, weathering and erosion.

The formation of Uluru is explained by the fact that the area was once covered by sea. Over time the area became covered with marine sediments and was folded. When material covering the area was eroded Uluru started to appear above the ground (see pages 122–123).

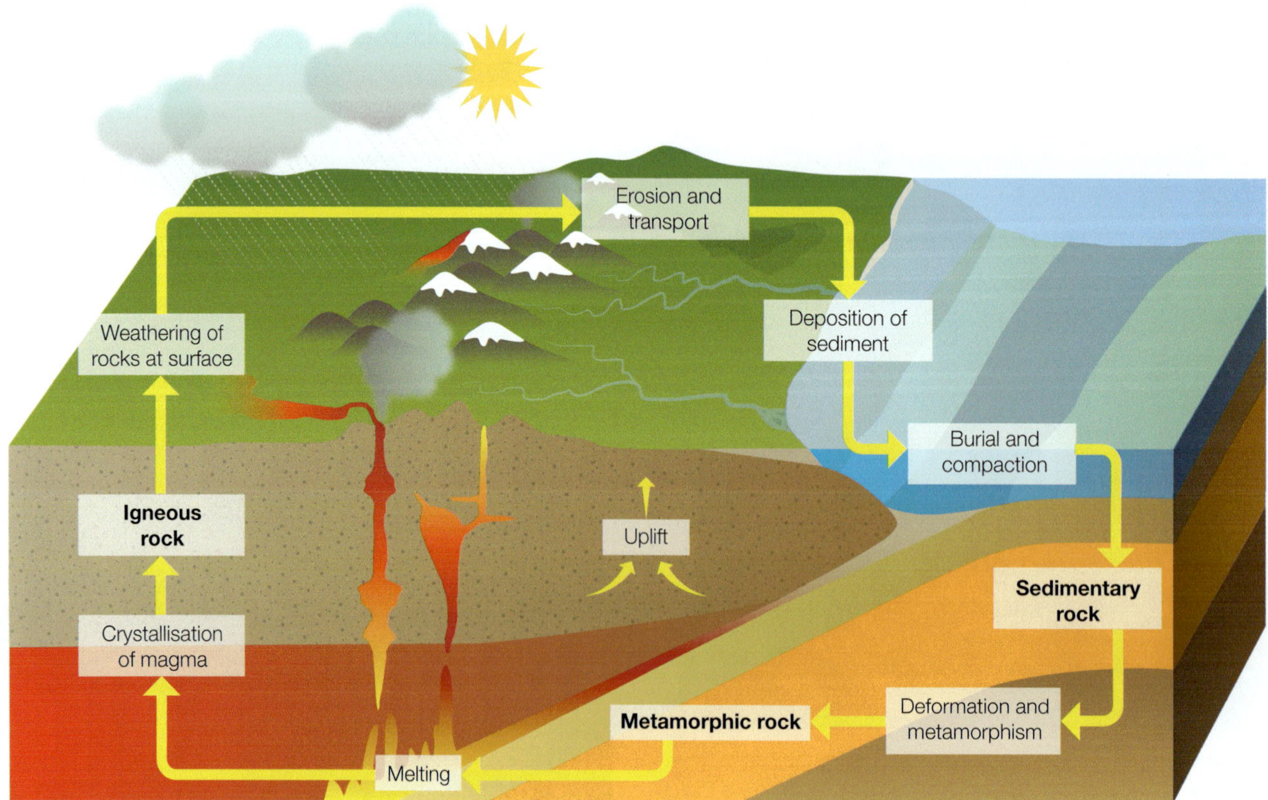

2.1.1 Geomorphic processes operate within the rock cycle

ISBN 978 1 4586 6277 4

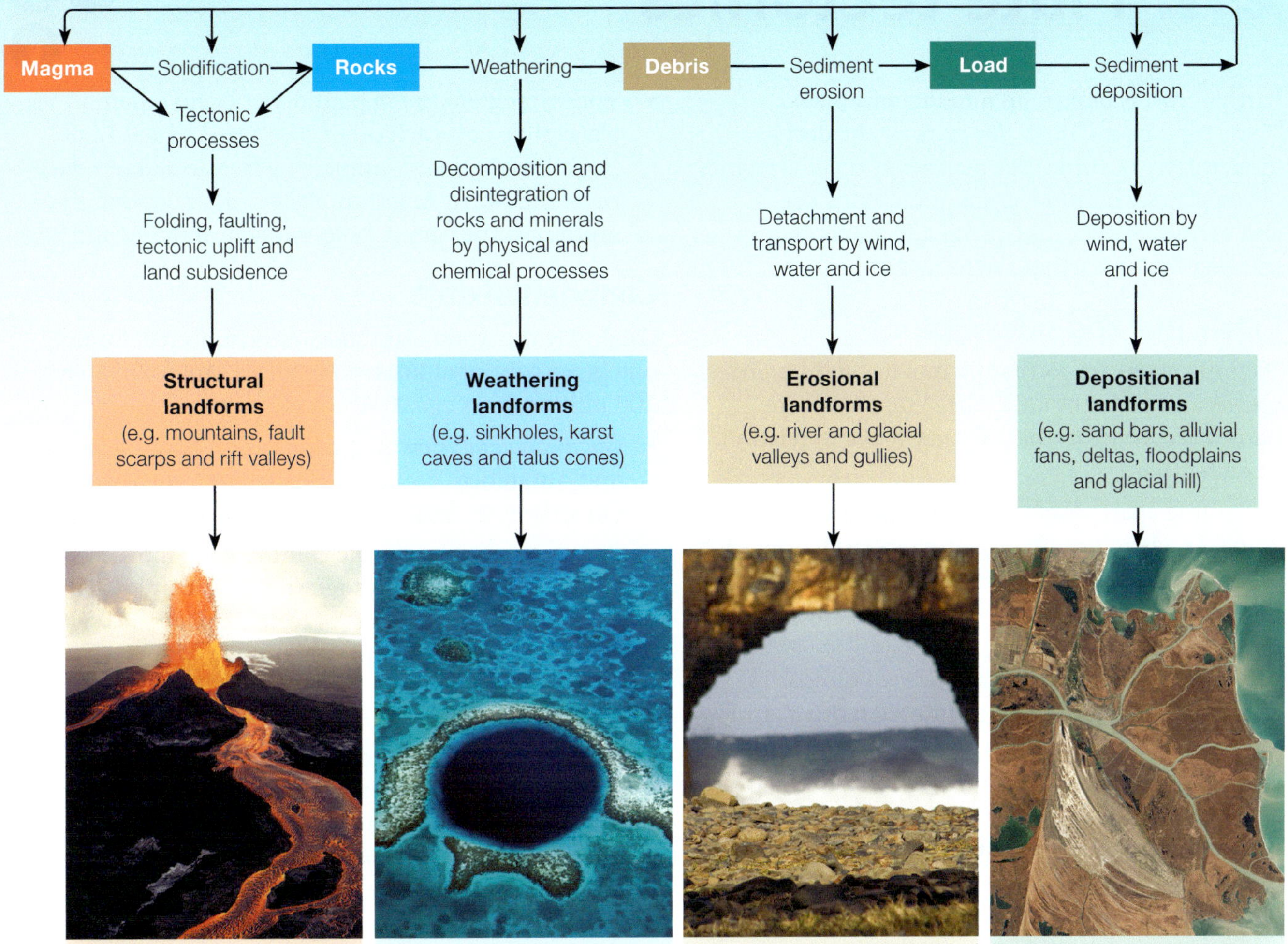

Magma — Solidification → Rocks — Weathering → Debris — Sediment erosion → Load — Sediment deposition

Magma → Tectonic processes → Rocks

Tectonic processes → Folding, faulting, tectonic uplift and land subsidence → **Structural landforms** (e.g. mountains, fault scarps and rift valleys)

Weathering → Decomposition and disintegration of rocks and minerals by physical and chemical processes → **Weathering landforms** (e.g. sinkholes, karst caves and talus cones)

Sediment erosion → Detachment and transport by wind, water and ice → **Erosional landforms** (e.g. river and glacial valleys and gullies)

Sediment deposition → Deposition by wind, water and ice → **Depositional landforms** (e.g. sand bars, alluvial fans, deltas, floodplains and glacial hill)

Volcanic eruptions in Hawaii, USA, are tectonic processes that create landforms

Blue Hole, Belize, is the world's largest sinkhole. The 300 m wide by 120 m deep solution hole was created by chemical weathering

Hole in the Rock outside Coffee Bay, Transkei, South Africa, is an example of a coastal erosion landform

Danube River Delta, Ukraine, shows depositional landforms when the Danube River empties into the Black Sea

2.1.2 Relationship between geomorphic processes and landforms

Geo**activities 2.1**

Knowledge and understanding

1 What are geomorphic processes?

2 List three main geomorphic processes.

Inquiry and skills

3 Refer to 2.1.1 and explain how geomorphic processes are part of the rock cycle.

4 Refer to 2.1.2 and compare the processes and features of structural and depositional landforms.

5 Discuss in a short report how landforms such as Uluru and the Sydney Basin are not static but change over time.

6 The study of geomorphology involves creative thinking. How do geologists know what something looked like millions of years ago? Do you think it is imagination or fact? Explain your answers.

7 Using the internet, research one of the following inselbergs in Australia: Mt Oxley, Pildappa Rock or Murphys Haystacks. Present your findings as a PowerPoint presentation.

ISBN 978 1 4586 6277 4

2.2 Plate tectonics

Earth's crust is broken up into **tectonic plates**. These rigid plates 'float' like rafts on the deeper, molten layer of the crust's mantle. At the boundaries of the plates, volcanoes erupt, mountains rise and earthquakes occur (see 5.6 and 5.7 for more information on earthquake hazards).

Crash, pull and sideswipe

Tectonic plates are constantly moving at different speeds and different rates. Plate boundaries are named according to the way they move in relation to each other.

- Pulling apart: *divergent* or *spreading* boundaries occur where tectonic plates move away from each other. For example, North Ethiopia is located where the Arabian Plate is drifting away from the African Plate.
- Crashing: *convergent* or *destructive* boundaries occur where Plates collide into each other, and one plate moves under the other plate. The collision of the Indian Plate and the Eurasian Plate 40–50 million years created the Himalayan Mountains. Northern Australia was once a separate landmass. Less than 2 billion years ago, it crashed into central Australia, near Alice Springs, and moved underneath it.
- Sideswiping: *transform* boundaries occur when plates slide past each other. The San Andreas Fault in California is an example of a transform boundary. Here, the North American Plate slides horizontally south past the Pacific Plate, which is moving north.

Continental drift

Originally the world's landmass was connected, forming one super-continent known as Pangea. Approximately 200 million years ago, Pangaea split into:

- *Gondwanaland*: now Australia, Africa, South America and Antarctica
- *Laurasia*: now Asia, North America and Europe.

Plate tectonics caused the continents to move away from each other—a process referred to as **continental drift**. Today, the movements of tectonic plates are detected by remote sensing satellite data.

Ring of Fire

The 40 000 km **Ring of Fire**, located around the Pacific Ocean, is where some of the plates collide. The horseshoe shaped ring is dotted with 452 volcanoes, and is the location of 81% of the world's earthquakes. Along the Ring of Fire are active volcanoes such as Mount Ruapehu in New Zealand and Mount Fuji in Japan. In 1883, a volcano in Indonesia, Krakatoa, erupted and sent volcanic ash and gas 80 km into the air.

2.2.1 Global pattern of tectonic plates

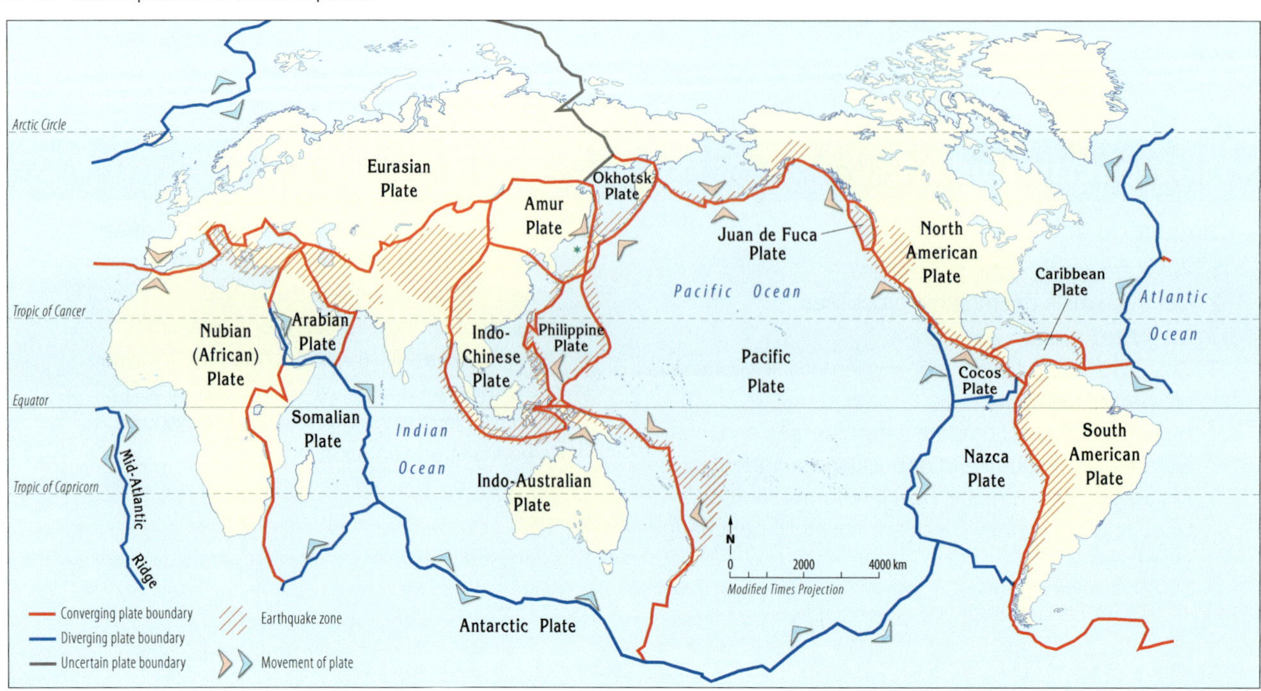

ISBN 978 1 4586 6277 4

2.2.2 Three types of plate boundaries

	Divergent boundary	Convergent boundary	Transform boundary
Movement	Spreading apart	Colliding and subducting	Lateral sliding
Result	Constructive	Destructive	Nil
Topography	Ridges and rises	Trenches	Small hills
Geological activity	Volcanic	Volcanic	Earthquakes
	A Ridge Lithosphere	B Volcanoes (volcanic arch) Trench Earthquakes	C Earthquakes within crust

2.2.3 In the future the Somalian and Arabian plates will break away from the Nubian Plate

Present

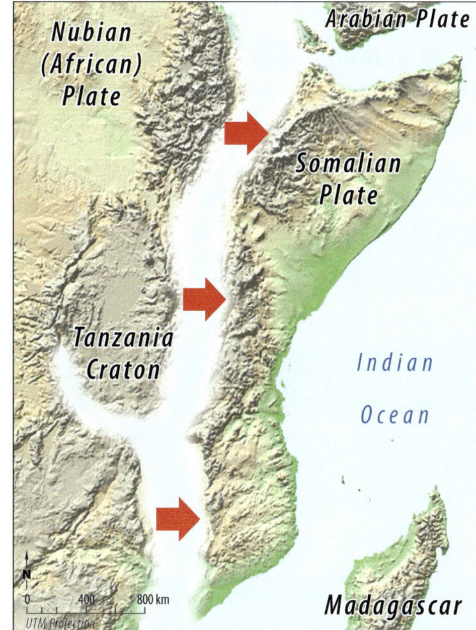

Future

Geoinfo

Tectonic plates move from 2 to 10 cm per year—about the speed at which fingernails grow.

Geoactivities 2.2

Knowledge and understanding

1 Explain the terms *plate tectonics*, *continental drift* and *Ring of Fire*.

2 List the continents that were part of Gondwanaland.

Inquiry and skills

3 Refer to 2.2.1.
   a Name the countries on the edge of the Pacific Plate.
   b State whether the following plates are converging or diverging: above India, Antarctica and Europe.

4 Refer to 2.2.2.
   a Describe the movements of diverging, converging and transforming tectonic boundaries.
   b Name two plate boundaries where volcanic activity occurs.

5 Refer to 2.2.3. Imagine you visited Somalia in a few million years. Describe what happened to the country. What were the geomorphic processes responsible for these changes? Present information using a web 2.0 tool.

ISBN 978 1 4586 6277 4

2.3 Wrinkles and breaks: folding and faulting

Folding: crumples and wrinkles

Tectonic forces compress and buckle the crust to create various types of folds. Folds are characterised by peaks (anticlines) and valleys (synclines). They vary from small (a few centimetres) to large (kilometres). These also vary from simple symmetrical and asymmetrical folds to more complex overturned and overthrust folds.

Folding has produced mountain ranges and peaks, including the Himalayan Mountains, Rockies, Andes and Alps. Fold mountains form at converging tectonic plates where they collide. As there is no place to go they fold themselves upward into a bump.

Fold mountains are used in a variety of activities. The Nepalese terrace hillsides to grow crops. Skiing, water rafting, abseiling and walking are popular with tourists in mountain ranges. Some fold mountain regions with fast-flowing rivers generate hydroelectric power.

Geoinfo

More resistant rock strata form ridges while less resistant strata often form valleys.

2.3.2 Folded rocks in the San Andreas Fault

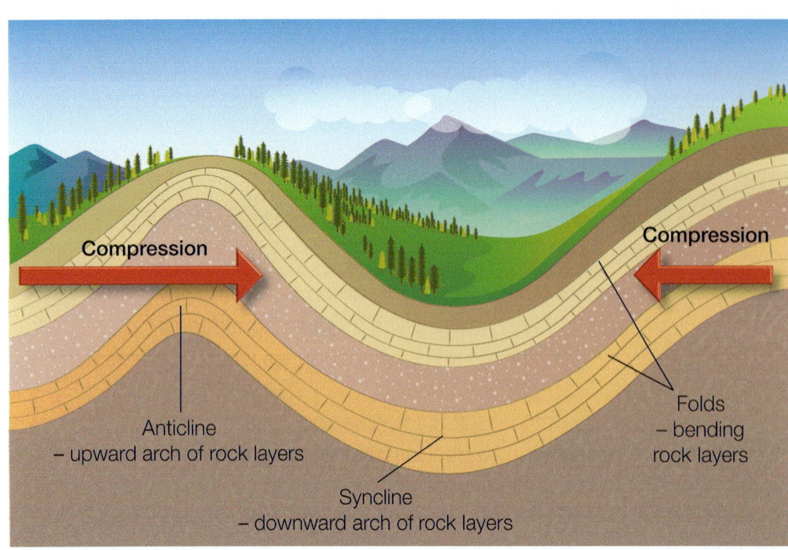

2.3.1 Folding: anticline and syncline

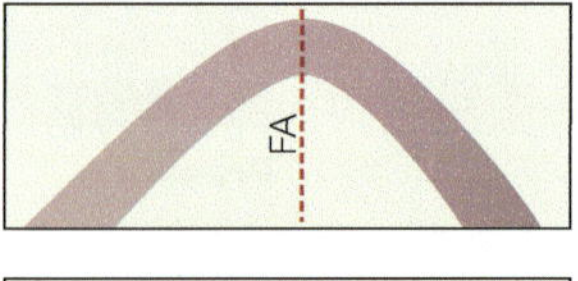

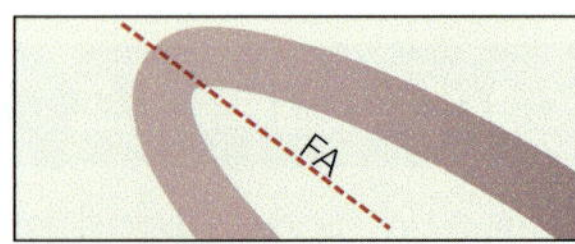

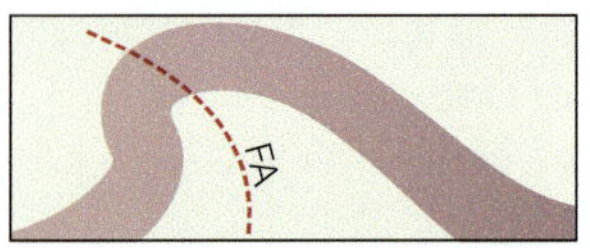

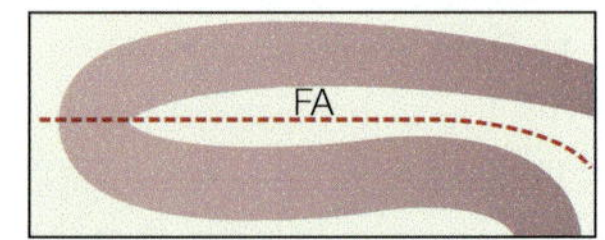

2.3.3 Major types of folds

ISBN 978 1 4586 6277 4

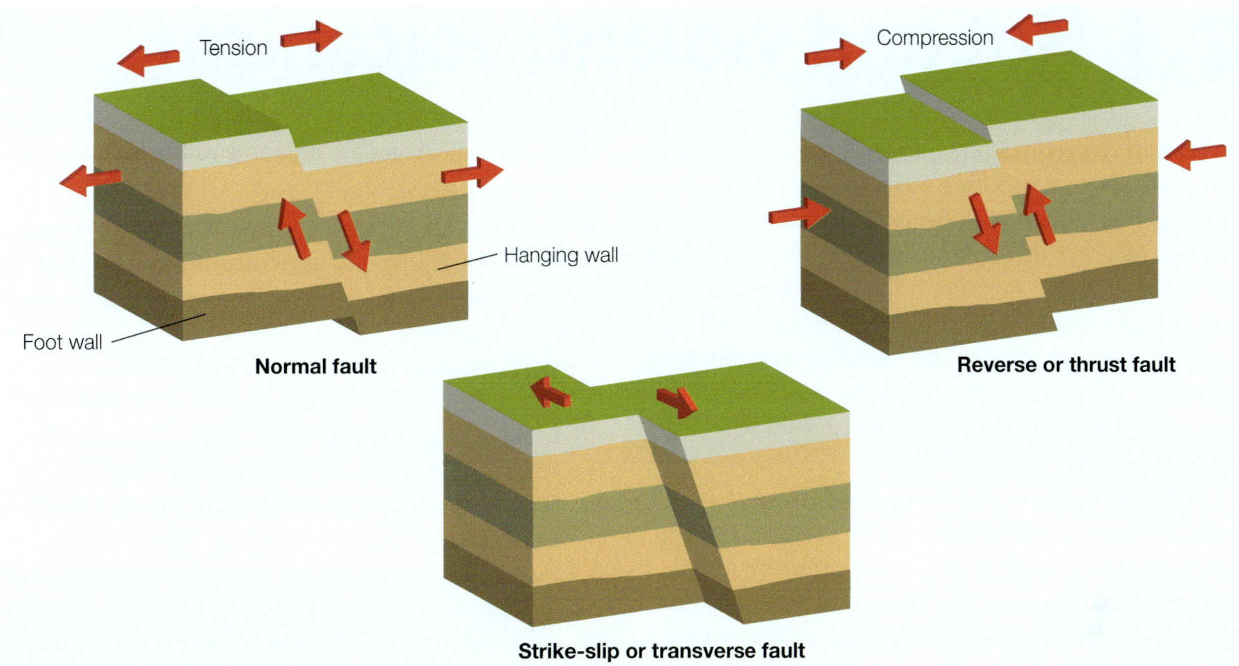

2.3.4 Three main types of faults

Faulting: breaks

When landforms are too hard to fold, **faulting** occurs. Compression forces the bedrock to fracture or break. Faults may result when land either:

- *moves downwards*, as in grabens or rift valleys (e.g. the East African Rift Valley, Gulf of St Vincent in South Australia and Narmada River Valley in central India)
- *moves sidewards* (e.g. the San Andreas Fault in California and the Alpine Fault in New Zealand).

In many cases, faults occur in groups called fault blocks. Depressed blocks are called grabens and raised blocks called horsts.

Folding and faulting cause earthquakes, which can impact significantly on humans. Scientists are therefore trying to gain a better understanding of these geomorphic processes.

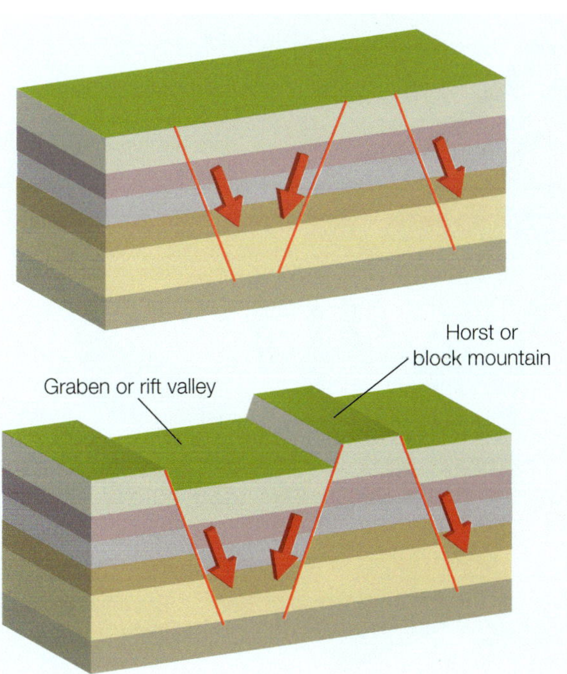

2.3.5 Faults forming block mountains (up) and rift valleys (down)

Geo**activities 2.3**

Knowledge and understanding

1. List two forces that change and displace Earth's crust.
2. What causes folding?

Inquiry and skills

3. Refer to 2.3.1. What is the difference between a syncline and an anticline?
4. Refer to 2.3.3 and list the four major types of folds.
5. Refer to 2.3.4 and name three types of faults.
6. Refer to 2.3.5 and distinguish between a graben and a horst.
7. Using geographical concepts prepare a poster that informs people of the early warning signs that an earthquake may be imminent.
8. Research using the internet and describe different human activities occuring in fold mountains, such as the Alps in Europe, in summer and winter. Present as a Tagxedo.

ISBN 978 1 4586 6277 4

2.4 Hot and violent: volcanism

Volcanism is a geomorphic process by which volcanic materials create landforms. A volcano forms when magma (hot molten rock from beneath Earth's surface) emerges through an opening or weakness in the ground. When pressure builds up the volcano erupts. Some volcanoes erupt continuously and are called active (e.g. Mauna Loa, Hawaii, USA) while others are dormant or extinct. Mauna Kea on Hawaii is classified as dormant as its last eruption was about 2460 BCE.

Volcanoes consist of a vent, where magma is extruded from underground, and a crater at the top where material (such as lava, steam, pumice and volcanic bombs) is ejected. Volcanic eruptions can occur on a massive scale causing death and destruction to properties, or they may only exude hot gases.

However, volcanoes are valuable to humans due to their:

- *magma*—abundance of minerals that form rich soils which increase agricultural productivity (e.g. grapes in Sicily, Italy, and cows in Taranaki, New Zealand)
- *sulphur*—used in the chemical industry
- *spring water*—source of drinking water, believed to have healing properties
- *basalt*—hard rock for building roads
- *tuff*—soft rock for carving statues in Bali and Easter Island
- *geothermal energy*—clean energy for industry and homes.

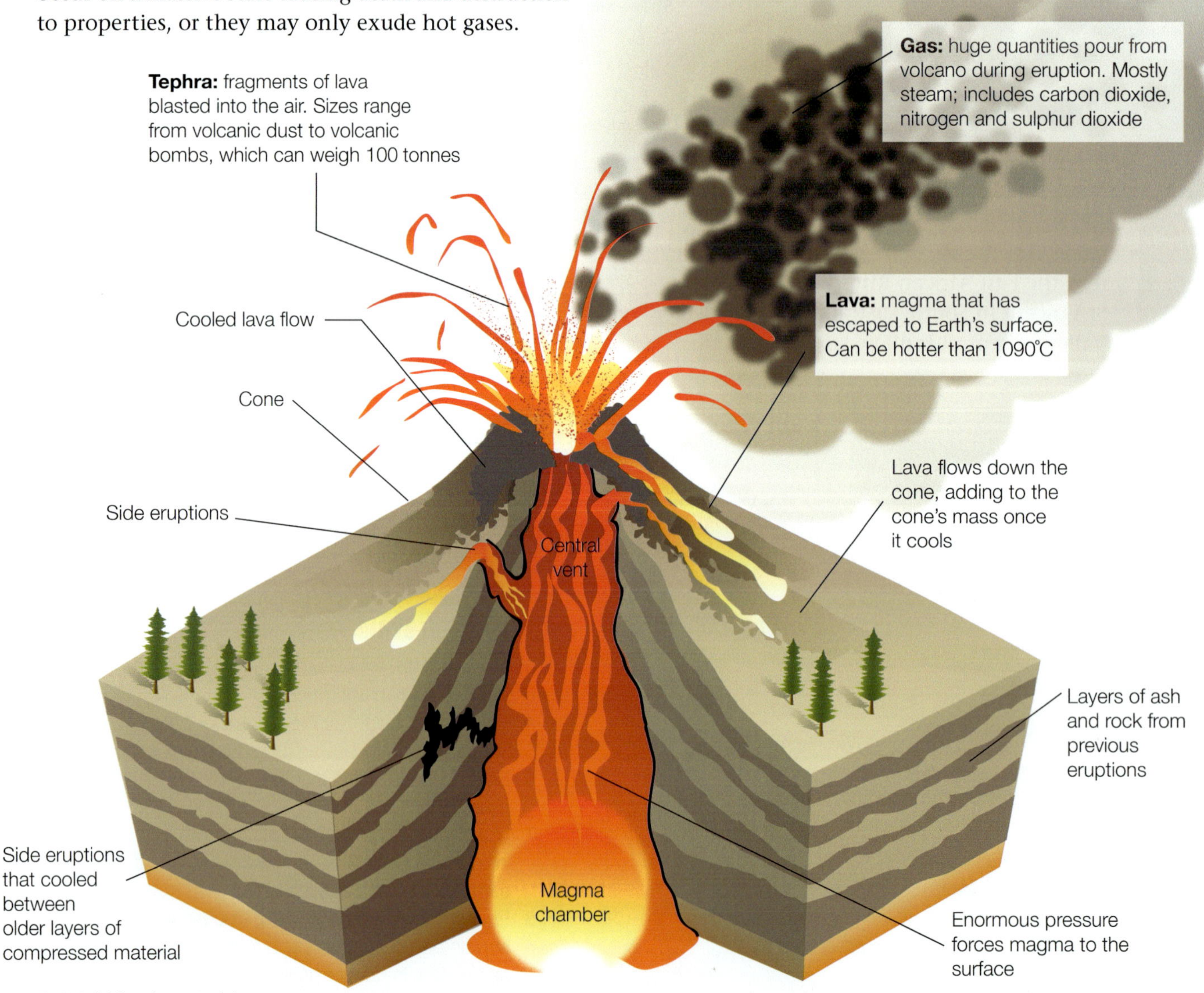

Gas: huge quantities pour from volcano during eruption. Mostly steam; includes carbon dioxide, nitrogen and sulphur dioxide

Tephra: fragments of lava blasted into the air. Sizes range from volcanic dust to volcanic bombs, which can weigh 100 tonnes

Cooled lava flow

Lava: magma that has escaped to Earth's surface. Can be hotter than 1090°C

Cone

Lava flows down the cone, adding to the cone's mass once it cools

Side eruptions

Central vent

Layers of ash and rock from previous eruptions

Side eruptions that cooled between older layers of compressed material

Magma chamber

Enormous pressure forces magma to the surface

2.4.1 Volcanic materials

ISBN 978 1 4586 6277 4

Geo**info**

- Earth's largest volcano is Mauna Loa on Hawaii. It's made up of 75 000 m^3 of volcanic material.
- Australia's only active volcano is Big Ben, a 25 km wide composite cone on Heard Island.

Multiple shapes

Most people think a volcano is the shape of a conical mountain. However, this is only one of many types of shapes of volcanoes. Others include shields, cinder cones, domes and composites, depending on the composition of material ejected from the eruption.

When pressure builds up in the vent and blows out the plug, a caldera forms. Blocks of hot lava and rock explode as lava bombs are catapulted from the crater. Once dormant, craters can fill with water to form crater lakes, such as Lake Batur in Bali.

(See 5.3, 5.4 and 5.5 for information on volcanic hazards.)

Geo**activities 2.4**

Knowledge and understanding

1 What is volcanism?

2 What type of landforms do volcanoes produce?

Inquiry and skills

3 Refer to 2.4.1. Describe the three types of volcanic materials.

4 Refer to 2.4.2.

 a Explain the three main types of volcanic eruptions.

 b Describe the shape of the five types of volcanoes.

5 Explain why volcanism is one of the most significant geomorphic processes on Earth.

6 Imagine you are in a helicopter hovering above an active (but not erupting) volcano. Create a five-senses picture or word chart to explain your experience.

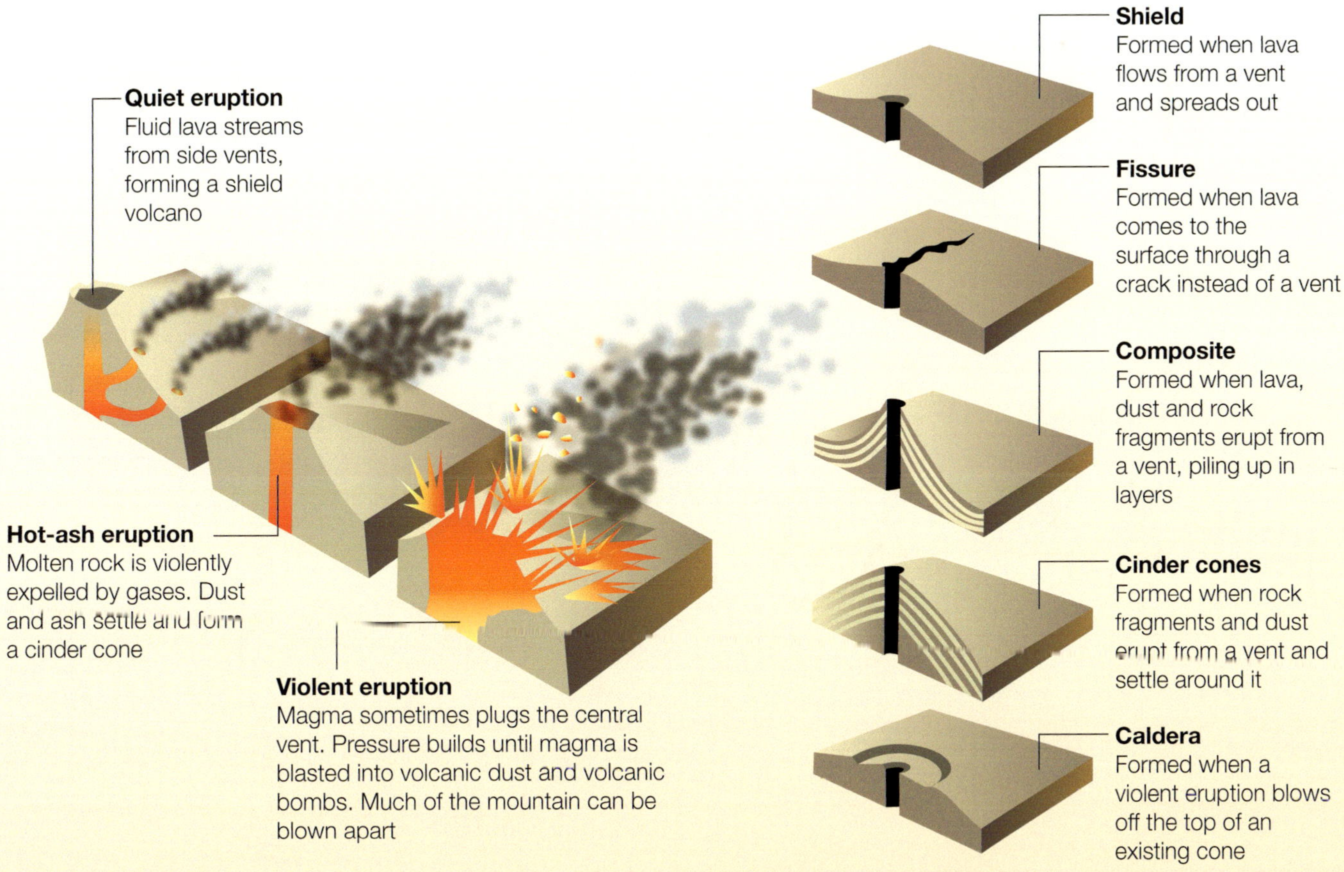

Quiet eruption
Fluid lava streams from side vents, forming a shield volcano

Hot-ash eruption
Molten rock is violently expelled by gases. Dust and ash settle and form a cinder cone

Violent eruption
Magma sometimes plugs the central vent. Pressure builds until magma is blasted into volcanic dust and volcanic bombs. Much of the mountain can be blown apart

Shield
Formed when lava flows from a vent and spreads out

Fissure
Formed when lava comes to the surface through a crack instead of a vent

Composite
Formed when lava, dust and rock fragments erupt from a vent, piling up in layers

Cinder cones
Formed when rock fragments and dust erupt from a vent and settle around it

Caldera
Formed when a violent eruption blows off the top of an existing cone

2.4.2 Types of volcanoes and eruptions

ISBN 978 1 4586 6277 4

2.5 Hot today, cold tomorrow

Volcanoes are generally found where tectonic plates are diverging or converging. They can also be found away from plate boundaries where Earth's crust is thin and stretched (e.g. in the East African Rift Valley) or in hotspots (e.g. in Hawaii), which rise from the molten core deep in the Earth and feed volcanoes. Hotspots are stationary but the Earth's crust moves, so over time these volcanoes become extinct. Old oceanic volcanoes cool and are eroded into atolls and seamounts. However, new volcanoes form over the hotspots and the cycle continues.

Hotspots

One of the world's most famous hotspots is the Hawaiian Islands. The oldest islands are mainly inactive volcanoes formed millions of years ago. The newest island is the big island of Hawaii, which contains two active volcanoes.

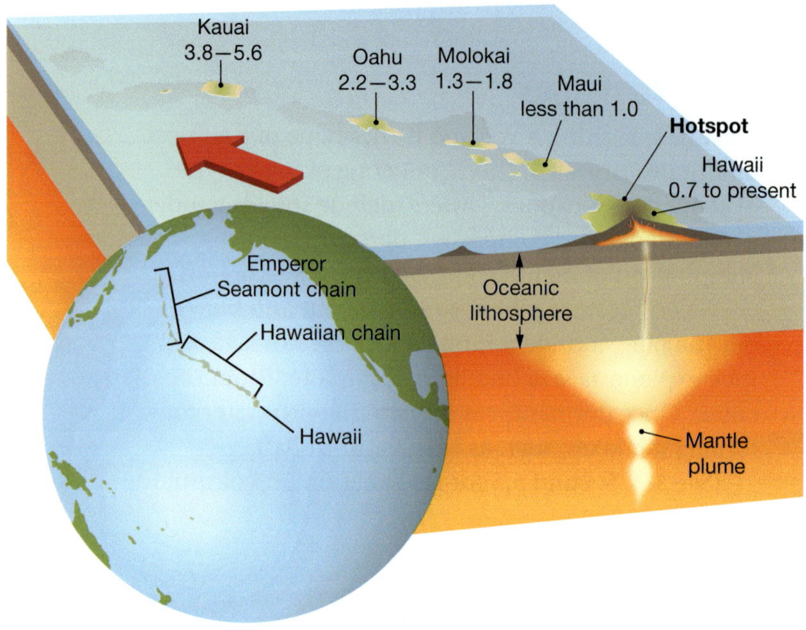

2.5.2 Formation of the Hawaiian Ridge–Emperor Seamount Chain. The oldest exposed rocks are up to 5.6 million years old and new volcanic rock is continually being formed

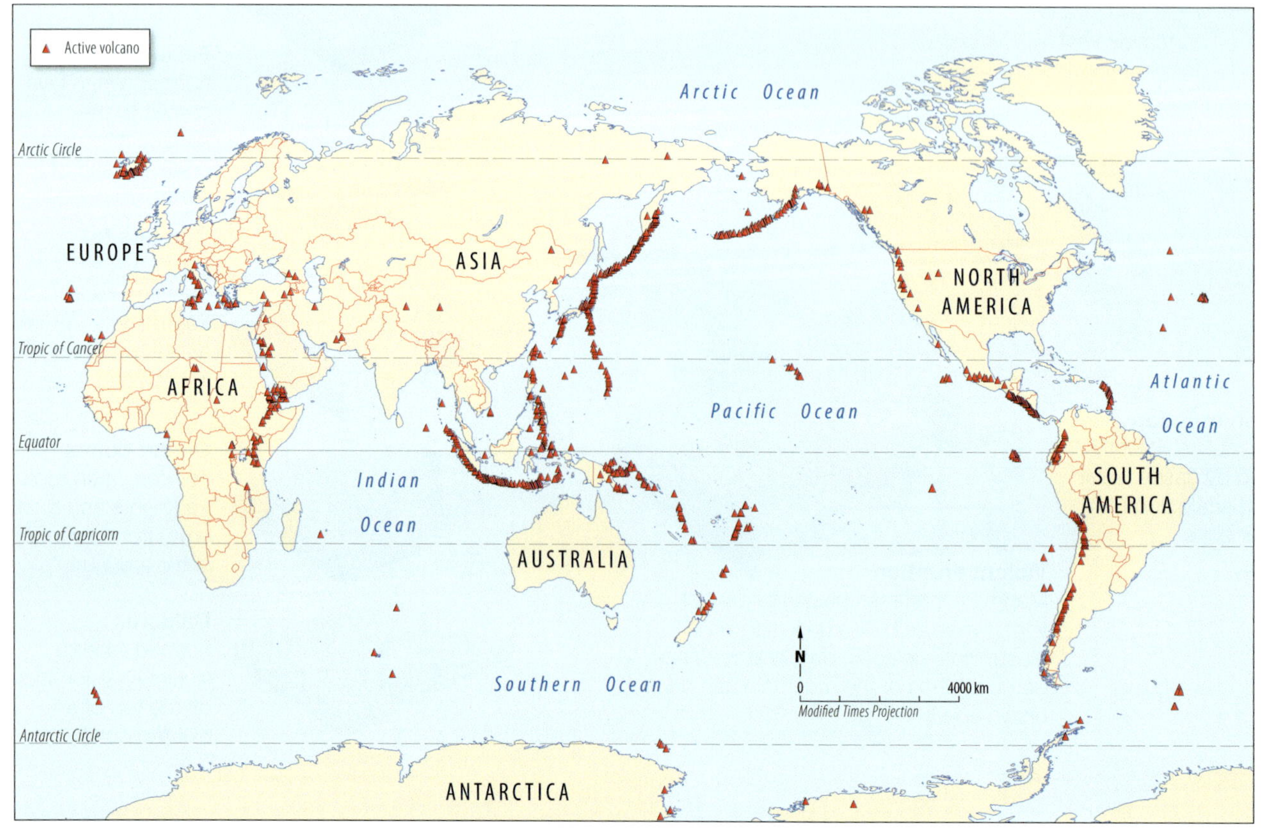

2.5.1 Spatial distribution of volcanoes around the world

ISBN 978 1 4586 6277 4

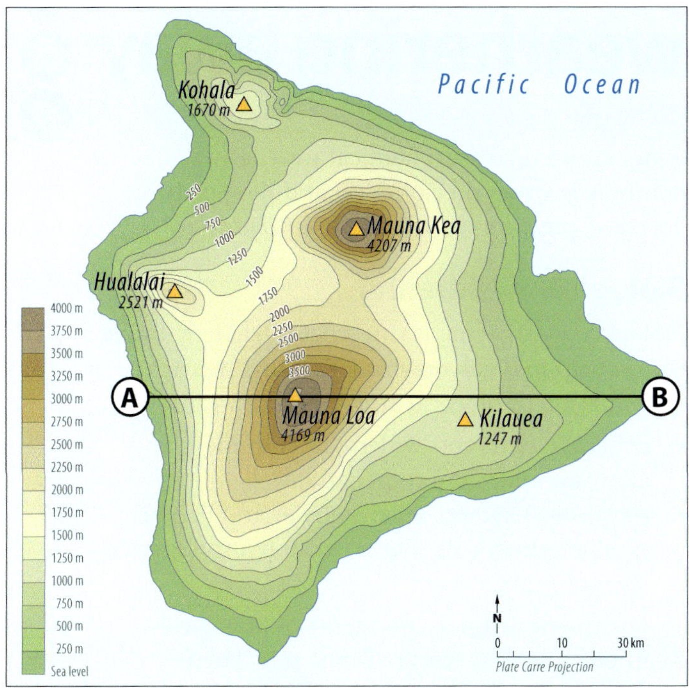

2.5.3 Topographic map of Hawaii's Mauna Loa shield volcano (see also 1.15–1.18 for topographic skills)

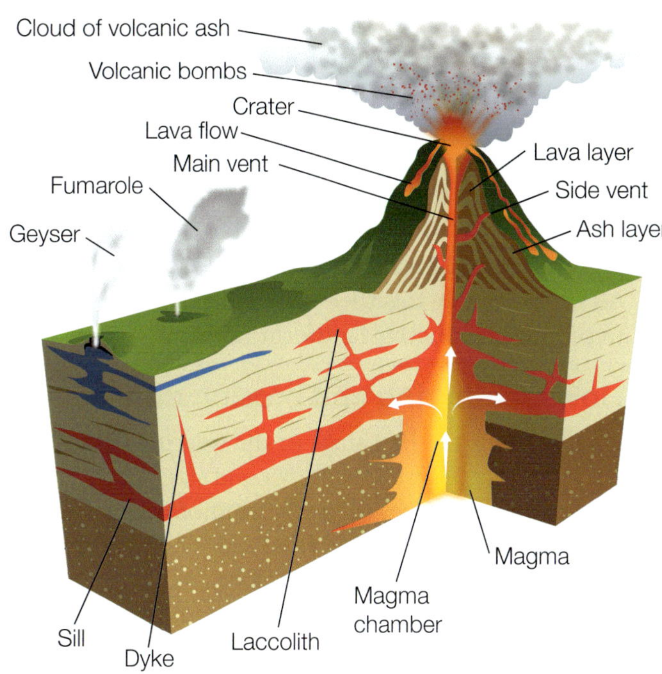

2.5.4 Intrusive and extrusive volcanism

Both old and new volcanoes have been fed by the same hotspot. As the Earth continues to move, additional islands will form over this same hotspot.

Yellowstone National Park, USA, is also located on a hotspot. The North American Plate is shifting in a south-west direction over a stationary hotspot. Over the last 17 million years a series of volcanoes formed a track, with the most recently active Yellowstone volcanoes at its north-east end. The track is referred to as the Snake River Plain.

Intrusive and extrusive volcanism

There are two basic volcanic processes, each named after its origin:

- *Extrusive*—includes volcanic materials ejected from Earth's crust. Thousands of square kilometres of Earth are covered by thick basalt lava flows, some more than 50 km thick.
- *Intrusive*—includes volcanic materials within Earth's crust. Most magma that enters Earth's crust solidifies deep underground. Magma solidifies slowly to form large, dome-shaped batholiths. When the batholiths reach the Earth's surface, erosion leaves **dykes** (magma that cuts through adjacent rock), **sills** (magma running parallel to rock strata) and **laccoliths** (magma between two layers of sedimentary rock, forcing the overlying strata upwards and giving a mushroom-like form at the top).

Geo**activities 2.5**

Knowledge and understanding

1. Where are volcanoes located along tectonic plates?
2. What are batholiths? Describe what happen after they solidify and are exposed at the surface.

Inquiry and skills

3. Refer to 2.5.1.
   a. Name two countries that have volcanoes.
   b. Why are volcanoes located inland in north-east Africa?
   c. Volcanoes are located in oceans. Name two of these oceans.
4. Refer to 2.5.2.
   a. What is a hotspot?
   b. Name the oldest volcano.
   c. Explain how hotspots cause volcanic eruptions in different places over time.
5. Refer to 2.5.3.
   a. Draw a cross-section from A to B.
   b. What is the contour interval?
   c. What is the approximate length and width of the Mauna Loa island?
6. Refer to 2.5.4. Research the Tower Hill eruptions and explain what Indigenous Australians near the eruptions would have witnessed.

ISBN 978 1 4586 6277 4

2.6 Getting older: weathering away

Weathering is the disintegration and decomposition of rock in situ (where the rocks are sited) at or near Earth's surface. There are three types of weathering: physical (mechanical), chemical and biological. All types operate together—not independently—to produce distinctive landform features.

How weathering and erosion differ

Weathering is the first stage in the wearing down of the landscape. Erosion is the second stage of the rock cycle, which involves the movement of weathered material away from the original site. The weathered sediments are transported and deposited by agents of erosion, such as running water, ice, wind and waves.

Factors controlling weathering

Several factors determine the type and rate of weathering. These include:

- rock type
- rock texture
- rock strength and hardness
- mineral composition of rock
- rock joints, which allow access for water.

Other influences include type of vegetation and soil, and steepness of topography. However, the biggest influence on the type and rate of weathering is climate. In general, when precipitation and temperature increase, chemical weathering dominates; when precipitation and temperature decrease, physical weathering dominates.

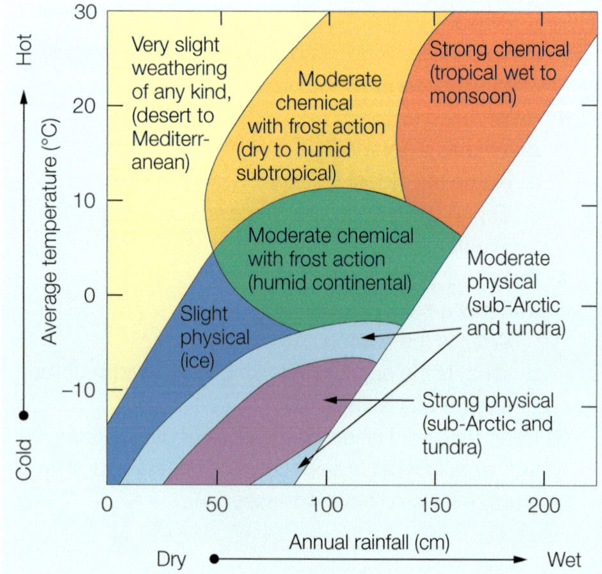

2.6.1 Influence of temperature and rainfall on physical and chemical weathering

Biological weathering

Biological weathering is the process whereby living things help to break down rock. It involves a combination of physical and chemical weathering:

- *physical*—tree roots widen joints in rocks and allow air and water to penetrate
- *chemical*—bacteria, algae, mosses and lichens produce chemicals that break down the surface layer of rocks.

Burrowing animals are agents of biological weathering as they turn over the soil. These animals range from small ants, termites and earthworms to larger moles and wombats. Additionally, sea creatures such as shellfish and

2.6.2 Quiver trees growing among rocks in Namibia, Africa. Roots from these trees prise rocks apart as they grow. This enables air and water to further break down the rocks through weathering processes

ISBN 978 1 4586 6277 4

2.6.3 This blue parrotfish feeding on algae on the rock surface also scrapes away the surface layer of the rock

parrotfish scrape algae from rock surfaces, slowly wearing away the surface. Trampling animals also expose rock to weathering.

2.6.4 Biological deposition: a termite mound in Mole National Park, Ghana

Geo**info**

Wombat burrows can be up to 30 m long and almost 2 m deep. Their burrows on the Nullarbor Plain are visible in satellite images.

Geo**activities 2.6**

Knowledge and understanding

1 Define weathering.
2 What are the three types of weathering?
3 How is weathering different from erosion?
4 List the factors controlling weathering.
5 What is biological weathering?
6 Give three examples of biological weathering.

Inquiry and skills

7 Refer to 2.6.1.
   a Describe the relationship between weathering and climate.
   b Which type of weathering dominates in tropical wet to monsoon climates and sub-Arctic and tundra regions?
   c Give the average temperature and rainfall ranges for humid continental climates.
8 Research the role of weathering in the formation of iron ore and bauxite. Write a report on these two minerals, outlining their location in Australia and their importance to the economy.

ISBN 978 1 4586 6277 4

2.7 Weathering changes landforms

Rocks formed deep in Earth's crust are slowly broken down by weathering processes at or near the surface. Chemical weathering breaks and changes rocks, whereas physical weathering breaks rocks apart but does not alter their make-up.

Physical weathering

Physical weathering is most effective in three places: deserts with a large diurnal (daytime) temperature range, mountains or high-latitude regions with temperatures that fluctuate around freezing, and areas with little vegetation. Examples of physical weathering include:

- *frost wedging*—very common in regions where temperatures are cold. Water infiltrates into cracks in rocks and expands when frozen. Over time, this alternating freeze–thaw process shatters the rock into pieces. The pieces form scree slopes on sides of mountains
- *onion skin weathering (exfoliation)*—occurs mainly in deserts. The rocks heats up and expand by day. At night the rocks cool and contract. The alternate heat and cold causes the outer layers of rocks to peel off in thin sheets just like an onion
- *salt*—when saline water enters rock spaces and dries, salt crystals expand and break up the rock. This is most common in deserts and coastal areas. On sandstones this process produces honeycombing.
- *pressure release (unloading)*—when rocks formed deep underground (e.g. granites) are exposed to the surface, they crack and split if the weight of the overlying material is removed.

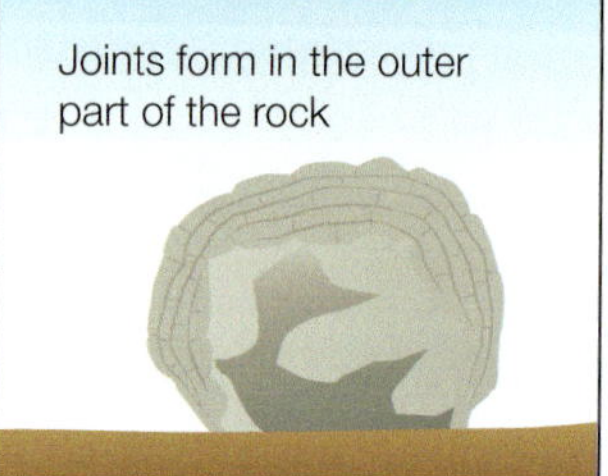

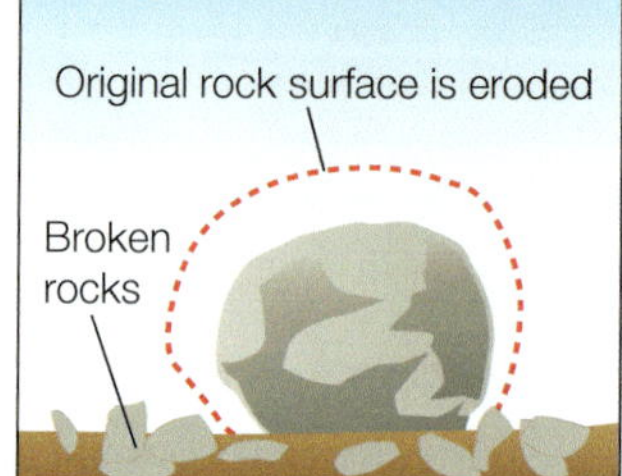

2.7.2 Onion skin weathering or exfoliation—rocks peel off in sheets rather than eroding grain by grain

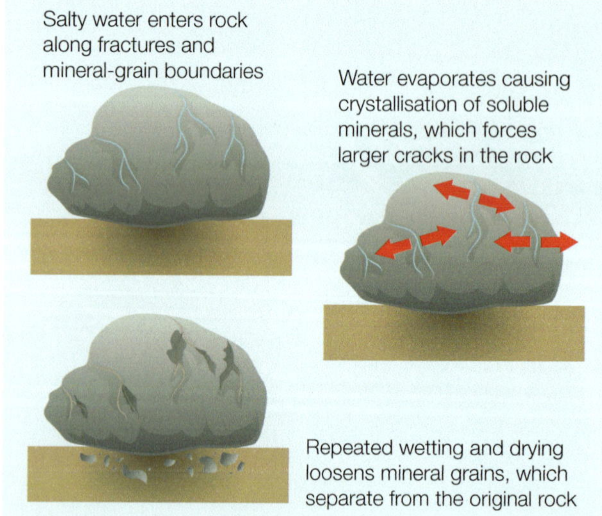

2.7.3 Salt breaks up rocks

2.7.1 Frost wedging process

ISBN 978 1 4586 6277 4

Chemical weathering

Chemical weathering breaks down or decomposes rocks and is the most important process in the formation of soil.

The rate of a chemical reaction increases as temperature rises. The most common chemical weathering processes are hydrolysis, hydration, oxidation, carbonation and solution.

- *Hydrolysis* and *hydration* occur when rocks react with split water molecules or absorb water, leading to changes in structures of new compounds.
- *Oxidation* occurs when rocks are exposed to oxygen in air and water. It makes iron minerals in rocks rust and take on a distinctive reddish colour.
- *Solution* is the removal of rock in solution by acidic rainwater. Rocks composed of calcium bicarbonate, such as limestone, can be weathered by rainwater that contains dissolved CO_2—the process of *carbonation*.

2.7.4 Formed by exfoliation 600 million years ago, Sugarloaf Mountain in Rio de Janeiro, Brazil, is one of the world's best-known granite domes

2.7.5 Frost wedging helped shape the Hoodoo rock formations in Bryce Canyon National Park, Utah, USA

Geo**activities 2.7**

Knowledge and understanding

1. Describe two types of weathering.
2. Distinguish between chemical and physical weathering.
3. Where is salt crystallisation most common?
4. How does oxidation operate?
5. Explain how carbonation weathers rocks.

Inquiry and skills

6. Refer to 2.7.1. Explain how frost wedging occurs and how it changes landforms.
7. Refer to 2.7.2. Explain onion skin weathering or exfoliation.
8. Refer to 2.7.3. How does salt break up rocks?
9. Investigate Sugarloaf Mountain in Brazil. How was it formed? How has the rock type contributed to the formation of the landform? Imagine you took a cable car to the top of the mountain. Describe the landscapes from a bird's-eye perspective using ICT.
10. Research the role of human activity in accelerating weathering processes, such as through increased carbon dioxide from enhanced greenhouse gases.
11. Investigate how resources produced from weathering, such as China clay, provide wealth to communities. Present as an annotated poster.

2.8 Rocks and sliding

Rocks are an important part of landforms. There are three main rock categories: igneous, sedimentary and metamorphic.

- *Igneous*—all rocks on Earth were originally igneous. They were formed as liquid magma when the Earth's upper mantle cooled.
- *Sedimentary*—these rocks are formed by the deposition of material on the Earth's surface and within bodies of water, such as lakes. The formation of layers of sediment is called bedding.
- *Metamorphic*—rocks that formed deep in the Earth, where heat and pressure caused the existing rocks to change their structure.

2.8.2 Main rock types and geomorphic processes

Rock landforms

The four major rock types produce distinctive landforms.

- *Granite*—granite forms when magma cools slowly deep underground. When water penetrates underground between the joints, it weathers the granite into rounded blocks. Once granite is exposed, it forms rounded boulders called tors. Granite also forms isolated resistant landforms known as inselbergs.

2.8.1 Exposed granite tors at the Devils Marbles Conservation Reserve, Northern Territory

- *Limestone*—rainwater, which is weak carbolic acid, dissolves limestone in a process called carbonation. This produces karst landforms, such as tower karsts; collapsed caverns, called dolines; limestone pavements; and cave depositional features, such as stalactites and stalagmites (see also chapter 1).
- *Sandstone*—sandstone is a sedimentary rock that is porous and permeable. Water passes along the horizontal bedding planes between layers and also through the rock. Sandstone forms distinctive landforms, such as the Three Sisters in the Blue Mountains.
- *Basalt*—basalt forms when magma flows to the surface and cools rapidly. It mainly produces volcanoes, mid-oceanic ridges and extensive lava flows. Basalt sometimes produces hexagonal, columnar jointing.

Sliding rocks

Sliding rocks refers to a rare geological phenomenon where rocks move along a smooth valley floor, such as Racetrack Playa, Death Valley, USA. The stones move every two to three years. Stones with rough bottoms leave straight tracks while those with smooth bottoms wander. The force behind their movement is uncertain.

ISBN 978 1 4586 6277 4

2.8.3 Basalt columns of Giant's Causeway, Ireland

2.8.4 Sliding rocks on Racetrack Playa, Death Valley National Park, California

Geoinfo

The karst landforms of Chillagoe, North Queensland, were formed millions of years ago when the Great Dividing Range uplifted an ancient Great Barrier Reef.

Aboriginal Peoples' use of rocks

Aboriginal and Torres Strait Islander Peoples made axes out of stone, and ground ochre to use as paint. In Tennant Creek in Central Australia, Aboriginal Peoples made blades out of quartzite and painted the handles with red ochre, white pipe clay and charcoal. Across Australia there are thousands of Aboriginal sites with rock art—especially in Sydney's sandstone area.

Grinding stones were among the largest stone implements used by Aboriginal Peoples. They were usually made from abrasive rocks, such as sandstone or basalt.

2.8.5 Aboriginal grinding stone and berries

Geoactivities 2.8

Knowledge and understanding

1 What characteristics of rocks influence landforms?
2 How does water create limestone landforms?
3 Describe the characteristics of sandstone that influences the creation of distinctive landforms.
4 What distinctive landforms are made of basalt?

Inquiry and skills

5 Refer to 2.8.1. Distinguish between sedimentary, igneous and metamorphic rocks.
6 Investigate where sedimentary, igneous and metamorphic rocks in Australia are located. Present as an annotated map.
7 Discuss how different types of rocks produce different landforms.
8 Reflect on your learning and investigate the main rock types in your local area and how they influence landforms.
9 Research the significance and use of rocks to Aboriginal and Torres Strait Islander Peoples and note the process involved in creating three examples of tools/weapons. Present your findings as a Prezi.

ISBN 978 1 4586 6277 4

2.9 Water and wind erosion

Erosion is a natural part of the rock cycle that occurs when sediments are removed by geomorphic agents, such as running water or wind. These agents loosen sediments, pick them up and transport them away. The erosion process is complete when sediments are deposited.

Water erosion

There are four categories of water erosion:

- *Sheet* erosion occurs when water flows over the ground and is not confined to streams or rivers.
- *Fluvial* erosion is carried out by streams and rivers and smooths rocks in rivers. The main fluvial processes are corrosion (or solution), attrition, abrasion and hydraulic action (see 1.20).
- *Coastal* erosion occurs when waves, tides and currents erode coasts producing wave-cut notch platforms and coastal cliffs (see 1.11).
- *Glacial* erosion occurs when glaciers with moraine (glacial sediment) erode valleys into a U-shape (see 3.12).

Wind erosion

Wind (or aeolian) erosion is more noticeable in arid regions where there is little vegetation (see 3.11). Wind is not as effective at moving large particles as rivers and glaciers, but it can pick up small particles and transport them long distances. There are two main processes of wind erosion:

- *Deflation* is the removal of fine sediments by winds in dry environments. As deflation progresses, all the fine sediment on the surface is removed, leaving behind the coarse sand and larger pebbles. Sometimes pebbles may form a hard surface called a desert pavement or reg.
- *Abrasion* is the process of smoothing and polishing rock surfaces by sandblasting. This happens when dry sand particles are moved by saltation (see 2.10). In general the wind cannot move sand more than a metre high, as erosion is concentrated close to the surface. Wind abrasion forms ventifacts (polished stones) and rock pedestals.

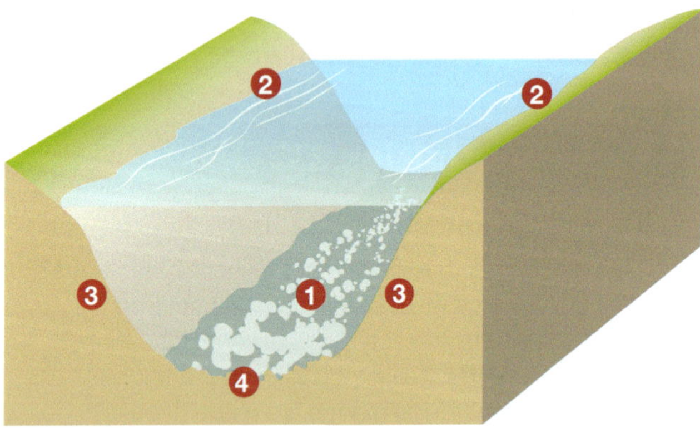

1 **Attrition:** collisions between sediment particle breaks them down and rounds their edges

2 **Hydraulic action:** the force of flowing water wears away the bed and the banks, e.g. outside bends of meanders and the base of waterfalls

3 **Corrosion or solution:** this continuous chemical erosion process dissolves rock minerals. It depends on the chemical composition (acidity) of the water

4 **Abrasion:** Water has a sandpaper effect on river channel, especially in floods. Pebbles in potholes act as grinders widening the holes and lowering the rock surface

2.9.1 Four main fluvial erosion processes

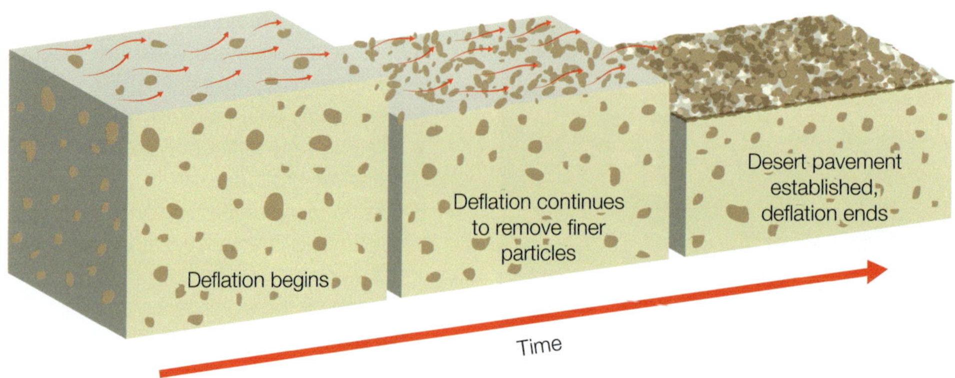

2.9.2 Formation of desert pavement by deflation over time

ISBN 978 1 4586 6277 4

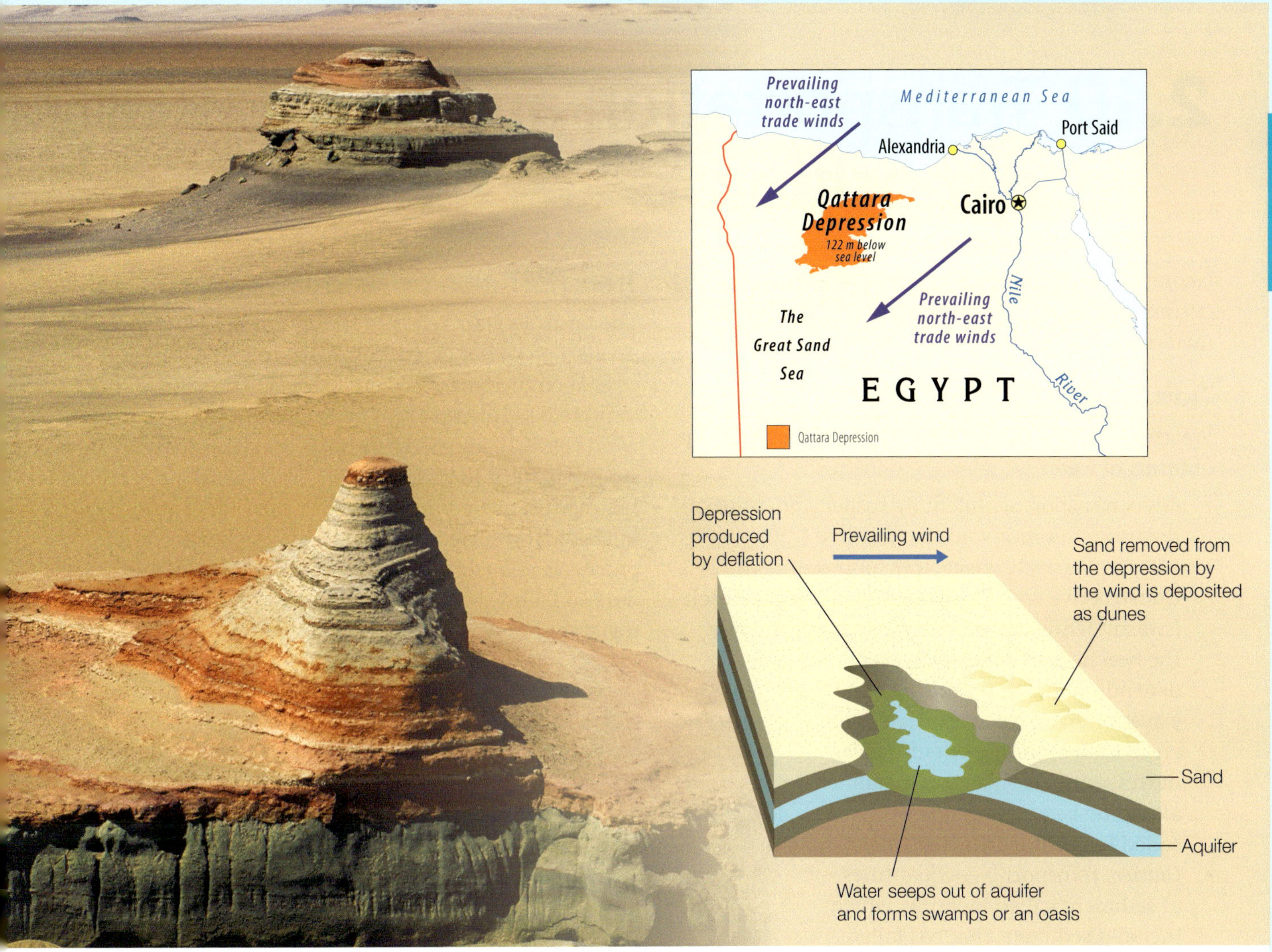

2.9.3 Deflation has formed the Qattara Depression in the Sahara Desert, Egypt

Sliding land

Landslides are also geomorphic processes that alter landforms. They can be a result of water erosion as well as human interactions, such as the deforestation of steep slopes. Landslides frequently follow heavy rainfall in the Andes Mountains in South America and Italian coastal areas around Cinque Terra (see 4.9).

Geo**info**

- Rivers carve V-shaped valleys whereas glaciers carve U-shaped valleys.
- The Qattara Depression is 80 km long and 120 km wide.

Geo**activities 2.9**

Knowledge and understanding

1. When and where does wind erosion happen?
2. What are the four categories of water erosion?
3. What is abrasion and what transportation process does it rely on?

Inquiry and skills

4. Refer to 2.9.1 and describe the four main fluvial processes.
5. Refer to 2.9.2 and discuss the formation of a desert pavement through wind erosion.
6. Refer to 2.9.2 and 2.9.3.
   a. Explain how the Qattara Depression was formed.
   b. Using secondary sources, research landuse in the depression, such as oil. Present your findings as a photo story using digital images.
7. Discuss how human activity can speed up the natural geomorphic processes that cause landslides. Report to your class on one example and the effects of this landslide on people. Apply geographical concepts to your report.

ISBN 978 1 4586 6277 4

2.10 Transportation and deposition

Geomorphic agents (such as rivers, waves, tides, currents, glaciers and wind) transport weathered materials.

Transportation by water

Weathered materials can be transported by a variety of forms of water:

- *Rivers* transport sediment by suspension, solution, flotation, saltation and traction (see 1.20).
- *Waves*, *tides* and *currents* transport sediment. Constructive waves move sediment up the beach (swash), while destructive waves move it down the beach (backwash). However, depending on the prevailing wind, waves often arrive at an angle. This creates a longshore current that moves sediment in a zig-zag manner along the beach. This process is called longshore drift (see 1.12).
- *Glaciers* transport large quantities of sediment called moraine, which includes sediment that has fallen onto a glacier or has been gouged from the sides and base as it moves slowly down a slope. Glaciers transport moraine in three ways: *supraglacially* (on its surface), *englacially* (within the glacier) and *subglacially* (beneath the glacier) (see 3.12).

Transportation by wind

The term 'aeolian' refers to the action of the wind. Aeolian processes play a dominant role in desert and coastal environments where there is little vegetation to protect sediments from being transported by the wind (see 3.10 and 3.11). Aeolian processes have a similar pattern to fluvial (river) processes as sediment is moved in three main ways: suspension, saltation and surface creep. The effectiveness of each process is related to sediment size, wind speed and duration, and vegetation cover. Dust plumes can travel for many thousands of kilometres.

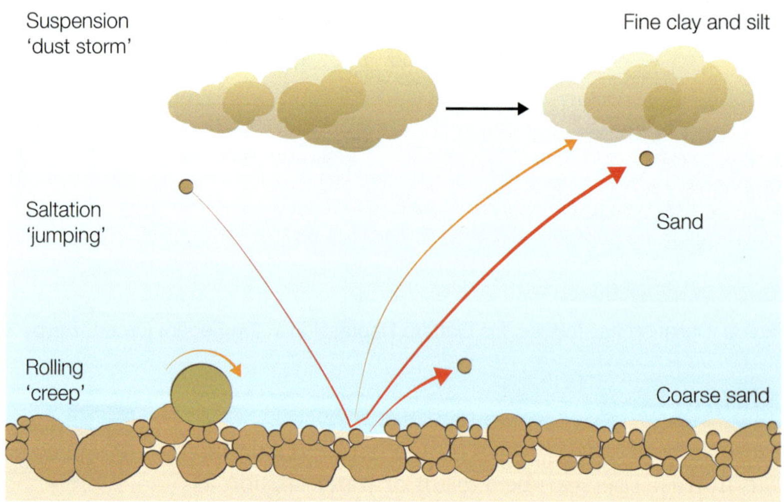

2.10.2 Transportation by wind

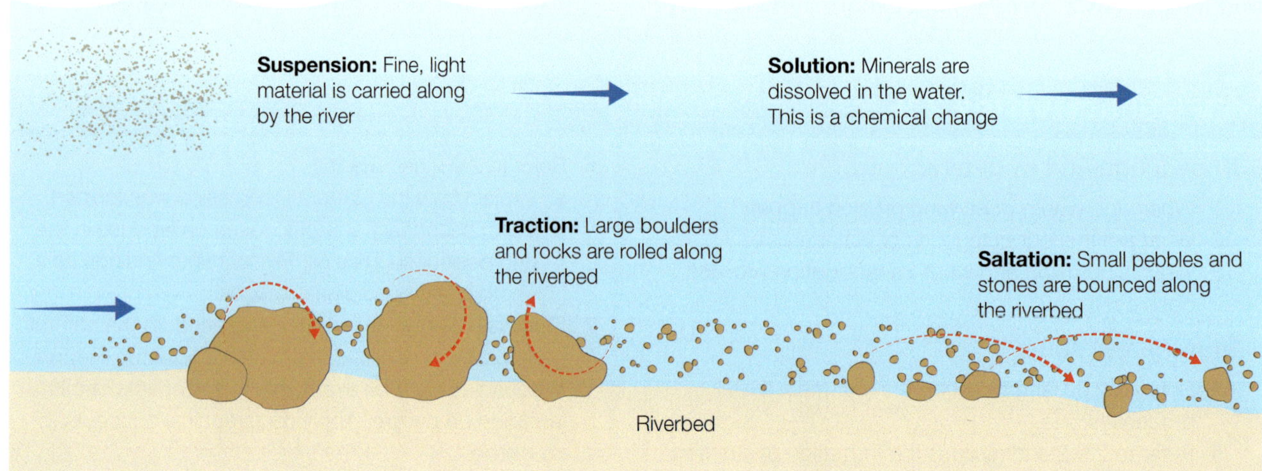

2.10.1 Transportation by running water

ISBN 978 1 4586 6277 4

2.10.3 Transportation by wind: a satellite image of a dust storm across Egypt and Saudi Arabia

Losing energy

Deposition occurs when the geomorphic agents that transport sediment lose their energy. Depositional features in different landscapes include:

- *rivers*—deltas (see 1.20 and 1.21)
- *coasts*—beaches, spits and offshore bars (see 1.12)
- *glaciers* and *ice sheets*—terminal moraines (see 3.12–3.14)
- *deserts*—sand dunes (see 3.10 and 3.11)
- *volcanoes*—ash and lava and deposits (see 2.4).

In some landscapes, such as arid lands and coasts, deposition can be temporary because erosion and transportation are continually at work.

Geo**activities 2.10**

Knowledge and understanding

1 How is sediment transported?
2 When does deposition occur?
3 What is longshore drift?

Inquiry and skills

4 Refer to 2.10.1.
  a What is the difference between solution and saltation?
  b Describe four processes involved in the transport of sediment by running water.
5 Refer to 2.10.2. What type of soil particles become airborne?
6 Draw and label the depositional features of a coast (see also 1.14).
7 Describe the depositional landforms of glaciers (see 3.12–3.14) and rivers (see 1.20 and 1.21).
8 Research the importance of dust plumes to marine ecosystems. Illustrate your report with maps and diagrams. Use maps which conform to cartographic conventions.

2.10.4 These volcanic ash deposits were laid down 11 000 years ago in Germany when the Laacher See volcano last erupted

ISBN 978 1 4586 6277 4

2.11 Skill: maps—geographical tools

Whether lost in a shopping centre, looking for the exit in a hotel, needing directions to a World Cup match, or locating important environmental resources (such as clean water), maps are your answer!

A globe is a true representation or model of Earth's surface, unlike a map, which is a visual representation of Earth's natural and human elements. Because Earth is round and maps are flat it is impossible to create a map with perfect scale. Maps are drawn to a smaller scale to fit the paper or the computer screen, and a curved shape has to be represented on a flat surface.

Which map will I use?

A variety of maps can be used to show the same place, from a simple sketch map to a complex topographic map or Google Earth map. Maps communicate information for different purposes. For instance, meteorologists refer to weather maps and charts, Indigenous Australians draw sketch maps to show the location of campsites, and engineers study topographic maps before building dams.

Today cartographers generate maps using computer programs, such as Google Earth, Geographic Information Systems (GISs) and the Global Positioning System (GPS).

Geoinfo

McArthur's Universal Corrective Map locates Australia at the top of a world map.

Sketch maps are also called mud maps as they are drawn in the dirt or sand during fieldwork. These maps are roughly drawn without scale. ▼

Geographic Information System (GIS) maps show layers of natural and cultural information on top of each other (e.g. landforms and population density). ▼

Topographic maps are used when bushwalking, planning roads or orienteering. They show height above sea level as well as physical (e.g. buildings) and natural (e.g. landforms and rivers) features. ▼

Google Earth maps provide fly-over access to nearly any spot on Earth. ▼

2.11.1 There is a huge variety of maps

ISBN 978 1 4586 6277 4

Diagram maps or network maps show positions of stations along a route and where they link up. ▼

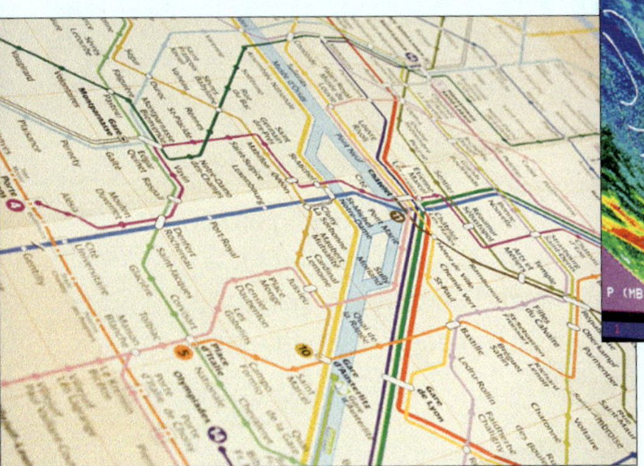

Weather maps, or synoptic charts, are seen on television and the internet. These maps show low and high air pressure systems, cold and warm fronts, temperature, rainfall, and wind direction and speed. They are useful for tracking cyclones and working out what we should wear and if the weather is suitable for our favourite sport. ◀

Global Positioning System (GPS) maps are on mobile phones and in cars. They are linked to satellites enabling people to see and adjust their route as they move. ▼

Thematic maps show a range of themes (e.g. political boundaries, climate, vegetation, mining and population density). ▼

Australian Indigenous maps show an aerial view of campsites and location of water and hunting grounds. ▼

Street directory maps show roads, shopping centres, sporting facilities, schools, hospitals and parks in a suburb or city. ▼

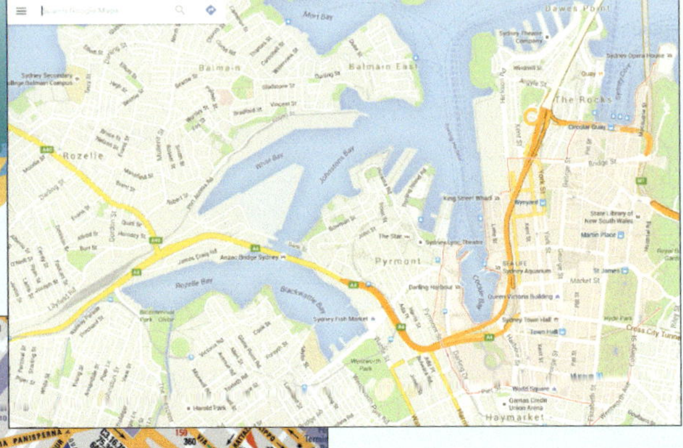

Pictorial maps are called illustrated maps. They provide a simple overview of a place. They are often used as tourist maps. ◀

Geo**activities 2.11**

Knowledge and understanding

1. Explain why maps are inaccurate pictures of the world.
2. In your own words explain the terms 'GPS' and 'GIS'.
3. List the advantages of Google Earth maps for military purposes, weather maps for sailors, and GIS maps for planning the location of homes.

Inquiry and skills

4. In groups brainstorm the types of maps you have seen and used. Present your findings as a Tagxedo.
5. Discuss the importance maps play in everyday lives and the increasing use of technology such as the mobile phone.
6. List 10 different types of maps used in the textbook. Compare two maps and explain why they illustrate different information.
7. Draw a pictorial map of your local area or shopping centre.

ISBN 978 1 4586 6277 4

2.12 Skill: map projections

For over 2000 years people have seen a distorted world through maps, as it is impossible to accurately represent Earth on a flat piece of paper or on a computer screen. Imagine peeling an orange and pressing the orange peel flat on a table. The peel would break as it was flattened. Making accurate maps of the world is also difficult, because it is mathematically impossible to flatten a sphere onto a piece of paper without distorting or cracking it. Geographers aim to choose the best map projection for a given dataset or theme, such as environmental resources studied in this topic.

Limitless maps

Map projections represent a three-dimensional surface of the Earth on a two-dimensional plane. The transformation includes distortion, such as to area, shape, direction and distance. Map projections are constructed to preserve one or more of these properties—though not all simultaneously. The challenge when selecting a map projection is whether the information requires accurate shape or size of objects, but not both. By blending maps or hybrid maps it is possible to create a map balancing the distortion of both size and shape.

There are limitless maps, each presenting its own point of view or perspective. As an Australian have you wondered why our island continent is generally located at the bottom or at the edge of world maps? Why not top and centre?

Old Mercator projection

Mercator published his map projection in 1569. It was a map for the sailor, navigator and world traveller. The scale and shape of regions near the equator are accurate but regions closer to the poles appear larger. The Mercator projection is unusable when latitudes are greater than 70° north or south of the equator, as it distorts countries. Greenland is as large as Africa, when in reality it is closer to Mexico in size, while Antarctica appears the largest continent on Earth. Despite these faults, Google Maps and Geographic Information Systems (GISs) use variations of Mercator projections for their map images.

Map	Projection	Use	Strengths	Weaknesses
	Mercator	Popular for navigation	Scale and shape of regions near the equator more accurate	Regions near the poles appear larger than in reality
	Mollweide	Thematic maps (e.g. distribution of rainforests and population)	Area more accurate	Distortion towards the poles; poles not shown
	Peters	Popular for determining flooded area or area under crops	Area more accurate; represents developing countries more fairly	Shape exaggerated in the southern hemisphere
	McArthur's	Southern hemisphere at top of map	Different perspective from other projections	Can be any projection type, with the same weaknesses of that projection

2.12.1 Common map projections

ISBN 978 1 4586 6277 4

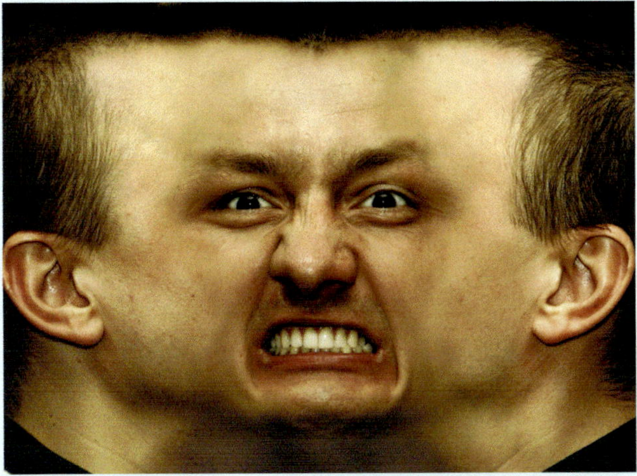

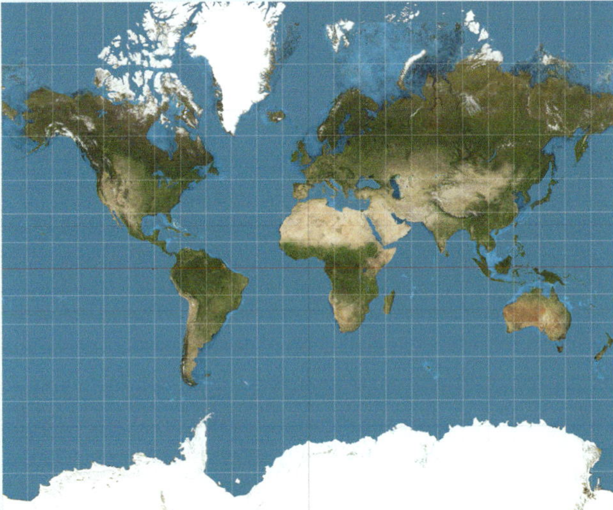

2.12.2 What a human head would look like if it were 'unwrapped' like a Mercator map projection. Which is more disturbing: the world presented this way or a human head?

Future maps

In the past the shape of Earth and its continents were constructed by map-makers using a pen while walking through isolated places, such as the rainforests in Papua New Guinea. Today remote sensing uses satellites and computers to measure Earth and identify physical and human features. Map-makers in hot air balloons used to use basic cameras to acquire aerial photographs, whereas today high-tech cameras in aeroplanes capture photographs at different angles to create detailed maps of different places.

Infrared sensing detects the temperatures of different objects on the ground and helps map living objects, such as the distribution of plant and animal species. Microwave rays are used to cook meals, but they can also be used for mapping what exists beneath the Earth's surface, such as the distribution of non-renewable minerals. Places can now be mapped using radar sensing when obscured by clouds and rain.

A variety of maps, photographs and satellite images help geographers map the world more accurately. By using both technology and fieldwork a clearer picture of Earth emerges, for instance the impact of humans on environmental resources.

Geo**activities 2.12**

Knowledge and understanding

1 'There is no perfect map.' Explain this statement.
2 Discuss why different maps are used for different geography topics.

Inquiry and skills

3 Peel an orange and press the skin flat on a table. Explain the relationship between this activity to globes and maps.

4 Refer to the photo and answer the inquiry questions.
   a What is a map projection?
   b What occurs when you flatten a globe onto a flat piece of paper?
   c Why are there different map projections?
   d What are the strengths and weaknesses of the four different map projections?
   e Do countries look their true size on all projections? Explain your answer.
   f Why is Australia at the bottom of most world projections? Which projection has Australia at the top?
   g Why does this textbook have Australia in the centre of many of its world maps?
   h How has technology enabled geographers to obtain a more accurate view of the world?
   i Imagine you were an alien: what does Earth look like from space?

ISBN 978 1 4586 6277 4

2.13 Skill: perspectives—the world you've never seen

Large and small maps

Maps can be drawn on paper, bodies, garments and sand. They are drawn using pens or using modern technologies, such as Google Earth and GIS. And they can vary greatly in size. Earth Platinum is the world's largest atlas. The book is 1.8 m high by 1.4 m wide and weighs 150 kg. As a contrast, Ghent University fabricated a world map on a scale of 1 trillionth, referred to as 'Terra' scale. The 40 000 km circumference of the Earth at the equator was scaled down to 40 micrometres. This is half the width of a human hair. The world map is placed in the corner of a silicon chip to enable complex telecommunications, high-speed computing and healthcare functions to operate.

2.13.2 World according to Facebook: the brighter the line the more friendships

Strange maps

There are an endless number of maps depicting geographical information from a variety of perspectives, as shown in the examples below:

- *Population size* rearranged according to country size. Monaco is the world's most densely populated country with 16 923 people per km^2, contrasting with Mongolia containing 1.7 people per km^2. On the map a country with a high population moves to a country containing a larger area. As a result China, with the world's largest population, is transferred to Russia, which has the largest area. Australia's 22.5 million inhabitants move to Spain, and Pakistan moves to Australia. The map illustrates how global population realignment would involve massive migration and changes to lifestyles.
- *Physical size* of country altered to reflect population size. This map leaves the country where it is geographically located and redraws it in proportion to the population size. Australia is a big country with a small population compared to India—a small country with a large population.
- A world map highlighting *Facebook friendships*. Visualisation of Facebook connections around the globe show that the USA has the highest concentration of Facebook friendships and Africa the lowest concentration. While most of Russia and Antarctica are not found on the map, the rest of the world is identifiable.

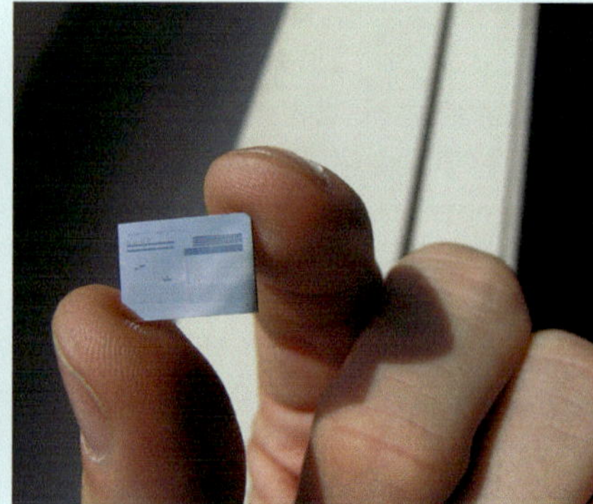

2.13.1 Map extremes: **a** World map on body; **b** Earth Platinum; **c** World's smallest map

ISBN 978 1 4586 6277 4

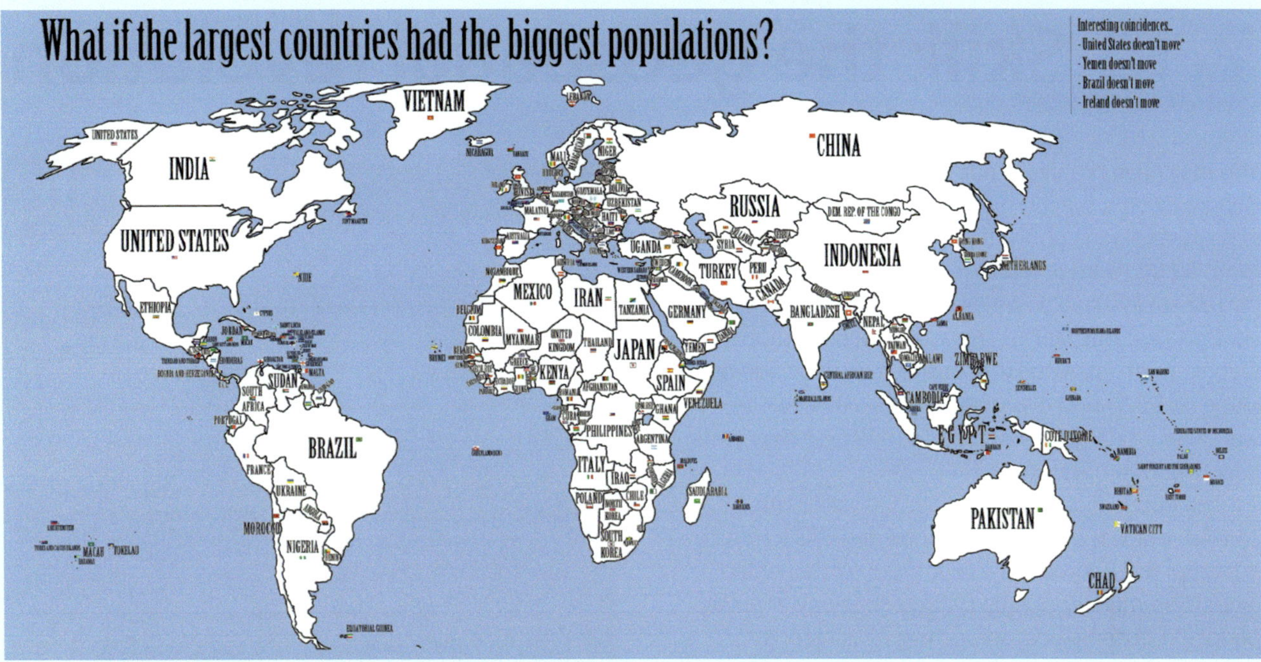

2.13.3 Map of the world's countries rearranged by population size

2.13.4 The size of countries reflects their population size

Geoactivities 2.13

Knowledge and understanding

1 Explain the changes to maps over time.

2 List the advantages of GIS over the Mercator map on page 79?

Inquiry and skills

3 Refer to 2.13.2.
   a List countries where Facebook is popular.
   b Locate the greatest Facebook links in Australia.

4 Refer to 2.13.3 and explain what happened to Australia. What country replaced Australia? What would be the impact of a larger population on Australia's environmental resources?

5 Refer to 2.13.4 and list five countries that appear larger on this map than on the World map on pages 382–3.

6 Not everyone in the world has access to 'all' the internet. These are referred to as 'black holes'—some countries have eliminated freedom of press and expression within cyberspace to control information and communication. Research online which countries have black holes (see Geolinks).
   a List five countries with internet black holes.
   b Discuss how limited information impacts on students.

ISBN 978 1 4586 6277 4

2.14 Skill: the geospatial revolution

You may have dreamt of riding camels across the Sahara Desert, surfing giant waves in Hawaii, floating in hot air balloons above the Grand Canyon or climbing Mt Everest. Most people, unable to participate in these activities, instead travel the world via Google Earth. With a finger tap on the computer, Google Earth lets you fly, spin and zoom down to any place on Earth and experience virtual fieldwork by wandering through the Amazon jungle exploring rivers, forests and remote villages.

The geospatial revolution uses Google Earth satellite images, aerial photographs and GIS to create geographical knowledge vital to the interconnected global community. Google Earth technology enables people to observe an approaching cyclone on a mobile phone, and a GPS in your car helps you find where to go for your sports match.

Google, a corporation specialising in internet searches, processes over 1 billion search requests a day. It provides geographical information from a variety of perspectives and includes geographical tools, such as photographs, graphs and statistics. The Google Earth Blog shares satellite images and the Google Earth Community adds place markers (such as mines, roads, forests, restaurants and hospitals onto these maps).

2.14.1 Google Earth showing Tropical Cyclone Yasi approaching Australia on 1 February 2011

2.14.2 A big herd of hippopotamuses swimming in a river in Tanzania caught on Google Earth

Earth observation satellites

Earth observation satellites show changes in environmental resources, such as deforestation in the Amazon Rainforest, melting glaciers, impacts of oil drilling in the Arctic and damage to coral reefs. Satellites track wildlife, such as polar bears and hippopotamuses, and show the impact of natural disasters, such as tsunamis and cyclones. Instant disaster information on an earthquake allows quick emergency responses from international organisations (e.g. United Nations), governments and non-government organisations (e.g. World Vision).

Google Earth monitors the environment. It shows before and after satellite images of changing environmental resources via the United Nations Environment Programme (UNEP) and is used by conservation organisations promoting sustainability, such as the WWF.

ISBN 978 1 4586 6277 4

Geoinfo

Some 6578 satellites have been launched into orbit since 1957.

High-tech tribes

Google Earth Outreach provides non-profit organisations with resources to visualise environmental problems and potential solutions. The Surui in Brazil and the Wayana and Trio in Suriname face threats to their forest and culture. These tribes have been trained to use GPS, map their land and locate resources, such as medicinal plants and hunting grounds. Additionally, GPS assists these tribes to guard against threats to their environmental resources from logging, drug lords and mining operations.

The Jane Goodall Institute uses Google Earth to monitor forest projects and chimpanzee populations in Tanzania and Uganda. Using satellite collars, the Save the Elephants organisation uses Google Earth to track and protect elephants from poaching. In 2010, Defenders of Wildlife used Google Earth to view millions of litres of crude oil gushing out of a BP well in the Gulf of Mexico and to show its impacts on marine and bird species.

Fieldwork: geocaching

Geocaching is an outdoor game using GPS coordinates (latitude and longitude) to find the geocache hidden at a location. Geocaches are placed in over 100 countries and there are more than 5 million geocachers worldwide. The geocache contains geographical questions for searchers to answer before moving onto the next location.

2.14.3 GPS helps Amazon tribes fight exploiters

Geoactivities 2.14

Knowledge and understanding

1 'Smile, you're on satellite imagery!' What do you think this means?

2 Explain how Google Earth can be used in the geography classroom.

3 'Google Earth helps geographers understand the use and misuse of environmental resources.' What does this mean?

4 Discuss the advantages of Google Earth Outreach to indigenous communities in isolated locations.

Inquiry and skills

5 Refer to 2.14.2 and estimate how many hippopotamuses are in the photograph. Explain why they are congregating at this place.

6 'The Geospatial Revolution examines the world of digital mapping and how it changes the way we think, behave, and interact' (National Geographic). Explain this quote.

7 Organise a geography geocache activity at your school.

8 Using Google Earth (see Geolinks):
   a Track the routes of chimpanzees in the Gombe Forest in Tanzania. What should be done to conserve the chimpanzee?
   b Measure the distance from your home to a mine in the Pilbara using a Google Earth ruler.
   c Plan a holiday to three countries. Collect three photographs of each country and write a summary of their main environmental resources.

9 Virtual fieldwork (see Geolinks):
   a Take a virtual trip in the Amazon and describe what you observed.
   b Take a virtual ride in the Tour de France. What environmental resources do you pass on your journey?

10 Refer to the Google Earth Blog (see Geolinks) and complete the following activities.
   a Tour with geographers on the job around the world. Describe the different jobs requiring geography.
   b Take a tour of Asian countries. Provide an overview of the environmental resources in three countries.
   c Complete the orientation quiz.

11 Why is the United Nations Environment Programme Atlas of our Changing Environment a useful site for environmental managers and geography students (see Geolinks)?

ISBN 978 1 4586 6277 4

Geo**think**

Topographic map of volcanoes

Tongariro National Park is located on the North Island of New Zealand. It was the first national park in the country and is also a World Heritage area, listed for its physical features and its cultural significance to the Maori people. The volcanic features in the park are unique because they are highly explosive, erupt frequently, and have a large number of active vents, lava flows and crater lakes.

The main volcanoes—Mt Ruapehu, Mt Tongariro and Mt Ngauruhoe—are among the world's most active composite volcanoes. Mt Ruapehu's crater lake, which contains 10 million m^3 of acidic water, violently erupts when glacial melt water interacts with the magma. This causes lahars (volcanic debris and mud flows). Since 1945, there have been more than 17 eruptions and at least 60 lahars recorded on Mt Ruapehu. Two lahar warning systems are in place for visitors to the park.

Inquiry and skills

1 Refer to 2.15.1.
   a List five volcanoes located in New Zealand.
   b Explain why New Zealand is the location of volcanic activity.

2.15.1 Volcanoes in New Zealand and location of Tongariro

2.15.2 Aerial photograph of Mt Tongariro in the foreground with Mt Ngauruhoe then Mt Ruapehu in the background, in Tongariro National Park, New Zealand, during winter.

2 Draw an annotated line drawing of the photograph.

3 Refer to 2.15.3.
   a What is the GR for each of the following features: Mt Ngauruhoe, Mt Tongariro, North Crater, Te Maari Crater and Mangatepopo Hut?
   b What feature is located at each of the following grid references: 300210, 308204 and 310208?
   c Explain why this would be a dangerous area during a volcanic eruption.
   d Rank the following craters according to their relative size: South Crater, North Crater, Red Crater and Central Crater.
   e Estimate the area of Blue Lake caldera.
   f Choose one of the walking circuits through the volcanic area and describe the main features you would see along the route. List the grid references for two of the features along the way.
   g Calculate the following distances: from Mt Ngauruhoe to Mt Tongariro as the crow flies, and from Mangatepopo Hut to Ketetahi Hut along the Tongariro Northern Circuit.
   h New Zealand is nicknamed 'The Shaky Isles'. Research the internet and write a report outlining the reason for this. Mention tectonic plates, major faults, volcanoes and earthquake activities. Use maps to illustrate your report.

ISBN 978 1 4586 6277 4

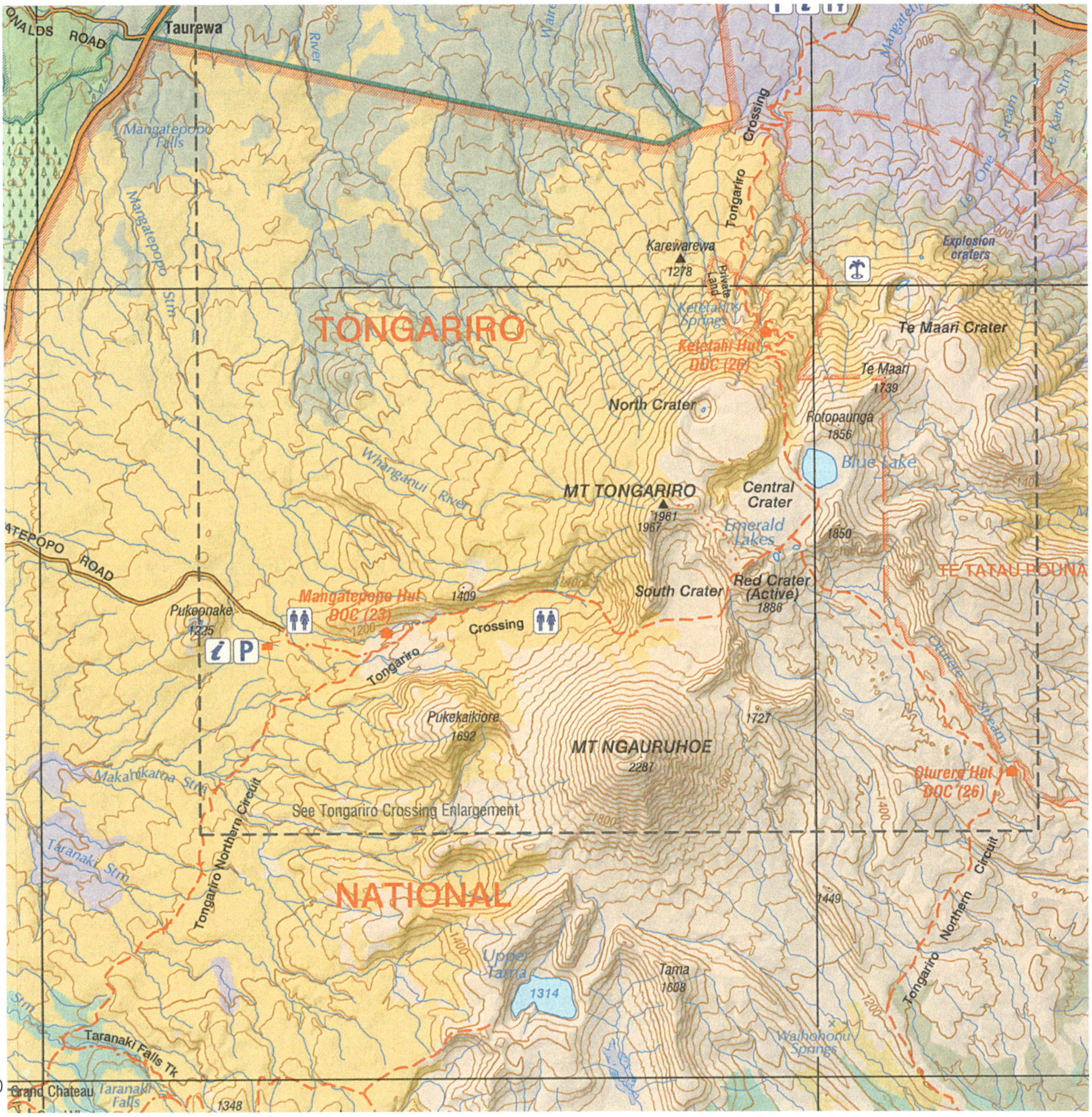

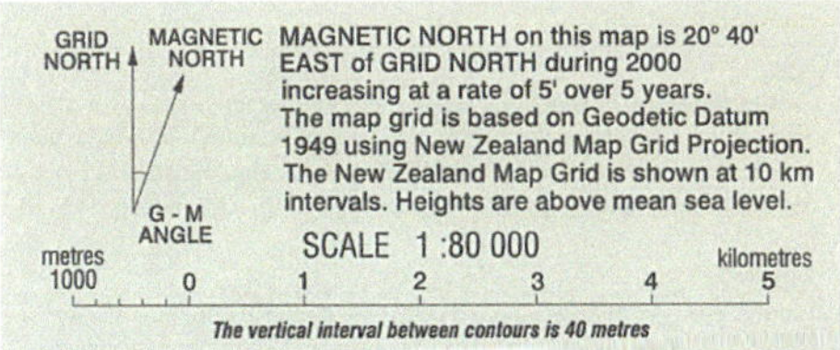

GRID NORTH
MAGNETIC NORTH
G - M ANGLE

MAGNETIC NORTH on this map is 20° 40' EAST of GRID NORTH during 2000 increasing at a rate of 5' over 5 years. The map grid is based on Geodetic Datum 1949 using New Zealand Map Grid Projection. The New Zealand Map Grid is shown at 10 km intervals. Heights are above mean sea level.

SCALE 1 :80 000

metres 1000 0 1 2 3 4 5 kilometres

The vertical interval between contours is 40 metres

REFERENCE

National park

Reserves/Conservation areas

Nature reserves/Restricted areas

Wilderness areas

	Road, track or route providing public access: sealed	unsealed	Access restricted; permission required, private property.
State highway	4		
Major road			
Other road			
Vehicle track			
Walking track			
Tramping track			
Route			

2.15.3 Topographic map extract of Tongariro National Park, North Island, New Zealand.

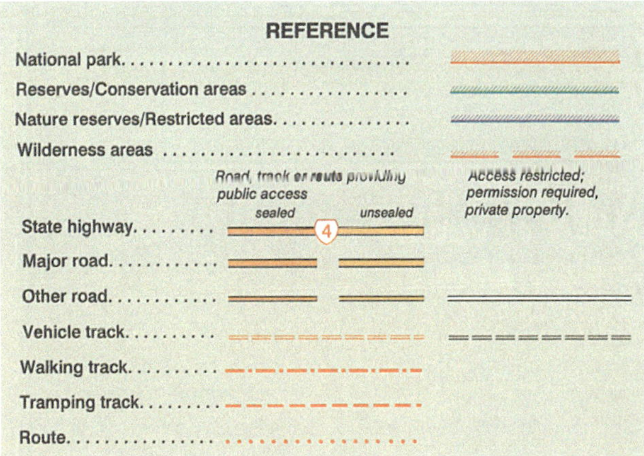

i Reflect on your learning and research individual and collective actions to reduce adverse impacts of future volcanic eruptions on New Zealand's people and places. Present research using web 2.0 tools.

ISBN 978 1 4586 6277 4

chapter 3 Landscapes: processes and values

'The landscape is what lies before our eyes. It is everywhere.'

Susan Bliss

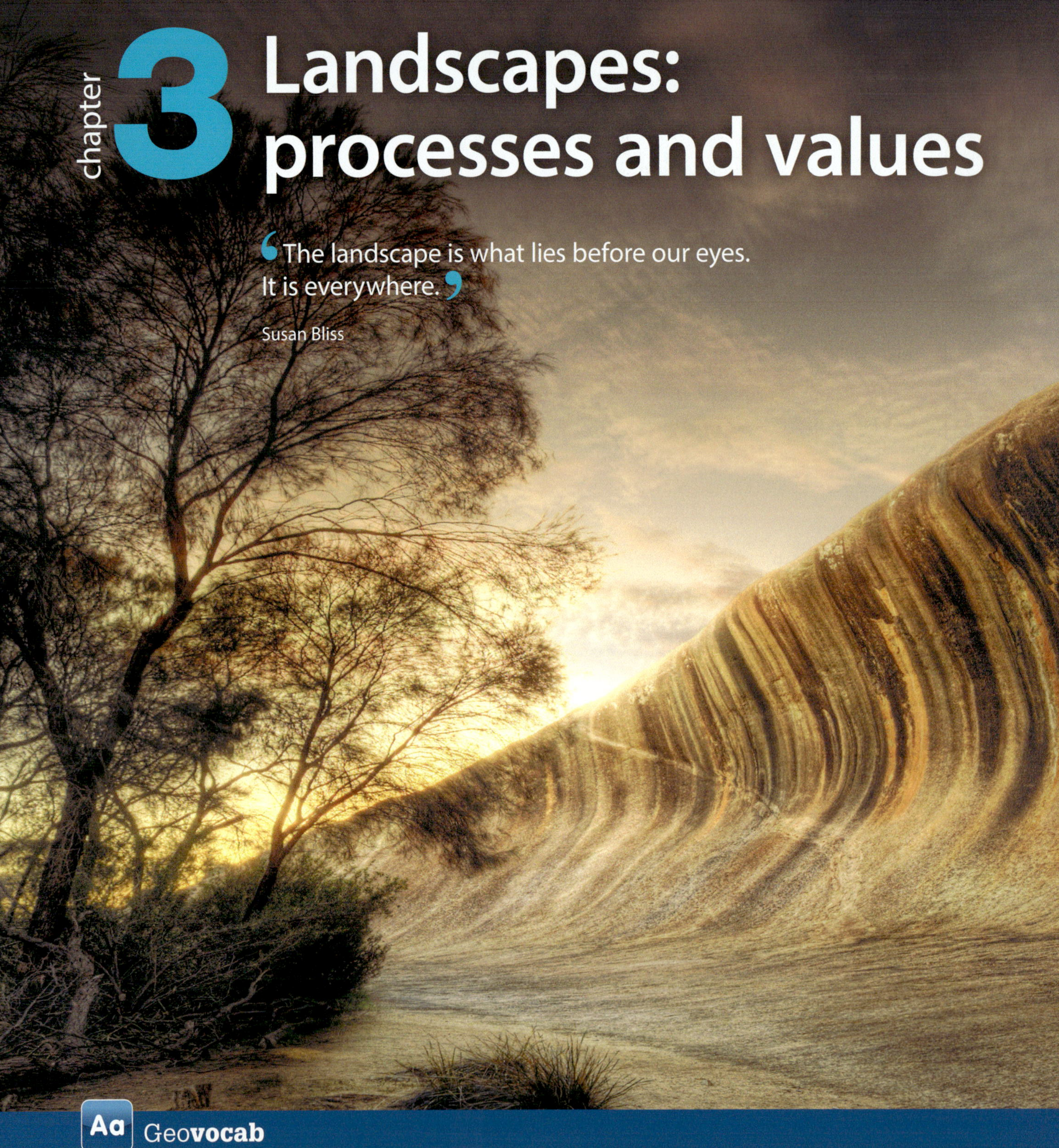

Geovocab

arid landscapes: very dry places where evaporation exceeds precipitation

bushfires: uncontrolled fires in areas containing combustible vegetation

chernozem: fertile black earth containing high percentage of humus

cultural landscapes: landscapes changed by humans

epicormic buds: shoots growing underneath the bark of a stem or branch of a plant that regenerates after fire

erg: sea of sand (e.g. sand dunes)

forest: area containing high density of trees

glaciers: dynamic, larger moving bodies of compacted ice that exist for long periods of time

grasslands: areas in which the natural vegetation is dominated by perennial grasses; referred to as 'savannas' in Africa, 'steppes' in Asia, 'prairies' in North America, 'pampas' in South America and 'rangelands' or 'savannas' in Australia

ISBN 978 1 4586 6277 4

Wave Rock in Hayden, Western Australia, was formed by chemical weathering while underground 60 million years ago

hamada: basins and plateaus in deserts
lignotubers: roots of plants that generates after fire
reg: stony pavements in arid environments
relict landforms: formed during a previous time
sustainable: (in terms of landscapes) practices conserving landscapes for present and future generations

ISBN 978 1 4586 6277 4

Our planet contains a diversity of landscapes and their landforms. Landscapes create the bigger picture, as they are a combination of natural and human features. Landscapes such as grasslands are important to the migration of animals in Africa, while forests are a source of 5000 products. The diversity of desert landscapes—from sand dunes and salt lakes to the white expanse of glaciers and ice sheets—provide beautiful vistas. As these landscapes change over time, an understanding of the geomorphic processes is important.

Landscapes and their landforms are perceived, valued and protected in different ways by different cultures. They hold great value for people, in particular for Aboriginal and Torres Strait Islander Peoples. As such, they need to be protected for future generations.

Think, puzzle, explore

- **Place:** Why do landscapes differ between places?
- **Space:** What is the spatial distribution of grasslands?
- **Environment:** Why do forest landscapes vary with environments?
- **Interconnection:** How are deserts landforms interconnected with dry and hot environments?
- **Sustainability:** How can the landscapes of World Heritage sites be conserved for future generations?
- **Scale:** Why are large ice sheets located in Antarctica and not in Australia?
- **Change:** How can glaciers change river valleys?

Geo**skills** in focus

- **Identifying** the variety of landscapes and their changes
- **Collecting** and **analysing** relevant geographical data on landscapes and their changes over time, from primary and secondary sources
- **Concluding** that landscapes are combinations of interconnected features that exist at a variety of scales
- **Reflecting** on the complexity of landscapes and their landforms and the need for protection

3.1 Photographs: landscape diversity

3.1.1 Hot, cold, natural and human landscapes.
A: Longitudinal sand dunes in the Oman desert. B: Glaciers and cirque in the Haute-Pyrenees Mountains, France. C: Rainforest vegetation in the Mount Field National Park, Tasmania. D: Cityscape beside the Huangpu River, Shanghai, China. The city's estimated population exceeds 24 million.

Landscapes are scenery or what lies before our eyes—a combination of natural and human features. Postcards, photographs posted on Facebook, television documentaries, movies, *National Geographic* magazines and Google Earth all provide glimpses and different perspectives of landscapes. We look, and dream of visiting one day!

The big picture: landscapes

Landscapes surround us. We live in a landscape and pass different landscapes on our daily journeys to school. Today, all landscapes contain the fingerprints of humans and can be classified into:

- **cultural landscapes**—natural landscapes change into cultural landscapes when humans intervene (e.g. when wetlands are cleared for settlements)
- *moving landscapes*—these landscapes involve moveable objects (e.g. traffic congestion in Mumbai, India, and roaming lions in grasslands in Kenya, Africa)
- *changing landscape*—landscapes change with seasons, weather, time of day, tides, and after disasters such as floods.

3.1.2 The big picture: components of landscapes

Landforms	Plants and animals	Human interactions
• Hills, plains, mountains and plateaus • Rivers • Coasts • Glacial (fiord) • Oceans • Volcanoes • Caves and sinkholes • Lakes	• Native and introduced plants and animals • Plants (e.g. rainforests, grasslands) • Animals (e.g. camels, llamas, whales)	• Urban, mining and industry • Agriculture • Roads, railway lines and ports

ISBN 978 1 4586 6277 4

3.1.3 Lauterbrunnen Valley in the Swiss Alps: landscapes contain both natural and human features

- *atmospheric landscapes*—landscapes provide different atmospheres (e.g. sitting alone on top of a mountain or amongst people in a busy city).
- *imagined landscapes*—people invent thought-provoking futuristic landscapes for movies (e.g. *Avatar*, 2009) and video games.

Geo**info**

Earth's highest landform is Mt Everest (8850 masl); the deepest is the Mariana Trench in the South Pacific Ocean (11 034 m deep).

Most individuals and non-government organisations aim to develop landscapes that are environmentally, socially and economically **sustainable**.

Indigenous landscape—their Country

Through their continual relationship with the land, Aboriginal and Torres Strait Islander Peoples have developed a vast body of knowledge of landscape resources. There is no landscape not expressed through song and dance—traditional owners can see the imprint of sacred creation in every landscape.

Geo**activities 3.1**

Knowledge and understanding

1. What is a landscape?
2. Why are landscapes more than simply landforms?
3. *Landscapes contain the fingerprints of humans.* What does this statement mean?

Inquiry and skills

4. Refer to 3.1.1.
   a. Why are photographs only snapshots of landscapes? How would time lapse provide a different perspective of the same landscape?
   b. Design a collage of different landscapes in Australia. Label each landscape with its location and a description.
5. Refer to 3.1.2.
   a. List five landforms.
   b. What are the living elements of a landscape?
   c. Mind-map the human elements of a landscape and present using web 2.0 tools.
6. Refer to 3.1.3 and create a labelled line drawing of the photograph, distinguishing between natural and human features. (Refer to 1.14.)

ISBN 978 1 4586 6277 4

3.2 Seas of grass

Grasslands cover 30% of Earth's land. They encompass 50% of the land area in Africa, 33% in South America and 75% in Australia and Kenya. Grasslands are not the same across the world; they vary according to climate, altitude, landform and soil. The majority of grasslands are located between forests (wetter environments) and deserts (drier environments). Some have the ability to tolerate saline and acid soils.

Grasslands are home to a rich biodiversity of species including elephant grass, feather grass, tigers, kangaroos, snakes and ants.

Hardy grasslands

Grasslands thrive on being eaten, burnt and trampled upon. Because the grasses grow from deep in the soil rather than the top of the soil, they are protected from fires and constant nibbling by animals. Traditional hunting communities started fires to encourage new growth of grasses, increase nutrients in the soil and prevent fire-intolerant trees and shrubs from taking over the landscape. Native Americans used fire to assist in hunting buffalo and Indigenous Peoples in Australia used firestick farming to herd animals such as kangaroos into particular areas.

Grasslands are the primary source of food for many animals in the food web, but they provide little protection—animals being preyed upon by lions in the African grasslands have nowhere to hide. In this 'catch me if you can' environment some animals rely on speed for survival. The cheetah can reach speeds up to 120 km/h when trying to catch its prey.

3.2.1 Variety of grasslands across Earth according to climate and altitude

Tropical and subtropical	Temperate	Mediterranean	High altitude	Periodic floods
Serengeti in Tanzania; Arnhem Land in northern Australia	Great Plains of USA; Eastern Australia mulga shrublands; Texas prairies; Patagonian grasslands, Argentina; Middle Eastern steppes in Iraq and Syria	Oak tree savannas in California; South-west Australia Mallee; Mediterranean Basin in Spain and Greece	Australian Alps; Tibetan plateau; Western Himalayan alpine shrubs and meadows in Nepal	Bogs, marshes, mangroves, fens and peats (e.g. Pantanal in South America and Viru Bog in Lahemaa National Park, Estonia)

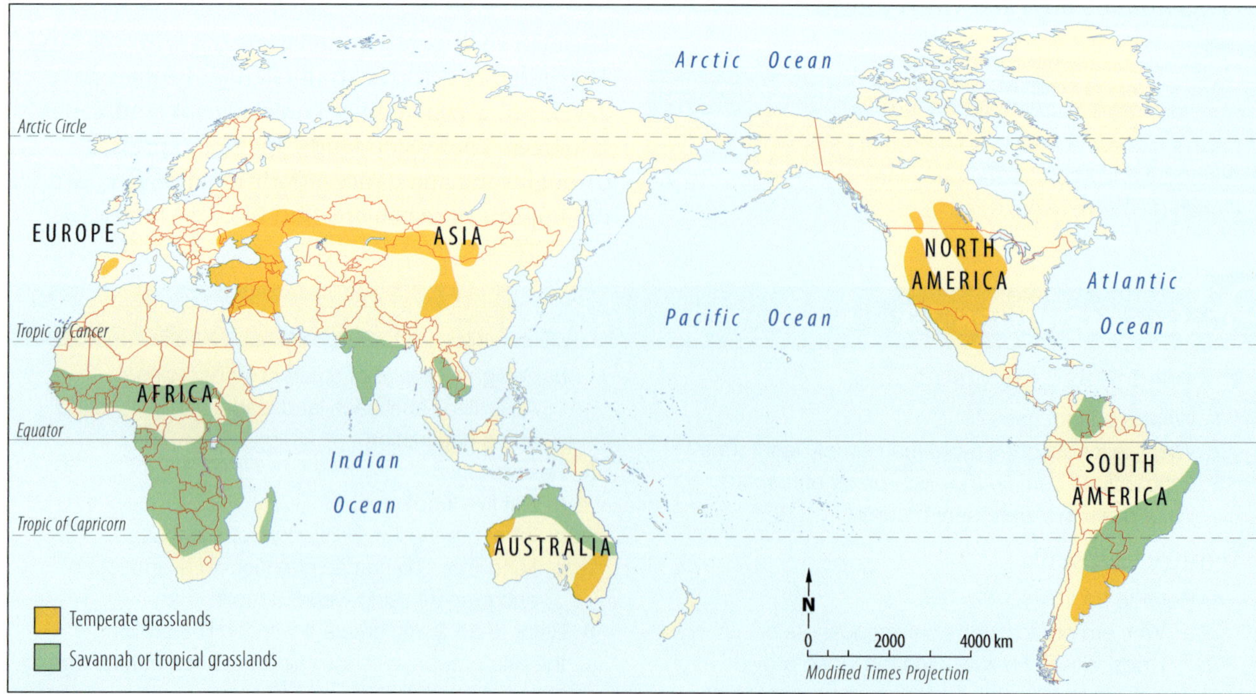

3.2.2 The location of the world's grasslands (spatial distribution)

ISBN 978 1 4586 6277 4

3.2.3 Threats to grasslands

Humans change grasslands

Grasslands are one of the most endangered environments on Earth as they are constantly being altered by human activities. Where soils are rich and the land is flat and treeless, areas have been turned into farms to grow crops or graze animals—the Prairie Soils (USA) and the **chernozem** soils in Russia have been labelled the 'bread baskets' of the world. Approximately 160 million ha of North America was once an unbroken sea of grass; today it has almost disappeared.

Many communities have developed distinctive ways of thriving in grasslands:

- *Nomadic herders* in Mongolia and Kenya (the Maasai) graze animals on grass and move when pastures have been exhausted. Recently, the nomadic way of life has been threatened by soil erosion and growth of farms, making it difficult to earn a living.
- *Argentine gauchos* are famous for their large cattle ranches. However, 30% of the pampas has been converted to agriculture, threatening their way of life.

The World Wildlife Fund (WWF) is working to ensure the world's remaining grasslands are available for endangered animals, such as elephants and tigers.

Geo**info**

- Three big, flightless birds found in grasslands are the rhea (South America), ostrich (Africa) and emu (Australia).
- The North American grasslands were home to millions of bison before most were slaughtered.

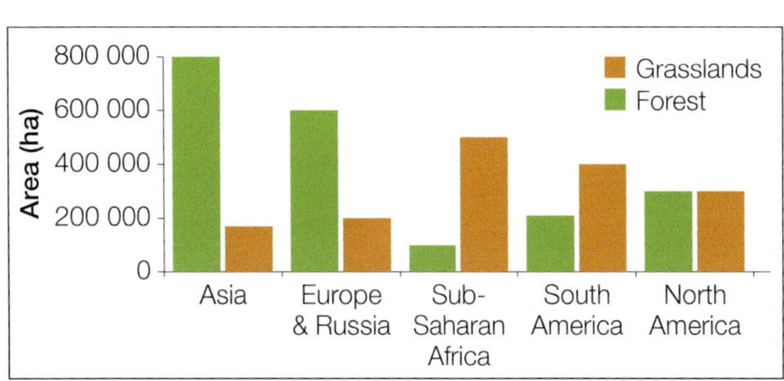

3.2.4 Amount of forest and grassland converted to farmland

Geo**activities 3.2**

Knowledge and understanding

1. What are grasslands?
2. What proportion of Earth and Australia is covered in grasslands?
3. Describe how animals have adapted to grasslands.
4. Discuss how grasslands have been shaped by humans.

Inquiry and skills

5. Refer to 3.2.1.
   a. Use Google Earth and locate these places on a map.
   b. Explain how grasses differ according to climate.
6. Refer to 3.2.3.
   a. List the threats to grasslands now and in the future.
   b. Imagine you are an environmentalist. In groups design a plan to reduce the threats to grasslands.
7. Refer to 3.2.4.
   a. Describe how many hectares of grasslands have been converted to farms.
   b. List two reasons for the conversion of grasslands in sub-Saharan Africa.
8. Using secondary sources, design a collage of different grasslands and their importance to humans. Present as a Prezi.

ISBN 978 1 4586 6277 4

3.3 On the move in the Serengeti

The name Serengeti is derived from the Maasai word meaning 'sea of grass on plains'. It is located in Tanzania and Kenya between latitudes 1–3°S and longitudes 34–36°E. The landscape hosts the largest mammal migration in the world. Over 24 million hooves pound the plains, with 200 000 zebras and 500 000 gazelles joining the two million wildebeests' trek for fresh grazing land.

Along the ancient 500 km migration route, lions feast on gazelles, giraffes eat the top leaves of acacia trees, and elephants trample down trees in their way. At the same time leopards lie quietly on acacia tree branches waiting for their prey to pass, and zebras have mud baths to camouflage their black and white stripes.

Diverse landscapes

The Serengeti region encompasses the Serengeti National Park (15 000 km^2), the Ngorongoro Conservation Area and numerous game reserves. Volcanic activity has added mountains, craters and granite outcroppings (kopjes) to the flat landscape. Rivers such as the Seronera River flow through the area.

The climate of the Serengeti is warm and dry with temperatures varying between 15 °C and 25 °C. Most precipitation falls between October and May, followed by drier months. Rainfall varies across the region from 400–1000 mm pa (mm per annum).

Rainfall determines when animals migrate as well as their route. Leading the migration, the wildebeest have the ability to locate rain over 50 km away. The Serengeti depends on the wildebeest to fertilise the soil with 380 tonnes of dung every day, to bite back grass to produce new growth and to trample it to prevent uncontrolled fires.

One common feature of savannas is fire. In the past natural wildfires helped maintain the savanna landscape. Now, controlled burning by park management occurs at the beginning of the dry season to reduce the risk of larger wildfires at the end of the dry season.

Long grass, unfriendly trees

The Serengeti contains a continuous cover of perennial grasses 1–2 m tall, which are constantly clipped by grazing animals, such as the African buffalo. Most of the trees appear unfriendly as they are covered in spines and needles to defend themselves from herbivores. Some plants have developed long taproots to reach the water table while others have thick fire-resistant bark and trunks to store water. The Maasai plant the poisonous candelabra trees as fences to stop predators taking their cattle—one drop can blind or burn the skin.

3.3.1 Annual migration of wildebeest around the Serengeti grasslands

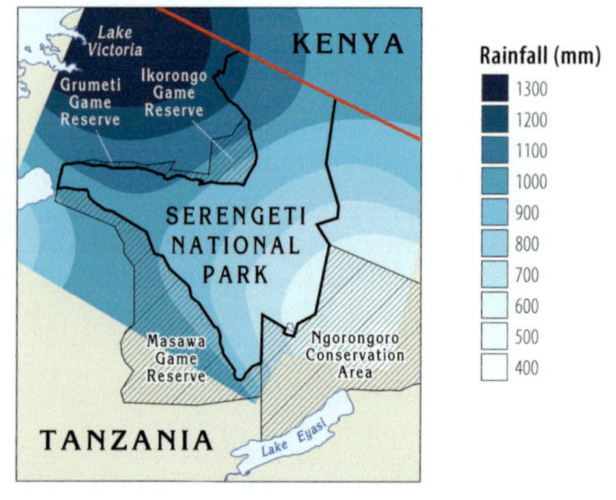

3.3.2 Annual rainfall: animals follow the rain, which means following the grass

ISBN 978 1 4586 6277 4

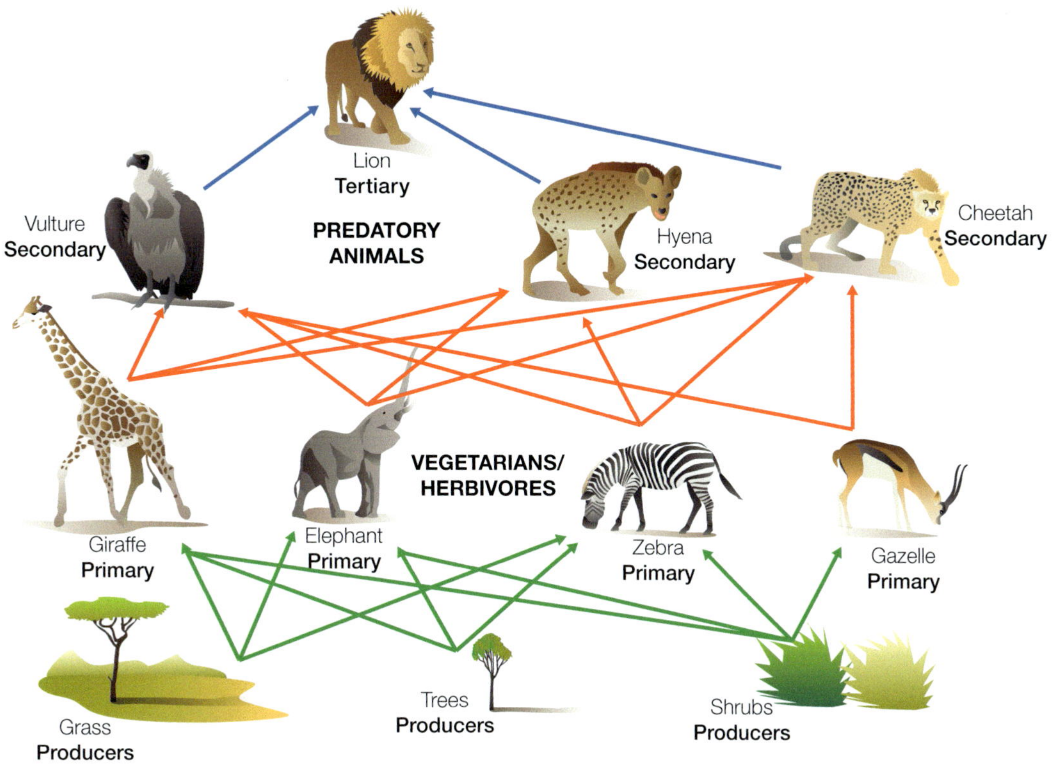

3.3.3 Serengeti food web depends on the availability of grasses

Humans have changed the landscape. Scarce wood has been cut for fuelwood for cooking and providing heat. To repair damaged grasslands the Serengeti and Ngorongoro Conservation Area were declared a World Heritage site in 1979. Today, human habitation is prohibited in the area and 90000 tourists a year are confined to specific areas aimed to minimise their impact on the landscape. At present there is heated debate over a proposed highway to be built across the animal migration path, and fears the absence of wildfires could see the development of dense woodlands over the next 30–50 years.

Geoactivities 3.3

Knowledge and understanding

1 What does the Maasai word 'Serengeti' mean?

2 Describe the Serengeti landscape.

3 The mass migration of animals is an important part of the landscape. Describe the annual migration and why they move.

Inquiry questions

4 Refer to 3.3.1.
 a Wildebeest follow the rain throughout the year. Describe their movements from July to June.
 b Why are the wildebeest important in the grassland environment?

5 Refer to 3.3.2 and describe the variation of precipitation across the Serengeti.

6 Refer to 3.3.3.
 a Describe how grass is important to the lion.
 b Draw a flow diagram illustrating the food web interconnections between animal species that enable them to survive.

7 The African buffalo is one of the migrating animals in the Serengeti. Complete the inquiry questions.

How does rainfall effect the movement of animals?

Why does the Serengeti need to be protected?

How is the Serengeti protected?

8 Explain the popular Maasai quote: *There is a place where the grass meets the sky, and that is the end. What we must face, all of us—poachers, tourists, farmers, conservationists and pastoralists—is the difficult truth that the land does not go on forever.*

9 Refer to the internet and organise a safari to the Serengeti. Explain the importance of the grassland landscape. Present your findings as a tourist brochure, including a map and climate graph.

ISBN 978 1 4586 6277 4

3.4 Skill: climate graphs

Climate data of average monthly temperatures and precipitation for a location is presented as a table or a graph. A climate graph is a combination of column and line graphs.

How to interpret climate graphs

Climate graphs, or climographs, provide important information about the location of a place. Here are some clues for interpreting them.

Temperature graphs

- *Hemisphere*—The maximum and minimum temperatures show whether the location is north or south of the equator. If the temperature is lowest in the middle of the year (a 'valley' pattern), it is in the southern hemisphere. Winter in the northern hemisphere is at the end and start of the year—a 'hill' pattern. If the temperatures are relatively constant throughout the year, the location is near the equator.
- *Latitude*—The difference between the monthly minimum and maximum temperatures near the equator is small because there are no seasons. The further away from the equator, the greater the temperature range and the more pronounced the seasons. Inland deserts, located in the centre of continents, experience a large daily (diurnal) range in temperature—heatwaves at midday and freezing temperatures at night.

Geoinfo

Satellite technology provides accurate measurements of temperature, precipitation, air pressure, wind speed, humidity and ice thickness.

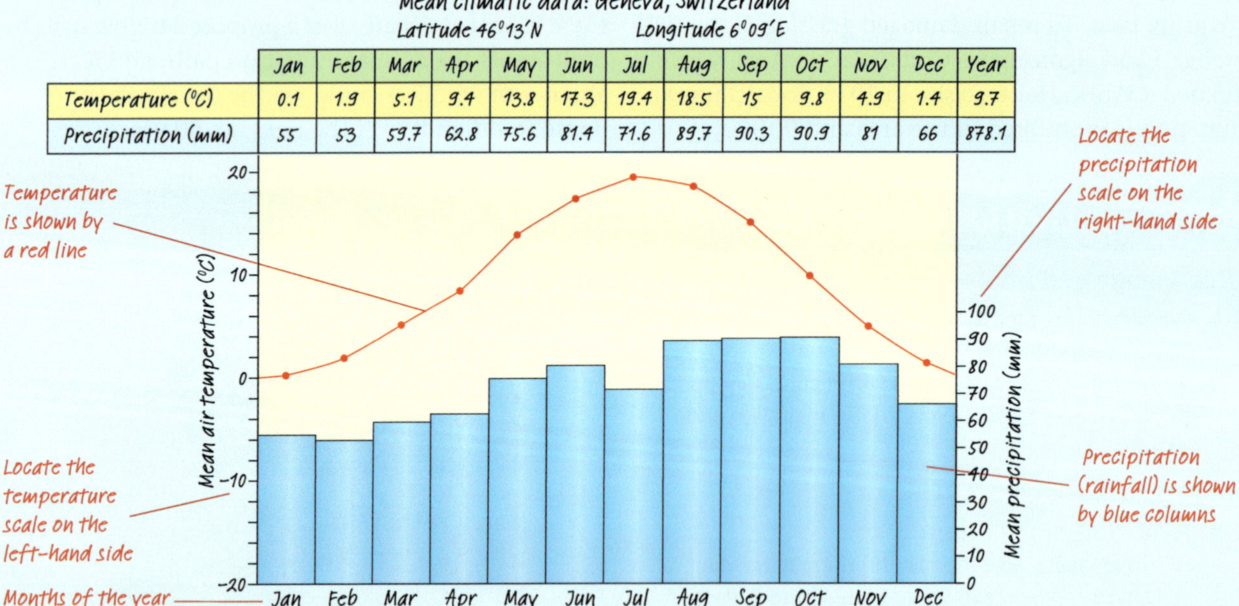

Mean climatic data: Geneva, Switzerland
Latitude 46°13′N Longitude 6°09′E

	Jan	Feb	Mar	Apr	May	Jun	Jul	Aug	Sep	Oct	Nov	Dec	Year
Temperature (°C)	0.1	1.9	5.1	9.4	13.8	17.3	19.4	18.5	15	9.8	4.9	1.4	9.7
Precipitation (mm)	55	53	59.7	62.8	75.6	81.4	71.6	89.7	90.3	90.9	81	66	878.1

3.4.1 Sample climate graph of Geneva explaining main features

	Beijing, China 39.93°N 116.20°E												
	Jan	Feb	Mar	Apr	May	Jun	Jul	Aug	Sept	Oct	Nov	Dec	Year
Temperature (°C)	–4.6	–1.8	4.7	13.6	20	24.4	26	24.7	19.8	12.6	3.9	–2.6	
Precipitation (mm)	3.9	4.7	8.2	18.4	33	78.1	224	170	58.4	18	9.3	2.7	635.3

3.4.2 Climate data for Beijing

ISBN 978 1 4586 6277 4

Precipitation graphs

- *Seasonality*—If more than 60% of precipitation occurs in the warmest months, there is a summer maximum pattern. This is the precipitation pattern for Beijing. If precipitation mostly occurs in the coldest months, there is a winter maximum pattern (e.g. Adelaide). If there is no distinct seasonal pattern, then precipitation is said to be evenly (or uniformly) distributed throughout the year (e.g. Hobart). Some places (e.g. London) receive uniform rainfall throughout the year. Many Asian regions have distinct monsoonal wet summer seasons and dry winter seasons. Regions with Mediterranean climates have the opposite: dry summers and wet winters (e.g. Adelaide and Perth).
- *Dry conditions*—If less than 25mm of precipitation falls in a month, then that month is classified as 'dry'. If the yearly total precipitation is also low, it is an arid climate.

Australian climate graphs

The climate graphs for nine Australian cities vary from a dry desert climate to a wet and dry monsoon climate. Sydney receives an East Coast temperate climate.

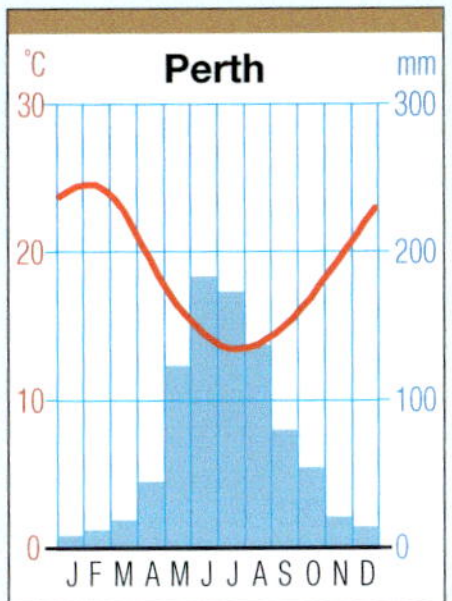

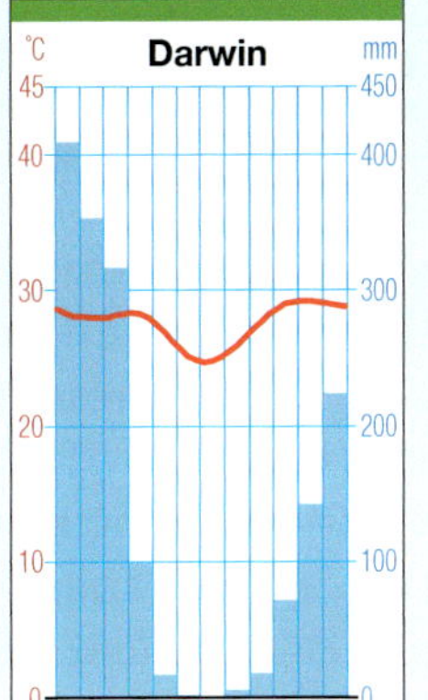

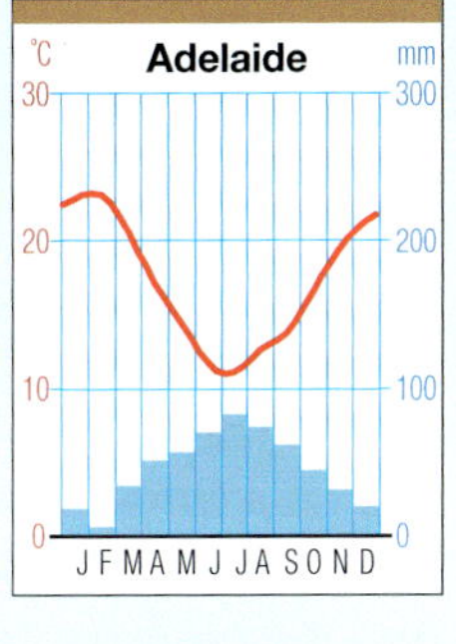

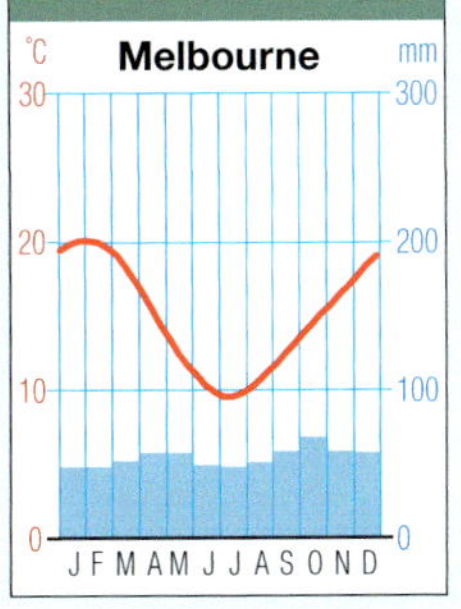

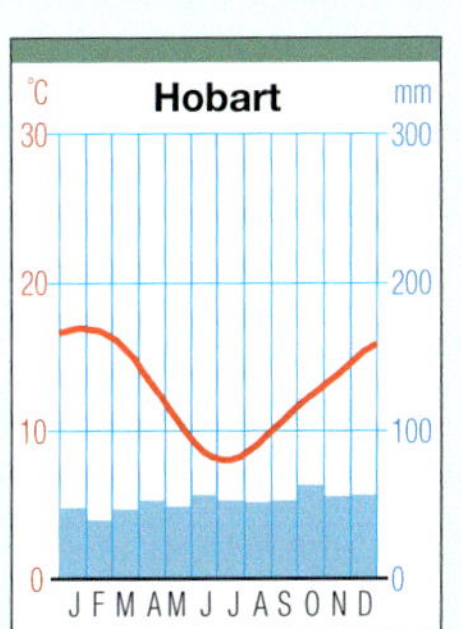

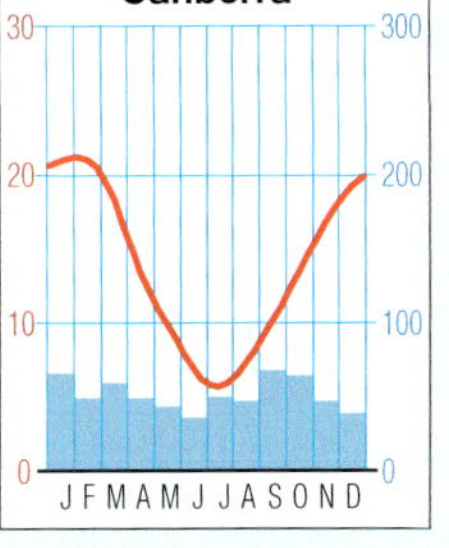

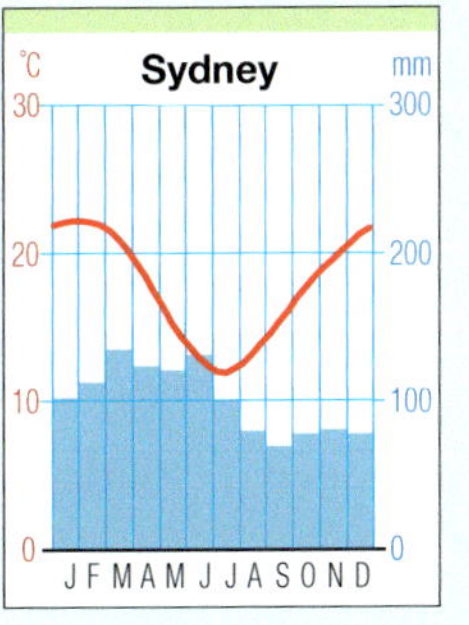

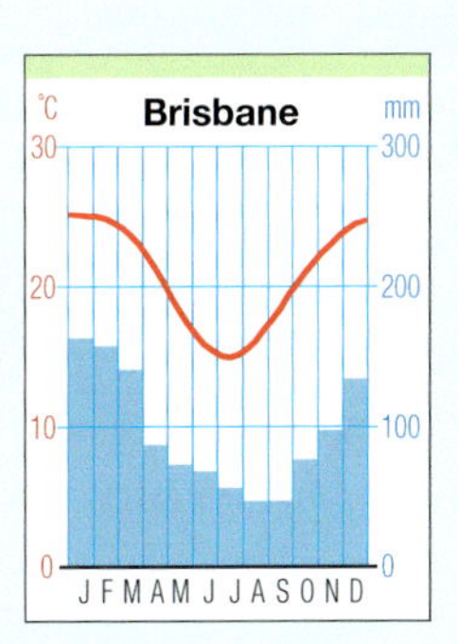

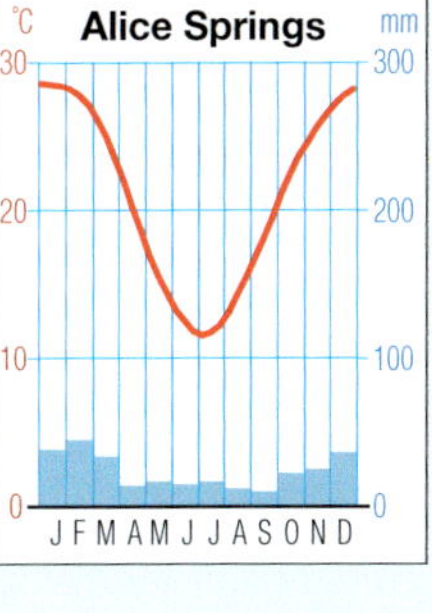

3.4.3 Climate graphs for Australia

Geo**activities 3.4**

Inquiry and skills

1 Refer to 3.4.1 and 3.4.2.
  - a Using graph paper, draw and label a climate graph for Beijing.
  - b Name the hemisphere for Beijing.
  - c Describe the temperature and precipitation patterns in Beijing.

2 Refer to 3.4.3.
  - a Add the rainfall totals for each month to work out the total annual rainfall in Sydney.
  - b In which season does most rain fall in Darwin?
  - c Why does Alice Springs have the greatest temperature range and the smallest precipitation?
  - d Compare the rainfall and temperature of Darwin and Adelaide.
  - e Compare the rainfall and temperature of Alice Springs and Melbourne.
  - f Investigate the effect of ocean currents off Australia on precipitation.

3 What are the consequences of the uneven distribution of climate on people and places such as Darwin and Perth?

ISBN 978 1 4586 6277 4

3.5 Australia's tropical grasslands

Tropical grasslands, or tropical savannas, are located in warmer regions of the world. In Australia they cover 25% of the land, stretching across northern Australia from Broome to Townsville. This landscape is important for biodiversity, indigenous culture, pastoral industry, tourism, defence and mining, for instance gold, uranium, diamonds and petroleum. Large proportions of Aboriginal and Torres Strait Islander Peoples live in this landscape and possess traditional knowledge of land management.

Two or six seasons?

Tropical grasslands experience two seasons—a hot wet summer and a dry warm winter—or six seasons, from an Indigenous perspective. Average temperature is above 18 °C each month, with precipitation diminishing from 2000 mm pa on the coast to less than 600 mm pa inland. Vegetation is dominated by the acacia, which produces gum used in drugs and food, and the eucalyptus tree, known for its oil and timber. The area is home to hundreds of species, including koalas, wallabies, crocodiles, termites, ants and spiders. Many species that are endemic (native) to the country are threatened by feral animals, such as pigs and cane toads, and introduced plant species, such as mimosa. Balancing the value of income received from agriculture and mining with the need to conserve biodiversity is a management problem.

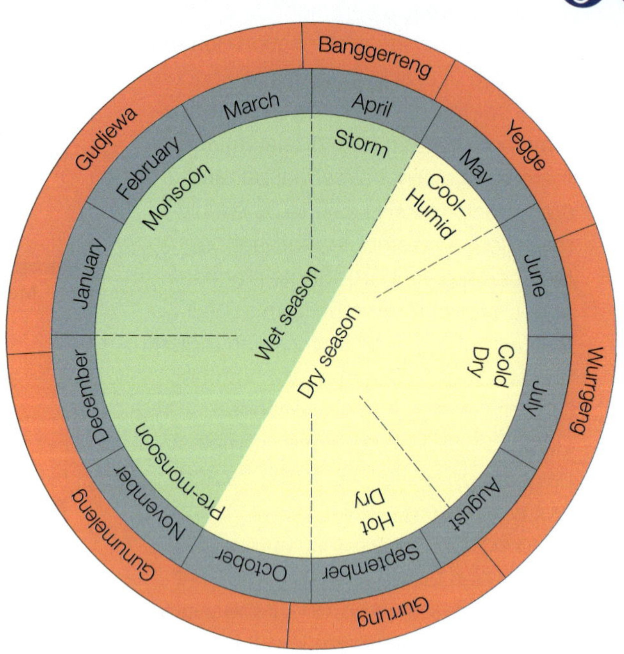

3.5.2 Two or six seasons of the tropical grassland landscape

Fire landscape and management

Over 98% of large bushfires occur outside the densely populated south-east and south-west sections of Australia. The largest and most frequent fires occur in northern Australia. Grass grows in the wet season and dries out in the dry season, making it vulnerable to fires each year. Managing the landscape means managing fire.

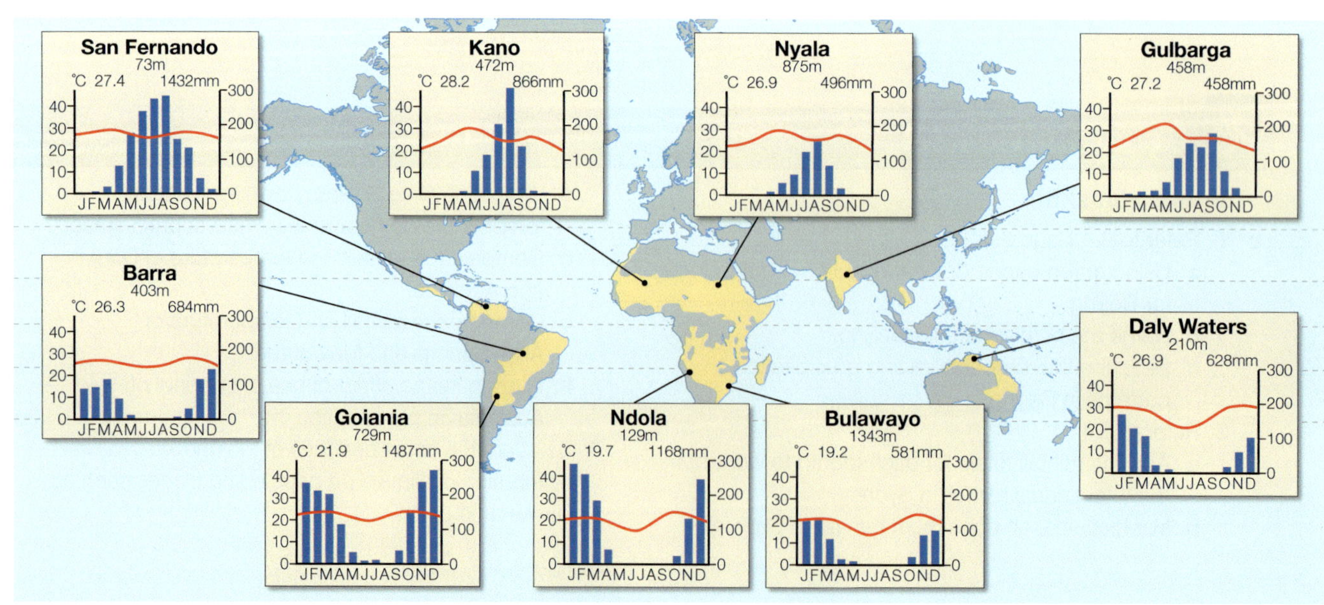

3.5.1 World map of tropical grasslands showing climate graphs

ISBN 978 1 4586 6277 4

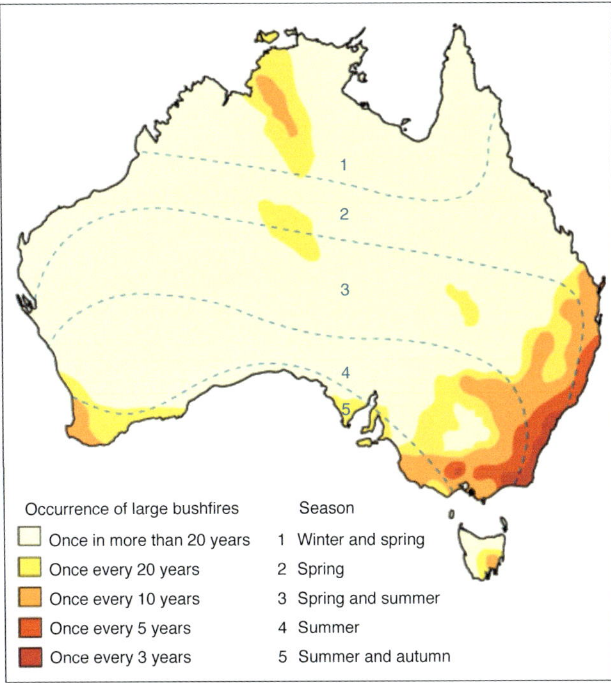

© IAWJ 2007

3.5.3 Extent and seasonal distribution of fire-affected areas

Fire is part of the natural cycle in tropical grasslands. Fires generate the growth of food plants, such as yams, providing nutrients to the soil. Some trees possess **epicormic buds** and **lignotubers**, which regenerate after fire. Less intense fires have been regularly burnt by Aboriginal Peoples for thousands of years, for hunting and protecting important food sources—a process known as firestick farming. However, introduced grasses have invaded the grasslands and provided fuel for more intense fires. As a result, plant and animal species that are an important part of Aboriginal bush tucker declined.

The West Arnhem Land Fire Abatement Project (WALFA) works to reduce the incidence and severity of wildfires that kill animals, damage properties and produce greenhouse gases. Having controlled fires in the cooler part of the dry season has reduced greenhouse gases in the area by 100 000 tonnes a year—a wildfire management solution that has been practised by Aboriginal Peoples for generations.

Termites and cattle

Most of the grasses are eaten and recycled by species such as termites, which have lived in the tropical grasslands for millions of years. Termites build huge mounds that can withstand fire. They also aerate the soil and create fertile soils containing nitrogen and phosphorus. Acacia trees growing near termite mounds contain 60% more shoots and bear more fruit than those growing elsewhere.

Grass is important to the cattle industry, but too many cattle in an area leads to weeds, a decline in grass species, erosion and an inability to cope with hot fires. Perennial grasses for cattle grazing include Mitchell grass, bluegrass and ribbon grass, while introduced species, such as gamba grass, compete with native grasses and require careful control.

Geo**info**

52% of the tropical grassland population live in Darwin and Townsville.

Geo**activities 3.5**

Knowledge and understanding

1 Where are tropical grasslands landscapes located?
2 Explain why tropical grasslands are important for the Australian economy.
3 How do plants adapt to a fire environment?
4 Describe the threats to the tropical grassland landscape.
5 Discuss the difference between firestick farming and wildfires.
6 Explain WALFA and its advantages to the environment.

Inquiry and skills

7 Refer to 3.5.1.
   a List the tropical grasslands located in the southern hemisphere.
   b Which months are the wettest in the southern hemisphere?
   c Which months are the driest in the northern hemisphere?
   d List the three places that receive the highest precipitation in one month.
8 Refer to 3.5.2.
   a Name the six seasons used by Indigenous Australians.
   b Explain why the Indigenous classification of seasons provides more information than the European two-season model. Which model do you prefer? Provide reasons for your answer.
9 Refer to 3.5.3. Describe the spatial distribution of large bushfires in Australia and their seasonality.
10 Using secondary sources research the following inquiry questions: Why is fire part of the grassland environment? What are the risks of uncontrolled fires? What are the benefits of using fire as a land management tool? Present the information as a PowerPoint presentation or a pamphlet.

3.6 Forest landscapes

Forests provide habitats for animals and are the source of more than 5000 products, including fuel, food, medicine, furniture and clothing. They prevent soil erosion, regulate climate (by absorbing CO_2) and provide the oxygen we breathe. However, most of the world's forests have been logged. Native species have been cleared for pine plantations in Australia, for oil palms in Indonesia and for beef ranches and soybeans in the Amazon Basin in South America. Forests are also threatened by climate change and wildfires.

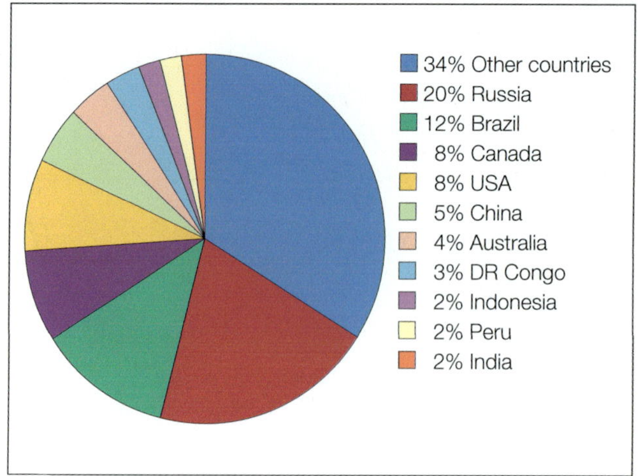

3.6.2 Global forest cover

Forest landscapes and distribution

A forest is an area containing a high density of trees. Today they cover 30% of Earth's land area, although they once covered 50%. There is a diversity of forests across Earth, such as rainforests (Brazil), coniferous forests (Switzerland) and sclerophyll forests (Australia). Forests are unevenly distributed across regions, countries and climates:

- *Regions*—27% in Europe; 23% in South America; 17% in Africa; 14% in Asia; 14% in North and Central America; 5% in Oceania.
- *Countries*—approximately 66% of the world's forests are distributed among 10 countries, including Russia, Brazil and Canada.
- *Climates*—the largest proportion of the world's forests is located in topical climates (47%). This is followed by sub-arctic (33%), temperate (11%) and sub-tropical (9%) climates.

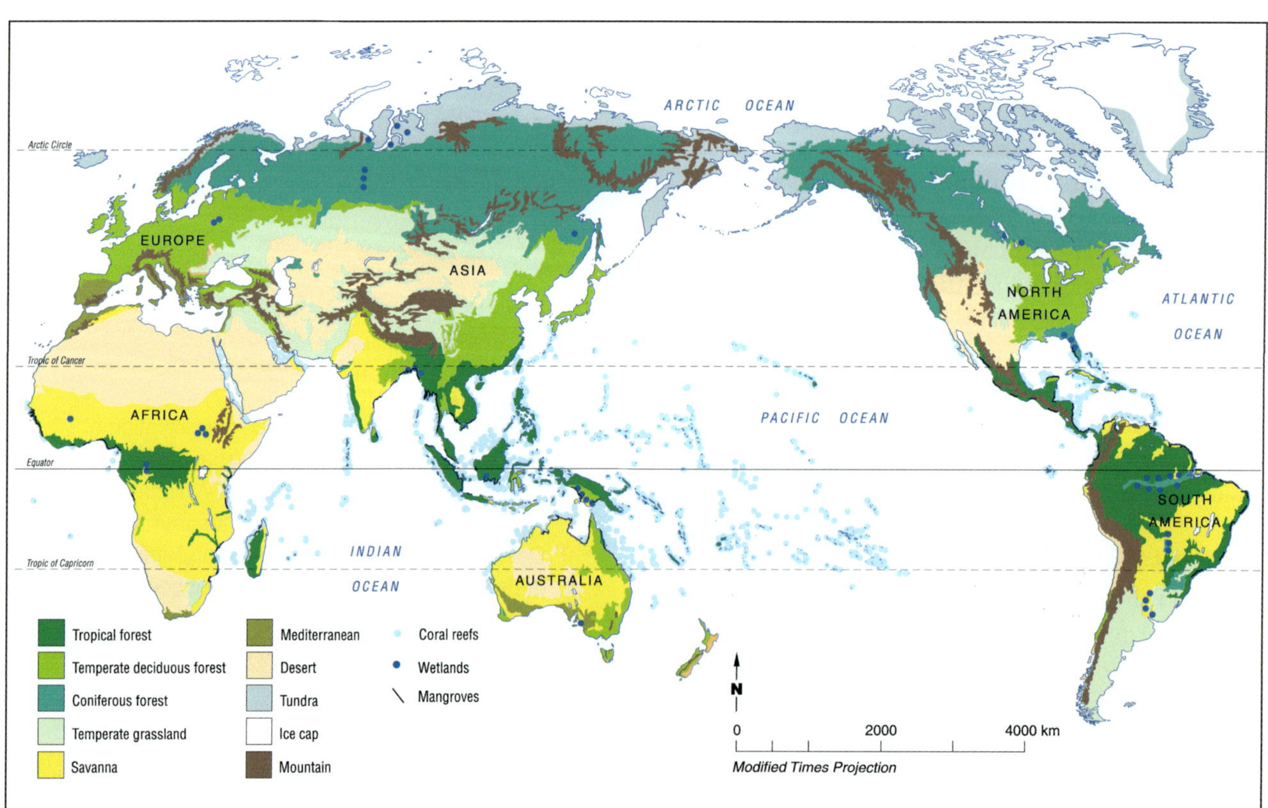

3.6.1 Spatial distribution of forest regions around the world

ISBN 978 1 4586 6277 4

Climate, soil and gradient

The factors determining the distribution of forests are largely climatic. Trees require moisture, warmth and nutrients. As a result, trees are generally absent from areas that are too dry (desert) or too cold (Antarctica), lack soil nutrients (nitrates) and contain inadequate soil cover (steep mountain slopes). Other factors that limit the distribution of forests includes fire, flooding, toxic minerals in the soil and water, and large herbivores, such as elephants. These large animals kill trees by uprooting and breaking branches, making them susceptible to infestation by wood-boring insects.

Australia: Gondwanaland species

Approximately 19% of Australia's land use consists of native forests. There are 457 forest communities distributed across Australia. Notable forest landscapes include:

- *Wollemi National Park* in New South Wales, which contains species thought to have become extinct 30 million years ago
- *Daintree Rainforest*, Queensland
- *Mt Warning National Park* in New South Wales is a World Heritage site with subtropical rainforests
- *Gloucester National Park* with karri trees in Western Australia
- the *Gondwana Rainforests*, consisting of 50 reserves located between Newcastle and Brisbane.

Temperate rainforest, Tasmania

Tropical rainforest, Polynesia

Redwood forest, California

Coniferous forest, Swiss Alps

Spiny forest, Madagascar

Sclerophyll dry forest, Western Australia

3.6.3 Diversity of forest landscapes

Geo**info**

1.6 billion people depend on forests for their livelihoods.

Geo**activities 3.6**

Knowledge and understanding

1. How would you define a forest?
2. What is the importance of forests to people and the environment?
3. List the changes to forests over time.
4. What climates contain the highest percentage of forests?
5. Explain how forests are unevenly distributed across Earth.
6. What are the main factors determining the distribution of forests?

Inquiry and skills

7. Refer to 3.6.1 and an atlas. What are the main forests in Australia, India, Scandinavia and the Amazon and Congo basins?
8. Refer to 3.6.2 and list the four countries with the largest percentage of forest cover.
9. Refer to 3.6.3. Design a collage of Australian forests using ICT. Label the types of forests and their locations.
10. Research past and present vegetation in your local area. Discuss why the vegetation has been cleared and whether native species remain or have been replanted.

ISBN 978 1 4586 6277 4

3.7 Changing forest landscapes

Over the last 8000 years, 45% of the Earth's original forest cover has disappeared. The approximate rate of deforestation is 1 ha/second—equivalent to two football fields per second. Logging leads to loss of biodiversity, soil erosion and greenhouse gases. Media campaigns and laws have contributed to a reduction in deforestation.

Illegal wood trafficking has increased, especially in countries where poverty is prevalent. The United Nations Environment Programme states that illegal logging accounts for 30% of the global logging trade. Approximately 80% of timber exported from Peru is illegal. One mahogany tree is worth $1936, which is why it is referred to as 'green gold'.

Deforestation and reafforestation

Madagascar, an island off the coast of Africa, has lost 90% of its original forests since the arrival of humans 2000 years ago. Forests have been cleared for timber, mining, slash-and-burn farming, pastures for animals, and fuelwood for cooking and heating. In 2009 political turmoil saw thousands of illegal loggers ransacking national parks for endangered species, such as ebony and rosewood. Currently, 98% of Madagascar's timber is exported to China to supply ebony for guitars and rosewood for beds.

Papua New Guinea (PNG) has the world's third largest tropical forest. However, logging for timber and forest fires have contributed to a deforestation rate of 362 000 ha per year. At the present rate of deforestation, 80% of PNG's forests will be lost in the next 10 years unless reafforestation strategies are implemented.

In Australia, wood products are worth $22 billion a year and 66 000 people are employed in the industry. Some eucalypt forests have been logged and replaced by pine plantations, which are later converted to wood chips for paper. Reafforestation programs have been introduced to reduce environmental problems.

India's forests

In 1847, the King of Jodhpur in Rajasthan ordered his army to cut trees in the Bishnoi forest to provide wood for his palace. The forest's inhabitants protested, offering their bodies as shields for the trees. The army killed 363 people before the king halted the logging and declared the Khejarli region a preserve—off-limits for logging and hunting.

The use of forests for fuelwood and the cremation of Hindus has exacerbated deforestation. Each year, more than 7 million Hindus die. For cremation, each body requires 400–500 kg of wood, resulting in 50–60 million trees logged to burn the dead each year. Most Indians have rejected electric cremations for cultural reasons. The new Green Cremation System may be the answer as it only requires 150 kg of wood to burn a body.

3.7.1 Highest deforestation rates of primary forests over last 10 years

Rank	Country	Deforestation rate (%)
1	Nigeria	56%
2	Vietnam	55%
3	Cambodia	30%
4	Sri Lanka	15%
5	Malawi	15%
6	Indonesia	13%
7	North Korea	9%
8	Nepal	9%
9	Panama	7%
10	Guatemala	6%

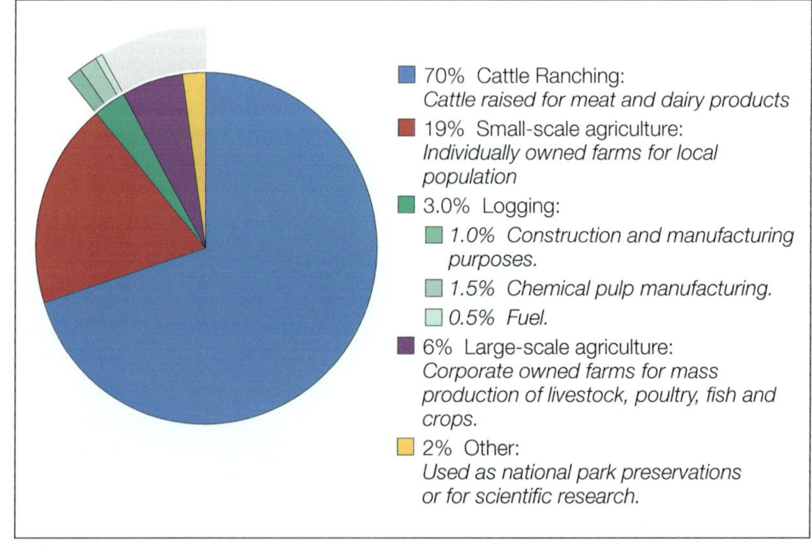

3.7.2 Causes of deforestation in the Amazon rainforest

ISBN 978 1 4586 6277 4

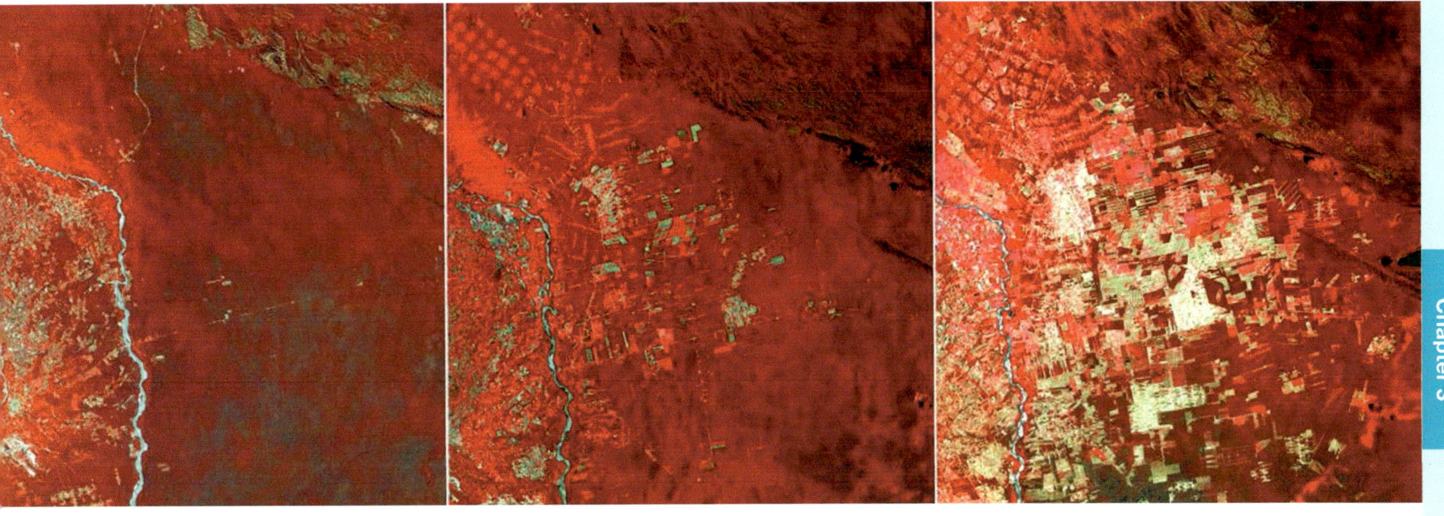

3.7.3 Satellite imagery showing deforestation in Bolivia, 1975–2013

Haiti's lack of forests

In 1923, 60% of Haiti's land was forested but by 2013 only 1.5% remained. Thousands of hectares of forests have been cleared for sugar cane and to power the sugar mills, and impoverished Haitians have cut trees for fuelwood and making charcoal. In 2010, a hurricane pounded the remaining forests and triggered mudslides. Haiti illustrates the consequences of valuing forests for the commodities they provide (food and fuel), while ignoring environmental services. Recently, grassroots projects planted thousands of trees and introduced solar cookers as an alternative to using wood and charcoal.

Recognition of the importance of protecting rainforests has led to the development of national parks and wildlife reserves. Bolivia, Brazil, Congo, Peru and Venezuela have devised projects to rehabilitate forest landscapes. The World Resources Institute uses geographic information systems (GIS)and satellites to strengthen sustainable forest management.

Geo**info**

- Deforestation accounts for 20% of global greenhouse gas emissions.
- 80% of logging operations are illegal in Bolivia and 42% in Colombia.

Geo**activities 3.7**

Knowledge and understanding

1 Why are forests disappearing in Madagascar?
2 What is the approximate rate of deforestation?
3 List the environmental problems of deforestation.
4 Why does illegal logging occur?
5 Why are illegally logged forests called 'green gold'?
6 Describe the deforestation problems in Papua New Guinea.
7 How has culture contributed to deforestation in India?
8 What was the good news concerning deforestation in Rajasthan, India?
9 Explain how humans changed the forest landscape in Haiti.
10 Suggest strategies to protect forests.

Inquiry and skills

11 Refer to 3.7.1. Why do you think deforestation is high in developing countries?
12 Refer to 3.7.2.
  a What are four main reasons for deforestation in rainforests?
  b What are the links between Western consumerism and deforestation?
13 Refer to 3.7.3.
  a Describe the changes over time to the Bolivian forests.
  b Why have Bolivia's forests been cleared?
14 Reflect on your learning and investigate how activists and non-government organisations (e.g. the Wilderness Society, the Australian Conservation Foundation and Greenpeace) contribute to the reduction in deforestation.
15 Research the importance of forests to the Australian economy and how they are managed in your state/territory. Explain whether the practice is sustainable? Present your findings using web 2.0 tools.

ISBN 978 1 4586 6277 4

3.8 Bushfires shape landscapes

Bushfires shape the landscape and are part of the natural process in some ecosystems, ignited naturally through phenomena such as lightning. Burning contributes to the decomposition of organic matter and plant growth. Many Australian plant species have adaptations to fire, such as epicormic buds (fire-resistant seeds) and lignotubers (shoots that sprout after fire). Eucalyptus trees stimulate fires as eucalyptus oil is flammable.

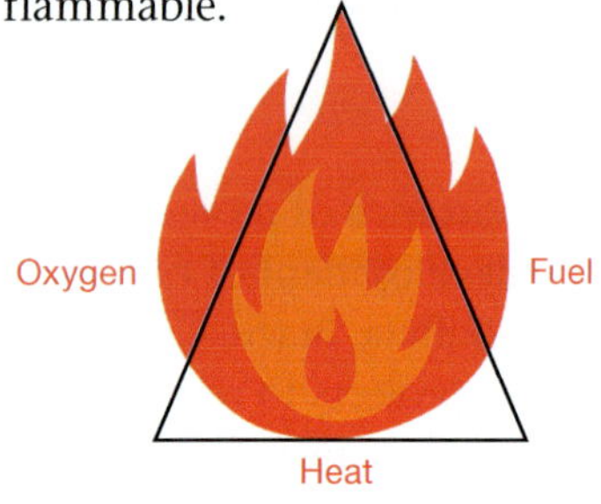

3.8.1 The fire triangle

3.8.2 How a fire storm starts

From trees to ashes

Bushfires, wildfires, forest fires and grass fires are all uncontrolled fires in areas containing combustible vegetation. These fires are extensive, travel quickly across landscapes, and have the ability to change direction and jump gaps, such as rivers and fire breaks. They occur on every continent except Antarctica. In the USA, 60000–80000 wildfires occur each year, causing death and damage to properties and adding carbon dioxide and ash to the atmosphere. Bushfires are common in Australia when:

- areas experience hot, dry summers
- winds blow from the dry interior
- vegetation dries out after a drought.

Fire requires heat, oxygen and fuel to burn. Three factors determine whether the fire fizzles out or becomes a raging blaze:

- *fuel*—long, dry grass burns faster than moist leaf litter. A continuous 'ladder' of dry fuel from the ground to the top of trees increases the incidence of 'crown fires'
- *weather*—strong, dry winds when temperatures are high and humidity is low encourage fire
- *topography*—fire moving up a slope travels faster than fire on a plain. For every 10° of upslope the rate at which the fire spreads doubles, and vice versa when going down a slope.

An inferno can start on a hot summer day with dry and windy conditions, ignited by a spark, unextinguished cigarette butt or lightning. The fire quickly spreads, consuming dry vegetation and anything in its path.

3.8.3 Five volunteer firefighters were killed when the wind suddenly changed direction during the Linton bushfire in 1998

ISBN 978 1 4586 6277 4

Forest fire types

Brush fire
A brush fire spreads along low-lying vegetation, moss and lichen, while trees remain unaffected. A small brush fire can spread at a rate of one meter per minute, while an intense fire can spread at over three meters per minute

Crown fire
A crown fire burns the crowns of trees and tall shrubs. A mild crown surface fire can spread at up to three meters per minute, while a massive fire can reach over 100 meters per minute

Peat fire
A mild peat fire can start at depths of approximately 25 centimeters, while a massive fire will burn at a depth of over 50 centimeters

Steps in extinguishing massive wildfires

1. Surveillance
This step includes locating the burn area, its type and intensity at the perimeter and its characteristics during the day and night. The results are used to forecast a fire's rate of expansion, its type and strength

2. Containing the wildfire
At this step, a fire is contained with control breaks and ditches in its probable path

3. Extinguishing the fire and suppressing ignition sources

4. Fire patrol, focusing on systematic inspection of affected areas along a fire's perimeter

Source: RIA Novosti, 2013

3.8.4 Forest fires: types and steps in extinguishing

Geo**info**

- 97% of fires in Mexico are lit for agricultural needs.
- Lightning accounts for 30% of fires in China's northern forests.

Prevention, detection and suppression

Prevention, detection and suppression programs aim to reduce the chance of a fire hazard becoming a disaster. Some countries back-burn the vegetation, which involves the controlled ignition of under-growth to prevent catastrophic fires.

Most fire-prone areas have fire-fighting services, water-spraying trucks and helicopters that douse water over inaccessible areas. However, urban sprawl in cities like Sydney and Melbourne has resulted in homes being built in the middle of highly flammable forests. In fire-prone areas, people can take precautions, such as constructing homes using flame-resistant materials and avoiding placing wood piles located close to home.

Fanning the flames

The extent, frequency, intensity and timing of a fire that prevails in a specific area is called the fire regime. Some scientists say climate change is changing the fire regime and 'fanning the flames', as large fires are increasing in intensity and frequency around the world. There are questions to be asked: Are people disrupting natural fire cycles by allowing fuel to build up? Has vegetation been converted to a type that produces frequent fires? Are more people living in fire hazard areas?

Geo**activities 3.8**

Knowledge and understanding

1 What is a bushfire?
2 Why are bushfires important to the natural environment?
3 How has Australian vegetation adapted to bushfires?
4 What is a fire regime?
5 What are the causes of bushfires?

Inquiry and skills

6 Refer to 3.8.1.
  a What is the fire triangle?
  b Suggest how reducing one side of the triangle reduces fires.
7 Refer to 3.8.2 and describe how a fire storm is created.
8 Refer to 3.8.3. Describe the extent and movement (i.e. direction, distance and time) of the Linton fire in Victoria in 1998.
9 Refer to 3.8.4.
  a List three types of forest fires.
  b What are the potential threats of crown fires?
  c Suggest steps to take to extinguish massive fires.
10 Most firefighters in Australia are volunteers. Research the training and expectation of a volunteer firefighter. Prepare an informative article for your school website.

ISBN 978 1 4586 6277 4

3.9 Bushfires: Australia burning

The 2009 Black Saturday bushfires in Victoria left 173 people dead, 500 injured, and thousands of homes destroyed. Many feared a repeat of Black Saturday in January 2013 when a heatwave hit Australia, affecting around 70% of the country. Soaring temperatures, strong winds and dry vegetation all combined to create a catastrophic fire threat, and on 8 January 2013, bushfires burned in all states and territories.

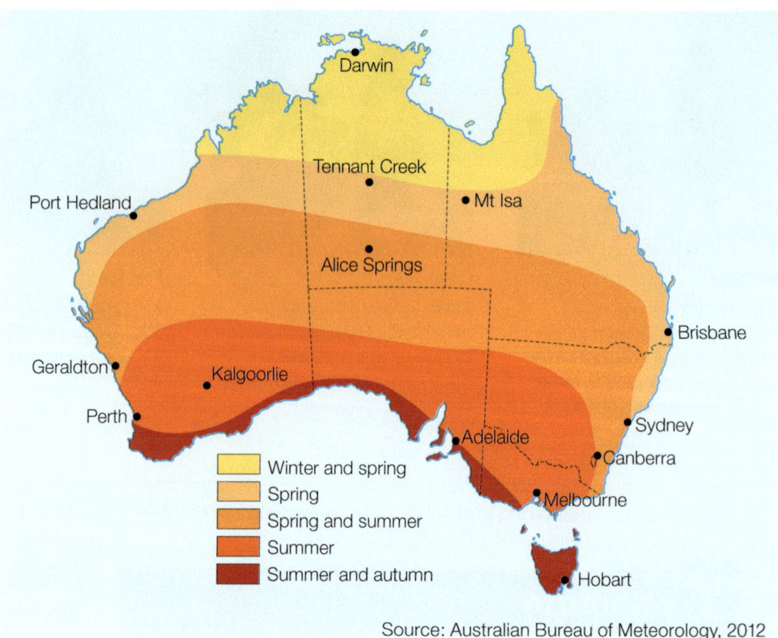

3.9.2 Australian fire seasons

Bushfires are part of landscapes

Fire has shaped the Australian landscape. Aboriginal Peoples used firestick farming for thousands of years, and early European explorers noted bushfires burning across the continent. The original settlers used fire to clear vegetation for agriculture, mining and urban settlements, frequently during the hottest and driest weather. As a result, the natural fire regime was altered by the arrival of humans. Fires became more frequent and intense and fire-tolerant species, such as eucalypts, expanded their range.

Australian bushfires can be divided into two categories:

- *mountain and hill fires*—the uncontrolled burning of dense forests. The steep terrain increases the speed and intensity of a firestorm
- *plain fires*—the uncontrolled burning of scrub and grasslands. Strong winds fan the fires, posing threats to settlements.

Managing bushfires

Bushfires are caused by natural and human forces, such as lightning, arson, campfires, cigarettes and the use of machinery (e.g. welding). The Fire Danger Rating (FDR) provides fire forecasts that consider temperature, humidity, wind speed and dryness

3.9.1 Impacts of bushfires in Australia

Bushfire	Location	Area burned (ha)	Summer	Deaths	Damages
Black Thursday	Victoria	~5 000 000	1851	12	1 million sheep; thousands of cattle
Black Friday	Victoria	2 000 000	1938–39	71	3700 houses
Ash Wednesday	South Australia and Victoria	418 000	1983	75	2400 houses
Black Saturday	Victoria	450 000+	2009	173	2029 houses, 2000 other structures
Eastern seaboard	NSW	~400 000	1993–94	4	225 houses
Margaret River	Western Australia	4 000	2011	0	31 houses
Black Tuesday	Tasmania	1 264 000	1967	62	1293 houses
Canberra	Australian Capital Territory	160 000	2003	4	500 houses

ISBN 978 1 4586 6277 4

OUR NEW NORMAL

The science is clear, our world is warming. Globally, it has now been 27 years since the world experienced a month that was colder than average.[2]

“The current heatwave – in terms of its duration, its intensity and its extent – is now unprecedented in our records... Clearly, the climate system is responding to the background warming trend.”

David Jones, the Bureau of Meteorology's manager of climate monitoring and prediction.

THE BUREAU OF METEOROLOGY ADDED A NEW COLOUR TO ITS HEAT MAPPING INDEX THIS WEEK TO REPRESENT TEMPERATURES ABOVE 50°C[4]

40.3°C[4] AVERAGE MAX DAILY TEMPERATURE RECORD FOR AUSTRALIA BROKEN JAN 8TH

7 HOTTEST DAYS EVER[4]

ON JAN 8TH AUSTRALIA BROKE THE RECORD FOR CONSECUTIVE DAYS ABOVE 39°C AVERAGE TEMPERATURE

7 DAYS IN A ROW

312 | 3 DARWIN

21 | 1 BRISBANE

14 | 3 SYDNEY

32 | 5 CANBERRA

44 | 17 ADELAIDE

27 | 9 MELBOURNE

3 | 1 HOBART

72 | 27 PERTH

MAP KEY

DAYS OVER 35°C THIS CENTURY

IN 2008 BY 2100

FIRES BURNING ON JAN 8TH

HOT AND DRY

THE CLIMATE COMMISSION PREDICTS A LARGE INCREASE IN THE NUMBER OF DAYS OVER 35°C THIS CENTURY[3]

‘CATASTROPHIC’ FIRE DANGER

From Queensland to Tasmania, vast tracks of Australia are burning and the threat is far from over. Over 100 fires are still burning across the country, with NSW reaching “catastrophic” danger levels.

4TH JANUARY 2013 WAS TASMANIA'S HOTTEST DAY IN ONE HUNDRED AND TWENTY YEARS

250 HOMES WERE DESTROYED

Source: Get Up! Action for Australia

3.9.3 The new normal: we're being told to ‘get used to it’

of vegetation. To alert the community, Fire Danger Ratings are broadcast via radio, TV and the internet.

Many scientists regard climate change as having contributed to the current extreme heat conditions and resulting bushfires.

The danger and destruction of bushfires to people and properties is undeniable. Unless effective preventative plans are implemented, it is anticipated climate change will increase the intensity and number of bushfires in Australia.

Geo**activities 3.9**

Knowledge and understanding

1 Describe what occurred in January 2013.
2 What was the impact of Black Saturday in Victoria?
3 *Fire has shaped the Australian landscape.* Explain what is meant by this phrase.
4 List the two catagories of bushfires in Australia and how they differ.

Inquiry and skills

5 Refer to 3.9.1. Describe the impact of bushfires on people, animals and properties.
6 Refer to 3.9.2.
  a What is the fire season in your local area?
  b *Every state in Australia suffers from bushfires but at different times of the year.* Explain this statement.
  c Write a fire plan to give people living in fire-prone areas.
7 Refer to 3.9.3.
  a Describe what is called ‘our new normal’.
  b What are the links between hot weather and catastrophic fire danger?
  c Compare the temperature in inland Australia with Darwin and Adelaide.
8 Heat waves, strong dry winds and dry vegetation are ideal conditions to generate fierce and uncontrollable bushfires across Australia. Research one bushfire in Australia and explain its impacts on people and properties. Reflect on your learning and suggest strategies to ensure a bushfire hazard does not become a disaster.

ISBN 978 1 4586 6277 4

3.10 Desert landscapes and landforms

Deserts are very dry places where evaporation is higher than precipitation. They cover nearly one-third of Earth's land surface and are found on every continent except Europe.

Classifying desert landscapes

Arid landscapes can be broadly classified into three types, according to precipitation:

- extremely arid (no precipitation for 12 months)
- arid (less than 250 mm pa)
- semi-arid (250–500 mm pa).

About 70% of mainland Australia receives less than 500 mm pa and is classified as arid or semi-arid.

Types of desert landscapes

There are five major types of desert: subtropical deserts, interior (or mid-continental) deserts, rainshadow deserts, coastal deserts and polar deserts. Each type has a wide variety of landforms. Only about 15% of the world's deserts are sand dunes (**ergs** or sand seas). Most deserts are mountains, basins and plateaus (**hamada**); stony pavements (**regs**); and dry salt lakes (playas).

- *Subtropical deserts*—located along the Tropics of Cancer and Capricorn between 15°N and 30°N and between 15°S and 30°S. They are dominated by high-pressure systems where dry air descends. Subtropical deserts include the Kalahari Desert and the world's largest desert, the Sahara.
- *Interior deserts*—interior (or mid-continental) deserts are located far from moisture-bearing winds and oceans. Winds that reach these deserts have little moisture left. The Gobi Desert in China and Mongolia is an example of an interior desert.
- *Rainshadow deserts*—occur behind mountain ranges where prevailing winds are forced to rise and drop their precipitation. The wind that descends on the leeward side of the mountains

3.10.2 Sand dunes in the Grand Erg Occidental, Sahara Desert, western Algeria

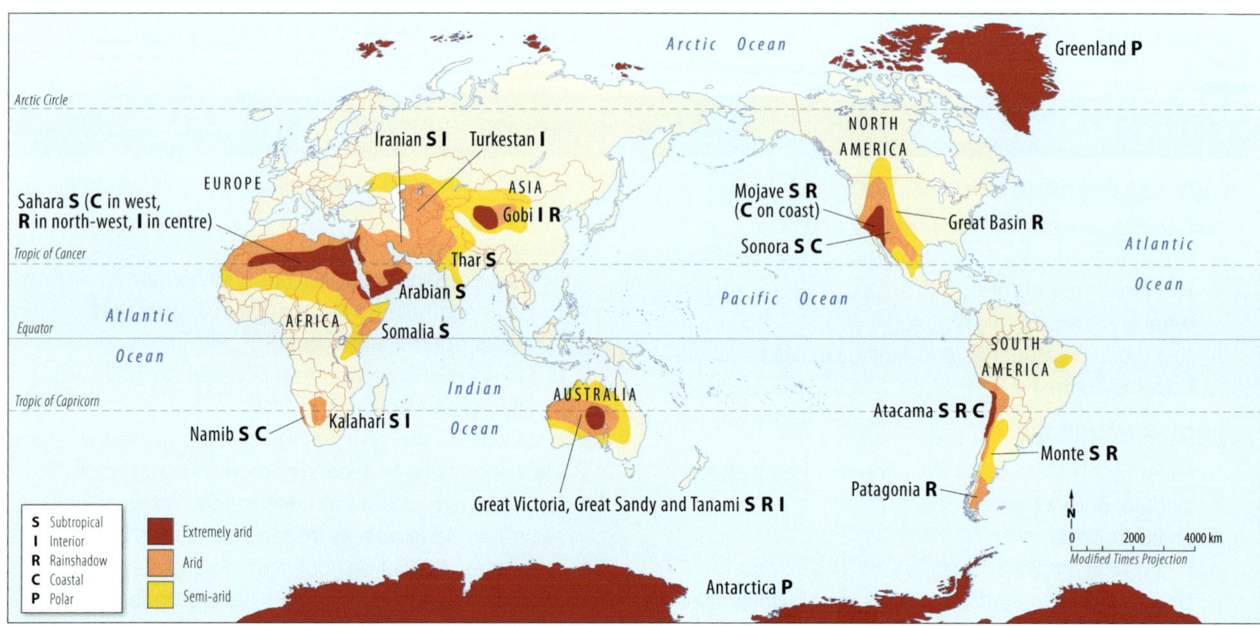

3.10.1 Spatial distribution of desert landscapes around the world

ISBN 978 1 4586 6277 4

has little moisture left. The driest place in North America, Death Valley in the Mojave Desert, is in the rainshadow of the Sierra Nevada Mountains.

- *Coastal deserts*—have cold offshore ocean currents, which produce less rainfall than warmer ocean currents. They are mainly found on the west coast of continents.
- *Polar deserts*—Antarctica is the world's driest and largest desert. Parts of the Arctic, including Greenland, are also polar deserts. Cold, dry air dominates with little precipitation. Even though there are large amounts of water at the poles, it is locked away as ice and is therefore unavailable for use by plants and animals.

3.10.3 Eroded ridges created by sandstorms blowing from one direction in the Gobi Desert, China (note the scale by comparing the yardangs with the vehicle in the foreground)

3.10.4 Playa and mountains, Death Valley, Mojave Desert, USA

3.10.5 Badlands in the Atacama Desert, Chile

3.10.6 Tourists in Taylor Dry Valley frozen desert, Antarctica

Geoinfo

Severely eroded arid hills are sometimes called 'badlands'.

Geoactivities 3.10

Knowledge and understanding

1 What are desert landscapes?
2 Name the five types of deserts according to the causes of their aridity.
3 Why are deserts often found in the centre of continents?
4 What are rainshadow deserts?
5 How do cold ocean currents create coastal deserts?
6 Explain why there are polar deserts.

Inquiry and skills

7 Refer to 3.10.1. List places which contain subtropical and rainshadow aridity.
8 Explain why it is geographically more accurate to classify deserts into the five major types rather than a simplistic classification into hot and cold deserts.
9 Using Google images, compile a collage of desert landforms from each of the five types of deserts. Comment on any similarities or differences between the types of deserts.

3.11 Deserts: weathering processes

Most people think desert landforms consist of sand dunes and oases. However, this stereotype is extremely limited. Desert landforms vary enormously according to geology (rock type), climate and the geomorphic processes that have produced them over time. There are five main types of desert landforms in Australia (3.11.2). Both water and wind erosion have worn away and removed surface materials to create distinctive landforms.

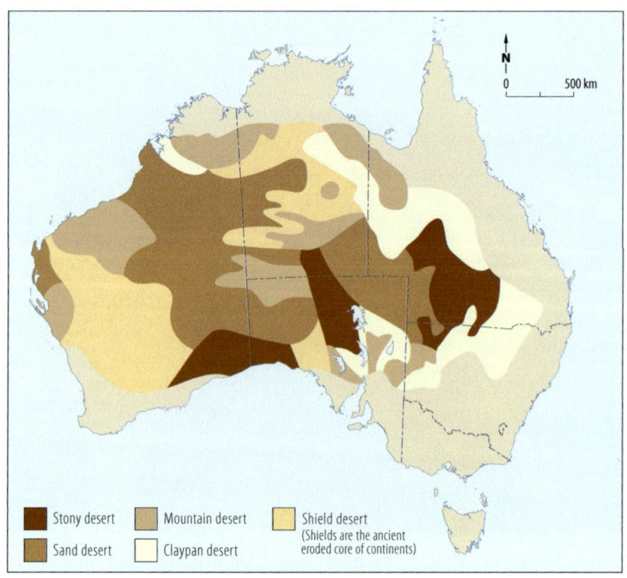

3.11.2 Spatial distribution of arid landforms in Australia

Water erosion and deposition

Although water is scarce in desert landscapes, it plays a significant role in creating landforms, especially in mountainous areas. Where vegetation is sparse, loose surface sediments are easily washed away by sudden downpours. In addition, precipitation infiltrates slowly in arid landscapes and most of it runs off quickly, causing sheet erosion. Over time, the erosive power of fast-flowing water and flash floods forms gullies and wadis. Eroded sediments are deposited as alluvial fans and playa lakes (salt lakes).

Geoinfo

About 70% of Australia is desert. Most deserts occur in Western and Central Australia.

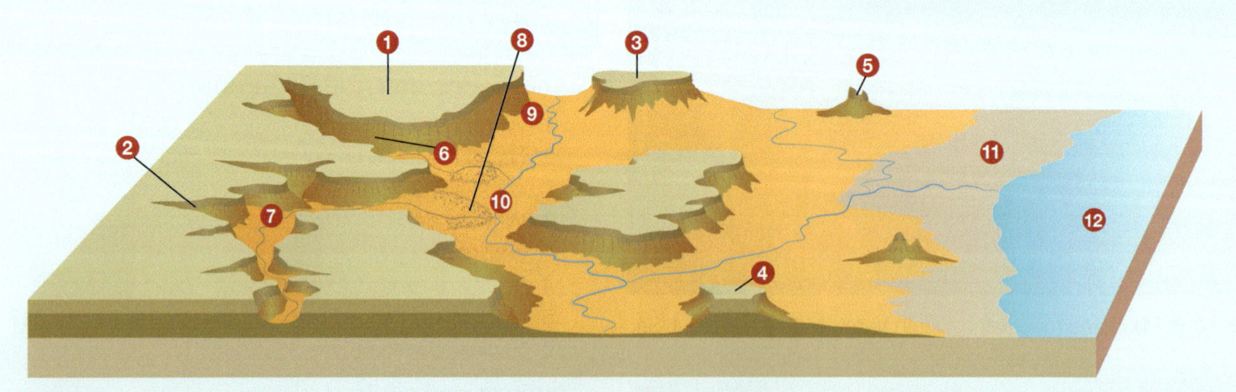

1 Plateau—wide, fairly level, uplifted landform with steep sides
2 Gully—surface channel eroded by running water, especially after heavy rain
3 Mesa—smaller, more resistant part of a former plateau with a flat top and steep sides
4 Butte—smaller isolated mesa
5 Chimney rock—the eroded remnants of a butte
6 Canyon—steep-sided gorge formed by water erosion
7 Wadi—dry river bed except during occasional flash floods
8 Alluvial fan—semicircular deposit of eroded materials washed out of the wadis
9 Pediment—deposit of eroded materials at the base of steep slopes
10 Bajada—alluvial fans that have joined together to form an alluvial plain
11 Pediplain—level surface that has been eroded to the lowest possible level
12 Ocean

3.11.1 Desert landforms eroded by water

ISBN 978 1 4586 6277 4

Wind erosion and deposition

In low-lying desert landscapes with a plentiful supply of sand, winds sandblast rocky surfaces in a process called abrasion. Over long periods of time, wind erosion creates many desert landforms, such as mushroom rocks and stony plains (gibber plains). Wind deposition creates loess (wind-blown silt) and sand dunes, which may be mobile or stabilised by vegetation. Major sand dune systems called ergs (sand seas) are often formed by wind patterns over many tens of thousands of years. Four of the most common dune forms include barchan, transverse, longitudinal (or linear), and parabolic.

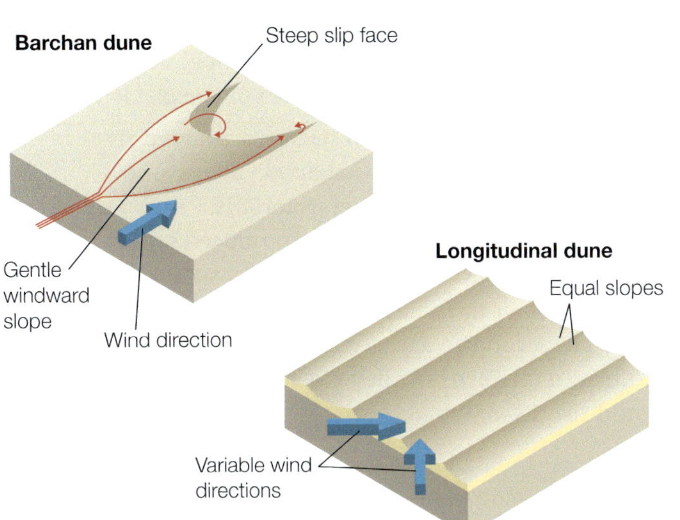

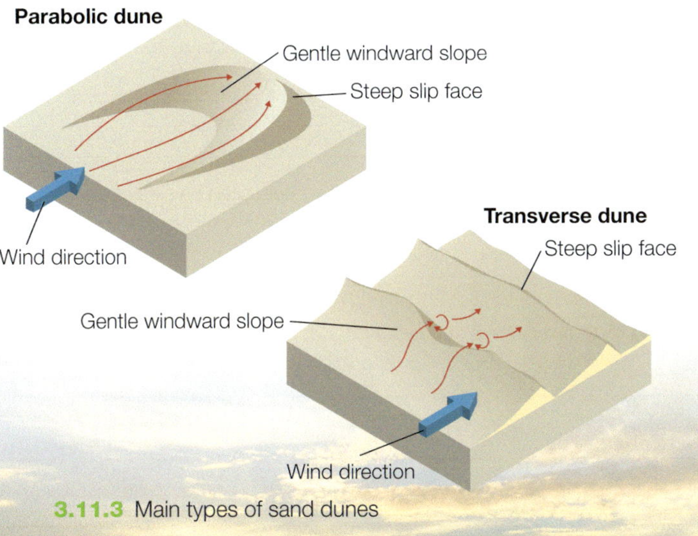

3.11.3 Main types of sand dunes

Geoactivities 3.11

Knowledge and understanding

1. Why do desert landforms vary?
2. Explain why water plays an important role in desert landforms.
3. In which two ways does wind create desert landforms?

Inquiry and skills

4. Refer to 3.11.1.
   a. Explain how a plateau is eroded into a mesa.
   b. Distinguish between a mesa and a butte.
   c. What are depositional features, such as alluvial fans and playa lakes?
   d. Describe why it is dangerous to be caught in a sand storm in the desert or to camp in a wadi if precipitation is anticipated in the area.
5. Refer to 3.11.2 and list the different types of deserts in Australia.
6. Refer to 3.11.3. Explain how winds move sand dunes across the landscape forming a variety of shapes.
7. Using the internet, describe what happened to form the channelled Scablands of the Colombia Basin in North America. Explain how this example shows the power of water erosion to sculpt landforms in short periods of time.

Butte (smaller)

Mesa

3.11.4 Mesas and buttes in Monument Valley, Colorado Plateau, USA

ISBN 978 1 4586 6277 4

3.12 Glacial valley processes

Glaciers are dynamic, large, moving bodies of compacted ice that exist for long periods of time. Glaciation is the process whereby glaciers create distinctive landscapes and landforms through erosion as they move slowly downhill carving valleys, under gravity and their own weight. When their journey is completed they deposit the transported material, leaving landforms such as terminal moraine and drumlins.

Rivers of ice

Valley glaciers exist in high mountains on most continents except Australia. They form in a cirque or armchair-shaped feature where snow collects. When the glacial ice fills the cirque, it begins to move. As glaciers progress down the valley, they scour the valley sides and transport sediment and rocks called moraine.

Moraine takes many forms:

- *lateral moraine*—side of the glacier
- *medial moraine*—where two glaciers merge
- *ground moraine*—bottom of the glacier
- *terminal moraine*—end or mouth of the glacier.

Valley glaciers transform narrow V-shaped river valleys into broad glacial U-shaped valleys, and produce spectacular hanging waterfalls.

Recently, climate change has produced rapid melting of over 90% of valley glaciers, such as the Aletsch Glacier in Switzerland, which is the largest valley glacier in the Alps. According to the United Nations, if all Earth's valley glaciers melt, sea levels would rise 15–37 cm.

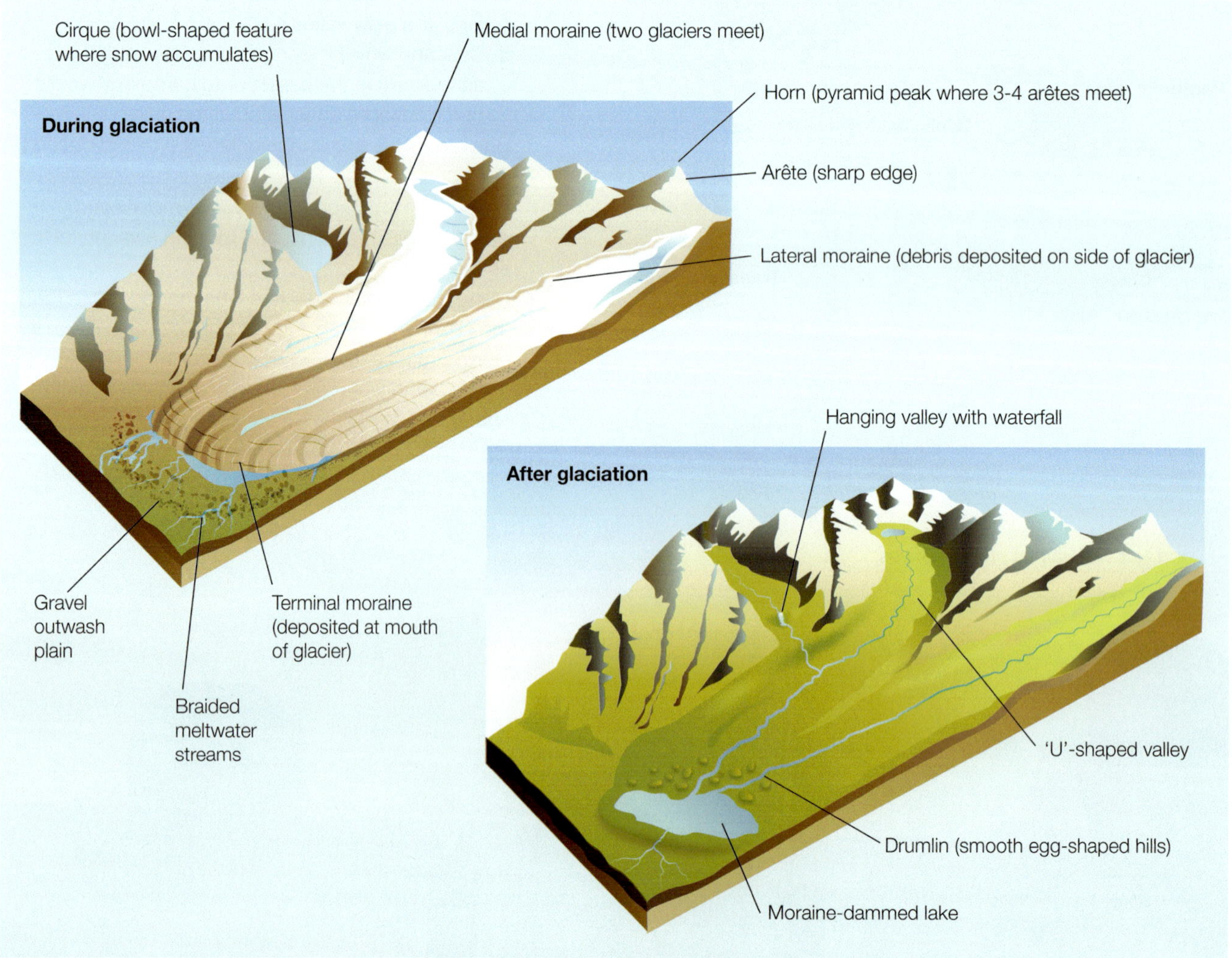

3.12.1 Landforms before (left) and after (right) glaciation in a mountain landscape. If there are multiple cirques on the top of a mountain, they create pointed pyramidal peaks or horns

ISBN 978 1 4586 6277 4

lateral moraine

medial moraine

3.12.2 Aerial view of Barnard Glacier, Alaska, USA, showing lateral and medial moraine

Geo**info**

- Glaciers cover roughly 10% of Earth's land surface.
- Glaciers cover 510 000–540 000 km^2 and have a volume of 50 000–130 000 km^3.

3.12.3 This fjord at Geiraner, Norway, is a flooded former glacial valley. Fjords are located on the western coasts of glaciated lands, such as Norway and New Zealand

Geo**activities 3.12**

Knowledge and understanding

1 Distinguish between a glacier and glaciation.
2 Why do many landscapes today have relict glacial landforms?

Inquiry and skills

3 Refer to 3.12.1.
   a Describe what happens during and after glaciation.
   b Explain the origins of the various types of moraine.
4 Refer to 3.12.2. Explain how lateral moraine develops into medial moraine.
5 Refer to 3.12.3.
   a Draw and label a sketch of the fjord (refer to 1.14).
   b Explain how fjords form.
6 Using Google Images, copy three images of valley glaciers from different parts of the world and label their landform features.
7 Refer to valley glacier photographs on the BBC website (see Geolinks). Describe the changes to valley glaciers shown in the photos.
8 Using ICT, research the Third Pole. Describe its location and the forecast problems if glaciers melt. Reflect on your learning and suggest actions to reduce glacial melt.

ISBN 978 1 4586 6277 4

3.13 Skill: glacial topographic map

The Southern Alps in New Zealand's South Island contain examples of glacial valley landscapes. The Tasman Glacier is New Zealand's largest and longest glacier, measuring about 28 km in length and 600 m thick. It flows south from Mt Tasman (3497 masl) and Aoraki/Mount Cook (3724 masl) and ends at Lake Tasman, which is referred to as a terminal lake. Along its downward journey it wears away the valley sides and carries eroded material, called moraine.

The Tasman Glacier provides evidence of climate change, as it has been retreating 180 m a year since the 1990s. As it retreated it left Lake Tasman. In the 1970s the lake consisted of a few small ponds but today it has grown to 7km in length and is 2km wide.

Scientists predict that New Zealand's glaciers will shrink and the Tasman Glacier will eventually disappear if Earth continues to become warmer. Glacial melting is likely to reduce tourism in Aoraki/Mount Cook National Park and tourism to nearby towns such as Queenstown.

The following map extract shows the Tasman River flowing in the glacial valley in Aoraki Cook alpine area—a World Heritage site.

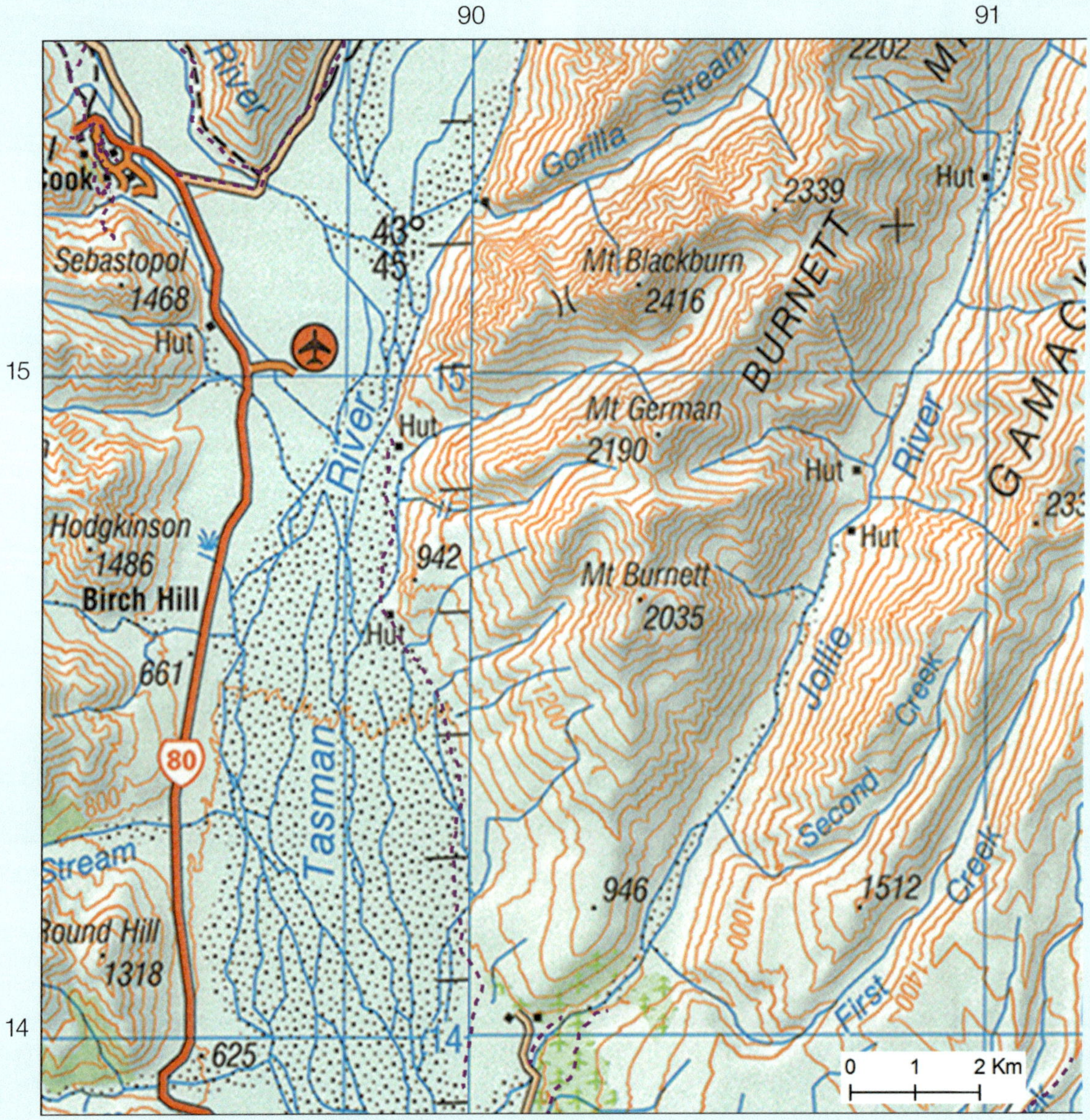

3.13.1 Topographic map of the Tasman Glacier

ISBN 978 1 4586 6277 4

Geoinfo

The Tasman Glacier will eventually disappear if current melting rates continue.

3.13.2 Upper Tasman Glacier and Malte Brun Range, Aoraki/Mt Cook alpine area, New Zealand

3.13.3 Queenstown is located on Lake Wakatipu, which was formed by glacial processes

3.13.4 Milford Sound—a fiord produced by glaciers

Geoactivities 3.13

Knowledge and understanding

1 What is the glacial response to climate change in New Zealand's South Island?

2 New Zealand's frozen assets are slowly melting away. What does this mean?

Inquiry and skills

3 Refer to 3.13.1
   a The contour interval is 100 m. Before you start, in groups write in the heights of the other contours. Remember numbers get larger going up a hill and smaller going down a hill to a river. This will make the cross-section easier to draw.
   b What is the scale of the map?
   c What is the height of Mt Burnett?
   d What is the distance from spot height 1318 masl to 946 masl?
   e Name one hill and one river.
   f The Tasman Glacier has carved a wide valley in which the Tasman River now runs. The small black dots are moraine deposited by the glacier. Why do you think the glacier dropped the moraine?
   g List two types of transport located in AR 8915 (top left).
   h Draw a cross-section from Hodgkinson 1486 m to Mt German 2190 m.
   i The walking track is represented by purple dots. Imagine you are walking along the track. What would you see?

4 In groups, investigate the impact of climate change on the Aoraki/Mt Cook alpine area using satellite imagery, photographs, diagrams and data. Present the information as a media report for television.

5 Milford Sound is a fiord located in New Zealand's South Island. Using the internet, download a topographic map of Milford Sound (see Geolink). In groups discuss how Milford Sound was formed and why it is a popular tourist destination.

ISBN 978 1 4586 6277 4

3.14 Ice sheets: processes and landscapes

Unlike valley glaciers, ice sheets (also called ice caps) are masses of land ice covering more than 50 000 km^2. They are not confined to mountains, but extend over vast areas of land, often burying mountains, lakes and other landforms. Ice sheets flow slowly downhill (like glaciers, due to gravity and their own weight), drastically changing the landscape and producing distinctive landforms.

Ice ages

During the last ice age, polar ice sheets expanded to cover a third of Earth's land surface. When they retreated they left distinctive glacial landforms, such as eskers and kettle lakes in Europe and Asia. These are referred to as **relict landforms**.

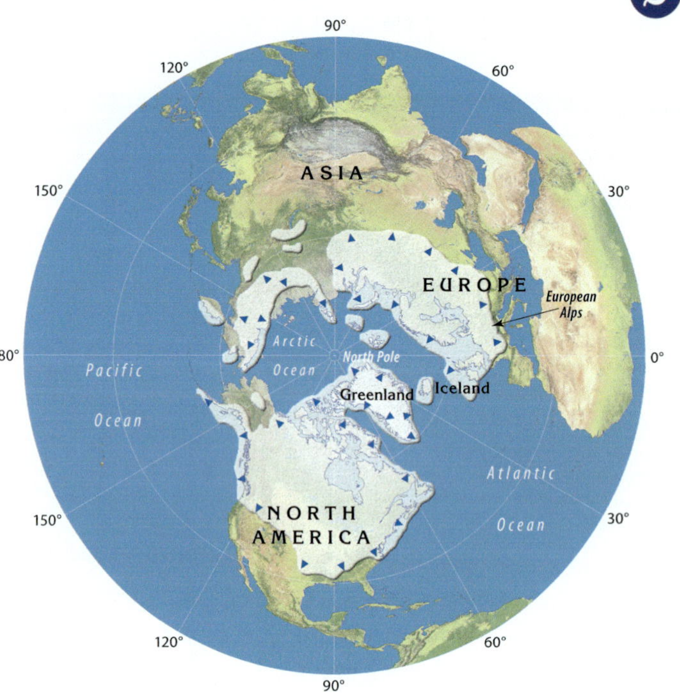

3.14.2 Spatial distribution of the northern hemisphere affected by continental glaciation during the last ice age

Before

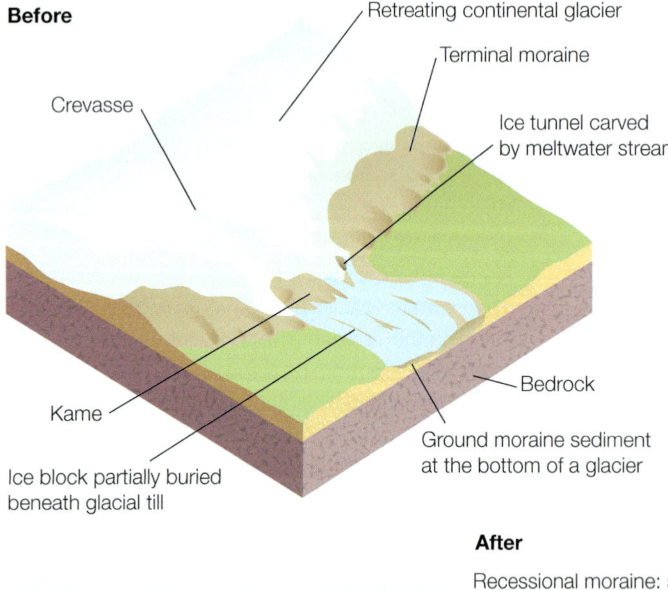

Polygons and kettle lakes, Arctic National Wildlife Refuge, Alaska, USA

After

Recessional moraine: sediment deposited as glacier retreats

Esker (long ridge of sediment)

Terminal moraine

Ground moraine (ground underneath eroded by glacier)

Outwash plain

Bedrock

Kame

Kettle lakes

Eskers in Alaska, USA

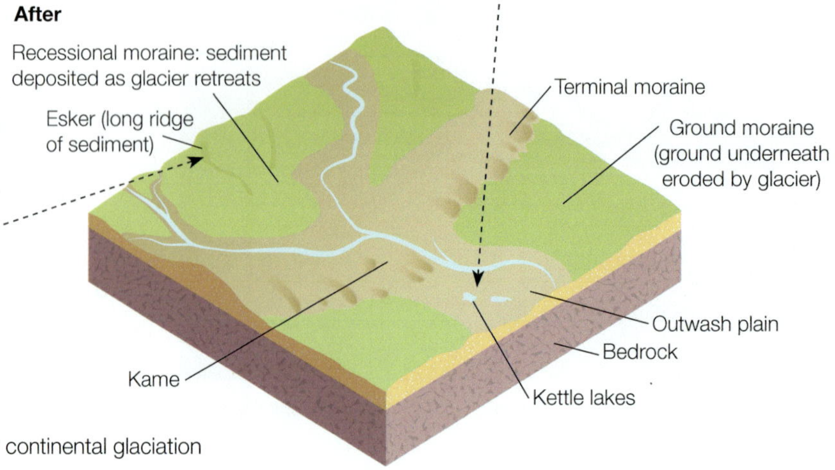

3.14.1 Features of a landscape before and after continental glaciation

ISBN 978 1 4586 6277 4

At the peak of the last ice age around 20 000 years ago, sea levels dropped 120 m below their current level, because glaciers stored vast amounts of water on land. As ice sheets retreated:

- *sea levels* rose, creating rias (drowned V-shaped river valleys, such as Sydney Harbour) and fjords (drowned U-shaped glacial valleys found in Norway and New Zealand)
- *land* slowly moved upwards, in a process called isostasy. This is still happening in Scandinavia today
- *melted ice* created huge quantities of water in rivers. This caused moraine dams to fail, resulting in floods. For example, glacial Lake Missoula, in northern USA, released about 3.74 million times the volume of water in Sydney Harbour in major flood episodes. This occurred about 40 times from 10 000–100 000 years ago. This created the famous Scablands landscape, which has an area of around 5200 km^2.

Icebergs: climate change

An iceberg is a large piece of ice that has broken off from a glacier or ice sheet. They range from 1 masl (metres above sea level) to 168 m tall. The largest is in the North Atlantic, and is around the height of a 55-storey building. There are concerns about global warming causing the break-up of icebergs. For example, the 11 000 km^2 Iceberg B15 broke off from the Ross Ice Shelf in 2000.

Geo**info**

About 15 000 years ago, Lake Bonneville in Utah, USA, broke through a moraine dam and released around 4200 km^3 of water (almost 7.5 million times the volume of Sydney Harbour) in one monstrous flood.

Geo**activities 3.14**

Knowledge and understanding

1 What is the difference between valley glaciers and ice sheets?

2 Explain why ice sheets have left a legacy on many landscapes today.

3 Explain how post-ice age catastrophic flood episodes have affected Earth's landscapes.

Inquiry and skills

4 Refer to 3.14.1.
   a Describe the main landforms created by ice sheets.
   b List two landforms after glaciation.

5 Refer to 3.14.3. Describe the changes to North America after the ice sheets retreated. Why could these floods be referred to as hazards?

6 Using a variety of sources, research in groups relict glacial landforms in Australia, such as the New England area (NSW), Inman Valley (SA) or Bacchus Marsh (Vic.). Locate five landforms on a map. Explain how they were formed and how they are protected. Present your findings using web 2.0 tools (see Geolinks).

Chapter 3

3.14.3 Lake Missoula and the Scablands 18 000 years ago

ISBN 978 1 4586 6277 4

3.15 Ice sheet landscapes of Greenland and Antarctica

The twin ice caps of Greenland and Antarctica, located at the North Pole and South Pole respectively, contain 99% of Earth's fresh water. As the world's largest ice sheets, they exert a powerful influence on climate. However, their existence is threatened by global warming.

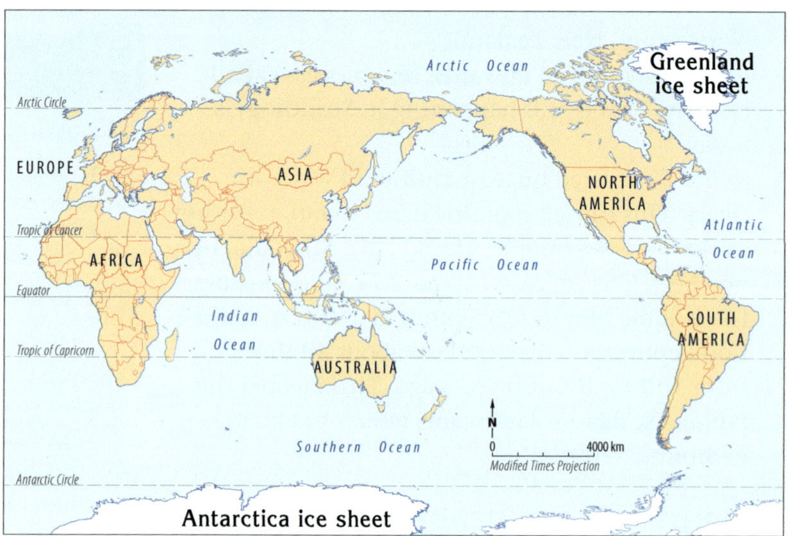

3.15.2 Location of Greenland and Antarctica ice sheets

World's last great ice sheets

Near the coasts of Greenland and Antarctica, ice sheets sometimes squeeze through narrow gaps in mountains acting as relatively fast-moving glaciers. In Antarctica, they flow out as ice shelves and ice tongues that project over the sea. Icebergs eventually break off the edges in the ablation zone.

The Greenland ice sheet contains around 12% of the world's ice and the Antarctic ice sheet contains approximately 87%. The ice sheets on the island of Greenland and the continent of Antarctica are the last remnants of the last ice age. The two ice sheets have different characteristics, relating to the depth of underlying bedrock and the thickness of the ice.

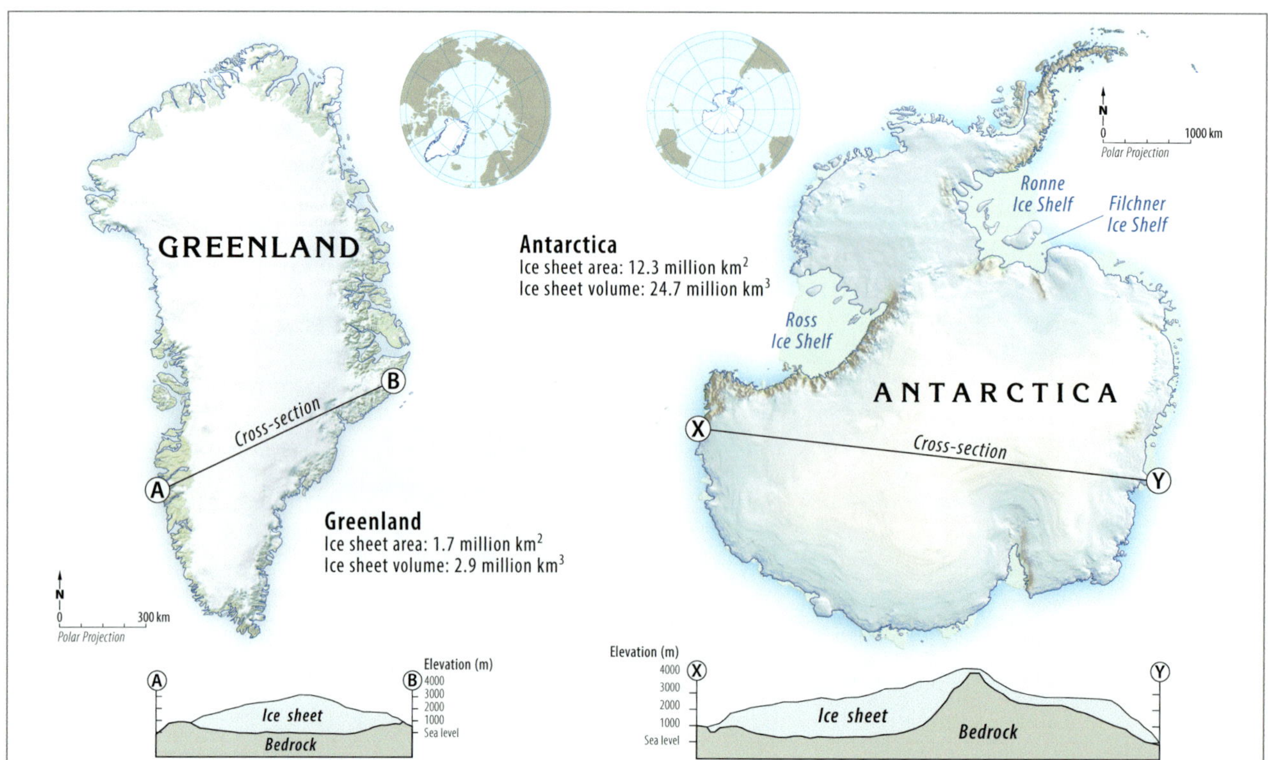

3.15.1 Ice sheets in Greenland and Antarctica

ISBN 978 1 4586 6277 4

Global warming's threat to ice sheets

The impact of global warming on Greenland and Antarctica has raised fears that the melting ice sheets will trigger sea level increases that will inundate low-lying countries. According to the United Nations Environment Programme, if the Greenland and Antarctic ice sheets melt, sea levels would rise 7.3 m and 56.6 m, respectively. The mass of the Greenland ice sheet is declining rapidly as glacial melting increases near the coast. In Antarctica, the Antarctic Peninsula has warmed about 2.5 °C since 1950. A large part of the West Antarctic ice sheet is losing mass, but the situation in East Antarctica is relatively stable.

Greenland
Melting on the lower parts of the surface, icebergs calve off from ice sheet edges into ice fjords and the sea.

Antarctica
Ice shelves with subglacial melting. Icebergs calve off from ice shelves.

Snow accumulation
Equilibrium line
Ablation (removal of ice)
Iceberg calving
Ice flow
Ice sheet
Ice flow
Ice shelf
Iceberg calving
Ocean
Sub-glacial lake
Ocean
Bedrock
Sub-glacial melting
Terminal moraine

3.15.3 Comparing ice sheets on Greenland and Antarctica

Geo**info**

Antarctica is the world's largest glacial landscape. The Antarctic ice sheet feeds thousands of glaciers, including the world's longest—Lambert Glacier.

3.15.4 Satellite image of Jakobshavn glacier and ice fjord, Greenland, showing the calving ice front from 1850–2014

Geo**activities 3.15**

Knowledge and understanding

1. Describe the significance of Antarctic and Greenland landforms and landscapes.
2. What are ice sheets? How are they interconnected to ice shelves and ice tongues?
3. Explain how global warming is threatening the Greenland and Antarctic ice sheets.
4. What impacts will global warming have on sea levels and low-lying countries?

Inquiry and skills

5. Refer to 3.15.1 and 3.15.3. Describe the differences between the Greenland and Antarctic ice sheets in a two-column table.
6. Research the cryosphere (ice sphere). Prepare a PowerPoint presentation on the glacial landscapes of the cryosphere and the anticipated impacts of global warming on them.
7. Using ICT, research and explain how Australia was part of Gondwana and subjected to valley and ice sheet glaciation. Present your findings as a photo story.

ISBN 978 1 4586 6277 4

3.16 Aesthetic, cultural and spiritual values

Landscapes and landforms are valued by different people around the world. In general, geographers divide the values into four categories:

- aesthetic
- cultural
- spiritual
- economic.

Some landscapes, such as national parks and coral islands, have economic, environmental, recreational and scientific values. Other landscapes are a source of conflict, such as the economic value of land for mining, versus the spiritual value of the same land to Aboriginal and Torres Strait Islander Peoples.

Aesthetic values

The aesthetic value, or attractiveness, of landscapes and landforms is based on the relationship between the landscape and the human observer. Different landscapes have the capacity to elicit pleasure (positive value) or displeasure (negative value). Unfortunately, many places of exceptional natural beauty, such as the Great Barrier Reef, are threatened by cyclones and coastal development.

3.16.2 With about 40 000 hexagonal basaltic prisms, the World Heritage listed Giant's Causeway, Antrim, Northern Ireland, has mythological value to the Celtic people. The prisms are excellent examples of volcanic landscapes

3.16.1 Different values of the Lizard Island group and Blue Lagoon of the Great Barrier Reef, which are within a World Heritage Area and the Great Barrier Reef Coast Marine Park in Queensland

ISBN 978 1 4586 6277 4

Cultural values

Cultural values are linked to landscapes and landforms through creative means such as stories, art, films and poetry. Indigenous Australians express the importance of land through Dreamtime stories, dance and rock carvings. In the 1880s, painters captured the colours and mood of the Australian 'bush' landscape. Arthur Boyd and Sidney Nolan are well known for their landscape paintings.

3.16.3 Mt Fuji and Lake Kawaguchi have great cultural significance to the Japanese. Mt Fuji is a World Heritage site and one of Japan's three holy mountains

3.16.4 Aerial view of the entrance to a cenote—one of more than 2200 mapped underwater caves in the Yucatan Peninsula, Mexico. It is a popular cave diving site

Spiritual values

Most cultures assign spiritual values to different landscapes and their landforms. Aboriginal and Torres Strait Islander Peoples hold spiritual values to landscapes, as they believe ancestral beings created the landform features in the Dreaming. The Jewish and Christian faiths hold Mt Sinai in Egypt to be sacred, because they believe this is the site where God gave Moses the Ten Commandments. The Mayans of the Yucatan Peninsula in Mexico believe that cenotes (vertical sinkholes filled with fresh water) are spiritually significant, as they are the entrances to the underworld where the gods and ancestors can be contacted.

Geo**info**

The Ganges River in India is revered by millions of Hindus as a symbol of spiritual purity.

Geo**activities 3.16**

Knowledge and understanding

1. For geographers, what are the main types of values?
2. What are aesthetic values and how do they vary?
3. What are cultural values and how do they differ from other values?
4. Discuss the spiritual values of landscapes and landforms held by a variety of communities.

Inquiry and skills

5. Imagine you visit a spiritually or aesthetically significant site. Write a 200-word report for a newspaper on your experience.
6. Refer to the internet and answer the following questions: Where is Lizard Island located? What are its significant landforms and landscapes? Why is Lizard Island valued? List the guidelines for caring for the national park on Lizard Island.

ISBN 978 1 4586 6277 4

3.17 The Dreaming: spiritual and cultural values

Aboriginal and Torres Strait Islander Peoples tell of a Dreaming, or Dreamtime, when all things began. This oral history is passed down by word of mouth as Dreamtime stories or songlines. They inform of journeys across Earth by ancestral spirits who created humans, plants and animals. On these journeys, the spirits left behind landforms, such as Uluru and the Devils Marbles.

Landforms became interconnected as the routes taken by creators in their Dreamtime journeys linked their sacred sites. Some Dreaming tracks ran for thousands of kilometres through desert, rainforest and coastal landscapes.

The Dreaming and interconnection

The Dreaming describes the period when spirits emerged from the ground, sky and sea to create landforms and living things over a featureless Earth. Paul Morin's Aboriginal Dreamtime story stated that:

> The Australian Aborigines believe that long ago the Earth was soft and had no form. The features of the landscape were created as the result of the heroic acts of ancestral spirits, who often assumed the form of animals. The origins of land shapes—mountains, deserts and water holes—echo these events, which the Aborigines refer to as Dreamtime. For at least fifty thousand years, the Aborigines have maintained the traditions of Dreamtime through stories, music, dance, art, and ceremony.

3.17.1 Wandjina figures are a feature of rock art sites scattered throughout the western Kimberley

ISBN 978 1 4586 6277 4

Dreaming is preserved in Aboriginal paintings, which resemble maps depicting different landscapes and events from an aerial perspective.

Multilayered meanings

Aboriginal and Torres Strait Islander Peoples hold multilayered meanings to different landscapes and landforms, including:

- *physical meanings* e.g. food and shelter
- *cultural* e.g. ceremonies
- *spiritual* e.g. burial sites

For example, a rock shelter may contain bone fragments, stone artefacts, rock art, and remains of meals and ceremonial activities.

Landforms such as caves were popular sites for Aboriginal Peoples in NSW:

- *Wirradjuri* and *Gundungurra* groups frequented Abercrombie Caves, which were a source of shelter, food and water.
- *Njunawal* people in Bungonia National Park left remains in caves that indicate the making of tools, and bone deposits that signify a burial site.
- *Bidewal* and *Ngunnawal* groups in the Snowy Mountains used caves for shelter and burial places.

Geo**info**

Karlu Karlu is the Aboriginal term for the Devils Marbles. They are believed to be the eggs that the Rainbow Serpent laid during the Dreaming.

3.17.3 Devils Marbles (Karlu Karlu), Northern Territory

Some landscapes that have significant meanings to Aboriginal and Torres Strait Islander Peoples have been declared World Heritage sites; for example, the Willandra Lakes Region, the Tasmanian Wilderness and Fraser Island.

Willandra Lakes Region (pictured)—arid lake landscape
Contains ancient lakes formed over the last 2 million years. The region illustrates how Aboriginal communities adapted to local resources and changing natural environments. The remains of a 40 000-year-old female were found in the dunes of Lake Mungo (believed to be the oldest ritual cremation site in the world) and 18 000-year-old grindstones used to crush grass seeds for flour.

Tasmanian Wilderness—forest landscape
Contains 37 cave sites occupied between 30 000 and 11 500 years ago. Stone artefacts, quarries and rock shelters in the Tasmanian highlands indicate Aboriginal adaptation to a subalpine environment.

Purnululu National Park—sandstone landscape
Has Aboriginal cultural heritage spanning back 20 000 years. Today, many of these people retain traditional land management practices and use the landscape for harvesting wild food.

Uluru-Kata Tjuta National Park—desert landscape
UNESCO states that *Tjukurpa*, which is traditional religious philosophy, 'provides an interpretation of the present landscape, flora, fauna and natural phenomena in terms of journeys and activities of ancestral beings and consequently binds people socially, spiritually and historically to the land'.

Wet Tropics of Queensland—rainforest landscape
Aboriginal occupation between Cooktown and Cardwell dates back 40 000 years. The Barrineans or northern tribes are said to represent the first wave of the Aboriginal occupation in Australia.

Fraser Island—coastal landscape
Indicates remains of Badtjala and Kabi Kabi settlements, such as middens and canoes, dating back 1500–2000 years ago.

3.17.2 UNESCO World Heritage landscapes of Aboriginal significance

ISBN 978 1 4586 6277 4

Uluru—a sacred landform

Uluru is probably Australia's most recognisable landform. It is a 348 m high sandstone inselberg located 465 km south-west of Alice Springs. Uluru was originally part of a mountain range that has resisted 300–400 million years of erosion. Today, this majestic red rock sits isolated on a sandy plain.

Spiritual and cultural values

Many people regard Uluru as a national symbol of Australia. To the traditional Pitjantjatjara and Yankunytjatjara owners (known as Anangu), it is a sacred site and their home. In Anangu culture:

- *Tjukurpa* is the creation period when the world was made
- *Tjukuritja* are the features of the landscape that the ancestors created.

Three ancestors are considered important to the Uluru region:

- *Mala*—hare wallaby
- *Kuniya*—woma python
- *Liru*—poisonous snake.

There are many sacred sites around the base of Uluru where significant Tjukurpa events occurred. For example:

- *caves* represent the open mouths of dead warriors
- *water stains* represent the blood of warriors.

Traditional *Tjukurpa* stories, ceremonies and dances have been passed down for tens of thousands of years to keep Anangu culture alive. These can be traced for hundreds of kilometres, where the ancestors journeyed across the vast Australian landscape. The journeys were called *iwara* or songlines.

Management

Today, the Anangu believe that the landscape is inhabited by ancestral beings and their spirits. As a consequence, they understand their responsibility to look after their 'Country' and its sacred sites.

The Anangu believe that the path to climb Uluru is the traditional route taken by the Mala men during the *Tjukurpa*. Because of its spiritual significance, the Anangu prefer that visitors respect their laws and culture and not climb Uluru. Out of respect for the Anangu's culture, between 1990 and 2014 the number of climbers fell from 75% to 18%.

The climb also poses a risk to personal safety, as many people have died in falls. In 2015, a Taiwanese tourist was trapped in a 20-metre rock crevice until a helicopter winched him out.

Uluru–Kata Tjuta National Park was inscribed on the World Heritage list for its outstanding natural values in 1987, and for its cultural values in 1994. A major responsibility of having a World Heritage listing is to protect and sustainably manage the landscape. This involves combining traditional Anangu land management with modern scientific management methods, including the tradition of fire management, called *waru* (patch burning).

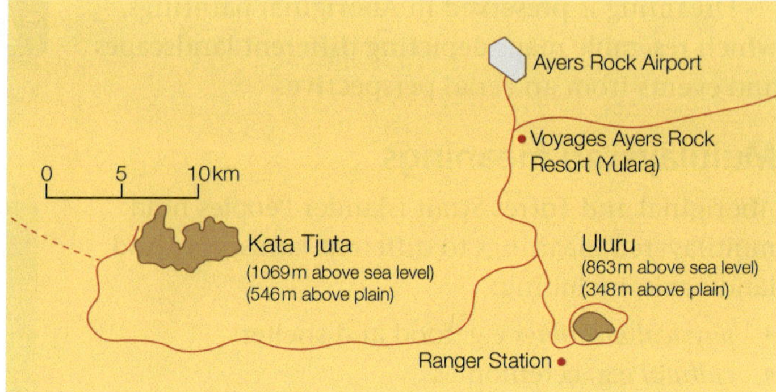

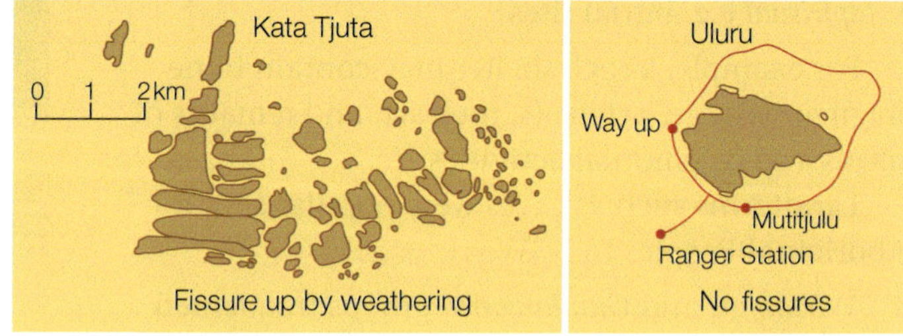

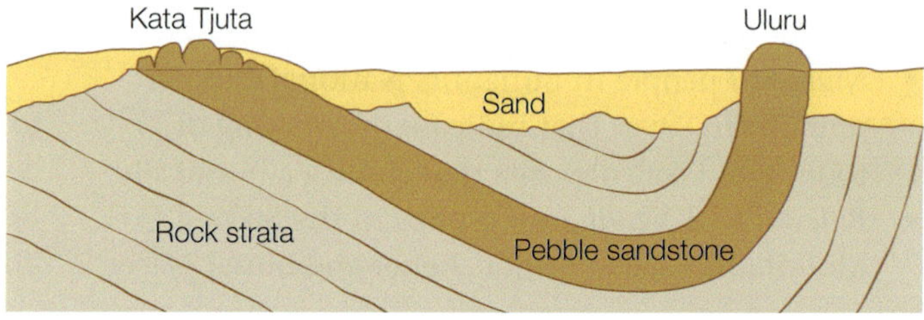

3.17.4 Location and formation of Kata Tjuta and Uluru

Geo**info**

Dreamtime stories reference climate change, extinct megafauna, sea level changes, volcanic eruptions and even glaciations.

ISBN 978 1 4586 6277 4

3.17.6 Cross-section of Uluru

3.17.5 Oblique aerial view of Uluru showing its distinctive shape

Geo**activities 3.17**

Knowledge and understanding

1 What is meant by the Dreamtime? Discuss the connections between the Dreamtime and the landscape.
2 Explain how landscapes and landforms possess multilayered meanings for Aboriginal and Torres Strait Islander Peoples.
3 Describe how traditional Aboriginal communities use natural resources, such as rocks and soil.
4 Describe why Uluru is sacred to the Anangu.
5 Explain the difference between *Tjukurpa*, *Tjukuritja* and *iwara*.
6 Describe the Anangu tradition of *waru*. What are two advantages of this practice?
7 Why don't the Anangu want visitors to climb Uluru?

Inquiry and skills

8 Refer to 3.17.1.
   a What is distinctive about these figures?
   b What are the common colours?
   c In groups, design a Wandjina tour and include: What would you see? What are the connections between culture and the land? Why do Mowanjum Peoples repaint the images each year?
9 Refer to 3.17.2. Explain why these World Heritage sites are of importance to Aboriginal Peoples.
10 Refer to 3.17.4.
   a What is the estimated height above sea level of Kata Tjuta and Uluru?
   b What do Kata Tjuta and Uluru have in common?
   c Explain how Kata Tjuta became separate landforms.
11 Refer to 3.17.6.
   a Draw a cross-section from A to B and from C to D.
   b Compare the different perspectives from the cross-sections.
   c Using Geocontext-Profiler on the internet draw a topographic profile of Uluru.
12 Investigate two Aboriginal Dreaming stories concerning the formation and meaning of landforms in your local area. Present your findings as a Prezi.
13 Aboriginal Peoples celebrate their culture at many events and festivals. In groups research three of the following festivals: Ord Valley Muster (WA), Laura Dance (Qld), the Dreaming Festival (Qld), Walking with the Spirits (NT), Yeperenye Dreaming (NT), Barunga Festival (NT), Yabun (NSW), Message Sticks (NSW), Spirit Festival (SA), Yalukit Willam Ngargee Festival (Vic.). Discuss how these festivals are linked to the Dreaming and the landscape using web 2.0 tools. Propose individual and collective actions to conserve the sites of the festivals.

ISBN 978 1 4586 6277 4

3.18 Landscapes and cultural identity

The importance of landscapes and landforms to Australia builds on the ancient traditions of Aboriginal and Torres Strait Islander Peoples. The Three Sisters in NSW's Blue Mountains is a World Heritage Site. According to Aboriginal Dreamtime stories, these rock formations represented three sisters who were turned to stone. People from the Gundungurra and Darug tribes consider the Three Sisters (which were once seven) to be sacred.

From another perspective, geologists state that over time the soft sandstone of the Blue Mountains was eroded by wind, rain and rivers. This caused the cliffs to slowly break up—one of the remaining formations is the Three Sisters.

Literature

Australian landscapes are often generalised as the 'bush' or 'outback'. Late nineteenth century poets and novelists, such as Henry Lawson and Banjo Paterson, wrote about the bush as a landscape that needed to be tamed by resourceful pioneers. Henry Lawson identified the struggle with the 'bush' as crucial in the development of a distinctive Australian identity. Traditional Australian values of treating people equally and being strong in the face of hardship are believed to have been influenced by the landscape.

Dorothea Mackellar's *My Country* describes her feelings towards Australia through landscape images.

I love a sunburnt country,
A land of sweeping plains,
Of ragged mountain ranges,
Of droughts and flooding rains.
I love her far horizons,
I love her jewel-sea,
Her beauty and her terror –
The wide brown land for me!

3.18.1 The Three Sisters in the Blue Mountains is an iconic Australian landscape—the blue colour in the distance comes from vaporised oil from eucalyptus leaves

3.18.2 How seven sisters became three—sandstone eroded by wind and water formed the seven sisters. The other four have been eroded and their remnants lie in the valley below

ISBN 978 1 4586 6277 4

3.18.3 Set on the banks of a billabong, Banjo Paterson's *Waltzing Matilda* is Australia's unofficial national anthem and the subject of this 1980 series of stamps

Music, art, dance and film

The different landscapes and their distinctive landforms have been central to the development of Australian culture. For example:

- *Music:* Bush ballads with references to landscapes were popular in the pioneering phase of Australia's history. They told stories of life droving cattle and sheep-shearing in outback landscapes. Since the mid-twentieth century, bush bands and country music festivals have popularised bush ballads. More modern music is also filled with references to Australian landscapes. 'It's raining on the Rock' by John Williamson is about a storm over Uluru–Kata Tjuta. 'My Island Home' by Christine Anu is about Elcho Island off Arnhem Land, but is a metaphor for Australia.
- *Art:* Russell Drysdale was famous for his series of outback paintings depicting the harsh relationship of its inhabitants with the landscape.
- *Dance:* The Bangarra Dance Theatre *Rites* production explores the natural forces that determine the ancient Australian landscape, capturing the spiritual essence of Earth, Wind, Fire and Water.
- *Film:* Director Baz Luhrmann depicts the harsh Australian landscape in *Australia*, while *Crocodile Dundee* features Paul Hogan in the hot, wet tropical landscape in northern Australia. *The Adventures of Priscilla, Queen of the Desert* follows a journey from Sydney to Alice Springs, across the Australian desert.

Geo**activities 3.18**

Knowledge and understanding

1 Outline how landforms and landscapes have been represented in Australian literature and art.

2 Give examples of references to landscapes in Australian music, film and dance.

Inquiry and skills

3 Refer to 3.18.1 and 3.18.2.
   - a How were the Three Sisters formed according to Aboriginal Dreamtime?
   - b How do geologists describe the formation of the Three Sisters?

4 Brainstorm more examples of Australian music that mention landscape. Present as a Wordle.

ISBN 978 1 4586 6277 4

3.19 Economic values

The economic value of landscapes and landforms is measured by how much money they generate and the number of people employed. Mining and tourism are lucrative Australian industries. In Australia, the land contains valuable minerals like coal, iron ore and uranium. Over the last 10 years the Australian mining industry invested $125 billion in the country. About 187 000 people are employed in the industry and 600 000 are employed in support industries.

Economic value of tourism

Tourism generates wealth in all countries, for example, in Tanzania where visitors come to go on safaris to view the annual migration of animals in the Serengeti National Park.

In 2014, international visitors spent $31.1 billion in Australia. Iconic landscapes (such as Kakadu National Park) and landforms (such as Uluru) not only possess cultural, spiritual and aesthetic values, but are of economic value to the tourism industry. More than 300 000 tourists visit Uluru-Kata Tjuta National Park each year, generating $400 million.

In 2015, the economic value of the Great Barrier Reef World Heritage Area was about $6 billion, and it generated 69 000 jobs.

Economic value of National Parks and World Heritage Sites

Landscapes and landforms considered to possess 'outstanding universal value'—naturally or culturally—have been classified as World Heritage Sites, to be protected and managed for future generations. World Heritage Sites such as Petra, in Jordan, and Fraser Island, in Australia, attract tourists and their valuable dollars, which help sustain the management of these sites.

The NSW National Parks and Wildlife Service (NPWS) manages 850 protected areas, which includes over 225 national parks. It focuses on environmental values (such as its Kosciusko alpine adventures) and cultural values (such as the conservation of Aboriginal sites). The organisation manages five of the six World Heritage Sites in NSW, including the Willandra Lakes Region—an ancient landscape formed by wind and water.

3.19.1 Economic value of tourism to Australia 2014. Photo shows Standley Chasm, near Alice Springs

ISBN 978 1 4586 6277 4

Mountains	Plateaus	Plains	Beaches/coasts
• generates hydro-electricity for industry • source of water for industry • tourist attraction – picturesque scenes, skiing, climbing • forests utilised for timber	• large grassy areas on plateaus used for grazing animals such as sheep and cattle. These animals produce meat, milk, wool and leather that can be sold	• rivers deposit fertile soil on the plain for agriculture • flat land easier to build cities, and construct roads, railway lines and runways to move goods and people	• beaches are popular tourist attractions, generating money from food, accommodation and transport • income from whale watching and hiring sailing boats

3.19.2 Economic value of major landforms. Photo shows tourists at the grassland plains of Serengeti National Park, Tanzania—a World Heritage Site. People come to see the 'big five animals': lion, elephant, buffalo, leopard and rhinoceros.

World Heritage Sites and the NPWS are of economic value, because tourists pay entrance fees and purchase souvenirs and food, as well as spending money on transport and accommodation. This stimulates employment in local businesses.

Australia's national parks not only possess aesthetic, cultural and spiritual values, but 25% of tourism expenditure is spent visiting Australia's national parks. This amounts to $16 billion a year.

3.19.3 Some of the landscapes and landforms that have been declared World Heritage Sites and attract tourists. Photo shows Petra, in Jordan

Geo**info**

- World Heritage sites in NSW employ 22 000 people.
- In 2014 there over 1.1 billion travelled abroad.

Geo**activities 3.19**

Knowledge and understanding

1 What is the economic value of mining to Australia?
2 Explain why national parks possess many values.

Inquiry and skills

3 Refer to 3.19.1.
   a Explain the economic value of tourism to Australia.
   b Investigate the current economic value of global tourism. Present research as a short report.
4 Refer to 3.19.2. Create a digital poster on Glogster showing the economic values attached to different landforms in Australia or an Asian country.
5 Refer to 3.19.3. In groups select one popular tourist site in Australia. Explain how this landscape or landform has different values.

ISBN 978 1 4586 6277 4

Geothink

Inquiry and fieldwork—value of forests

Before deforestation

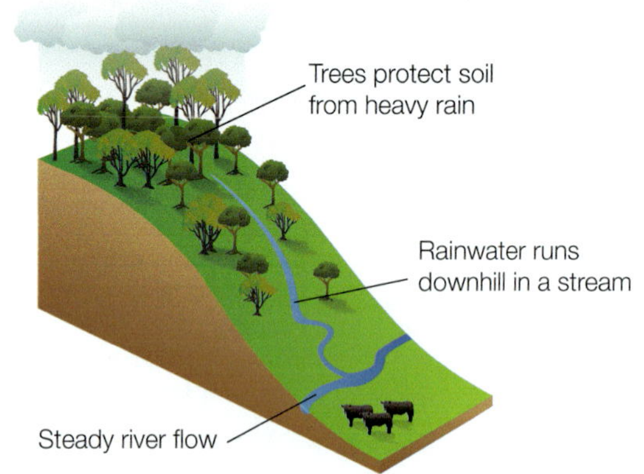

After deforestation

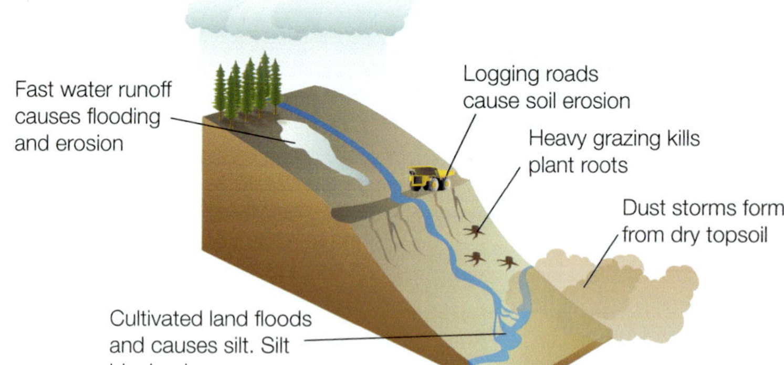

3.20.1 Before and after deforestation

Trees absorb carbon dioxide

Place for recreation and enjoying nature

Habitat for a wide range of species

Produce berries, mushrooms, wildlife, plants for food and medicines

Renewable source of timber and paper-based products

A sustainably managed forest contributes far more than just timber

3.20.2 Value of sustainably managed forests

1 Refer to 3.20.1.
   a Discuss the changes to landforms before and after deforestation.
   b Explain how these changes cause landscape degradation.

2 Refer to 3.20.2.
   a Explain what is meant by a sustainably managed forest.
   b Research where your used paper ends up once you have placed it in the garbage bin.
   c List five recycled paper items in your home or in the supermarket.

3 Refer to 3.20.3.
   a Explain how most of the biomass affected by logging is left to rot or burn or ends up as wood chips and processing waste in Tasmania.
   b Suggest strategies to reduce the waste.

4 *Geographical inquiry*: Refer to 3.20.4.
   a Answer the key inquiry questions.
   b Complete the inquiry process:
      i Develop a geographical question—such as why are forest landscapes disappearing?
      ii Plan a geographical inquiry—explain the cause and effect relationships of deforestation in developing or developed countries.
      iii Collect and evaluate information using a variety of sources.
      iv Communicate your research using PowerPoint.
      v Reflect on the investigationand suggest sustainable actions.

5 a Mind map your favourite memories and feelings about forests and present as a Wordle.
   b Explain why you think forests should be protected.

6 *Fieldwork*: Conduct actual or virtual fieldwork in a forest. Identify and label the main species. Explain how they adapted to the environment. Describe the threats from exotic species. Discuss how humans have changed the landscape and protection strategies to make the forest sustainable.

7 Research the Tasmanian Forests Agreement and the proclamation of the World Heritage Area extension in 2013. Include key questions such as What? Where? Why? Present findings using ICT.

ISBN 978 1 4586 6277 4

On average, every 100 tonnes of trees that are felled make three tonnes of sawn wood and wood-based panels. A further four tonnes out of the 100 becomes domestic paper products, and 23 tonnes are exported as woodchips. The remaining 70 tonnes is waste: 65 tonnes in every 100 is left on the forest floor as harvest residues (harvest slash) and five tonnes becomes wood processing waste that is burnt for energy or turned into mulch … With only 3% of the biomass affected by harvest events being converted into long-lived wood products, most of the remainder is waste and much of the carbon in that waste ends up in the atmosphere.

Source: www.crikey.com.au

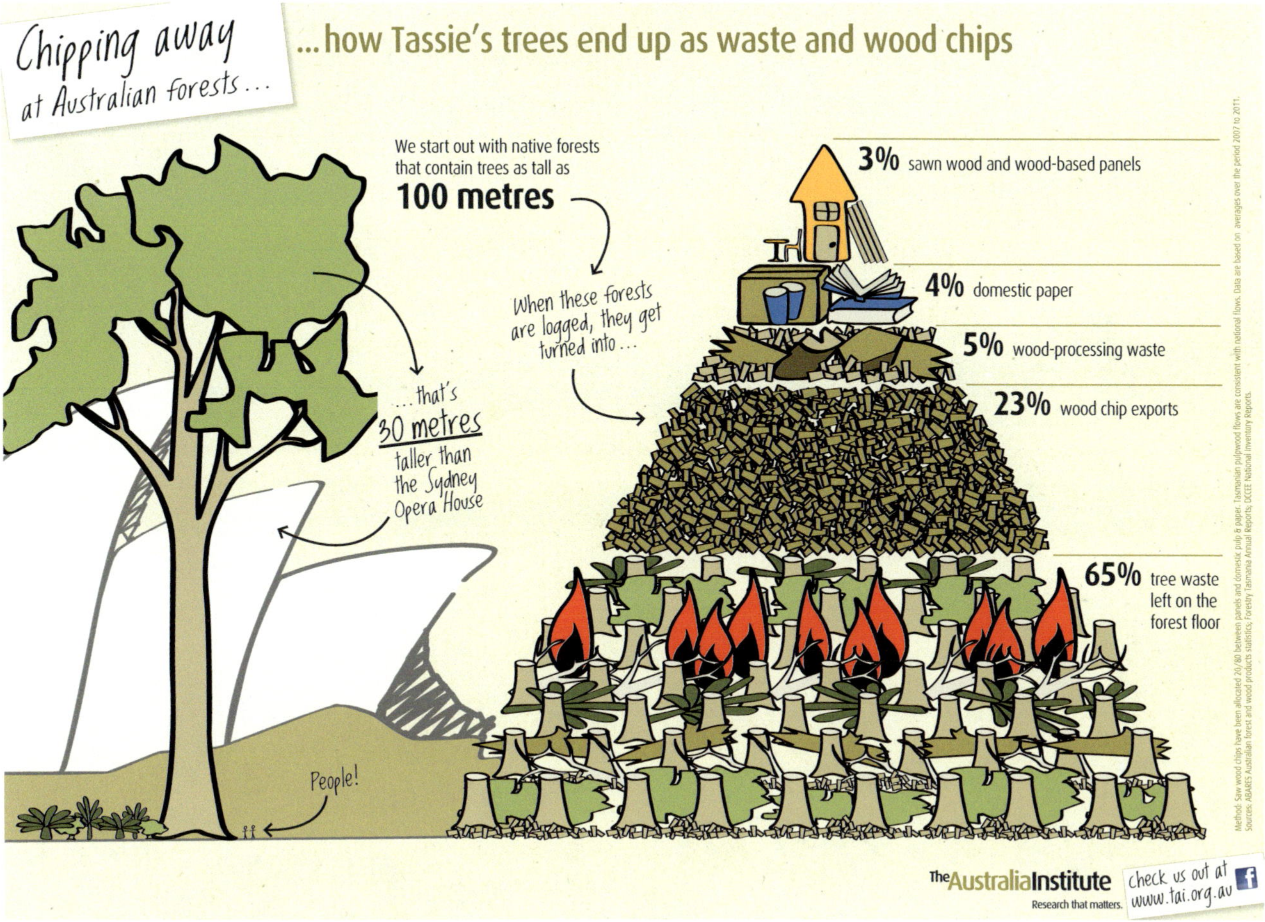

3.20.3 Waste in Tasmania's forests: economic versus environmental values

3.20.4 What did I learn?

ISBN 978 1 4586 6277 4

chapter 4 Changing landscapes: manage and protect

'The real voyage of discovery consists not in seeking new landscapes, but in having new eyes.'

Marcel Proust

Geovocab

acid rain: rainfall that has become acidic from atmospheric pollution; causes environmental harm to soils, forests and lakes

anthropogenic: caused by human activity

canal: artificial waterway constructed to allow the passage of boats or irrigation of crops

climate change: long-term change in weather patterns that range from decades to millions of years

cultural landscapes: natural landscapes changed by humans

dykes: earth walls that hold back water

estuarine: where the sea extends inland to meet a river mouth

gross domestic product (GDP): total value of goods and services produced in a country in a given year

heritage: our legacy from the past, what we live with today, and what we pass on to future generations (UNESCO definition)

Indigenous People's Knowledge (IPK): refers to the knowledge, innovations and practices of Indigenous communities

land reclamation: creating new land from the sea or rivers

landscape: comprises landforms, including the sea; living elements, such as vegetation and animals; and human elements, such as land use and buildings

landscape degradation: decline in natural features (e.g. mountains, lakes, the sea and vegetation) and human features (e.g. land use and buildings)

ISBN 978 1 4586 6277 4

Humans have changed sea into land to grow flowers in the Netherlands

mountaintop removal (MTR) mining: alteration and removal of a summit of a mountain or hill to obtain geological material, such as coal

polder: low-lying land reclaimed from a sea or river, protected by dykes

riparian: situated on the banks of a river

storm surge: rising sea as a result of a tsunami or cyclone

sustainable: a pattern of resource use that aims to meet both present and future needs; consists of environmental, social and economic factors

tundra: flat, treeless Arctic region in which the subsoil is permanently frozen

wetland: land consisting of marshes or swamps

All landscapes contain the imprint of human use and hold aesthetic, cultural and spiritual values for different people. Over time, rivers have been dammed to reduce flooding, mountain tops removed for minerals, and coastal areas enclosed within seawalls. Roads are dug through mountains, while forests are cleared to make way for farms and wetlands for settlements. Land is reclaimed from the sea for agriculture, cities and holiday resorts.

Human activities have resulted in landscape degradation—tourism and mountain biking has caused soil erosion, industry has contributed to acid rain, and mining has polluted water sources. As a result, these landscapes need to be protected for future generations to enjoy.

Aboriginal and Torres Strait Islander Peoples have managed their landscapes for over 40 000 years. These original Australians have a spiritual and cultural attachment to particular landforms, and have had a lasting impact on the landscape through the use of fire. Their knowledge has contributed to the sustainable management of landscapes and their distinctive landform features.

Think, puzzle, explore

- **Place** What are the effects of human changes on landscapes in different places?
- **Space** What is the spatial distribution of major dams?
- **Environment** How can Earth support an increasing global population without landscape degradation?
- **Interconnection** What are the interconnections between recreation and land degradation?
- **Sustainability** How can landscapes be managed sustainably?
- **Scale** How can humans protect landscapes from the local to the global scale?
- **Change** Why do humans change landscapes?

Geo**skills** in focus

- **Identifying** changes to landscapes through fieldwork, such as visiting areas degraded by recreation, mining, farming or urban activities
- **Collecting, recording** and **analysing** primary and secondary sources to investigate causes and effects of landscape degradation by humans, such as interviewing people
- **Evaluating** alternatives to improve landscape degradation
- **Communicating** information using web 2.0 tools such as infographics and Prezi, and incorporating maps, graphs, statistics and photographs
- **Reflecting** on the values of landscapes and ways to protect landscapes, from the local to the global scale

ISBN 978 1 4586 6277 4

4.1 Land degradation: causes, effects, management

All life on Earth depends on landscapes. Soils produce food, plants provide oxygen, and minerals and timber contributes to the production of goods. Over time, humans have changed the land. In some locations they have caused a decline in output that has affected the land's ability to provide resources, such as food. This is called land degradation, which affects 1.5 billion people globally.

Land degradation is mainly driven by a large and growing population, poor land-use practices, deforestation, contaminated soil and mining. Governments, farmers, city planners and individuals work to protect the land for future generations.

Tunisia: land degradation

Tunisia is located in North Africa. Land degradation affects 60% of Tunisia's total area. Approximately 50% of the degraded land is dry scrubland, 30% cropland and 12% forests. Changes to the country's landscape is due to the growing rural population, which has:

- *cleared forests* to make way for farms, and for use as wood for cooking and heating
- *overgrazed land,* caused by cattle eating too much grass, resulting in soil erosion
- *overcropped land*, a result of farmers continuously cultivating crops on one plot of land, causing a decline in soil fertility.

Over 8000 ha of arable soil (suitable for farming and that can be cultivated to grow crops) is lost every year, with desertification threatening agricultural land. Poor land management has led to salinisation, water and wind erosion, and loss of production from the land.

4.1.1 Global causes and effects of land degradation. Photo shows degraded landscape in Tunisia, North Africa

ISBN 978 1 4586 6277 4

Stabilising dunes in Tunisia. Terracing prevents further gully erosion and stores surface water for olive trees.

Goats climb an argan tree planted to reduce land degradation.

Draining inland Tunisian salt lakes shows small naturally coloured salt hills at Chott El Jerid.

Ksar consists of attached houses and a collective granary. It was a location in the film *Star Wars*, which has contributed to tourism. Money raised has been used to manage land degradation.

4.1.2 Managing land degradation in Tunisia

Managing changed landscapes

Anti-desertification initiatives involve planting wind breaks and stabilising dunes. The argan tree is planted to stabilise sand dunes, and is spaced in rows and intercropped with barley. However, the tree is threatened by the grazing of animals, intensive crop cultivation, and goats that stunt the trees by climbing up the branches to eat the leaves and fruit.

In the past, Tunisians overused water for irrigation in dry environments. As a consequence, little water entered inland lakes. Water was then subjected to high evaporation, leaving salt behind. These lakes are drained, and the natural and coloured salts sold to salt companies or tourists.

Remote sensing technology has been used to identify degraded areas in Tunisia, and organisations have implemented reafforestation programs, drip irrigation, and soil and water conservation to manage the landscape to manage the landscape.

Geo**info**

World Day to Combat Desertification is observed globally on 17 June each year.

Geo**activities 4.1**

Knowledge and understanding

1 What is meant by land degradation?
2 List the causes and effects of land degradation in Tunisia. Present as a two-column table.

Inquiry and skills

3 Refer to 4.1.1 and summarise land degradation in the format of a TV report.
4 Refer to 4.1.2 and discuss how Tunisians are managing land degradation.
5 Investigate how tourism to dry environments in Tunisia could increase land degradation. Suggest strategies to reduce the tourists' impacts.

ISBN 978 1 4586 6277 4

4.2 Humans change landscapes

All **landscapes** contain the imprint of human use. People live in landscapes, and can alter the landscape with little understanding of their impacts on the natural environment. Rivers are dammed to create reservoirs and wetlands drained for settlements. Forests are logged for paper and hillsides terraced to grow crops. Mountain tops are removed for minerals and coastal areas enclosed with seawalls to reduce the impacts of rising sea levels. Roads and railways are dug through rugged mountains, and monasteries are perched on steep cliffs. Land is dotted with quarries and oil rigs.

Additionally, many human activities have increased the rate at which natural processes shape landscapes, for example:

- *erosion*—from overcropping, overgrazing animals, logging forests for settlements, bush walking, mountain biking and driving off-road vehicles
- *air pollution*—from **acid rain** speeding up the weathering of Earth's surface, as seen on the Taj Mahal's marble facades.

Satellites monitor **landscape degradation**, for instance toxic mining waste in South African rivers, or irrigation leading to the decline of the Aral Sea area in Central Asia.

Anthropogenic landscapes

Humans constantly change natural landscapes. **Cultural landscapes** are the combined efforts of nature and humans, such as the polders in the Netherlands and the Three Gorges Dam in China.

Landscape changes can be permanent (e.g. mountain-top mining), semi-permanent (e.g. roads and power transmission lines) or temporary (e.g. commercial forests returned to previous natural vegetation). These changes are often a source of conflict, such as the sculpture of four presidents carved into the granite face of Mt Rushmore. The carvings are controversial among Native Americans because the USA seized the area from the Lakota tribe.

The impacts of human changes can be either positive, such as seawalls constructed to save a community from a dangerous tsunami, or negative, such as **anthropogenic** landscape degradation with a decline in wetlands and seagrasses.

4.2.1 Extracting oil changes landscapes in California, USA

4.2.2 Roads constructed through mountains alters landforms at St Gotthard Pass, Switzerland

4.2.3 Sculpturing landforms at Mt Rushmore, USA

ISBN 978 1 4586 6277 4

Changing the Tuscan landscape

Across centuries, people have left their imprint on the Tuscan landscape. Tuscany, in Italy, is famous for landscape paintings, fields covered with olive groves and vineyards, and medieval castles and fortresses perched on hill tops.

The hills were originally vegetated with chestnut, pine and cypress trees, but over time humans have changed the Tuscan landscape with farms and urban settlements. In order to protect the landscape, six Tuscan localities were designated as World Heritage sites and over 120 protected nature reserves were established. However, the natural environment is under stress from the 3.8 million inhabitants and growing tourism—the capital city of Tuscany, Florence, attracts 10 million tourists a year.

The wine maker, farmer, tourist, shop keeper, painter, town planner, urban dweller and conservationist each have different views on managing the Tuscan landscape. Of increasing concern is whether the resources in the area are used in a **sustainable** way.

Geoinfo

Many famous people used the Tuscan landscape as inspiration for their work (e.g. Leonardo de Vinci).

4.2.4 Cultural landscape of Tuscany

Geoactivities 4.2

Knowledge and understanding

1. What is a landscape?
2. How have humans changed the landscape over time?
3. List human behaviours that degrade landscapes.
4. *Landscapes are constantly changing.* What does this statement mean?

Inquiry and skills

5. Refer to 4.2.1 and 4.2.2.
   a. Why did humans change these landscapes?
   b. Do you think these human activities have degraded the landscapes? Provide reasons for your answer.
   c. Discuss how technology has changed landscapes.
6. Refer to 4.2.3.
   a. Why is Mt Rushmore a source of conflict?
   b. Explain whether you consider these statues landscape degradation.
7. Refer to 4.2.4.
   a. Where is Tuscany located?
   b. Describe how humans have changed the Tuscan landscape over time.
   c. Explain whether you consider these changes cause landscape degradation.
8. *All landscapes contain the imprint of human use.* Explain this statement in a short report.

ISBN 978 1 4586 6277 4

4.3 Past and future landscapes

Since the beginning of the Neolithic Revolution about eight millennia ago, humans have shaped and changed landscapes. These landscapes vary from Mongolian yurts to New York's high-rise apartments, and from small Indonesian rice farms to large Argentinean cattle ranches.

Landscapes have changed around the world to accommodate over 7 billion people. With over 50% of the global population living in urban areas, the world's wetlands, forests and farms have been converted into urban landscapes. Enormous landscape changes have occurred in China, with the urban population growing from 26% in 1990 to 53% in 2015. Changes continue as the country aims to create the largest megacity in the world containing 42 million people, and plans to build 20000 to 50000 new skyscrapers over the next 20 years.

Manila, the capital of the Philippines, changed from a fortified settlement of a few hundred people on the shores of the Pasig River in the 13th century to a population of 12 million today. As it is one of the most densely populated cities in the world, the city expanded and engulfed rural landscapes and coastal landscapes by clearing wetlands along waterfronts.

4.3.2 Location of the San Andreas Fault

San Francisco's changing landscape

San Francisco is located on the east coast of the USA. The area was once sand dunes and marshland, but after the 1849 gold rush it developed into a large city that accommodates 8 million people today.

In the 1840s, the San Francisco area was desolate with only stunted vegetation on sand dunes. By 1849 sand hills were levelled and tidal marshes filled. Thousands of tonnes of sand were dumped into the bay to create additional land. Located near the San Andreas Fault, there are concerns San Francisco could be affected by an earthquake and tsunami as well as the impacts of rising sea levels in the low-lying Bay Area.

4.3.1 San Francisco: sand dunes c. 1900 and today's urban landscape

ISBN 978 1 4586 6277 4

Blaenavon industrial landscape

The Industrial Revolution in 1700s saw the growth of manufacturing in the UK followed by the rest of the world. These industrial landscapes degraded the environment with air and water pollution. Today these old industrial sites have become areas of urban decay. Some have been rejuvenated with new functions, referred to as urban renewal, such as the Docklands in London, the Rocks Area in Sydney and Southbank in Melbourne.

Frequently original landscapes are forgotten, such as that of the Blaenavon Industrial Landscape World Heritage site in Wales. During the Industrial Revolution, Blaenavon was pitted with mines and quarries. Eventually, mining and associated industries declined, resulting in a degraded and desolate landscape. After some time, vegetation colonised mining wastes and wildlife returned. Concerns for the protection of this degraded landscape led to the Forgotten Landscapes Project, which aims to protect the area's industrial heritage and conserve its heather moorland.

4.3.3 Blaenavon's 19th century industrial and mining landscape

Geo**info**

3000 people died in the 1906 San Francisco earthquake (magnitude 7.8) and subsequent fires.

4.3.4 Map of Blaenavon

Geo**activities 4.3**

Knowledge and understanding

1. How have humans shaped landscapes?
2. How does urban growth change landscapes?
3. What is the cause of major landscape change in China?
4. Explain how the Industrial Revolution changed landscapes from the 1700s.
5. Describe the impacts of human-induced changes on future landscapes.

Inquiry and skills

6. Refer to 4.3.1. Explain the effects of urban growth on the landscape.
7. Refer to 4.3.2.
   a. Imagine San Francisco's landscape after an earthquake. Suggest strategies to reduce landscape degradation
   b. Describe the location of the San Andreas Fault.
8. Refer to 4.3.3, 4.3.4 and the text.
   a. Describe the changes to the Blaenavon landscape over time.
   b. Explain why the area was declared a World Heritage site.

Chapter 4

ISBN 978 1 4586 6277 4

4.4 Land reclamation creates islands and cities

For over 800 years, the Uros tribe of Lake Titicaca, Peru, weaved floating isles of totora reeds. Recently in South Korea, three floating islands were anchored in the Han River with concrete blocks. Should a flood eventuate, these islands rise and fall with the water levels.

4.4.1 Seoul Floating Island on the Han River

Land reclamation

Have you fantasised about living on your own island paradise? Today this can be a reality as engineers have created habitable islands by land reclamation.

Land reclamation is the process of creating new land from oceans, rivers or lakes, and as a result new landscapes have emerged. In the Netherlands 18% of the country is reclaimed from the North Sea and marshes, and 25% of Hong Kong is built on reclaimed land. Mumbai originally consisted of seven islands, which were united through land reclamation projects. Limited land availability for growing populations is expected to result in increased land reclamation around the world, causing degradation of wetlands, sea grasses and coral reefs.

Land reclamation has been extensive in the following places:

Dubai and Bahrain

In West Asia, artificial islands have been created, for instance, Dubai's Palm Islands and Bahrain's Amwaj Islands. The construction of the three Palm Islands adds 520 km of beaches to Dubai. The islands are composed of rock and sand dredged from the bottom of the Persian Gulf. Construction damaged the marine habitat by burying coral reefs, oyster beds and sea grass. As ocean currents have been disrupted, beaches are eroding.

Singapore

Located off the southern tip of the Malay Peninsula, Singapore consists of 63 islands. Limited land and a growing population led to the country reclaiming land from the sea. Since the 1960s, land reclamation has increased the original land area by 20%. Coastlines have been altered and mangroves reduced from 13% in the 1820s to 0.5% today. Singapore's population of over 5 million people is expected to reach 6.5–6.9 million by 2030. To accommodate the population, land reclamation is anticipated to increase by another 100 km^2 by 2030.

Hong Kong

Mountainous Hong Kong in China has a limited supply of useable land. The country commenced reclaiming land from the sea from 206 BC when beaches were turned into fields for salt production. Major land reclamation projects have been conducted since the mid-19th century.

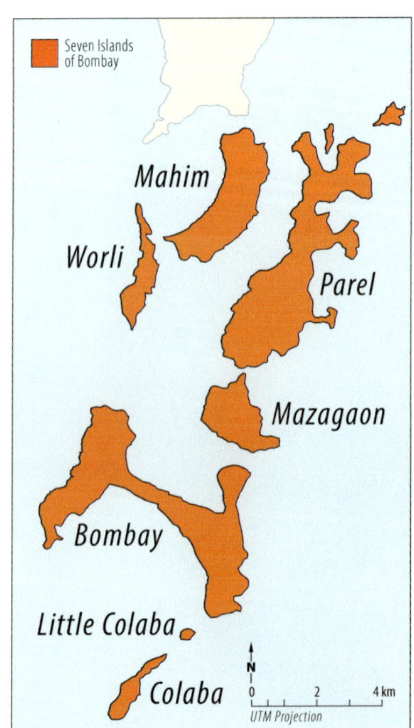

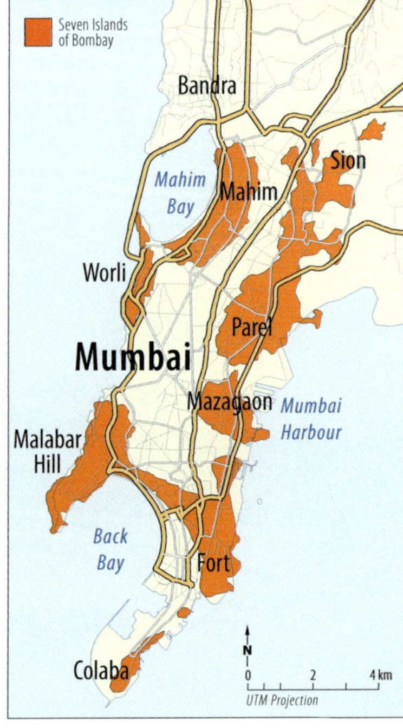

4.4.2 Seven Islands of Bombay merge over time

ISBN 978 1 4586 6277 4

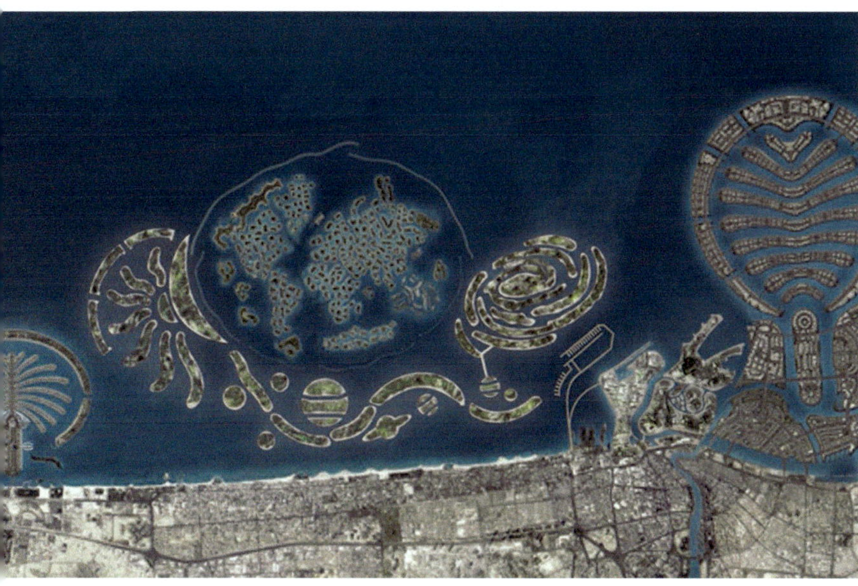

4.4.3 Satellite image and map showing Dubai's land reclamation from the sea

Today Hong Kong Disneyland Resort is located on reclaimed land, while Hong Kong International Airport involved levelling two isles and adding an additional 6000 km^2 via land reclamation. As the city expanded, new towns, (e.g. Sha Tin) were built on reclaimed land. Reclamation has impacted on seaweed beds, marine environments and the endangered Chinese white dolphin. In 1996 Hong Kong passed a law to safeguard the threatened Victoria Harbour against encroaching land development.

The Netherlands

The Dutch pushed back the water of the Zuiderzee by building **dykes** and creating **polders**. Once dykes were built, canals and pumps drained the land to keep it dry for agriculture and settlements. Further protective dykes were built, reclaiming IJsselmeer. The new land led to the creation of Flevoland. There are future threats from rising sea levels, as 27% of the Netherlands is below sea level.

Geo**activities 4.4**

Knowledge and understanding

1 Why do people change landforms? List the answers as a mind map.
2 What are floating isles and what are their advantages?
3 What is land reclamation?
4 List the advantages and disadvantages of land reclamation in two columns.
5 Describe how Singapore and Hong Kong changed landforms to accommodate their growing populations.

Inquiry and skills

6 Refer to 4.4.1. What are the advantages of South Korea's floating islands?
7 Refer to 4.4.2.
 a Describe how Mumbai connected islands into one landmass
 b What were the effects of the landform changes?
8 Refer to 4.4.3 and the text.
 a What are the advantages of using a map and a satellite image to identify areas where land has been reclaimed in Dubai?
 b Describe the impacts of artificial islands on the environment.
9 Refer to the chapter opening photo on pages 130–131 and secondary sources.
 a Explain how the Netherlands changed from being a landscape under the sea to one above the land.
 b Draw and label a line drawing of the Netherlands photo. Indicate changes by humans.
 c What are the advantages of windmills in the Netherlands? How do they differ from modern wind farms?
10 Polders also exist in countries other than the Netherlands. Select a polder in another country. Discuss why it was built and its effects on the landscape.
11 Research the artificial Zoran Island off Phuket in Thailand. Include answers to questions such as: Where will it be located? How will humans change the landscape? What will be the function of the new island? What are the environmental issues? Present as a short report.
12 Research Eden Island in the Seychelles or Kansai Airport island in Japan.
13 Design a collage of land reclamation of countries in the Asia region.

ISBN 978 1 4586 6277 4

4.5 Humans construct canals

Humans have changed landscapes by constructing features to reduce the impact of sea level rises (e.g. seawalls and dykes), decrease floods (e.g. dams and levees) and improve transport (e.g. canals). From one perspective, these human-built structures are aesthetically unattractive and cause landscape degradation, and from another perspective they are marvels of human engineering that have improved people's lives.

Rivers have been changed by dredging and damming their journey from their source in mountains to their mouth at oceans. Some meandering rivers have been straightened, resulting in degradation of **wetlands** and **riparian** vegetation.

4.5.1 Straightening rivers

Advantages
• Rivers are more suitable for navigation by larger vessels • Larger and deeper river channels reduces floods • Reduces natural erosion
Disadvantages
• Loss of wetlands causes a decline in habitat for wildlife and the loss of an important filter for fresh water • Straightening causes the rivers to flow more rapidly, which can increase soil erosion and flooding downstream from the channelised area • Causes decline in river fish populations

4.5.2 A canal in Amsterdam moves people and goods

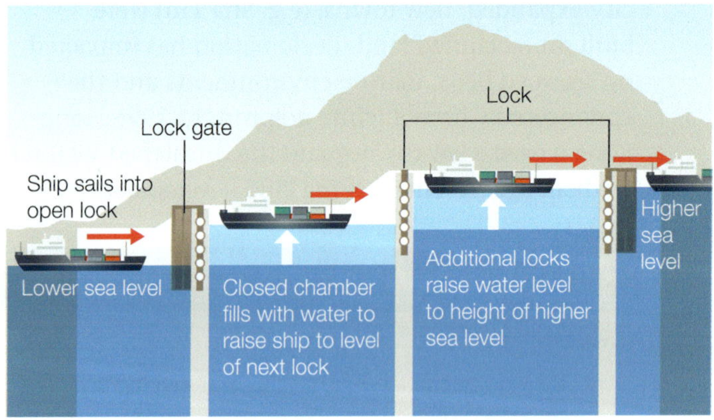

4.5.3 Operation of the Panama Canal. Ships move via locks, which increase or decrease their water level to raise or lower the ships

Canals

Canals are human-made channels for water that serve a variety of functions:

- connecting lakes (Erie Canal, USA), rivers (Lenin Volga–Don Shipping Canal, Russia) and oceans (Suez Canal in Egypt and Panama Canal)
- connecting city networks e.g. the canals in Venice.

Amsterdam has been called the 'Venice of the North' with over 100 km of canals and 1500 bridges. Bangkok is referred to as the 'Venice of the South', with canals, or klongs, transporting people and products. Today many canals in Bangkok have been drained or filled and replaced by roads.

The Suez Canal connects the Mediterranean Sea and the Red Sea, and all seaborne trade between Asia and Europe passes through the canal. Marine species located in the Mediterranean Sea are threatened by the movement of 300 species from the Red Sea. In Central America, the Panama Canal connects the Pacific and Atlantic oceans, enabling faster transport of goods and people. At present there is a proposal for the construction of Canal Istanbul in Turkey to link the Black Sea with the Sea of Marmara by 2023. However, Greenpeace Turkey warns that the canal will change the native habitat, excavations will cause erosion, and agriculture and water resources will be negatively affected.

i Geoinfo

- More than 200 million tonnes of cargo move through the Great Lakes Seaway System annually.
- The Grand Canal of China is the longest canal in the world.

ISBN 978 1 4586 6277 4

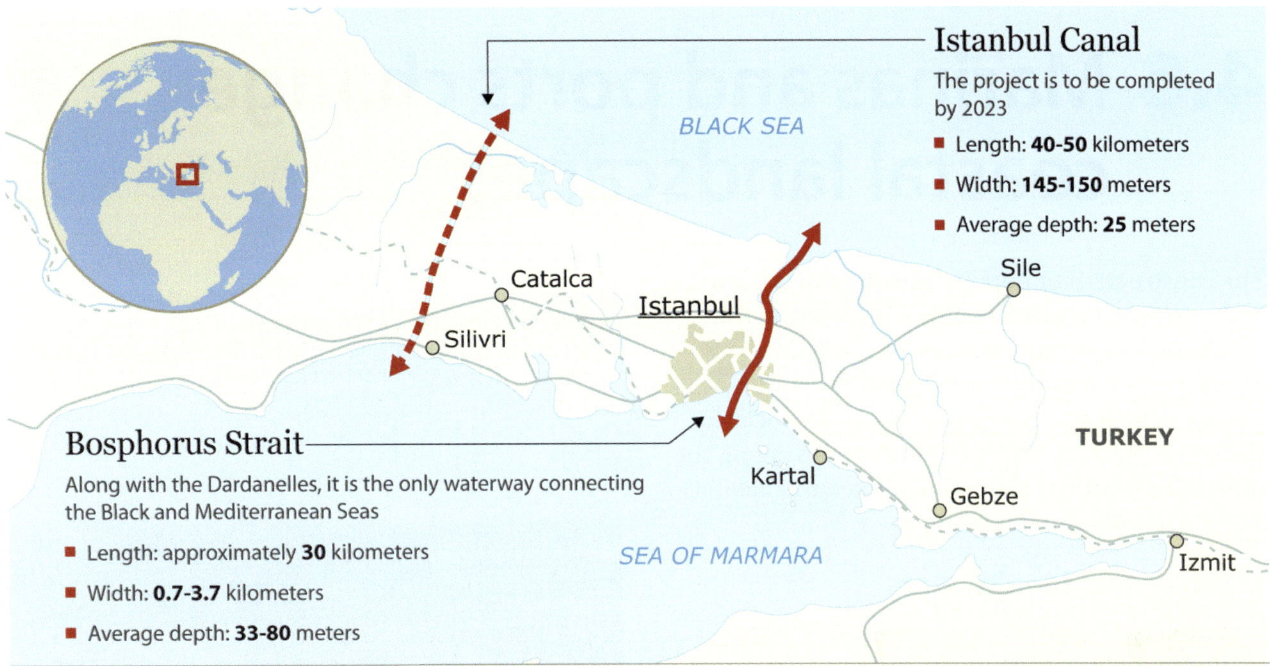

Congestion in the Bosphorus Strait

- **45 000** vessels pass through the strait each year
- **15 000** ferries cross the straight daily
- **1.5 mln** people transported by ferries each day
- **140 mln** tons of oil transported through the strait each year
- **4 mln** tons of liquefied gas transported through the strait each year

Source: RIA Novosti

4.5.4 Istanbul Canal project

Restoring rivers

The global trend to straighten rivers has recently changed its course. For example, in Europe some constructed canals have been restored to the natural meandering courses of the original rivers. In the USA a 'no wetlands loss policy' is popular, which means canals constructed in one place are to be offset by creating a new wetland in another place.

Geoactivities 4.5

Knowledge and understanding

1 List the features constructed by humans that change the flow of rivers and ocean currents.

2 Name two different types of human-made canals.

3 What is meant by a 'no wetlands loss policy'?

4 Give an example of a landscape restored to its natural state.

Inquiry and skills

5 Refer to 4.5.1. Explain your perspective on whether rivers should be straightened.

6 Refer to 4.5.2.
- a Discuss why Amsterdam is called the 'Venice of the North'.
- b Design a day excursion travelling on the canals of Amsterdam or Bangkok. Identify the different landscapes you would pass. Present as a Prezi.

7 Refer to 4.5.3.
- a Where is the Panama Canal located?
- b How does the canal work?
- c What are the advantages of the canal?
- d Investigate plans for future expansion of the canal.

8 Refer to 4.5.4.
- a Describe the location of the Istanbul Canal project.
- b Explain why it will be important for the flow of people and goods.
- c Discuss how the canal could degrade landscapes.

9 Using ICT, research the Suez Canal. Explain the reasons for the construction of the canal and its impacts on the global movement of people and goods.

10 Where is the St Lawrence Seaway located? What are the advantages of the canals for North Americans?

11 Plan a holiday on a barge in the UK or Europe. In your response include a map showing the canals. Present your information as an eco-tourist pamphlet.

ISBN 978 1 4586 6277 4

4.6 Marinas and ports change coastal landscapes

The construction of port facilities modifies coastal and estuarine habitats by physically altering the shoreline. Frequently, wetlands and seagrass beds are destroyed, which reduces aquatic habitats. The increasing number of marinas and associated structures, such as jetties and boat ramps, along the Australian coast has altered shorelines and habitats, and degraded natural landscapes.

Marinas and breakwaters

A marina is a dock with moorings and supplies for yachts and small boats. They are located beside oceans, rivers and lakes, such as Salt Lake, USA. San Francisco Bay is a mecca for sailors as it contains 40 marinas and over 11 000 slipways. In Greece there are 20 marinas under construction and 50 in operation, such as at Piraeus—the starting point for boat trips to the Aegean islands for millions of tourists every year.

Australia's coastline stretches approximately 60 000 km. Two-thirds of the population live in towns and cities located near the coast. Marinas have been constructed to accommodate affluent lifestyles, such as in the Docklands in Melbourne, Darling Harbour in Sydney, the Gold Coast, Hamilton Island, and Hillarys Boat Harbour, north of Perth.

At present the Western Australian Government aims to build a multi-million dollar marina and port facility in Port Hedland. The Spoilbank Marina Precinct is part of the Pilbara Cities initiative, which aims to revitalise Port Hedland into a city of 50 000 people. However, dredging during construction will cause loss of species in the intertidal zone associated with mangroves, salt marshes and cyanobacterial mats. These mats are rich in organic matter and phosphorus, and reduce erosion by binding and stabilising land. The Spoilbank Marina project has strategies in place to reduce loss of biodiversity, trap nutrients, maintain water quality, protect against storm surges and erosion, and promote the recreational value of fishing.

4.6.1 The planned Spoilbank Marina, Port Hedland, WA

Landscape degradation

Excessive boating activity in river mouths and estuaries leads to erosion of riverbanks and destroys the vegetation that is important for the preservation of biodiversity. Boats and marinas cause water pollution from oil spills and sewage disposal. Additionally, they require wharves, navigational markers, dry-docks, slipways, fuel storage, pump-out facilities and parking access.

Before a marina is constructed, an environmental statement of its effects is prepared. Questions to be answered include:

- Is a breakwater or channel diversion required?
- Can clearing of natural riparian vegetation be avoided?
- Will wetlands or seagrass be affected?
- Is bank erosion likely?
- Will nearby aquaculture and fishing grounds be affected?
- Will there be significant visual impact?
- Is the proposal likely to affect the significance of Indigenous or non-Indigenous heritage items found on the sites?

ISBN 978 1 4586 6277 4

To improve navigation, natural channels have been widened and deepened by removing soil from the bottom of waterways—a process called dredging. This destroys the habitats of benthic, or bottom-dwelling, organisms that live in the sediment. To reduce the impact of waves, such as tsunamis or storm surges after cyclones, stone breakwaters are constructed around ports and marinas. This interferes with natural erosion and seagrass species and their habitats.

4.6.3 Photo of Gainsville Marina on Lake Lanier, USA, depicting the facilities a marina requires. A–G and 1–10 are docks offering various facilities

4.6.2 Images of Mantoloking, New Jersey, before and after Hurricane Sandy in 2012

Geo**activities 4.6**

Knowledge and understanding

1 What is a marina?

2 Describe how marinas can cause landscape degradation.

3 Why are cyanobacterial mats and riparian vegetation important?

Inquiry and skills

4 Refer to 4.6.1.
   a What is a spoilbank?
   b What is the aim of developing the Spoilbank Marina?
   c How will the construction of the Spoilbank Marina impact on the environment?
   d How will the project reduce landscape degradation?

5 Refer to 4.6.2.
   a What were the impacts of Hurricane Sandy on New York marinas?
   b Describe what could occur in marinas with rising sea levels and increasing intensity of storms.

6 Refer to 4.6.3.
   a What facilities does a marina require?
   b Explain how the natural environment surrounding a marina changes over time.

7 Plan a class debate on the following statement: *Marinas are the environmentally destructive playgrounds of the rich.* Evaluate your sources and summarise conclusions.

ISBN 978 1 4586 6277 4

4.7 Seawalls and barriers: saving Venice

Climate change and the associated projected rise in sea levels could result in the catastrophic flooding of coastal cities. Thirteen of the world's 20 largest cities are situated along coastlines, including Mumbai (India), New York (USA), Shanghai (China) and Alexandria (Egypt). Their protection requires the construction of seawalls or sea barriers.

The Thames Barrier prevents London from being flooded by high tides and **storm surges**. The flooding of low-lying parts of New York after Hurricane Sandy in 2012 saw calls for the construction of a barrier.

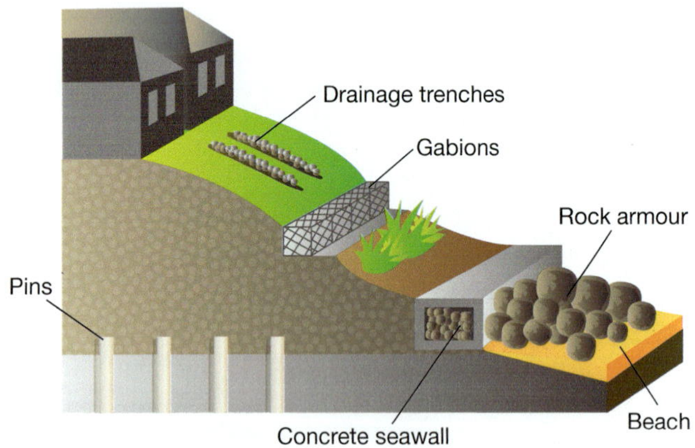

4.7.1 Structures to protect coastlines

Flooded Venice

Venice is a city partially flooded by water. A network of canals is used in place of roads as a means of transport. High tides flood the streets from September to December forcing residents to walk on wooden planks to stay above the floodwaters. There is an increase in flooding in Venice due to climate change causing sea levels to rise slightly every year. Furthermore, the city is also sinking 1–2 mm a year due to:

- *pumping of groundwater* from beneath the city (a practice that stopped in 2000)
- *compaction of sediments* beneath Venice from centuries of building
- *plate tectonics*, as the Adriatic Plate, on which Venice sits, is subducting beneath the Apennine Mountains.

In 2013, the first successful tests of the new electro-mechanical flood barrier system for Venice were trialled. The Moses project (as it is called in English) is scheduled for completion in 2016.

Seawalls hold back the sea

A seawall is a form of coastal defence constructed to protect humans and their settlements from tides and waves. Seawalls are constructed from a variety of materials, such as reinforced concrete, boulders and fibreglass. On the downside, seawalls stop the exchange of sediment between land and sea and alter ocean currents. Seawalls built across the world have achieved positive and negative results:

- *Pondicherry*, on the coast of India, escaped the 2004 tsunami due to a massive stone seawall constructed by the French 300 years ago. The only people killed were those who lived in villages beyond the seawall.
- *The Maldives*, located in the Indian Ocean, is building seawalls to protect the islands from sea level rise. In 2004 the seawall protecting the capital, Male, prevented half the city from being destroyed by the tsunami.

The Maldives

Gulf of Maine, USA

Isle of Wight, UK

4.7.2 Different types of seawalls

ISBN 978 1 4586 6277 4

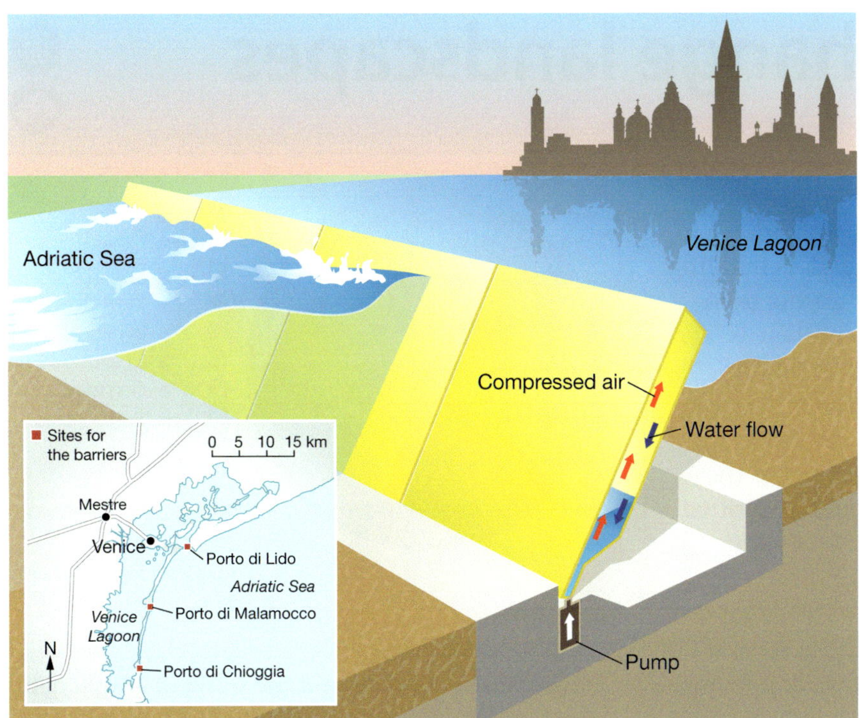

4.7.3 Design and sites of the proposed sea barrier for Venice

- *Japan* has 43% of its coastline lined with concrete seawalls and breakwaters, for protection from earthquake-generated tsunamis. However, these did not prevent the 2011 tsunami surging over seawalls and flooding coastal areas.
- *Phuket* is located off the coast of Thailand. A seawall was built to protect the sea gypsy community at Laem Tukkae. However, it did not last the first monsoon season.

Other countries preparing for sea level rise include:

- *the Netherlands*, which has constructed the largest seawalls (dykes) in the world. Rising sea levels continue to breach its floodgates. In response, the Dutch are designing floating houses, roads and cities.
- *Australia*, which has cities and towns built on coastal waterways that are vulnerable to sea level rises, high tides and cyclones. Coastal protection projects include breakwaters and seawalls aimed to guard against erosion and sea level rise.

Geo**info**

- The biggest mega-structure seawall is in the Netherlands. It is twice as long as the Great Wall of China.
- The St Petersburg's barrier in Russia consists of 26 km of levees and gates, providing protection from the Baltic Sea and the Neva River.

4.7.4 Effects of seawalls

Positive
• Provides a hard coastal defence • Minimises deaths and injuries • Reduces damage to homes and infrastructure along coasts • Used for recreational purposes (e.g. walking and cycling) • Long-term solution compared to continually replenishing eroded beach with sand
Negative
• Scars the landscape and can be an eyesore • Destroys coastal habitats (e.g. wetlands) • Disrupts natural shoreline processes (e.g. wave movement) • Reflects wave energy, which could erode the seawall base

Geo**activities 4.7**

Knowledge and understanding

1. Describe a seawall and a barrier.
2. Name cities affected by future sea level rises.
3. Venice suffers from rising sea levels and sinking ground. Explain why the land is subsiding.
4. Describe how seawalls cause environmental degradation.
5. Discuss whether seawalls are ineffective constructions.

Inquiry and skills

6. Refer to 4.7.1.
   a. List two structures humans build to hold back the sea.
   b. Describe how the coastline has been altered.
7. Refer to 4.7.2 and describe the different seawalls.
8. Refer to 4.7.3.
   a. Where are the sea barriers to be located?
   b. Explain the operation of the sea barriers.
9. Refer to 4.7.4. Summarise the positive and negative effects of seawalls on the landscape, in the form of a diagram.
10. Research one coastal city in Australia and one coastal city overseas. Include the impacts of increasing sea levels and what is being done to reduce the impacts (e.g. seawalls or barriers).
11. Investigate how low-lying islands (e.g. the Torres Strait Islands and the Maldives) can be protected from the sea and what the potential impacts are on landscape degradation. Present your findings as annotated photographs.

ISBN 978 1 4586 6277 4

4.8 Dams change landscapes

Of the world's 228 largest rivers, approximately 60% have been changed by the construction of dams, weirs and canals along their journey from source to mouth. In Australia, after 150 years of dam building some people wonder if the problems arising from dams outweigh the benefits. At present there is a debate over the construction of a dam on the Mary River in south-east Queensland after the river experienced flooding in 2010, 2011 and 2013. The traditional Gubbi Gubbi people object to the construction of a dam as the river touches spiritual places, birth places and sacred pools.

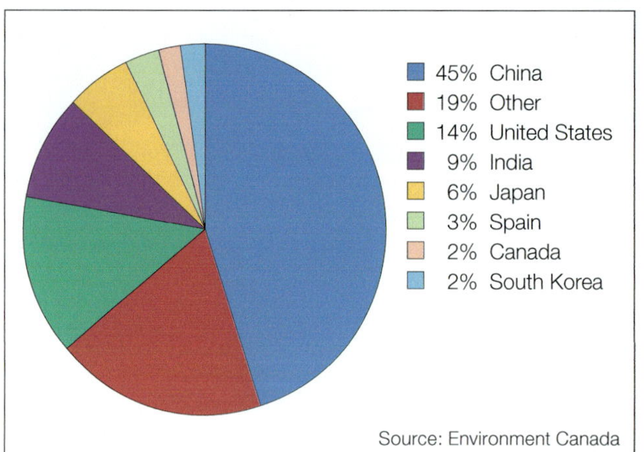

4.8.2 Global distribution of large dams

Damming the world

A dam retains water, while floodgates, levees and dykes are used to manage or prevent water flowing onto the land. The ideal location for building a dam is the narrow part of a deep river valley where the valley sides act as natural walls. Behind the dam wall a reservoir of water collects, frequently enjoyed for recreational activities, such as fishing. Dams are used to irrigate farms in the Murray–Darling Basin, provide water to urban settlements in London and supply electricity for Chinese industries. Additionally, dams reduce the impacts of floods downstream.

The construction around the world of 48 000 dams, over 15 m high, has displaced natural landscapes, such as forests, as well as agricultural lands and settlements. Dams have caused the degradation of wetland and riverine ecosystems, and their reservoirs have changed biodiversity, because water temperature is colder behind the dam wall where the water is deeper. Fish such as salmon are unable to migrate upstream to breed, although in many cases fish ladders have been built to assist their migration.

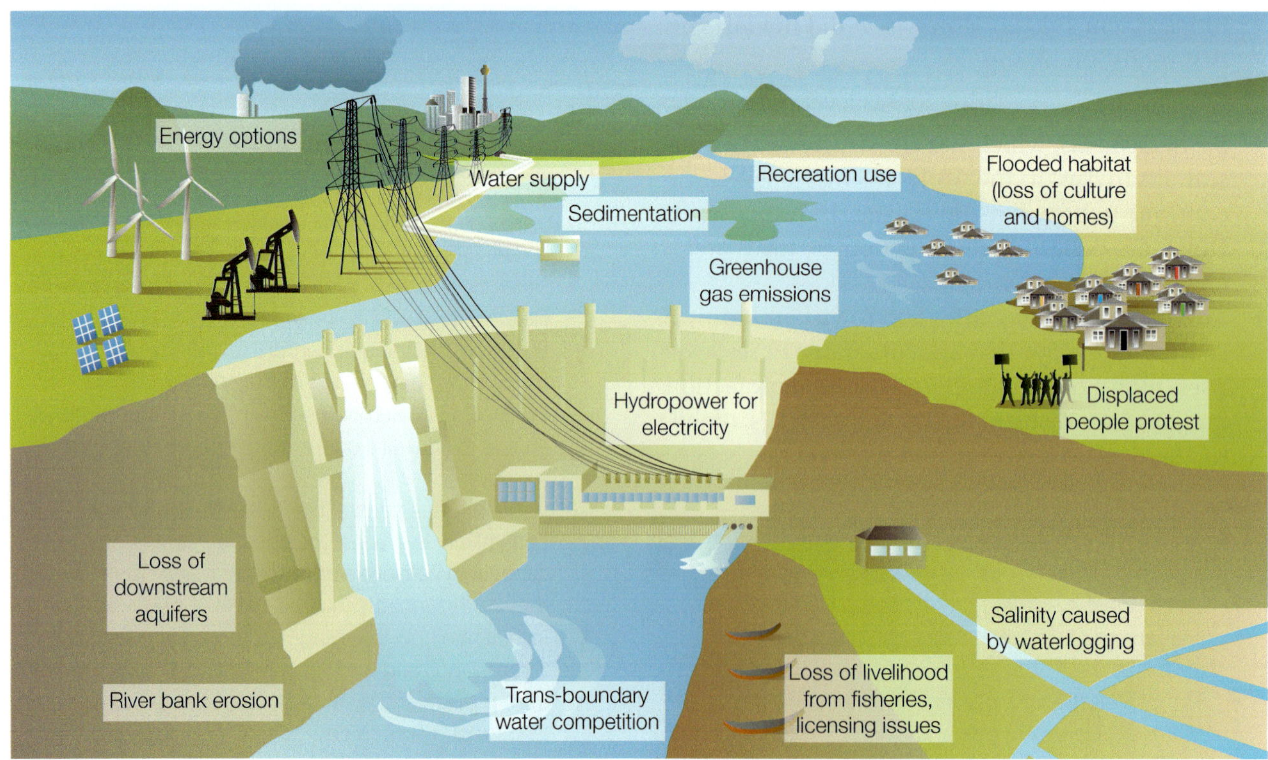

4.8.1 Dams cause changes to people and landscapes

ISBN 978 1 4586 6277 4

A dam collapse from an earthquake or faulty construction increases the chance of floods downstream. Dam failures are rare, but in 1975 the collapse of Banqiao and Shimantan dams in China killed 171 000 people and displaced the 11 million who had lost their homes.

Three Gorges Dam, China

In 1954, the Yangtze River in Central China flooded, killing 33 169 people and forcing 18 884 000 people to relocate. The Three Gorges Dam built on the river reduces floods downstream, provides electricity, and increases shipping capacity by constructing locks and lifts.

4.8.3 Three Gorges Dam changes landforms

Changes in Yangtze River upstream of dam and towards the source	Changes in Yangtze River downstream of dam and towards the mouth
• Large storage of water created behind the dam • Gradient and speed of river reduced • Build-up of sediment behind the dam could cause water to overflow or break the dam over time—needs constant dredging	• Discharge of water downstream reduced and more regular • Less alluvium in the river downstream to provide fertile soil for agriculture
Advantages of Three Gorges Dam	**Disadvantages of Three Gorges Dam**
• Improved navigation and trade as far as Chongqing, one of the fastest-growing cities in the world • Flooding reduced • Water stored behind dam used for drinking and irrigation • Produces renewable energy	• Increased traffic and pollution placing stress on freshwater species (e.g. freshwater dolphin, sturgeon and alligator) • Dam is vulnerable to earthquakes • 1.3 million people relocated when reservoir was flooded

4.8.4 The Three Gorges Dam

However, the future of the dam is uncertain as it sits on a seismic fault and 80% of the land around the river experiences high rates of soil erosion. As a result, around 35 million tonnes of sediment is deposited into the Yangtze River annually, which means sediment builds up behind the dam instead of flowing downstream. The Yangtze carries the fifth-largest sediment discharge of any river in the world.

The Yangtze River basin is home to 361 fish species and accounts for 27% of endangered freshwater fish species in China. The Three Gorges Dam on the Yangtze River alters the water flows, adversely affecting freshwater fish.

Geo**info**

- Approximately 40–80 million people have been displaced from their homes as a result of dam construction.
- Australia's largest artificial lake by volume is Lake Argyle, Western Australia, completed in 1971.

Geo**activities 4.8**

Knowledge and understanding

1 What is a dam?
2 How does a dam differ from a dyke?
3 List the advantages of dams.
4 Explain how dams cause landscape degradation.

Inquiry and skills

5 Refer to 4.8.1.
  a What is the ideal location for a dam?
  b List the multiple uses of dams.
  c Explain how dams impact positively and negatively on people.
6 Refer to 4.8.2 and name the three countries where most large dams are constructed.
7 Refer to 4.8.3.
  a Where is the Three Gorges Dam located?
  b How has the dam changed the river upstream and downstream?
  c Explain anticipated future problems of the dam.
  d Determine whether the dam is an engineering miracle or an environmental disaster. Explain your answer.
8 Investigate the proposal for the construction of a dam on the Mary River or a dam in another country. List the different views of people on its construction (e.g. indigenous communities, people who lost properties during a flood, conservationists, town planners and displaced settlements). Present the different perspectives as a table and summarise conclusions.

ISBN 978 1 4586 6277 4

4.9 Mining explosions and mountain top removal

Mining areas are frequently in the news as they account for 12 000 deaths annually, caused by fires, explosions, ground subsidence, and mudslides from steep waste dumps (tailings). Explosions killed 1572 miners in 1942 at a Honkeiko coal mine in China, and 96 miners in 1902 at a Port Kembla coal mine in New South Wales.

Mining changes landforms

Mining causes scars on Earth that can be seen from space. The Mirny diamond mine in Siberia is the largest open-cut diamond mine in the world. Its mining hole is 1.2 km wide and 525 m deep. Wind currents inside the huge hole form a downdraft that can result in helicopters being sucked in and crashing.

Mining changes landforms as mountain tops are removed and underground explosions conducted. **Mountaintop removal (MTR) mining** involves the removal of a summit on a mountain or hill to obtain geological material, such as coal. Mountain tops have been removed for coal in the Appalachian Mountains in the eastern USA and the Dnepropetrovsk-Donets and Kuznets area in Russia. Sometimes the land is dumped back on the ridge and compacted to reflect the original contour of the mountain. In other cases, wastes are dumped into nearby streams, altering their flow and causing a decrease in aquatic biodiversity.

Landscape degradation

During the process of mining, landscapes become degraded, for example:

- *logging forests*—contributes to a decline in biodiversity
- *water pollution*—causes toxins to enter rivers and groundwater
- *air pollution*—emits dust and carbon monoxide.

There are health impacts as well, for instance lung diseases (e.g. asbestosis).

Madre de Dios occupies Peru's lowland and hosts the Manu National Park, which was designated a UNESCO World Heritage site. Mining activities in Madre de Dios contributed to deforestation faster than any other activity. Roughly 7000 ha of pristine

4.9.1 Mining hotspots change people and places

Aberfan, Wales	Collapse of coal mining sludge killed 144 people and 116 children in 1966
Red Dog, Alaska	World's largest zinc mine and Alaska's largest polluter (toxic chemicals)
Carlin Trend, Nevada	World's second-largest gold deposit damaged Western Shoshone lands (indigenous lands)
Bingham Canyon, Utah	World's largest open copper and gold mine. Caused several landslides in 2013
San Martin, Honduras	Open pit gold and sand mine is a significant burden on water resources (reducing availability for agriculture)
Yabacocha, Peru	Gold mine causes mercury spill in 2000 (local communities affected)
Amazon, Brazil	Tens of thousands of small-scale miners subject to mercury poisoning
Ranger Uranium mine, Australia	Radioactive waste pollutes the wetlands in Kakadu National Park
Ok Tedi Mine, Papua New Guinea	Copper and gold mine discharged waste into Ok Tedi River (50,000 people living downstream affected)
Samangan, Afghanistan	Tunnel collapses in a coal mine, killing 24 Afghans. (Many mines in Afghanistan operate illegally, with rudimentary safety standards.)
China	Government pledge to shut down 2,000 small coal mines and improve safety standards by 2015
Syama, Mali	Expansion of gold-mining industry in Mali (now the third largest producer in Africa, after Ghana and South Africa). Has caused contamination of groundwater in Syama
South Africa	Environmental pollution from over 100 years of gold mining has polluted water supplies (affecting people's health and food chains)
Kabwe, Zambia	Mining and smelting has poisoned soils with lead, cadium and copper (affecting people's health and food chains)

ISBN 978 1 4586 6277 4

Process of mountaintop removal

1 Vegetation is cleared

2 Giant earthmovers remove bedrock

3 Material is dumped into valleys – called valley fills which changes the course of rivers

4 The land can be rehabilitated by returning topsoil and vegetation

Adverse effects

- Decline in biodiversity – vegetation and animal species
- Destroys rivers when buried by valley fills
- Blasting damages foundations of homes and infrastructure
- Flooding downstream from loss of vegetation and valleys

4.9.2 The processes and environmental effects of mountaintop removal

forest and wetlands were cleared at two mining sites between 2003 and 2009.

Over a period of 20 years the Ok Tedi open-cut copper and gold mines in Papua New Guinea (PNG) discharged over 70 million tonnes of mining waste into the Ok Tedi River, annually. The pollutants harmed the livelihood of 50000 people living downstream. Fish species declined and contaminated mud destroyed plantations of taro and bananas—the staple foods of the local people.

Conflicts: perspectives

Conflicts concerned with mining take place across the world, such as between mining and agriculture interests, and between mining groups and indigenous people's attachment to the land. There are also conflicts between values held by the mining industry and conservationists. Mining companies promote economic benefits, such as employment and an increase in **gross domestic product (GDP)**, while many have signed contracts agreeing to protect the environment. Conservationists believe the cultural values of an area may be more important than money gained by mining.

Geo**info**

Ok Tedi's mining lease expired at the end of 2013 and the PNG Government has taken ownership.

Geo**activities 4.9**

Knowledge and understanding

1. How does mining change landforms and landscapes?
2. What is mountain top removal?
3. Explain the effects of mountain top removal on landforms.
4. Discuss the effects of mining on the landscape of Madre de Dios.
5. How did mining degrade the environment in PNG?

Inquiry and skills

6. Refer to 4.9.1.
   a. Provide examples of the impacts of mining on land, water, biodiversity, air and people.
   b. Research two mines and discuss their effects on landscape degradation. Present you information using web 2.0 tools.
7. Refer to 4.9.2.
   a. Explain the process of MTR.
   b. What are the problems of valley fills on steep slopes after heavy rain?
   c. Describe how MTR degraded the landscape.
8. In groups study one mining conflict from around the world. Answer questions such as: Where is the mine located? What are the major interest groups (e.g. mining company and local community)? Reflect on your learning and suggest individual and collective actions to resolve the conflict.

ISBN 978 1 4586 6277 4

4.10 Gold mining and rehabilitation: Johannesburg

Johannesburg, the largest city in South Africa, contains 4.5 million people. The original landscape has been described as a waterless, treeless, dust bowl. It was inhabited by the San tribes, who were hunters and gatherers. In the late 1880s, the gold rush changed the landscape into mine pits and waste heaps. Mines brought wealth to a few, who were known as Randlords, while hundreds of thousands of black Africans worked deep underground, earning little money.

4.10.2 The original landscape featured sparse vegetation and was inhabited by the hunter–gatherer Sans

From mining to an urban landscape

The Witwatersrand is a large sedimentary range of hills rich in gold. Over the years the landscape changed as shaft gold mining and waste dumps spread across the 80 km mining belt. The Western Deep Level Mine is the deepest mine in the world, with shafts to depths of 3900 m. It is problematic as the temperature rises by 1 °C every 33 m, with rocks too hot to touch. Cave-ins, shaft collapses, explosions and land subsidence occur.

Most underground gold mines ceased operations in the 1970s, leaving large tracts of vacant land. Informal settlements encroached on unused mining land with people living beside toxic waste dumps.

At Krugersdorp, acidic lakes dot the landscape and hippopotamuses at a nearby nature reserve are going blind due to acid water runoff.

4.10.1 Location of South Africa and Johannesburg

Future landscape: protection

Johannesburg faces environmental challenges from mining operations, including contaminated soils, Acid Mine Drainage (AMD) and subsiding land. Gold mining companies recognise the environmental challenges they face, and are actively pursuing sustainable mining practices. Large quantities of mining land have been treated for contaminants and released for reuse. Water from mining activities is treated to reduce pollutants entering water courses. Waste dumps are lined to reduce the leaching of toxic waste (tailings) through the soil and into groundwater.

There are more than 5700 derelict mines in South Africa. The nation has endorsed international treaties to protect wetlands from mine operations, as well as to ensure the safe disposal of mining wastes.

Geoinfo

Johannesburg's Zulu name is eGoli, which means 'City of Gold'.

ISBN 978 1 4586 6277 4

Soccer City stadium, the main venue for the 2010 FIFA World Cup, is located in the midst of three large mine dumps.

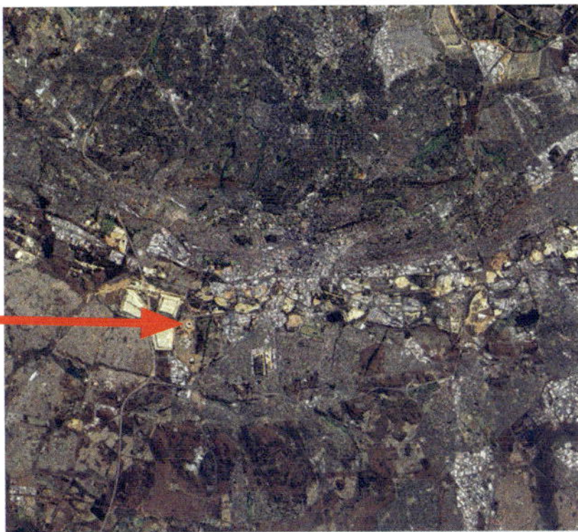
Satellite image of the soccer stadium surrounded by mine dumps.

AMD contains heavy metals harmful to humans and ecosystems, and is released into Johannesburg's water system.

The gold industry has shifted from deep-shaft mining to reprocessing waste and much of the original mining land has been redeveloped. The Gold Reef City theme park now covers 12 ha built on the site of a mine.

Informal settlements in the mining belt consist of shacks built of scrap metal and wood. Water is available at communal standpipes or delivered weekly by truck, and electricity is tapped illegally or provided by car batteries. Communal latrines provide sanitation. Waste disposal is usually non-existent.

In the early 1900s, Johannesburg introduced shade trees to alleviate airborne dust from mining. Originally a semi-arid grassland ecosystem, the city now boasts the largest urban canopy in the world. Large quantities of water are required for irrigation—an added burden for a region with limited water resources.

4.10.3 Mining landscapes in Johannesburg

Geo**activities 4.10**

Knowledge and understanding

1 Where is Johannesburg located?
2 Describe how mining changed the landscape.
3 What is the Witwatersrand?
4 List the problems of gold mining in Johannesburg.
5 Describe one procedure to rehabilitate mining landscapes and prevent future contamination.
6 Discuss how humans changed the landscape of Johannesburg over time.

Inquiry and skills

7 Refer to 4.10.3.
   a Describe the location and size of the soccer stadium in relation to the mine dumps.
   b The gold industry shifted from shaft mining to waste processing. How has this changed the landscape?
   c Describe how the benefits of gold mining were not distributed equally among the population as observed from different shelters.
   d What is Acid Mine Drainage (AMD)? How does it impact on the environment?
8 List the largest gold mines in the world and locate them on a map.
9 Using secondary resources, briefly outline how mining has changed landscapes in Australia since the gold rushes. Present as a photo story.

ISBN 978 1 4586 6277 4

4.11 Sustainable rehabilitation: mining landscapes

An increasing number of mining companies have implemented sustainable rehabilitation projects to restore mines to their pre-mine environment. Sustainable rehabilitation consists of environmental, social and economic factors:

- *Environmental* (e.g. reduce mining wastes entering groundwater, replace mountain tops and stabilise waste (tailings) slopes to reduce landslides).
- *Social* (e.g. return the land to the community and provide an aesthetically attractive landscape).
- *Economic* (e.g. sell the land for housing estates, graze animals on it or grow trees).

Today many degraded mining landscapes are in the process of rehabilitation. Underground disused mines are repaired to reduce land subsiding and mine dumps capped to stop toxins entering groundwater and rivers. Topsoil is removed before mining and later spread on top of the waste dump, then revegetated with native species.

4.11.1 The sustainability three-legged stool

Red Mountain Park in Birmingham, USA, was originally a mining site. Rehabilitation of the site aims to convert the land into the largest urban park in the USA. The 486 ha park is located along a ridge on the edge of the city. When finished it aims to connect the older African American communities on one side of the mountain to newly developing communities on the other side.

4.11.2 Red Mountain Park: from mine to urban landscape

Environmental sustainability
• Reclamation of mined areas • Long-term native forest and habitat management • Removal of invasive species • Commitment to Sustainable Sites Initiative standards for new construction
Social sustainability
• Enhances public health through active recreation • Secures social equity • Preserves cultural heritage—area still inhabited by original miners and their descendants
Economic sustainability
• Hotel or conference/retreat within the park • Gate fee, rentals, tour and food concessions • Employment training program

4.11.3 Syncrude's Mildred Lake oil sands mine in Alberta, Canada (left). After rehabilitation (right).

Creative solutions

Some unused mines are unable to be rehabilitated to their pre-mine environment as it would be too difficult and too expensive. However, many have

4.11.4 The Wieliczka Salt Mine in Poland is now a wedding venue

ISBN 978 1 4586 6277 4

been converted into recreational facilities, such as swimming pools and indoor ski centres. In Poland, the former Wieliczka Salt Mine is now a wedding venue and a World Heritage site. In Moldavia a former limestone mine is a wine cellar, in Italy a copper mine now stores cheese, and in Germany a former coal mine contains an ice-skating rink. Parts of the film *Mad Max* were filmed inside a disused mine, and the underground opal mines at Coober Pedy are popular on the tourist circuit. Above the ground, open-cut mines can become amphitheatres. Most of these new functions create employment (economic), provide enjoyable activities (social) and improve the environment.

Nauru: a new landscape

Over thousands of years, birds flying over the Pacific island of Nauru left their droppings on the island. Over time these droppings, called guano, fossilised and became a source of phosphates, which are used as fertilisers on farms and for making explosives.

Nauru, a phosphate rock island, is the world's smallest island nation, covering 21 km^2. After decades of phosphate mining, the island resembles a moonscape rather than a tropical paradise because 80% of the land has been degraded from mining activities. The runoff of silt and phosphates into the surrounding waters killed 40% of marine life, and most of the surrounding coral is black.

Rehabilitation projects aim to reverse the landscape degradation. Jagged limestone pinnacles will be crushed into rocks and used to level the land. Topsoil and organic matter will be added and land revegetated with native plants. A catchment area will be constructed for a freshwater reservoir, and a conservation area established.

4.11.5 Satellite image of Nauru

Geoinfo

In 1962 Nauru had the highest per-capita income in the world when exports of phosphates were high. Today it is a poor country.

Geoactivities 4.11

Knowledge and understanding

1 List the three parts of sustainability.
2 What is meant by rehabilitating mines?
3 How can an unused mining area be rehabilitated sustainably?
4 Explain the environmental impacts of mining guano in Nauru.
5 Describe how degraded mining areas in Nauru are going to be rehabilitated.

Inquiry and skills

6 Refer to 4.11.2 and discuss how Red Mountain Park will sustainably rehabilitate an unused mining area.
7 Refer to 4.11.3 and describe how the Mildred Lakes mine site has been rehabilitated.
8 Refer to 4.11.4.
   a Brainstorm the different uses of old mine sites. Present your findings as a Wordle.
   b List how unused mines can be reused sustainably—environmentally, socially and economically.
9 Refer to 4.11.5 and an atlas.
   a What is the latitude and longitude of Nauru?
   b Using an atlas or Google Earth, calculate the distance from Nauru to your home.
10 In a two-column table:
   a list the economic and social values of a mine (e.g. source of materials used in mobile phones)
   b list the environmental damage of a mine.

Chapter 4

4.12 Conflicting management: coal seam gas mining

Coal seam gas (CSG) mining is one of the most controversial issues in Australia. The industry uses toxic chemicals and vast amounts of water in the fracturing process to extract natural gas trapped underground. CSG threatens wetlands, rivers, groundwater and agriculture

Australia has an estimated 100 years of CSG reserves. The exploration, extraction and export of these will have economic benefit for regional towns and the country. Critics of CSG claim the environmental and social cost of CSG mining is one that Australians should not allow. According to the National Water Commission, there is 'less-than-perfect information' about long-term impacts of CSG, and 'there are also significant potential risks to water and water management as a result of the scale of the development'.

4.12.1 CSG mining: stakeholder conflicts

Stakeholder	Opinion
Farmer	Landowners have no say in CSG mining on their properties because they do not automatically own the rights to gas and minerals under the soil. This puts farmland and water resources at risk. Farms provide food security for the country as well as valuable exports. Once the water table is damaged or polluted, our livelihoods are ruined. The health of our rivers, farms and communities are not worth losing for the sake of CSG companies' profits. Lock the gates to CSG!
 Resource scientist 	CSG does have huge potential in Australia's energy future and export markets, but there are many environmental risks. These include the vast demand for water (especially groundwater), the treatment and storage of contaminated wastewater, the contamination and permanent damage of aquifers, land subsidence, and the release of greenhouse gases into the atmosphere. CSG technology is new and poorly understood. It needs to be carefully regulated to minimise environmental damage.
Ecologist	Health impacts of CSG mining include water supply contamination and destruction; air and soil pollution; and the degradation of agricultural land, wetlands and rivers. Chemicals used in mining, and contaminants released from the coal seam, may increase the risk of cancer. The CSG industry will also increase mental health issues for individuals as it divides communities and adversely affects tourism and the local economy.
CSG company 	Using CSG to create electricity reduces greenhouse gas emissions by up to 70%, thus helping Australia to meet its greenhouse reduction targets. The CSG technology is subject to a wide range of regulatory safeguards by state and federal authorities. These bodies have judged the risks to be minimal and manageable. If CSG mining is stopped, economic investment in remote and rural areas will halt. Thousands of people's jobs will be lost along with millions of dollars in tax revenue for governments to build roads, schools and hospitals.
Environmentalist 	CSG threatens our farmlands and water resources. There are serious concerns about the impacts of CSG on the environment, public health and the economic sustainability of rural communities. Gas wells are leaking, water tables are falling, rivers are being polluted, trees are dying, and people and animals are getting sick. CSG companies are also not properly monitoring air and groundwater levels as they are supposed to. A University of New England study in 2013 found that emissions from leaked methane and argon around gas fields make CSG mining more polluting than coal mining.

ISBN 978 1 4586 6277 4

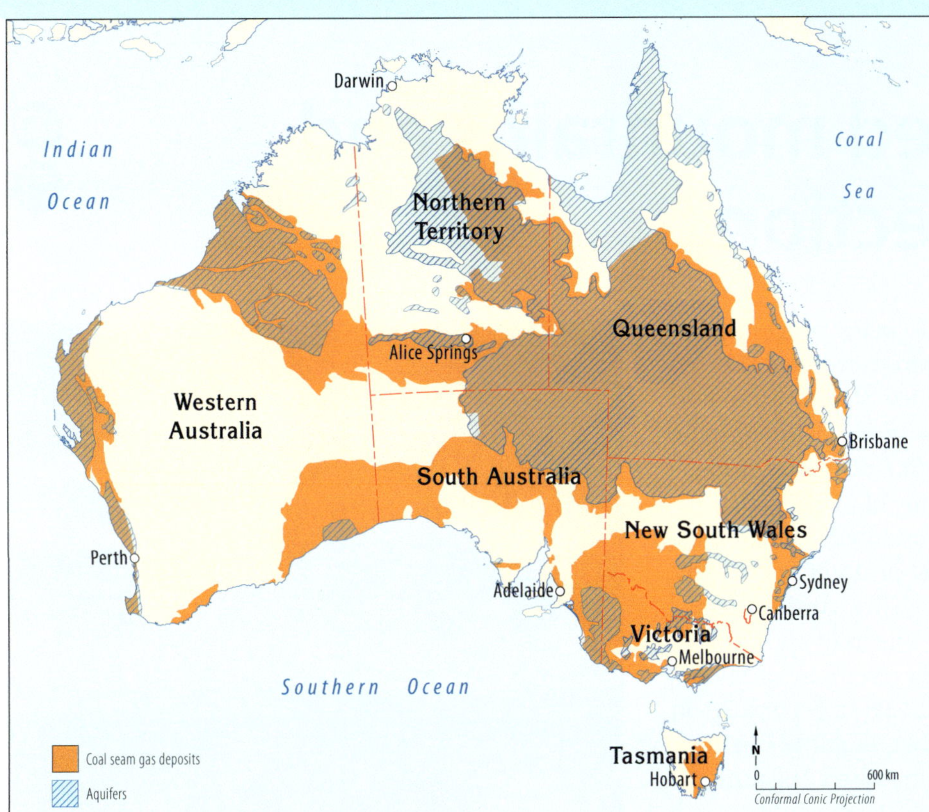

4.12.2 CSG and related aquifers in Australia

Geoinfo

Over the next 20 years, coal seam gas operations are expected to continue expanding. Up to 40 000 wells are projected for Queensland alone.

4.12.3 Coal seam gas fields, Tara, Queensland

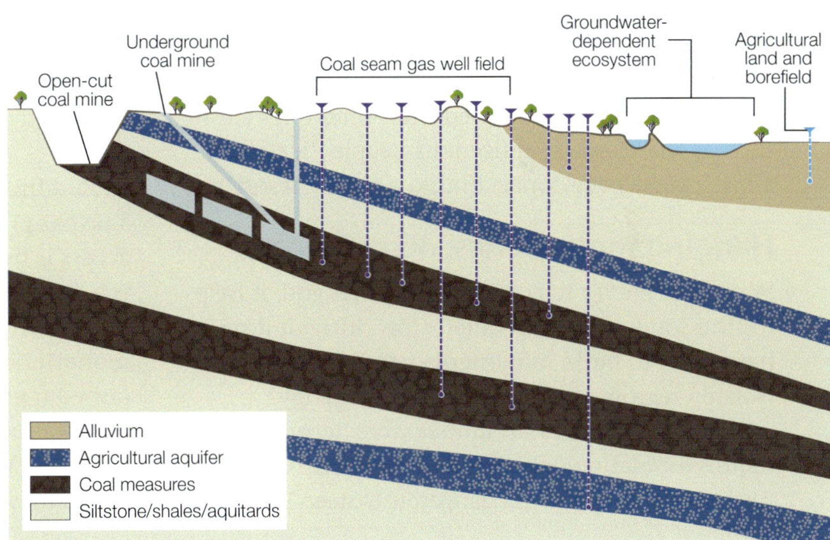

4.12.4 Bioregion featuring coal seam gas extraction

Geoactivities 4.12

Knowledge and understanding

1 Why is CSG mining controversial?

2 How will CSG mining change landscapes?

Inquiry and skills

3 Refer to 4.12.1.

a List the main arguments for and against CSG mining. Present as a table divided into advantages and disadvantages.

b Roleplay a meeting with five stakeholders in a rural community hall. What do you propose for the future of CSG mining and why? Present your findings as a report.

4 Refer to 4.12.2.

a Discuss the relationship between CSG and aquifers.

b What are the potential impacts of CSG on agriculture and water resources in Queensland? Justify your answer.

5 Refer to 4.12.4. Reflect on your learning and present a media file on CSG mining and how it alters landforms and landscapes.

ISBN 978 1 4586 6277 4

4.13 Sacred mountains and protection

Sacred mountains are central to many religions, for instance Mount Sinai, where Moses received the Ten Commandments according to Jewish, Christian and Islamic religions. Mt Etna in Italy was believed to be the home of Vulcan, the Roman God of Fire, and Mt Olympus in Greece the home of Greek gods.

In Peru, the ancient Inca honoured dead people by placing them on the highest peaks of mountains, such as Machu Picchu, to express the connections between the sacred mountain, the dead and the gods. In Bali, the annual Hindu ceremony of Mulang Pekelem occurs at Segara Anak Lake, a crater lake on 3700 m high Mt Rinjani. In Tibet, Buddhists perform sky burials on tops of mountains where the human corpse decomposes or is eaten by birds.

People's access to many sacred mountains is restricted, as they are viewed as a source of power and are to be treated with respect. Different views on the preservation of sacred sites sometimes clash, such as the views of indigenous people ('leave it alone') versus developers ('build a tourist resort').

Hanging monasteries: dizzy heights

When people try to connect with their god, it helps to find some peace and quiet. How difficult it must have been to build monasteries on steep cliffs in isolated regions without today's technology?

Originally, Christian monks were hermits and required isolation, which led to the construction of monasteries on high and dangerously steep landforms.

4.13.1 Mt Nebo, Jordan, is believed to be the mountain from which Moses viewed the Promised Land. It is an important Christian pilgrimage site

4.13.2 Hanging monasteries in Bhutan: Takshang or Tiger's Nest Monastery

In Buddhist Bhutan, the Tiger's Nest Monastery (or Takshang) is located on a 3120 m high cliff, and access is by foot or riding a mule. In Greece, Meteora is a complex of Eastern Orthodox monasteries built on sandstone pillars. In the past, access to the monasteries was difficult, requiring long ladders lashed together or large nets to haul up goods and people. Today the monasteries are inscribed on the UNESCO World Heritage List to protect and conserve the monasteries for future generations.

Mt Kailash, Tibet

Some mountains have never been climbed as they are sacred, such as Mt Kailash in Tibet. Instead people embark on a pilgrimage by walking around the mountain in a counter-clockwise direction for 52 km. The pilgrims bend down, kneel, lie prostrate, make a mark with their fingers, rise to their knees, pray, crawl forward to their finger marks and repeat the process. Buddhists consider the pilgrimage a major event in a person's spiritual life.

Geoparks: Mt Batur, Bali

A UNESCO Global Geopark is an area containing geological heritage of international significance. Geoparks promote awareness of geological hazards,

ISBN 978 1 4586 6277 4

including volcanoes and earthquakes, and develop strategies to reduce their impacts on people and places. To become a Geopark, the area must possess a plan for sustainable development that engages the local people and respects their traditional way of life.

Mt Batur is an active volcano in Bali containing Lake Batur—a vast freshwater lake. The area has been classified as a Geopark for both geological reasons and cultural values. The Goddess of the Lake, Dewi Danu, provides irrigation to central Bali to irrigate the crops.

For thousands of generations Balinese famers have transformed the landscape to grow irrigated rice on terraced slopes. Intricate network of shrines, temples and agricultural deities are located amongst the fields to aid agricultural productivity.

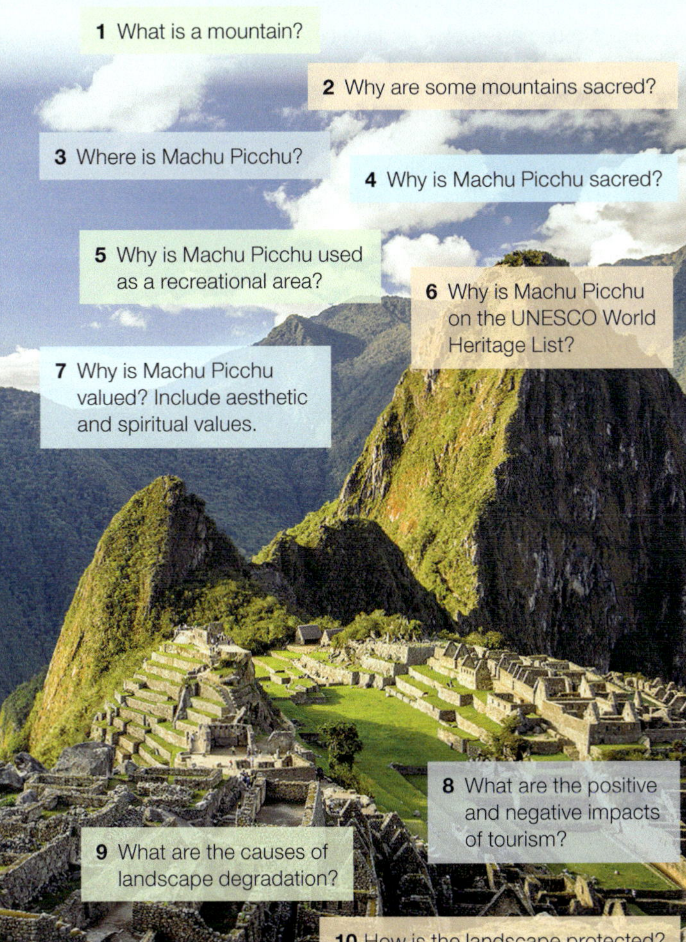

4.13.3 Inquiry questions

4.13.4 Spatial distribution of landforms showing Mt Batur, Bali

Geo**activities 4.13**

Knowledge and understanding

1. List mountains that are spiritually significant.
2. Explain why sacred mountains should be protected.
3. Describe how Tibetans and mountaineers show respect for Mt Kailash.
4. What is a Geopark?
5. Explain why Mt Batur is a Geopark.

Inquiry and skills

6. Refer to 4.13.1 and the text. Describe the value of mountains to different religions around the world.
7. Refer to 4.13.2.
   a. Why were monasteries constructed on dangerous landforms?
   b. Explain how you would reach this monastery.
   c. Debate for and against developing tourist accommodation and quality roads to sacred sites located in mountains.
8. Refer to 4.13.3 and complete the inquiry questions on Machu Picchu.
9. Refer to 4.13.4.
   a. Describe the location of Mt Batur and Lake Batur.
   b. Why are the mountain and lake of spiritual importance?
   c. What other mountains are located in Bali?
   d. Measure the length and breadth of Bali.
   e. What proportion of the island is above 1000 m?
10. Research monasteries on cliffs and how humans have changed the landforms to make them possible.

ISBN 978 1 4586 6277 4

4.14 Recreation: valuing and protecting mountains

Landforms and landscapes are an important part of people's recreational activities. For instance, we use:

- mountains for skiing, snowboarding, abseiling, climbing, white water rafting, trekking and viewing landscapes via cable cars
- karsts for caving
- coasts for surfing, rock climbing, scuba diving, sailing and fishing
- forests and wetlands for forest skyways and wetland boardwalks
- deserts for camel safaris
- grasslands for viewing animals, such as tigers
- rivers for boat trips and canal holidays.

The aesthetic, cultural and spiritual values of popular tourist sites require protection from the degradation of land, water and biodiversity. The establishment of local, national and global protected areas (such as national parks, wilderness areas and World Heritage sites) aims to protect these sites for future generations to enjoy.

Valuing mountains

Mountain tourism accounts for 15–20% of the world's tourism industry. International Mountain Day creates awareness of the importance of mountains as water reservoirs, sources of biodiversity and food, and the home for the majority of the world's indigenous people. For example, the Quechuas in the Andes Mountains in Peru value the land, water and forests. These resources should not be exploited as their wellbeing depends on careful stewardship of the landscape. For the Quechuas, spirituality and cultural identity is interconnected with the environment.

4.14.1 Recreational activities are carried out in geomorphic landscapes, such as cavers abseiling into the Greenland Ice Sheet

Protecting the Australian Alps: a significant geomorphic feature

The increased number of people skiing, hiking and climbing mountains is causing landscape degradation.

Kosciusko State Park was established in 1944 in order to protect the Australian Alps. Today 11 parks and reserves are collectively referred to as the Australian Alps National Parks with the purpose of protecting the unique mountain landscape and its natural and cultural values.

For Aboriginal Peoples the land holds spiritual values. For thousands of years, Aboriginal Peoples travelled hundreds of kilometres to meet on the highest peaks of the alpine region for trading, intertribal corroborees, marriages and the initiation of young men.

In 2008 the Australian Alps was included on the National Heritage List to recognise the region's landforms, biodiversity and recreational opportunities.

Management plans are developed through consultation with groups and individuals. These include plans to address climate change, predicted to reduce the extent of snowfields in winter and the threat of their eventual disappearance.

Protecting Himalayan Mountains

Each year 527 000 visitors, including over 100 000 trekkers and mountaineers, visit the Himalayan Mountains in Nepal. The number is high compared to other large mountains, which experience fewer than 1000 trekkers and mountaineers a year. In Nepal, most trekkers head to Annapurna (60 237) and Mt Everest (26 511). The Nepalese have a permit system for visiting their mountain summits, which brings revenue into the country.

Trekking tourism has contributed to environmental degradation. Rubbish and medical waste have created the most elevated junkyard in the world. Some conservationists want certain trekking tracks closed to reduce their environmental impact. As a result, a large proportion of the forests are allocated as reserves and national parks; for example, Chitwan National Park.

ISBN 978 1 4586 6277 4

Mountaineers, trekkers and tourists: they are attracted by recreational activities and aesthetically beautiful scenery

Sherpas: members of this ethnic group from high in the mountains are employed as trekking guides, improving their income and lifestyle. They have some control over tourism (e.g. running lodges, tea shops and trekking services). Mt Everest has strong spiritual meaning for sherpas

Government: money from tourist fees levied by the Immigration Office helped to:
- establish micro hydro-power plants and bottled gas, reducing the use of fuelwood
- construct latrines and establish rubbish collection systems
- improve roads, airport and hospitals

Conservationists are concerned about:
- declining forests leading to soil erosion and flooding downstream.
- firewood used for camp fires and lodges (e.g. cooking, heating and construction of tourist accommodation)
- the regenerative capacity of trees being poorer in higher altitudes
- the use of forest resources by local people—most use fuel wood

They have brought about change: trekking is limited and controlled in protected areas

Businesses: trekkers increase demand for businesses (e.g. restaurants, craft shops, mountain gear shops, cyber cafes and tour guides) and mean increased wealth for developers

Hill communities: trekkers spend money, enabling recipients to buy homes with modern facilities

4.14.2 Views of diverse groups with an interest in the Himalayan Mountains

Source: *Journal of Alpine Research*

4.14.3 Protected areas in Nepal

Geoinfo

- Over 1 million British people booked a snow-sports holiday last year.
- The Hindu–Kush–Himalayan region has some of the highest diversity of indigenous peoples in the world.
- The Universal Declaration of Human Rights declares leisure a human right.

Geoactivities 4.14

Knowledge and understanding

1 Landforms and landscapes are valued as recreation activities. Brainstorm the activities and present them as a Tagxedo.

2 What is meant by the aesthetic value of a landscape?

3 How do indigenous people view mountains?

4 Explain how the Australian Alps are being protected.

Inquiry and skills

5 Refer to 4.14.1 and explain the links between recreational activities and their landscapes.

6 Refer to 4.14.2.

 a Discuss how trekking in Nepal has a positive impact on diverse groups.

 b List some of the environmental problems arising from trekking.

7 Refer to 4.14.3.

 a Name one national park and one conservation area.

 b Measure the distance between Mt Everest and Chitwan National Park.

8 Design a collage illustrating recreational activities related to landforms. Evaluate sources for reliability and usefulness.

9 Investigate mountain-based sporting events (e.g. the Tour de France or cross-country skiing). List their impacts on mountain landscapes.

ISBN 978 1 4586 6277 4

4.15 Mountain bikes and off-road vehicles degrade landscapes

Bike riding is a fantastic form of exercise, and cycling on tracks is a great way to experience the diversity of nature and the aesthetically attractive landscapes found in parks.

Mountain biking

Mountain biking is riding bicycles off-road, often over rough terrain. Experienced riders pursue steep descents and aerial manoeuvres. Many riders belong to clubs and combine the sport with orienteering; others enter competitions. The International Mountain Bike Association (IMBA) states a mountain bike's impact on a trail surface is comparable to that of a hiker and less than that of an equestrian. However, the IMBA says: 'over 13.5 million mountain cyclists visit public lands each year. What was once a low use activity and easy to manage has become more complex'.

Landscape degradation and protection

Mountain biking causes trail erosion, which increases water runoff, destroys habitats and ruins the trails for other bikers as well as for runners and walkers. Unfortunately, some riders look for new challenges and venture into preserved lands and as a result damage habitats. The environmental impact of mountain biking is reduced by staying on the trail and by not riding on wet or sensitive trails and not skidding.

There are thousands of kilometres of managed trails open to cyclists in Australia, such as Garigal National Park and Barrington Tops National Park in New South Wales. The National Park Sustainable Mountain Biking Strategy and the IMBA have implemented strategies to protect trails to enable future generations of riders to enjoy biking.

Dakar Rally

An off-road vehicle (ORV) is a type of vehicle capable of driving off paved surfaces, such as on sand dunes, riverbeds, snow and rocks. Dune bashing is illegal in most places in Australia, but is a tourist attraction in Oman. People participate in cross-country safaris through uncharted African territory, while others participate in rallies such as the annual Dakar Rally. Cars slide sideways at adrenaline-pumping speeds as they criss-cross sand dunes and bump over rocks.

ORVs destroy habitats and kill plants, livestock and other animals. In semi-arid areas they stunt the growth of plants and increase the possibility of desertification. They spread seeds when they churn up vegetation. They also erode soils, which leads to sedimentation in rivers, causing a decrease in riverine species.

Geoinfo

The carbon emissions of the two-week Dakar Rally race are equivalent to a Formula One race.

4.15.1 Mountain bike race

4.15.2 The 1984 Paris–Dakar Rally

ISBN 978 1 4586 6277 4

4.15.3 4WDs on Fraser Island, Queensland

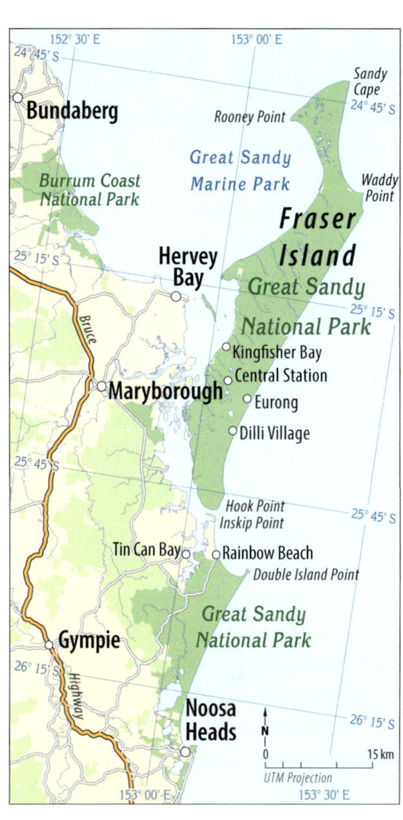

4.15.4 National parks in and around Fraser Island

Fraser Island

Fraser Island is a World Heritage site off the Queensland coast. It is the world's largest sand island and an area of natural beauty. Vehicle access is only by high clearance four-wheel drive (4WD) and a permit is required before embarking on the island. Walking tracks have been provided, such as the Fraser Island Great Walk, and mountain biking is permitted on certain beaches.

On the foredunes, 4WDs destroys vegetation, which increases the dunes' vulnerability to wind erosion. Today, some roads are metres below their original elevation. Eroded sand resulted in Yidney Lake being converted from a shallow, wet surface to a deep, dry surface. Suggested proposals to protect the landscape are replacing 4WDs with smaller, lightweight buses to reduce surface disturbance.

Wilderness nourishes the soul, but people don't need to visit a wilderness to be inspired to protect it. Most Australians have not visited Antarctica, yet 94% want the continent to be free from mining and commercial exploitation.

> We at Treehuggers International concede off-road vehicle and thrillcraft use has become a part of the outdoor experience for some … But we also stand by our belief the explosion of off-road vehicle use is the single-biggest environmental threat our special places face. If not properly regulated … the consequences will be regrettable and far-reaching.
>
> http://treehuggersintl.com/2011/herger-amendment-off-road-vehicles/

Geoactivities 4.15

Knowledge and understanding

1. What are mountain bikes and ORVs?
2. Explain the impacts of mountain bikes on environments from the perspective of the IMBA.
3. Describe how mountain bikes affect landscape degradation.
4. Suggest strategies to reduce the impacts of ORVs on the landscape.
5. Explain the impacts of ORVs during the Dakar Rally.

Inquiry and skills

6. Refer to 4.15.1 and the internet. What are Mountain Bike Australia's rules to protect the landscape?
7. Refer to 4.15.2. Imagine you were driving in the Dakar Rally this year. Use a photo story to describe your route and the changing landscapes.
8. Refer to 4.15.3 and 4.15.4.
   a. Where is Fraser Island located?
   b. What is the latitude and longitude of the island?
   c. What are the impacts of 4WDs on the environment?
9. Research the National Parks policies for mountain biking in one Australian state. Explain the increasing trend for protection policies in national parks.

4.16 Protecting landscapes

Jacques Cousteau, the famous French oceanographer and explorer, taught that 'People always protect what they love'. Yet many unique and beautiful landscapes and their landforms that are highly valued are not protected.

Protected areas

The International Union for the Conservation of Nature (IUCN) defines a protected area as:

> ... a clearly defined geographical space, recognised, dedicated and managed, through legal or other effective means, to achieve the long term conservation of nature with associated ecosystem services and cultural values.

According to the Protected Planet Report, 'Protected areas remain one of the cornerstones for promoting biodiversity, ecosystem services and human wellbeing'. In 2013, protected areas covered about 13% of Earth's terrestrial area and around 4% of the oceans. Since 1990, the number of protected areas has increased by 58% and their extent has increased by 48%. However, the report states that many protected areas face management challenges and half of the world's most important sites for biodiversity are still unprotected.

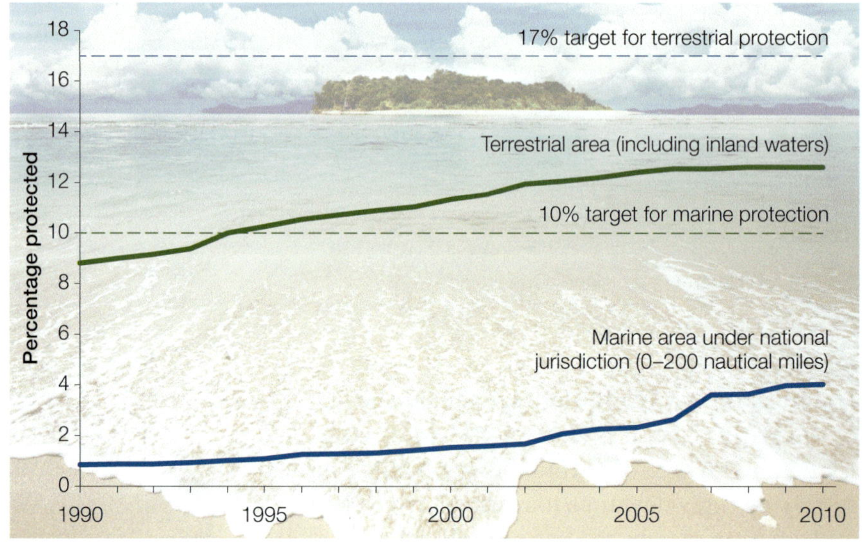

Source: Protected Planet Report 2012, UNICEF data 2013

4.16.2 Growth in the percentage of terrestrial and marine area protected, 1990–2013

4.16.1 IUCN protected area categories

ISBN 978 1 4586 6277 4

The Convention on Biological Diversity (CBD) calls for at least 17% of the world's terrestrial areas and 10% of its marine areas to be managed and conserved by 2020. With rapidly increasing global population and the threat of climate change, the need for well-managed protected areas is more important than ever.

4.16.3 Blue Lake, an extinct volcanic crater in Mount Gambier, and its surrounding volcanoes are part of Australia's most extensive volcanic province, the Kanawinka Geopark of south-west Victoria and south-east South Australia

Global Geoparks Network

UNESCO defines a Geopark as 'a nationally protected area containing a number of geological heritage sites of particular importance, rarity or aesthetic appeal'. In 2001, UNESCO established the Global Geoparks Network (GGN) to provide an international framework for the protection of outstanding geological heritage. In 2013, there were 91 Geoparks from 27 countries in the GGN. Kanawinka Geopark was Australia's only Geopark. UNESCO hopes to eventually have over 500 Geoparks in the GGN.

World Heritage sites

Through the 1972 Convention concerning the Protection of the World Cultural and Natural Heritage, United Nations Educational, Scientific and Cultural Organization (UNESCO) encouraged the 'identification, protection and preservation of cultural and natural heritage around the world considered to be of outstanding value to humanity'.

To be listed as a World Heritage site, an area must have outstanding universal value and meet at least one of the 10 selection criteria. The criteria that deal with landscapes and landforms are:

- *Criteria 7*—superlative natural phenomena or areas of exceptional natural beauty and aesthetic importance
- *Criteria 8*—outstanding examples of Earth's history, including significant ongoing geological processes in the development of landforms, or significant geomorphic features.

Places that have been listed as World Heritage sites include coasts, reefs, islands, mountains, arid landforms, karst, ice fields, glaciers, rivers, lakes, tectonic features and volcanoes.

Geo**info**

Nearly 25% of Australia's national reserve system is managed by Aboriginal and Torres Strait Islander Peoples.

Geo**activities 4.16**

Knowledge and understanding

1. What is the IUCN definition of a protected area?
2. Outline why protected areas are important.
3. Give statistics that show the percentage of protected terrestrial and marine areas and their growth.
4. What are World Heritage sites?
5. Outline how landscapes and landforms are included in the UNESCO selection criteria for World Heritage sites.
6. Explain how the GNN aims to protect geological heritage sites.

Inquiry and skills

7. Refer to 4.16.1. Identify the different types of protected areas used by IUNC.
8. Refer to 4.16.2. Describe the progress towards achieving the targets for terrestrial and marine protection.
9. Choose a biosphere reserve and write a report on its role in protecting landscapes. Reflect on the effectiveness of actions to protect landscapes. Present findings using web 2.0 tools.

Chapter 4

ISBN 978 1 4586 6277 4

4.17 Protecting World Heritage landscapes

> **Heritage** is our legacy from the past, what we live with today, and what we pass on to future generations
>
> UNESCO

Countries that adhere to the World Heritage Convention commit to preserving special landforms and landscapes declared as World Heritage sites. In 2012 there were 962 World Heritage sites across the world, including the Serengeti grasslands (Tanzania), Ha Long Bay limestone formations (Vietnam), Ifugao rice terraces (Philippines) and Australia's Uluru-Kata Tjuta National Park, Great Barrier Reef and Kakadu National Park.

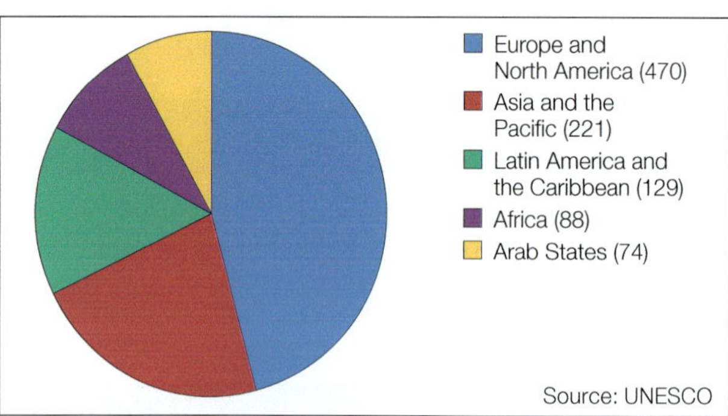

4.17.1 Number of World Heritage properties by region

Cultural landscapes

Cultural landscapes are natural landscapes changed by humans. The UNESCO World Heritage Committee defines a cultural landscape in three ways:

- deliberately shaped by humans (e.g. First Coffee Plantations in Cuba)
- evolving landscape shaped by humans (e.g. Orkhon Valley Cultural Landscape in central Mongolia, which has been used by nomads as pastureland since the 8th century)
- associative cultural landscape (e.g. Tongariro National Park in New Zealand, which has mountains that symbolise the link between Maoris and the natural environment).

4.17.3 Cinque Terre terraced hill and village

Cinque Terre: aesthetics and recreational

The Ligurian coast between Cinque Terre and Portovenere in Italy is an aesthetically beautiful landscape and important culturally. The five villages —Monterosso, Vernazza, Corniglia, Manarola and Riomaggiore—and surrounding hills are part of the Cinque Terre National Park, and are listed as a World Heritage site.

Over centuries, people built terraces on the region's steep hills. Recently the area has experienced a shift from farming to tourism, which has caused landscape degradation.

Today most of the area is not cultivated, with a decline from 1200 ha (1951) to 92 ha (2013). As a result bare ground and delapidated terraces have contributed to soil erosion. Sediment runoff into rivers and the ocean has caused a decline in marine

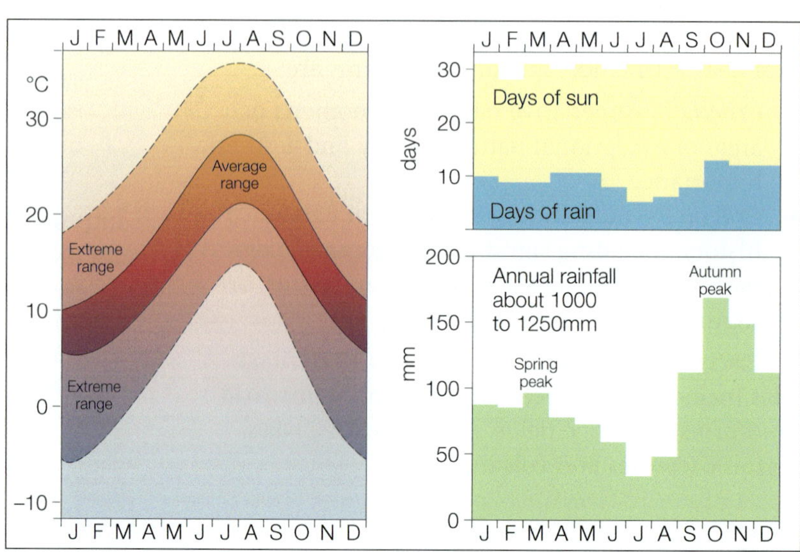

4.17.2 Climate of Cinque Terre

ISBN 978 1 4586 6277 4

species. Additionally, deadly flash floods rush down eroded steep slopes and mudslides are more frequent. The Uncultivated Lands LIFE Project plans to rehabilitate the old terraces and grow vineyards.

The area attracts tourists from around the world, who hike along trails observing spectacular scenery and traditional settlements. However, the large numbers of trekkers have caused soil erosion, requiring trail reconstruction and the building of stone retaining walls. The Park Organisation promotes sustainable tourism that considers the following factors:

- *environmental*—trekkers are encouraged to keep to designated tracks and reduce waste
- *social*—visitors are encouraged to experience the culture of the area (e.g. eating and drinking local food and wine)
- *economic*—local employment is boosted (e.g. accommodation and transport).

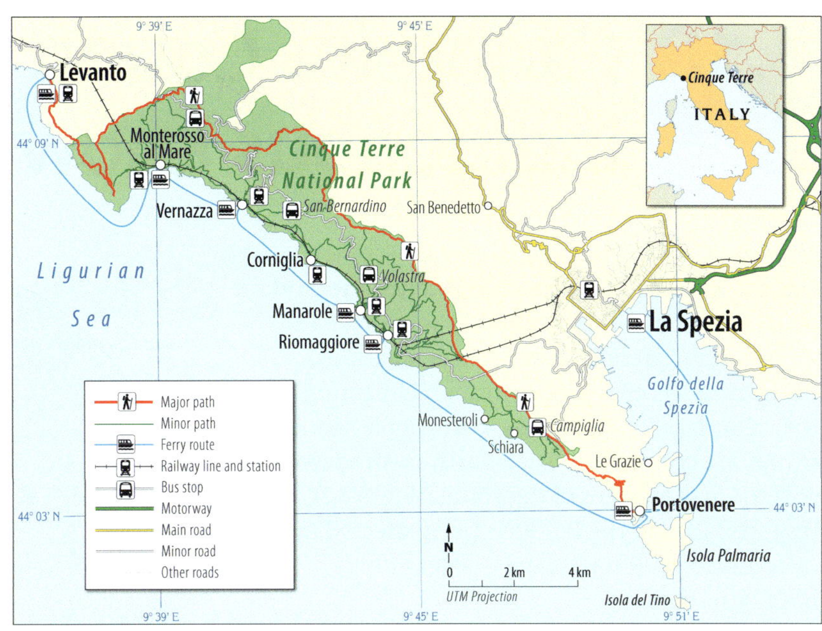

4.17.4 Map of Cinque Terre

Tongariro National Park: spiritual and recreational

In 1887, Tongariro became the first national park in New Zealand and the fourth in the world. It is also acknowledged by UNESCO as a natural and cultural World Heritage site. Tongariro National Park includes three volcanoes: Mt Ruapehu, Mt Ngauruhoe and Mt Tongariro. These volcanic mountains are culturally and religiously significant for the Maori people.

The park has a diverse range of ecosystems and spectacular landscapes attracting tourists. The annual number of visits to the park increased from 90 000 in 1960 to 800 000 in 2013. The main activities are skiing and snowboarding in winter, and hiking and climbing in summer. Developments are prohibited above 1500 m in the Tongariro and Ngauruhoe area, and above 2250 m on Ruhapehu (see pages 84–85).

Geoactivities 4.17

Knowledge and understanding

1 What is meant by the words 'heritage' and 'cultural heritage'?
2 Name five properties listed as World Heritage sites.
3 Explain three types of cultural landscapes defined by UNESCO.
4 Describe the main causes of landscape degradation in Cinque Terre.
5 List three main factors contributing to sustainable tourism in Cinque Terre.
6 Describe the significance of Tongariro National Park to past and present people.

Inquiry and skills

7 Refer to 4.17.1. Which region has the largest number of World Heritage sites?
8 Refer to 4.17.2.
  a Redraw the climate graph as required in Geography.
  b In which season does most rainfall occur?
  c Why are days of sunshine important for tourism and growing crops?
9 Refer to 4.17.3
  a Describe the scene in the photograph.
  b What is the impact of hillside erosion on the protected sea area?
10 Refer to 4.17.4.
  a Name the five main villages.
  b Describe how tourists travel into and out of the area.
11 Research on the internet and discuss how UNESCO contributes to the protection of significant landscapes.

ISBN 978 1 4586 6277 4

4.18 Caring for Country: Indigenous knowledge

Aboriginal and Torres Strait Islander Peoples consider Country a place that gives and receives life, and for which they have a responsibility to protect and maintain for present and future generations. Indigenous Peoples learnt through Dreaming stories that 'Our lands, Our Waters, Our People, All living things are connected'. Their earth-centred worldview governs management of their land. However, custodial responsibility to the land has been affected by loss of access to traditional lands, water and natural resources.

Indigenous People's Knowledge

Indigenous People's Knowledge (IPK), or Traditional Knowledge (TK), refers to the knowledge, innovations and practices of Indigenous communities, acquired through their long-term association with a place.

Earth-centred worldview

- Humans are part of nature and depend on it for survival
- Resources are limited and should not be wasted
- Humans should encourage Earth-sustaining forms of economic growth
- Human success depends on learning how nature sustains itself
- Humans have an ethical responsibility to be caring managers of Earth
- The success of the human race depends on how well humans manage Earth's life support systems for their benefit and that of the rest of the world's species and ecosystems

4.18.1 The Aboriginal and Torres Strait Islander People's earth-centred worldview guides management of the land

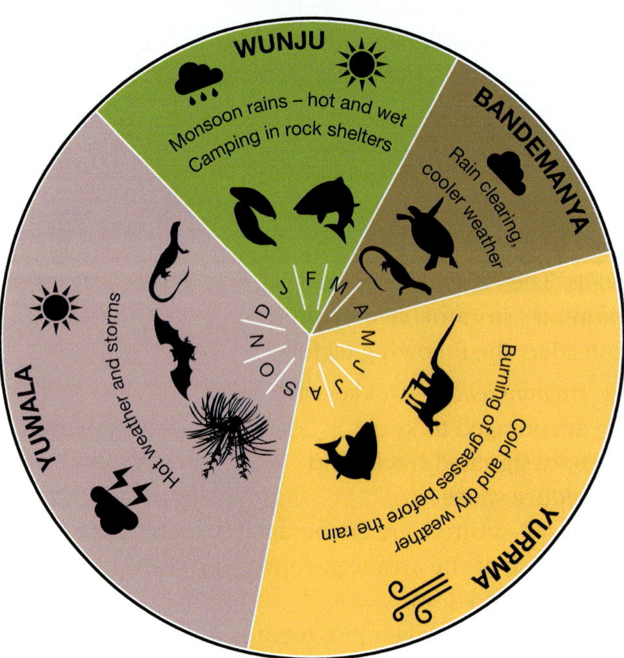

4.18.2 IPK: seasonal management of landscapes in the northern Kimberley area

This knowledge contributed to their survival in different Australian environments and landscapes, such as hot dry deserts. While IPK has sometimes been disregarded and undervalued it is now increasingly used in environmental projects across Australia. Investment in IPK programs is important as once knowledge is lost, it is gone forever.

IPK not only includes past knowledge, but also continuing knowledge. IPK is dynamic as environments change and so do people's relationships with the environment. IPK is living knowledge and is presently adapting to anticipated climate change.

Bush burns ease global warming

When Europeans first encountered Australia, they saw a continent ablaze as Aboriginal 'firestick farmers' lit up the bush with controlled burns that prevented destructive wildfires. Now a hi-tech version of the land management practice, which is possibly tens of thousands of years old, could form part of Australia's strategy to tackle a modern problem: global warming.

www.theaustralian.com.au

ISBN 978 1 4586 6277 4

North Australia: Indigenous knowledge

In north Australia, Indigenous Australians are responsible for managing over 45% of the land and have interests over the entire area. The North Australian Indigenous Land and Sea Management Alliance (NAILSMA) philosophy 'Looking after our country … our way' supports the practical application of IPK to sustainably manage land and sea resources. Combining IPK with science and research, NAILSMA delivers effective programs, such as:

- *Indigenous Tracker or I-Tracker*—natural resource monitoring, research and management activities using digital technology. Over 700 Indigenous rangers collect data which provides governments and environmental decision-makers with information.
- *Saltwater People Network*—connects traditional owners who share land and sea management knowledge (e.g. being responsible for the migration paths of marine turtles across north Australia).
- *Water Resource Management*—supports the Indigenous Community Water Facilitator Network, which works with traditional owners to ensure their interests are incorporated into water policy and water allocation decisions.
- *Livelihoods Program*—has resulted in a rise in the number of Indigenous-owned businesses based on IPK (e.g. bush plants for shampoos, harvesting of crocodile eggs to produce crocodile skins, and bark and wood for the arts industry).
- *Carbon Program*—promotes IPK as integral to land and fire management and supports Indigenous land managers to establish their own carbon abatement projects.

Indigenous Peoples living in the tropical savannas of north Australia burn the landscape to create better hunting areas and increase the production of resources. Aboriginal Peoples are aware that burning fires in the early, cooler part of the dry season prevents uncontrollable wildfires in later, hotter part of the dry season.

4.18.3 I-Tracker has been supporting Indigenous ranger groups to collect and manage information about sea turtles since 2009

Geo**info**

The Australian Department of Agriculture, Fisheries and Forestry is responsible for funding farms to reduce greenhouse gas emissions.

Geo**activities 4.18**

Knowledge and understanding

1. What is meant by 'Country' to Aboriginal and Torres Strait Islander Peoples?
2. Why does IPK include past and continuing knowledge?
3. What is meant by the NAILSMA philosophy 'Looking after our country … our way'?
4. Discuss how NAILSMA aims to deliver Indigenous knowledge to improve management of landforms (e.g. plains) and landscapes (e.g. savannas) in north Australia.

Inquiry and skills

5. Refer to 4.18.1.
   a. What is meant by an earth-centred worldview?
   b. In groups discuss whether landforms and landscapes should be mined, built on or left as wilderness. Explain your answer in an oral report.
6. Refer to 4.18.2.
   a. Indigenous Australians have their own seasonal calendars based on the local landscape. How many seasons are in this calendar?
   b. For hundreds of generations, Indigenous Australians have been using fire in the landscape. List the seasons when grass is burnt and the reasons behind this decision.
   c. Explain the use of resources in the local environment and why they should be managed sustainably.

ISBN 978 1 4586 6277 4

4.19 Management of Indigenous land: Kakadu

In the Northern Territory, 50% of the land and 85% of the coastline belongs to Aboriginal Peoples. The *Aboriginal Land Rights (Northern Territory) Act 1976* protects traditional owners of Aboriginal land, and the Northern Land Council (NLC) ensures that traditional landowners give informed consent before any action is taken that affects their land and sea.

Kakadu National Park in the Northern Territory was inscribed on the UNESCO World Heritage List in 1981. The park contains a diversity of landforms, such as Koolpin Gorge, Mamukala Wetlands, Yellow Water Billabong and Jim Jim Falls. However, within Kakadu National Park are also located uranium mines, such as Jabiluka and Ranger. As mining accounts for 80% of the Northern Territory's income, the mining of mineral-rich landforms on Indigenous land has led to heated campaigns concerning Indigenous rights to the land and their share in mining wealth. The Mirrar and other Bininj (Aboriginal) peoples have struggled over their cultural and spiritual attachments and legal rights to their land, which were covered by Jabiluka and Ranger mineral leases.

4.19.1 Location of Kakadu National Park showing mining projects

4.19.2 Ranger Uranium Mine lies within Kakadu National Park

Jabiluka Uranium Mine

Exploration for uranium at Jabiluka began in the late 1960s, with uranium discoveries in the 1970s. The land where Jabiluka was located is communally owned by the Mirrar people who lived as hunters and gatherers surviving on food such as wallabies, fruit, fish and lily bulbs.

In 1982 the Mirrar people were granted a claim to their traditional land, with the right to lease land to individuals, governments and corporations. However, in 1981, Indigenous leaders had agreed to allow a mining company (Pancontinental Mining) to construct an underground uranium mine at Jabiluka. Yvonne Margarula, the senior traditional owner of Mirrar land, declared that her father was coerced into signing the agreement.

The Mirrar people regretted signing the agreement, and argued that the Jabiluka mine is the location of culturally significant artefacts and rock art. Additionally, mineral leases cut across the Country containing Dreaming tracks of the Mirrar people, and were a threat to their living culture. The Mirrar people presented their arguments to the UNESCO World Heritage Committee and an agreement was made to halt mining.

ISBN 978 1 4586 6277 4

4.19.3 Yvonne Margarula is photographed near a monitoring station on Magela Creek, which flows into the Kakadu National Park in the Northern Territory. Millions of litres of radioactive water from Rio Tinto's Ranger uranium mine have flowed into Kakadu National Park wetlands

Traditional knowledge and management

Kakadu's wetlands are important hunting grounds for water birds and for gathering edible water plants. When feral Asian water buffalo were removed from the wetlands, the native grass mudja spread and choked other wetland plants. Aboriginal Peoples used fire to control mudja, using traditional ecological knowledge of vegetation, water, weather and fire behaviour. The CSIRO applied this knowledge to landscape management in its Burning for Biodiversity project.

4.19.4 View of the sandstone escarpment in Kakadu National Park

The Kakadu National Park Management Plan 2007–2014 says that Bininj (traditional owners), the Kakadu Board of Management and the Director of National Parks and Parks Australia will share knowledge to manage Kakadu. The plan stated that 'Maintaining healthy landscapes will help to maintain the World Heritage recognised conservation values of the Park'.

Geo**info**

The Gagudju people believe the sandstone escarpment that dominates the park's landscape was created in the Dreamtime by Ginga, the crocodile–man.

Geo**activities 4.19**

Knowledge and understanding

1 What percentages of land and coast in the Northern Territory belong to Indigenous Australians?

2 Describe the conflict between the mining company and the Mirrar people.

3 Discuss how traditional fire management can enhance biodiversity and cultural values of wetlands.

Inquiry and skills

4 Refer to 4.19.1, 4.19.2 and the text.

  a Name three uranium projects in Kakadu National Park.

  b What are the impacts of mining on Indigenous culture?

  c Uranium waste from mining has the potential to leak and pollute rivers and soil—a source of traditional Indigenous food (see Geolinks). Draw a diagram showing the links between mining and the environment.

5 Refer to 4.19.4.

  a Draw an annotated line drawing of the landforms.

  b Kakadu National Park contains a variety of landscapes that have been occupied by Aboriginal Peoples for over 40 000 years. Design and annotate a collage of different landforms and landscapes in Kakadu National Park.

  c Research the use and significance of landforms in Kakadu to Indigenous peoples.

6 Conflicts between Indigenous landowners and miners have increased since the Mabo High Court decision recognised the native title rights of traditional Aboriginal landowners. Using the internet, collect two newspaper articles covering this issue.

ISBN 978 1 4586 6277 4

4.20 Indigenous Peoples manage land and sea

In Dreaming stories, Aboriginal and Torres Strait Islander Peoples tell of the creation of Australian landforms and landscapes, such as coasts, wetlands, forests, coral reefs, deserts, grassland plains and mountains. With Indigenous People's continuous relationship with the land, IPK is incorporated into most current land management programs.

The Indigenous estate, covering over 20% of the continent, includes areas of cultural value and globally significant biodiversity. The condition of the estate has changed since colonialism and it now faces environmental threats from mining, tourism, agriculture, deforestation, land and water degradation, and climate change.

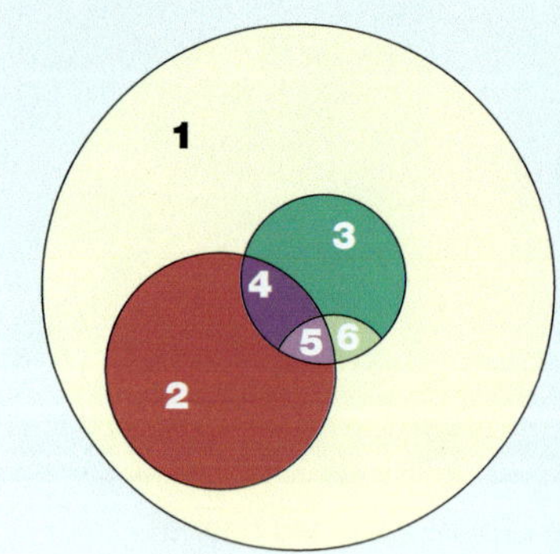

4.20.2 Relationship between Indigenous estates and conservation estates

Responsibilities for land

In 2009, the Australian Government supported the UN Declaration on the Rights of Indigenous Peoples. Article 25 states that 'Indigenous Peoples have the right to maintain and strengthen their distinctive spiritual relationship with their traditionally owned or otherwise occupied and used lands, territories, waters and coastal seas and other resources and to uphold their responsibilities to future generations in this regard'.

Indigenous Peoples have generally been excluded from decision-making processes that directly impact on their lives and livelihoods. As Aboriginal communities depend on different landforms and landscapes (e.g. Kakadu, Uluru, the Great Barrier Reef and Snowy Mountains) for their survival, a 'one-size-fits-all' environmental management approach is unsuitable. However, there has been a growth in incorporating Indigenous knowledge into modern management, such as in the Mt Kosciuszko National Park, where the Australian Alps Traditional Owners Reference Group (AATORG) advises the Australian Alps Liaison Committee on the management of the park.

Some Indigenous sites of international significance are protected as World Heritage sites, while others are protected as national and state heritage sites. Indigenous Protected Areas (IPAs)

Caring for Country programs engage Indigenous Australians to protect and conserve the land. Programs include Working on Country, Reef Rescue Indigenous Land and Sea Country Partnerships and the Indigenous Emissions Trading commitment.

Marine reserves—in 2012, the Australian Government established 40 new marine reserves, adding over 2.3 million km^2 to Australia's marine reserves estate, resulting in a total area of 3.13 million km^2 of ocean.

Sea Country Plan coordinates the priorities of Indigenous communities with other interested parties in sea country.

Indigenous Protected Areas (IPAs) are Indigenous-owned land and sea landscapes and landforms. There are 50 declared IPAs. The areas provide training and employment for Aboriginal and Torres Strait Islander Peoples and helps close the gap between Indigenous and non-Indigenous people.

Indigenous Land Management Facilitators support governments, Landcare and community groups in delivering the programs.

Working on Country programs engage Indigenous Australians to improve management of fire, feral animals, exotic weeds, threatened species and coastal and marine ecosystems.

4.20.1 Indigenous people's traditional knowledge is integrated into management of landforms and landscapes (photograph shows the Great Barrier Reef)

ISBN 978 1 4586 6277 4

have been established and Marine Parks aim to conserve natural resources in sea country.

Yolngu people

The Yolngu people living in north-east Arnhem Land have a worldview that connects people to the land and sea. They have traditionally used art, such as bark paintings, to articulate these interconnections. In 2013 about 4000 km^2 of ocean was added to the Dhimurru IPA in north-east Arnhem Land. This encompassed the Yolngu people's sea country.

In 1963, the government's grant of mining rights on land reserved for Aboriginal Peoples caused Yolngu people to dispatch bark petitions to the Australian Government. The bark petitions became a catalyst for the 1967 referendum, which amended the constitution to include Aboriginal Peoples in future censuses.

4.20.3 Yolngu Sea Country illustrated by different maps. Left: bark map of sea country showing flow of water and energy around the mouth of the Baraltja River. Right: conventional map showing clan ownership of water and where ancestral beings travelled

Great Barrier Reef

Australia boasts the world's largest marine park: the Great Barrier Reef Marine Park (GBRMP). While it is a big win for conservationists, it means that certain activities (including gas exploration, commercial fishing and recreation) are limited.

Indigenous peoples sustainably used and managed their traditional sea country, including the area covered by the GBRMP. Due to their long association with the region and their vast IPK knowledge of marine environments, their management expertise is included in current programs.

Today Indigenous Peoples comprise 5.7% of the Great Barrier Reef coastal population. Their activities are managed under Federal and Queensland legislation, Traditional Use of Marine Resource Agreements (TUMRAs) and Indigenous Land Use Agreements (ILUAs). Additionally the GBRMP aims to support traditional owners to maintain cultural and spiritual connections with their sea country.

Geoactivities 4.20

Knowledge and understanding

1. Explain the significance of the UN Declaration on the Rights of Indigenous Peoples.
2. Discuss why a 'one-size-fits-all' environmental management approach is not suitable for all landforms and landscapes across Australia.
3. How does polluted water in the Great Barrier Reef impact on Indigenous and non-Indigenous communities?

Inquiry and skills

4. Refer to 4.20.1 and discuss how Indigenous Peoples play an important part in managing land and sea landforms and landscapes.
5. Refer to 4.20.2.
   - a List the proportion of Australia's land that covers (i) Indigenous National Estate, (ii) IPAs and (iii) Indigenous-owned and jointly managed.
   - b The Register of the National Estate is a list of natural, Indigenous and historic heritage places. Why are the places significant and how should they be managed?
6. Refer to 4.20.3 and Geolinks.
   - a Who are the Yolngu people and where do they live?
   - b What have been the achievements of the Yolngu people over the past 50 years?
   - c Explain the differences between the two maps.

ISBN 978 1 4586 6277 4

4.21 Torres Strait: managing sea and climate change

The Torres Strait consists of a shallow body of water located between northern Australia (Cape York Peninsula) and Papua New Guinea (Western Province). Here, where the Arafura and Coral seas meet, are located different landforms such as islands, sandbanks, mudflats and coral reefs. Approximately 274 islands lie in the Torres Strait, ranging in size from Prince of Wales (Muraleg), which has a 23 km diameter, to tiny coral isles less than a hectare.

Sahul and climate change

During the last ice age, the landmasses of mainland Australia, Tasmania and New Guinea were joined, forming a single continent called Sahul, or Australia–New Guinea. About 12 000 years ago the sea level rose and covered the land bridge joining the two countries. As a result, Torres Strait was formed and high mountains protruding above water evolved into islands, such as Waibene (Thursday) and Muralag (Prince of Wales). Extinct volcanoes formed the Mer (Murray) and Erub (Darnley) islands. As sea levels continue to rise, islands such as Masig are already experiencing an increase in flooded areas.

Management

The coastal and marine environments located in Torres Strait are known as Saltwater country, and the traditional owners as Saltwater People. The links between these people and their sea country is governed by Ailan Kastom (Island Custom), which

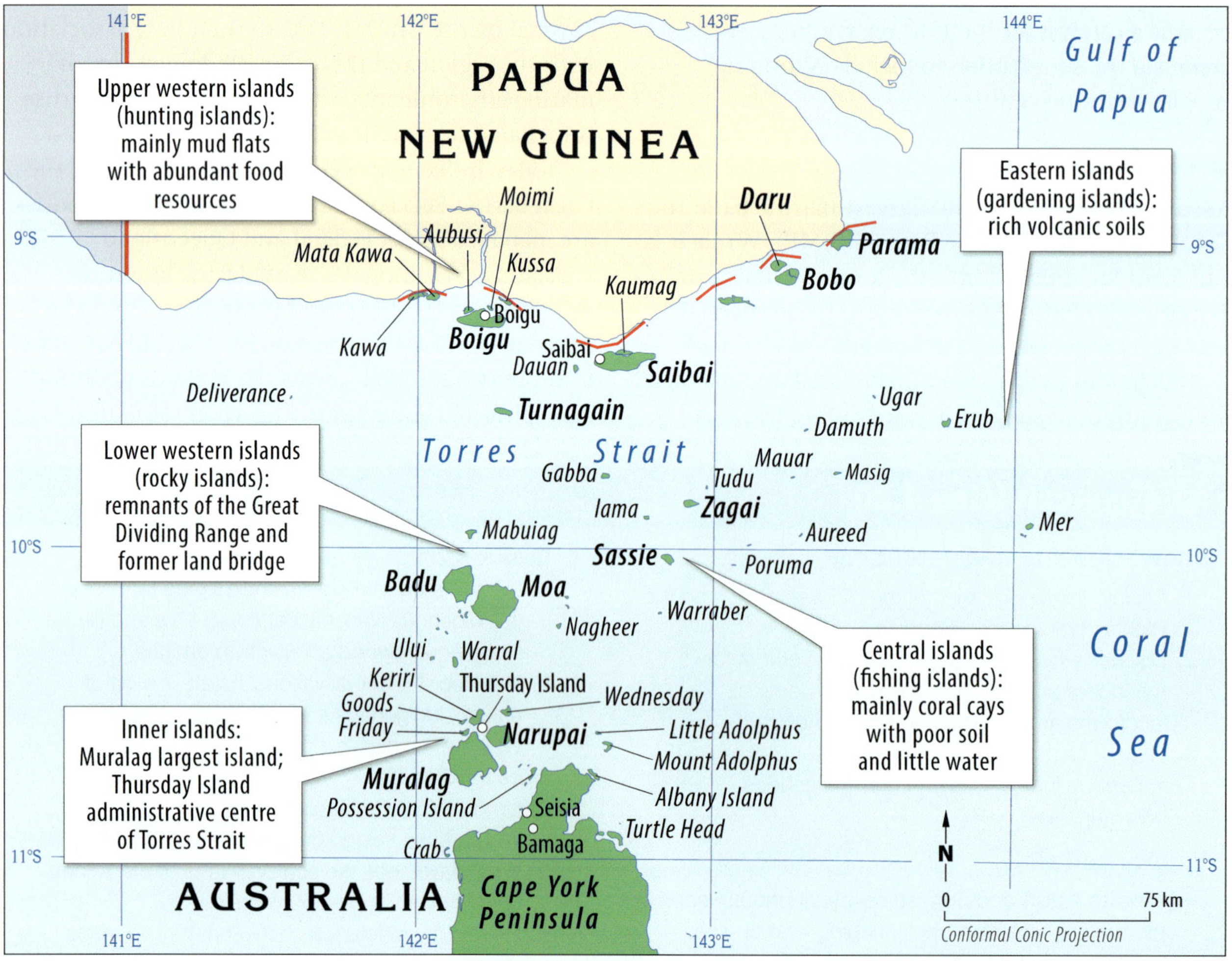

4.21.1 Map of Torres Strait illustrating a diversity of landforms and indigenous groups

ISBN 978 1 4586 6277 4

covers a range of customary practices, languages, traditions and beliefs. The management focuses on the interconnections of land and sea landscapes. For example, polluted water from land destroys fishing grounds, which are important for the Saltwater Peoples' material, cultural and spiritual wellbeing.

Fittingly, many Torres Strait indigenous groups possess totems of sharks, turtles and dugongs. A person's totem, worn as a pendant during ceremonies, acknowledges their responsibility to care for particular species, and hunting or eating a totem animal may be forbidden.

Torres Strait Islander Peoples consider the sustainable management of their land–sea territories essential to their survival. The management of the Torres Strait is covered in the 1978 Treaty, which allows Indigenous Peoples from northern Australia and Papua New Guinea to move freely within the area for traditional activities, such as fishing. The Indigenous Fisheries Advisory Committee (IFAC) recognises customary and limited commercial indigenous rights to marine resources in Torres Strait to ensure sustainable fish supplies.

4.21.4 Traditional clothing of the Sabai Island

4.21.2 A changing coastline: this image shows the Sahul coastline 25 000 years ago when the sea level was 135 m lower than today. The yellow lines show the present-day coast line

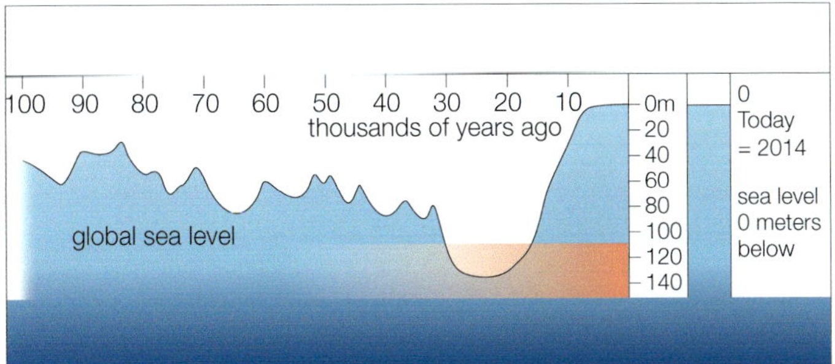

4.21.3 Sea level changes 100 000 years ago until 2014

Geoactivities 4.21

Knowledge and understanding

1 Where is the Torres Strait located?

2 What is Sahul?

Inquiry and skills

3 Refer to 4.21.1. Describe the activities of Torres Strait Islanders in different landscapes.

4 Refer to 4.21.2. Imagine you were migrating from mainland south-east Asia (Sunda) to Sahul (Australia–PNG). Explain the differences in your journey about 50 000 years ago compared with today.

5 Refer to 4.21.3.
   - a Estimate the years when the sea level was highest and lowest.
   - b As sea levels are constantly changing, outline how these changes have impacted on Torres Strait landforms in the past and in the anticipated future.

6 Reflect on your learning and design an annotated collage of three different Torres Strait islands. For each, include its name, a significant landform, and the indigenous groups managing the land and sea. Present findings using web 2.0 tools.

ISBN 978 1 4586 6277 4

4.22 Managing your footprint

Walking is beneficial for health and provides people with an opportunity to value the aesthetic beauty of the surrounding landscape. However, the increased popularity of walking has led to excessive numbers of people travelling along the same pathways, causing soil erosion, sedimentation in rivers and loss of biodiversity.

United Kingdom: National Parks

In 1951, in the UK 15 National Parks were created to protect the landscape and limit development. The Lakes District is England's largest National Park, and includes the country's highest mountain, Scafell Pike, and deepest lake, Wastwater. The park receives 12 million visitors a year. With so many feet pounding the ground, there are huge soil erosion scars, visible kilometres away. It is not only aesthetically unsightly but has led to habitat loss and sediment in lakes. The most popular areas in the Park are called 'honeypot' sites as they attract tourists like bears to honey. These areas experience greater footpath erosion because of the sheer numbers of people walking popular pathways.

Park authorities have implemented projects to repair erosion, such as revegetating paths with hardy plants and filling paths with stones. The challenge is to encourage sustainable tourism without damaging the landscape that visitors came to enjoy.

Boardwalks and skyways

Boardwalks and skyways are constructed to reduce the impact of erosion from walkers and runners in locally protected places. Boardwalks are located along intertidal zones, such as wetlands, and along coasts (oceanways) and rivers (riverwalks).

Boardwalks are found around the world, and they are particularly common along the East Coast of the USA. In Australia, the Gold Coast Oceanway is a 36 km foreshoreway stretching from the Gold Coast Seaway to Point Danger.

In the Yukon–Kuskokwim Delta, Alaska, USA, there are almost no roads and travel is by bush planes, and by river boats in summer and snow machines in winter. To travel between villages, 3 m wide heavy-duty boardwalks have been constructed to protect the **tundra**, which is part of the Yukon Delta National Wildlife Refuge.

4.22.2 Mangrove walkways in Australia

1 Area initially has good vegetation cover, and roots are keeping soil particles together.

2 Walkers cause soil to compact and rain infiltration is reduced. A small gully forms and vegetation starts to die.

3 Further walking causes more vegetation loss and the gully is enlarged. More soil is removed by wind and rain, exposing rocks.

4 The gully deepens further and walkers trample the grass beside the path, causing further damage.

4.22.1 Stages in footpath erosion

ISBN 978 1 4586 6277 4

4.22.3 Location of Australian mangrove boardwalks

Scenic skyways are a popular way of viewing forest habitats and difficult-to-access locations. Some people argue walkers are a minor cause of soil erosion compared with the braking and skidding of mountain bikes and the overgrazing by livestock, which have greater impacts on soil erosion. As a cause of soil erosion, walking has less impact on the environment than mining and deforestation.

4.22.4 A skyway over the tropical forests between Cairns and Kuranda in North Queensland

Geo**activities 4.22**

Knowledge and understanding

1 Describe how walking and running causes landscape degradation.
2 Explain the term 'honeypot sites' and how they affect soil erosion and sedimentation.
3 List different structures built to minimise erosion by walkers.
4 Explain the advantages of boardwalks and skyways in relation to soil erosion.

Inquiry and skills

5 Refer to 4.22.1 and describe how humans cause footpath erosion.
6 Refer to 4.22.2 and 4.22.3.
   a What is a wetland (see 1.22)?
   b Why are wetlands important?
   c How can boardwalks protect wetlands?
7 Refer to 4.22.4. What are the environmental advantages of skyways?
8 In groups list areas affected by erosion from walkers in your school and local area. Suggest strategies to protect these areas.
9 When walking in the forest or along shorelines, why should people stay on designated trails? What are the problems of short cuts?
10 Research the extent of boardwalk pathways on the Overland Track in Tasmania. What effect has this protection had on the environment and walkers?
11 Reflect on your learning and suggest individual and collective actions to reduce damaging human footprints.

ISBN 978 1 4586 6277 4

Geothink

Restructured landscapes

1 Refer to 4.23.1.
  a How have humans restructured landscapes?
  b Discuss how humans change landscapes by mining, industry, agriculture, urban activities and waste disposal.
  c What should be done to protect land and marine environments?

Nickel tailings (waste from mining) floating in northern Mindanao, Philippines

Discarded tyres stacked in piles in Phoenix, Arizona, USA

Ship breaking for scrap metal recycling in Bhavnagar, Gujarat, India

Favela in Rio de Janeiro, Brazil

Tulip fields in the Netherlands

Heavy industry in Teesside, England

4.23.1 Humans change landscapes

2 Refer to 4.23.2. Discuss how humans change landscapes to improve the environment.

The Netherlands contains an impressive display of over 600 wildlife crossings (including underpasses and ecoducts) that have been used to protect wild boar, red deer, roe deer and the endangered European badger.

Singapore's ecological EDITT tower. The 26-storey high-rise will increase biodiversity and rehabilitate Singapore's 'zeroculture' metropolis

4.23.2 Humans change landscapes to make them more sustainable

ISBN 978 1 4586 6277 4

Rural landscapes

3 Refer to 4.23.3.

a Using ICT research how different rural landscapes vary across Earth (see Geolinks).

Minnesota, United States

Gulf of Fonseca, Honduras

Queensland, Australia

4.23.3 Satellite and aerial views of rural landscapes across Earth

What did I learn?

4 Refer to 4.23.4.

a Answer the key geographical questions.

b Geographers recognise all elements on Earth are *interconnected*. How does this term shape our thinking?

c Using relevant sources, research a significant landscape threatened by human activities in your local area or in another part of Australia. What are the different views on its use and its protection?

d Discuss how Aboriginal and Torres Strait Islander Peoples use and manage landscapes. Present your findings as a Prezi, including a map and photographs.

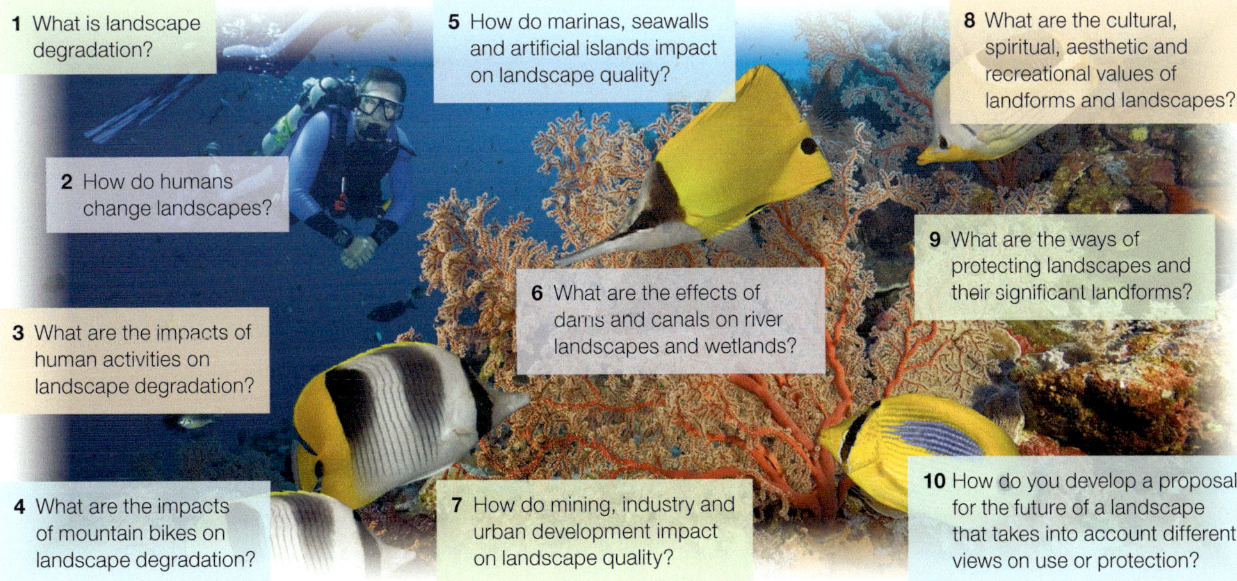

4.23.4 What did I learn?

ISBN 978 1 4586 6277 4

chapter 5 Geomorphic hazards: causes, impacts and responses

> 'No man is an island and everyone is a part of the whole.'
>
> John Donne, poet

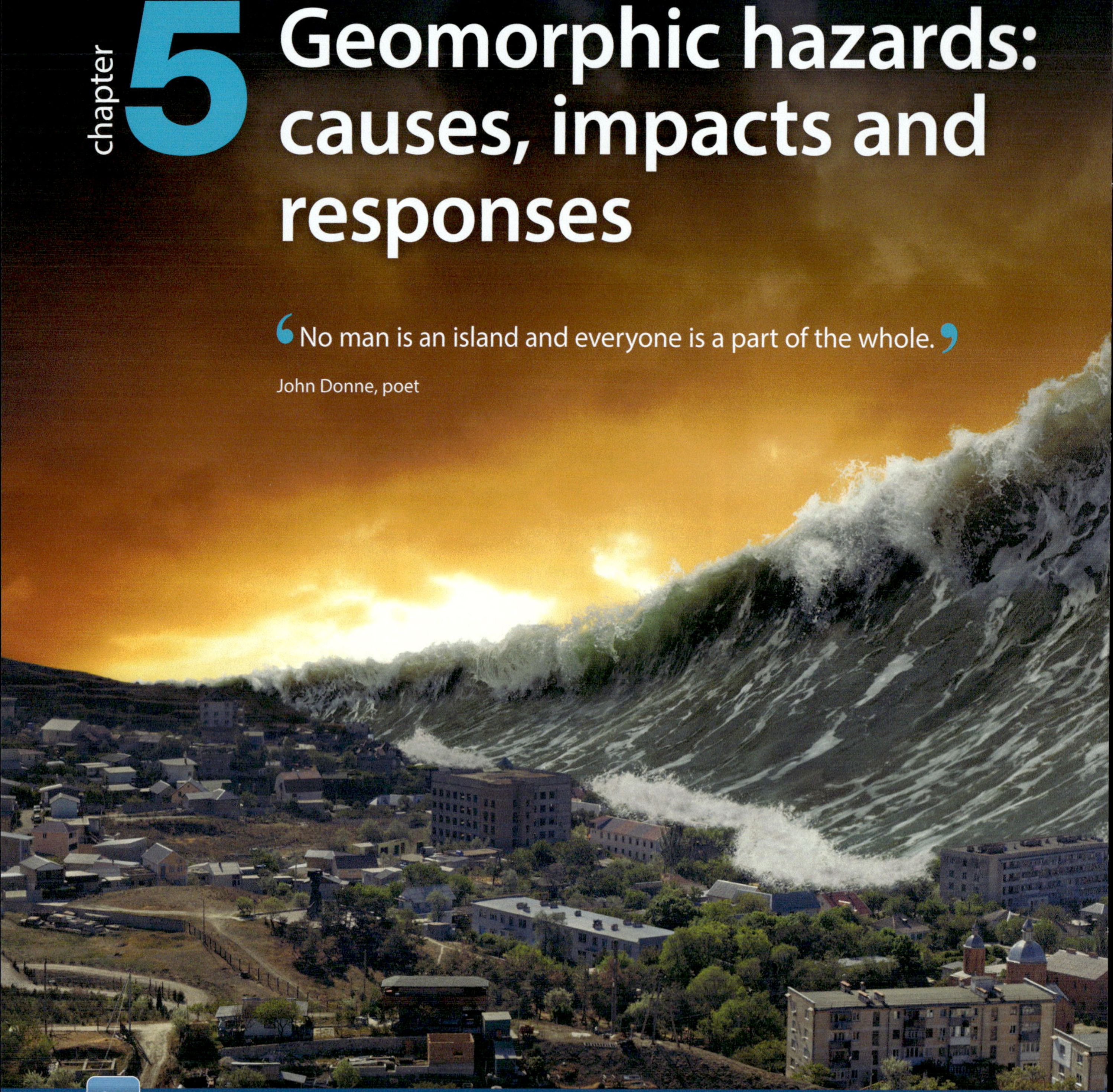

Geovocab

avalanche: snow, ice and rock falling rapidly down a slope

disaster: event caused by a hazard (e.g. a volcanic eruption) that results in loss of life and livelihoods and damage to property

earthquake: sudden shaking of the ground as a result of movements within Earth's crust

epicentre: point on the ground directly above the hypocentre of the earthquake

hazard: possible source of danger (e.g. an earthquake or landslide)

hypocentre: underground point or origin of an earthquake

inter-plate earthquakes: occur at plate boundaries

intra-plate earthquakes: occur in the middle of plates

landslide: mass movement of rock and soil down a slope or cliff

limnic eruption: dissolved carbon dioxide (CO_2) suddenly erupting from a deep lake

Richter scale: scale providing a measurement of the energy released during an earthquake

tsunami: series of large waves generally caused by an underwater earthquake, landslide or volcanic eruption

ISBN 978 1 4586 6277 4

Hazards (such as volcanic eruptions, earthquakes, tsunamis, landslides, avalanches and bushfires) occur across Earth every day. A hazard becomes a disaster when it causes death, loss of livelihood and damage to property. The impact of damage from hazards is influenced by social, cultural and economic factors. For instance, people in poor countries lack the money to build earthquake-resistant homes, and population pressure forces many to live in slums located on steep slopes subject to frequent landslides.

The impact of a disaster depends on the number of people and properties affected, the frequency of the event and whether recovery responses are quick and adequate. As disasters are unpredictable, preventative measures from the local to the global scale aim to minimise losses.

Think, puzzle, explore

- **Place:** What is the impact of an avalanche on a ski resort or a landslide on a town?
- **Space:** What is the spatial distribution of earthquakes and volcanoes?
- **Environment:** How can heavy rain from a cyclone contribute to a geomorphic disaster, such as a landslide?
- **Interconnection:** How can a volcanic eruption in Iceland impact on people living in other countries?
- **Sustainability:** How can technology contribute to economic and social sustainability by preventing and responding to geomorphic disasters?
- **Scale:** How can a tsunami impact on local communities as well as countries on the other side of the world?
- **Change:** How does an earthquake-generated tsunami change landscapes and people?

Geo**skills** in focus

- **Identifying** and **planning** investigations into the causes, impacts and responses of a geomorphic hazard
- **Collecting** information from a variety of sources
- **Recording, analysing** and **concluding** using a variety of maps; pie, line and column graphs; statistics; tables; cartoons; vertical and oblique photographs; simple and 3D diagrams; and satellite images showing changes over time
- **Communicating** information using web 2.0 tools
- **Reflecting** on management of hazards and **responding** by proposing collective action

Photo mock-up of a hypothetical tsunami overwhelming an urban area

Volcanic Explosivity Index (VEI): indicates how much volcanic material is thrown out of an erupting volcano, to what height, and how long the eruption lasts

volcano: mountain or hill with a crater or vent through which lava, rock fragments, gas and ash are expelled

ISBN 978 1 4586 6277 4

5.1 Geomorphic hazards

Hazards are a potential threat to life and property caused by natural and human forces. A **hazard** evolves into a **disaster** when it causes death, loss of people's livelihoods and destruction of properties. In 2003, a magnitude 6.6 earthquake shook Bam in Iran. It killed over 31 000 people, destroyed 85% of the buildings and caused landslides. More than 75 000 people were displaced.

5.1.2 Bam—a World Heritage Site—before and after the earthquake

Types of geomorphic hazards

Geomorphic hazards occur naturally or are caused by humans. In 1997 at Thredbo, in NSW, 10 000 tonnes of soil moved downhill from the road called the Alpine Way. As the soil moved, it killed 18 people and destroyed ski resorts in its destructive path. The causes of the landslide were a combination of natural and human causes:

- *natural*—heavy rain causing soil creep
- *human*—clearing the slope for the Alpine Way and leakage from a water main.

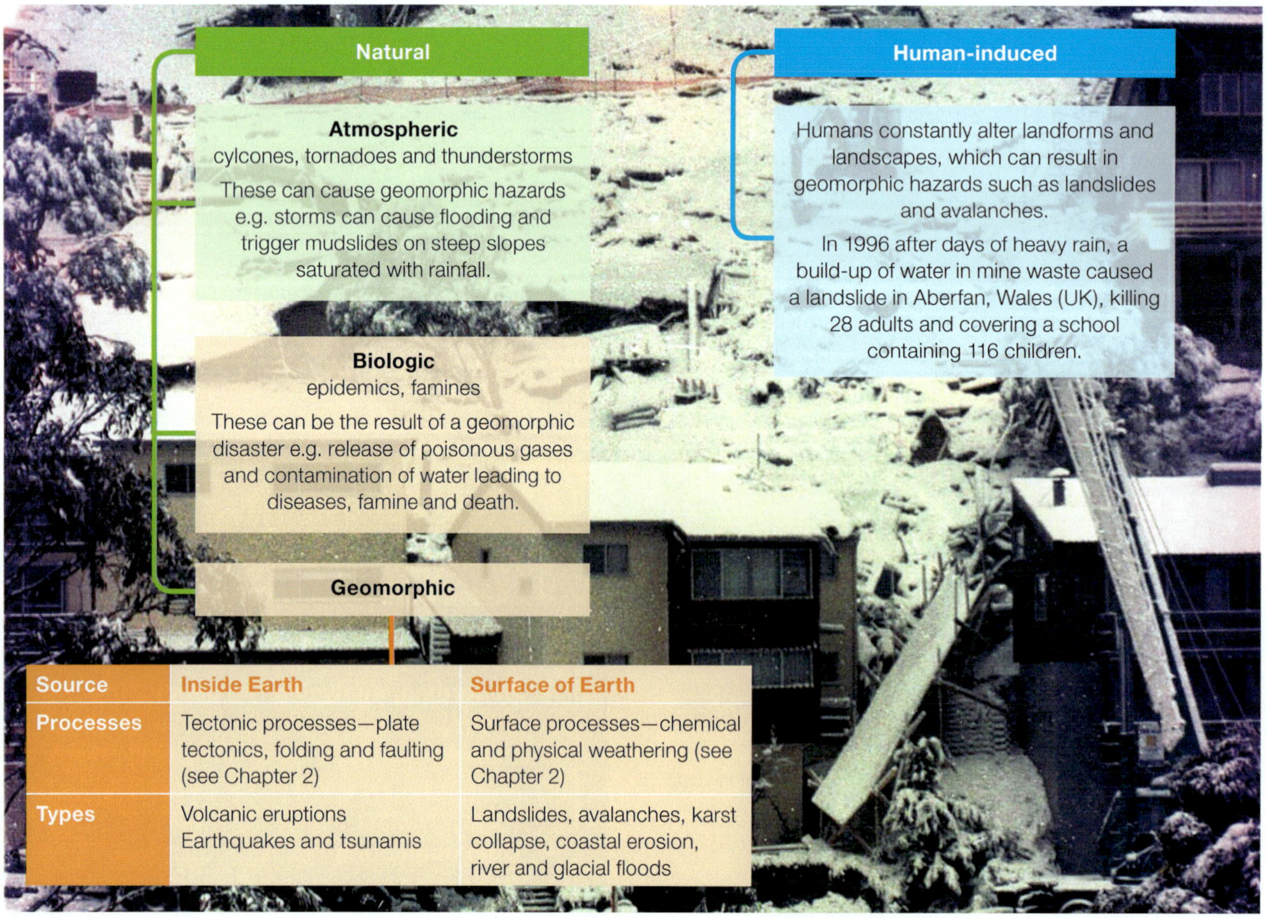

Natural

Atmospheric
cylcones, tornadoes and thunderstorms

These can cause geomorphic hazards e.g. storms can cause flooding and trigger mudslides on steep slopes saturated with rainfall.

Biologic
epidemics, famines

These can be the result of a geomorphic disaster e.g. release of poisonous gases and contamination of water leading to diseases, famine and death.

Geomorphic

Source	Inside Earth	Surface of Earth
Processes	Tectonic processes—plate tectonics, folding and faulting (see Chapter 2)	Surface processes—chemical and physical weathering (see Chapter 2)
Types	Volcanic eruptions Earthquakes and tsunamis	Landslides, avalanches, karst collapse, coastal erosion, river and glacial floods

Human-induced

Humans constantly alter landforms and landscapes, which can result in geomorphic hazards such as landslides and avalanches.

In 1996 after days of heavy rain, a build-up of water in mine waste caused a landslide in Aberfan, Wales (UK), killing 28 adults and covering a school containing 116 children.

5.1.1 Types of hazards. Photo shows Thredbo after the landslide in 1997

ISBN 978 1 4586 6277 4

In 1998, the Alpine Way was rebuilt with retaining walls and the road monitored to detect slope movement.

Deadly hazards

In the last ten years, more people have been killed by earthquakes compared to any other natural disaster. Earthquakes are sudden and their impact is immediate. The adverse impacts of earthquakes are anticipated to continue, given that eight of the ten most populated cities—including Tokyo, New York, Shanghai and Mumbai—lie on fault lines.

5.1.3 Selected deadly geomorphic disasters since 1900

Year	Event	Deaths (estimated)
1902	Mount Pelée eruption, Martinique	29 000
1908	Messina earthquake/ tsunami, Italy	123 000
1920	Haiyuan earthquake, China	234 000
1923	Great Kanto earthquake, Japan	142 000
1931	Floods and landslides, China	1 000 000–4 000 000
1970	Huascarán avalanche, Peru	20 000
1976	Tangshan earthquake, China	242 000–779 000
2004	Indian Ocean earthquake and tsunami, Indonesia	240 000
2010	Haiti earthquake, Haiti	316 000

It's hard to believe that living beside a beautiful lake could be hazardous. In fact it can lead to disaster: a **limnic eruption**, or an 'overturning lake' occurs when saturated CO_2 suddenly erupts from the bottom of a deep lake and kills marine species. The resulting CO_2 cloud then moves over the land, suffocating humans and animals. In Cameroon, Africa, a limnic eruption caused the asphyxiation of 37 residents living around Lake Monoun in 1984, and then 1744 people around Lake Nyos in 1986. Volcanic activity in these two crater lakes released gas into the lakes. At present, the lakes are being degassed to avoid future disasters.

Geo**info**

Scientists believe a big deep limnic explosion beneath the ocean could cause mass extinction.

5.1.4 The 1984 limnic disaster in Lake Nyos, Cameroon, turned once clear blue water into murky brown

Geo**activities 5.1**

Knowledge and understanding

1 What is the difference between a hazard and a disaster?
2 Describe the causes and management of the landslide at Thredbo in 1997.

Inquiry and skills

3 Refer to 5.1.2.
   a List the three categories of natural hazards.
   b State whether the following landforms originated inside Earth or above the ground: earthquakes, landslides, volcanic eruptions and river floods.
   c Explain how atmospheric and biologic incidents are linked to geomorphic disasters.
4 Refer to 5.1.3.
   a List three geomorphic hazards that turned into disasters.
   b Earthquakes remain a future hazard. What does this mean?
   c Research the internet for statistics on these geomorphic disasters and determine why the figures may be unreliable.
5 Refer to 5.1.4.
   a Describe how living beside a lake can turn from a hazard to a disaster.
   b Nyos, Monoun and Kivu Lakes in Africa accumulate deadly amounts of dissolved CO_2 at great depths. In groups research the causes and impacts of saturated CO_2 and the implementation of preventative measures.

ISBN 978 1 4586 6277 4

5.2 Hazards: risks and vulnerability

Humans live in many locations at risk of geomorphic hazards. They built settlements on low-lying coasts (e.g. New York) and on small coral islands (e.g. Maldives), which are vulnerable to rising sea levels. In China, famers live on fertile alluvial plains produced by the annual flooding of the Yellow River. This hazardous location caused 800 000–4 000 000 deaths in 1931.

People residing in San Francisco, USA, live on top of the San Andreas Fault. This hazard became a disaster in 1906, when an earthquake occurred along the fault. The earthquake and its accompanying fires killed 3000 people and destroyed 80% of San Francisco.

The risk of a hazard turning into a disaster depends on where people live and how vulnerable they are to the impacts of geomorphic hazards.

Risk
probability of harmful consequences—loss of life, property and livelihood

=

Hazard
natural or human-induced events that could lead to a disaster

Vulnerability
degree to which a society is susceptible to the impact of hazards e.g. dangerous locations and population

Dangerous locations
a. near erupting volcanoes
b. on the Pacific Ring of Fire, fault lines and low lying areas subject to tsunamis

Population
a. living in developing countries that lack resources to prepare and respond to the disaster
b. poor, elderly, disabled, women and children
c. densely populated areas e.g. cities

5.2.1 Risk = Hazard × Vulnerability

Lives at risk

About 500 million people live on or close to active volcanoes. These vulnerable people are at risk from exploding rocks and poisonous gases. In 79AD the eruption of Mount Vesuvius, in Italy, buried the city of Pompeii under six metres of ash and pumice. In 1902 on the island of Martinique, the eruption of Mt Pelée destroyed the town of Saint Pierre.

Most victims of the Indonesian tsunami in 2004 were unaware of the warning signs of an approaching tsunami (such as the water line disappearing into the far ocean). The tsunami was caused by a large earthquake (magnitude 9.1) in the Indian Ocean—the third largest earthquake recorded on a seismograph. The tsunami killed an estimated 240 000 people in 14 countries, and the cost of damages reached $10 billion. To reduce the impact of future tsunamis, the Indian Ocean Tsunami Warning and Mitigation System detects seismological changes and provides warnings of approaching waves.

5.2.2 Impacts of the 2004 tsunami and the recovery ten years later

ISBN 978 1 4586 6277 4

The Economic and Human Impact of Disasters in the last 12 years

$ 1.3 trillion damage (USD) | 2.7 million affected | 1.1 million killed

Year	Damage (USD)	Affected	Killed
2000	46 billion	174 million	16 666
2001	27 billion	108 million	39 496
2002	52 billion	659 million	21 342
2003	69 billion	255 million	113 513
2004	136 billion	161 million	24 880
2005	214 billion	160 million	93 076
2006	34 billion	126 million	29 893
2007	74 billion	211 million	22 424
2008	190 billion	221 million	242 191
2009	46 billion	199 million	15 957
2010	131 billion	261 million	308 152
2011	363 billion	162 million	32 816

2002 Europe, South Asia; *2002* China; *2002* Iran; *2004* Indian Ocean; *2005* Kashmir; *2005* *Katrina*; *2007* *Sidra*; *2008* China; *2008* *Nargis*; *2010* Pakistan; *2010* Haiti; *2011* Japan

Source: United Nations Office for Disaster Risk Reduction

5.2.3 Economic and human impacts of disasters in last 12 years

Impacts of disasters

Over the last 20 years, natural disasters cost US$2.6 trillion. Earthquakes and tsunamis were the deadliest types of disasters, causing 750 000 deaths. Over this period:

- Asian countries experienced 2778 disasters affecting 3.8 billion people
- lower-income countries suffered 44% of disasters and 68% of deaths, compared to higher-income countries, which experienced 56% of disasters and 32% of deaths.

Geo**info**

- In 2013, 22 million people in 119 countries were displaced by natural disasters.
- In 2014 Asian countries experienced 48% of the disasters and accounted for 85% of those killed.

Geo**activities 5.2**

Knowledge and understanding

1 List the people at greatest risk from geomorphic hazards.

2 Explain why the victims of the 2004 tsunami were vulnerable.

Inquiry and skills

3 Refer to 5.2.1.
   a What is meant by risk?
   b Who are the most vulnerable people?

4 Refer to 5.2.3.
   a What disaster had the highest cost?
   b What disaster affected the largest number of people?
   c Name the three disasters that killed most people.
   d Write a TV news report on the economic and human impact of geomorphic disasters.

5 Draw a mind map of geomorphic hazards that have developed into disasters. Present the mind map as a Wordle.

6 Research the San Andreas Fault and how it evolved from a hazard to a disaster in the past. Suggest management strategies to reduce future disasters along the fault line.

Chapter 5

ISBN 978 1 4586 6277 4

5.3 Volcanoes: from hazard to disaster

Volcanoes are one of nature's deadliest killers. Breathtakingly beautiful yet devastatingly fatal ... They stand as permanent reminders of the fragility of the human race—constantly at the mercy of the unpredictable ruthlessness of the natural world.

Random History

Volcanic eruptions are one of Earth's most violent agents of change. They create, alter and destroy landscapes, modify climates and change people's lives. In 1883 Krakatoa, a volcanic island in Indonesia, erupted and exploded. It destroyed 66% of the island and caused a massive tsunami, killing around 37 000 people. The following year the global temperature fell by 1.2 °C, because particulate matter blocked the sun's rays.

Volcanoes release 130 million tonnes of CO_2 into the atmosphere each year, contributing to changes in climate. Eruptions cause death, and destruction of towns, farms, transport systems and electrical grids. Fortunately, most volcanoes show signs of unrest before the event, allowing communities at risk to be forewarned.

When hazards become disasters

Approximately 2500 active volcanoes dot the globe, and most are located near the edge of tectonic plates and around the Pacific Ring of Fire (see 2.2). Within seconds, active volcanoes can change from a hazard to a disaster causing widespread destruction. Their explosiveness is measured by the **Volcanic Explosivity Index (VEI)**. The VEI indicates the quantity of volcanic material thrown out, to what height it is thrown, and how long the eruption lasts. The indices run from 0 to 8, with 0 being non-explosive or gentle, to 8 being a mega explosion as experienced in Yellowstone (USA) 600 000 years ago. For comparison, the 1980 eruption of Mt St Helens (USA) was a VEI 5 and in 1883 Krakatoa was VE1 6.

Geoinfo

In AD 180 the Hatepe eruption in New Zealand was 50 km in height rated 7 on the VEI.

5.3.1 Types of material extruded from volcanoes and their impacts

ISBN 978 1 4586 6277 4

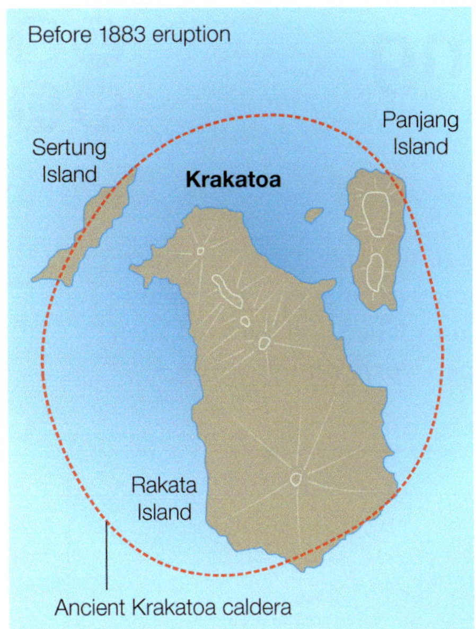

5.3.2 Krakatoa: volcano of destruction

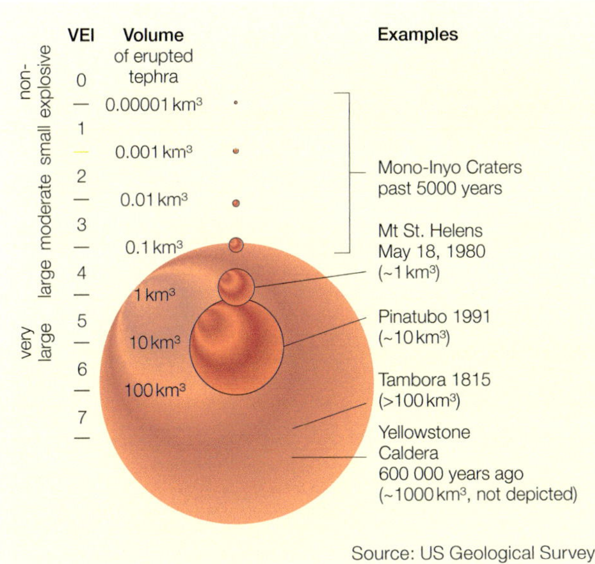

5.3.3 Volcanic Explosivity Index

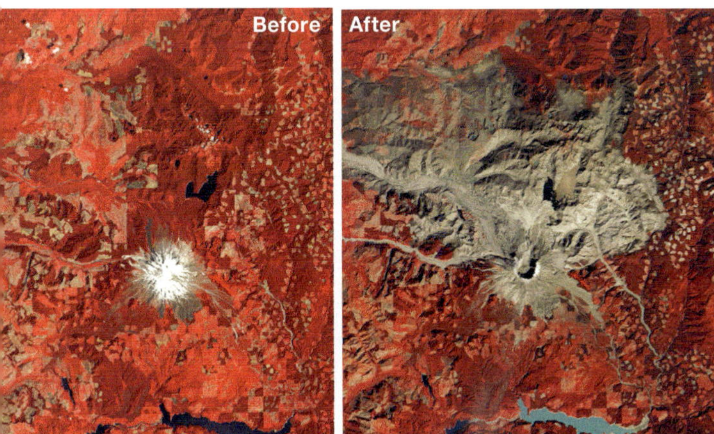

5.3.4 Satellite imagery: before and after the 1980 Mt St Helens eruption

Geo**activities 5.3**

Knowledge and understanding

1 What is a volcano?
2 Why are volcanoes a hazard?
3 Explain the Volcanic Explosivity Index (VEI).

Inquiry and skills

4 Refer to 5.3.1.
   a List five different types of material extruded from volcanoes.
   b List two materials that become air borne and are transported over long distances.
   c Name two extruded materials that cling to landforms.
   d What extruded material do you think is the most dangerous? Give reasons for your answer.
   e Active volcanos can change from a hazard to a disaster instantaneously. Discuss the impacts of volcanoes on people and places over time.
5 Refer to 5.3.2.
   a Compare the maps before and after the volcano.
   b Will subsequent undersea eruptions once again rebuild the island? Explain your answer
6 Refer to 5.3.3 and 5.3.4.
   a What was the VEI for Mt St Helens?
   b Describe how Mt St Helens changed after the eruption.
   c Imagine you lived near the volcano. Discuss its impacts on your life.
7 Research one recent volcanic eruption and discuss its causes and impacts on landscapes, people and places. Present using web 2.0 tools.

ISBN 978 1 4586 6277 4

5.4 Indonesia: our erupting neighbour

Indonesia consists of more than 13 000 islands, and is located near the junction of tectonic plates and the most volcanically active belt on Earth—the Ring of Fire. Indonesia contains around 400 volcanoes, of which 150 are active; this constitutes 75% of all active volcanoes on Earth.

Indonesia has a history of large eruptions, such as Krakatoa, but also:

- *Lake Toba, Sumatra*—a VEI 8 volcanic eruption occurred about 73 000 years ago. A caldera was formed which filled with water creating the largest volcanic lake in the world.
- *Mt Tambora, Sumbawa*—a VE1 7 volcanic eruption in 1815 killed 92 000 people. It also caused temporary changes to the climate, such as the volcanic winter in 1815, and the 'year without a summer' in North America and Europe in 1816. This was attributed to large quantities of volcanic particles, which decreased the amount of solar energy reaching Earth's surface. Crops failed and livestock died, resulting in Europe's worst famine in the 19th century.

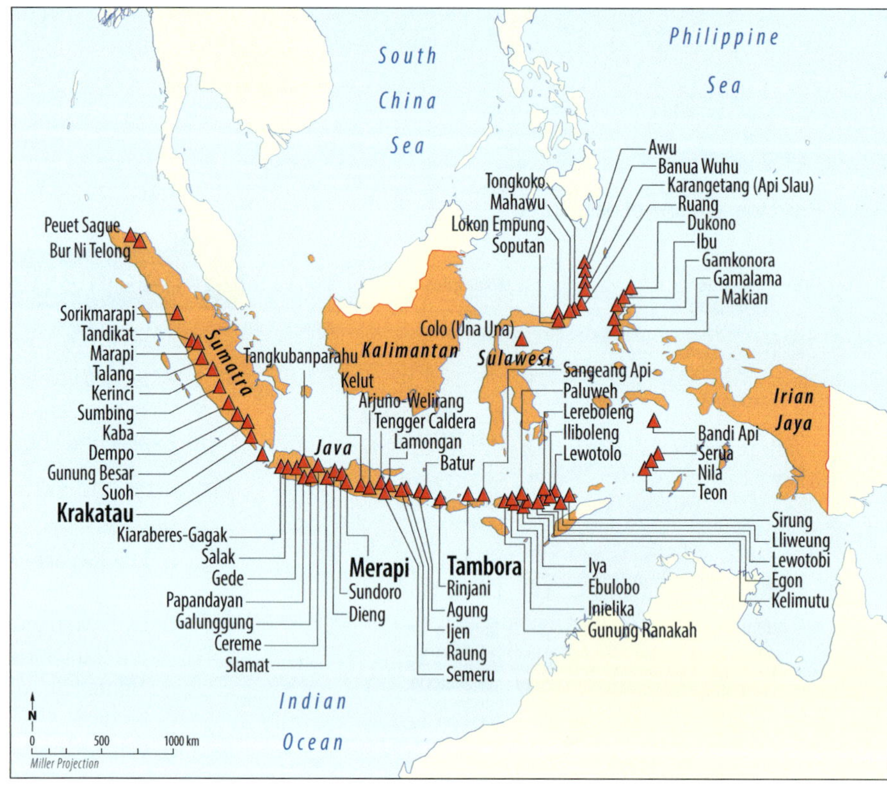

5.4.2 Location and pattern of major volcanic eruptions in Indonesia since 1900

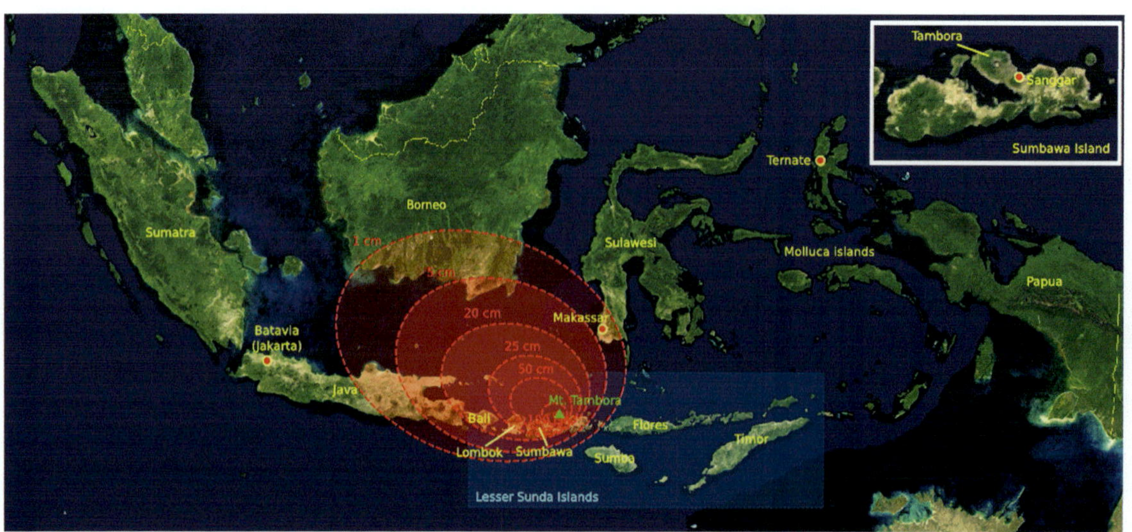

5.4.1 The calculated thickness of ashfalls following the 1815 eruption of Mt Tambora

ISBN 978 1 4586 6277 4

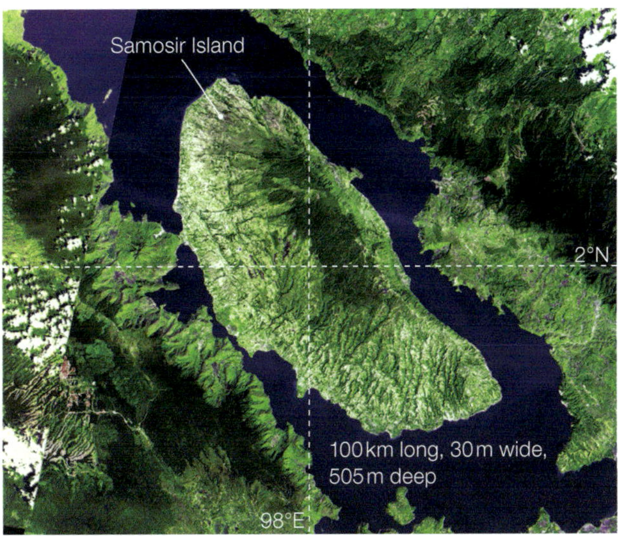

5.4.3 Satellite of the caldera of Lake Toba

Merapi: Fire Mountain

Merapi, or Fire Mountain, is an active stratovolcano located on the island of Java. It has erupted regularly since 1548. Its volcanic activity has spiritual significance for those living on its fertile slopes. Located near the large city of Yogyakarta, it is a threat to approximately 600 000 residents, as a new lava dome has been rapidly forming. Local authorities have banned the climbing of Merapi, halted sand mining at its foot and warned people living on its slopes to be prepared for evacuation.

5.4.4 Erupting Merapi

Australia's vulnerable flights

Beneath the aviation corridors linking Australia to Asia and Europe lies Indonesia. A cataclysmic eruption could cause major disruptions to international air traffic, tourism and trade. After Indonesia's 1982 Galunggung volcanic eruption, a British Airways flight from Kuala Lumpur (Malaysia) to Perth flew through the ash cloud. All four engines cut out and the plane descended more than 7000 m before the engines restarted.

Darwin has one of nine global ash monitoring centres that track volcanic activity and advise airlines of current risks around the world. In 2015, the Ruang volcano in Java erupted, ejecting clouds of dust into the atmosphere. The ash cloud disrupted air flights in and out of Bali.

Geo**info**

Mt Bromo (Indonesia) is sacred to the local Tenggerese people, who throw chickens and goats into the crater to appease the gods and bring good fortune.

Geo**activities 5.4**

Knowledge and understanding

1. Why does Indonesia have a large number of volcanoes?
2. Name three volcanoes in Indonesia.
3. Describe the impacts of the volcanic eruption of Mt Tambora.
4. Discuss why Australia is vulnerable to the impacts of Indonesia's volcanic eruptions.

Inquiry and skills

5. Refer to 5.4.1.
   a. Explain the extent of different ashfall thicknesses from Mt Tambora to Borneo.
   b. What are the impacts of ash on people and environments?
6. Refer to 5.4.2.
   a. List five major volcanic eruptions in Indonesia since 1900.
   b. Research the causes and impacts of one of these volcanic eruptions.
7. Refer to 5.4.3.
   a. What is the latitude and longitude of Lake Toba?
   b. How was Lake Toba formed?
8. Discuss the causes and impacts of volcanoes in Indonesia. Present as an annotated diagram.

ISBN 978 1 4586 6277 4

5.5 Ripple effect: Iceland's volcanic eruptions

Iceland lies on the Mid-Atlantic Ridge, which is the boundary between the Eurasian and North American tectonic plates. As these two plates move apart, magma rises along the ridge. This means Iceland is engaged in a geological race between the spreading motion that is moving the island apart, and volcanoes, which are building the island upwards.

Eyjafjallajökull

Eyjafjallajökull (pronounced ay-yah-FYAH-lah-yer-kuhl) or 'island-mountain glacier' is an Icelandic stratovolcano fed by magma. The volcano, which has been periodically active for the last 800 000 years, erupted on 14 April 2010, ejecting fine ash 10 km into the atmosphere. An estimated 750 tonnes of magma was ejected from the volcano every second.

The eruption took place under 200 m of glacial ice. By 20 April, most of the ice in the crater had melted. The local landscape was impacted as melting glaciers flooded low-lying agricultural land, and cattle inhaled ash, which caused internal bleeding. Lava destroyed crops and the thick layer of ash covering pastures affected the growth of crops and grazing of livestock.

The winds moved ash from Iceland towards Europe. By 16 April, thousands of flights were cancelled, leaving millions of passengers stranded around the world. Kenyans lost one million kilograms of fresh produce each night as they were unable to airlift their products to the European Union. This affected hundreds of thousands of Africans who depended on the export of horticultural products for their livelihood.

5.5.1 Satellite image of the ash cloud from Eyjafjallajökull

ISBN 978 1 4586 6277 4

Hazardous volcanic ash

Volcanic ash consists of small tephra, which are sharp-edged pulverised rock and glass fragments created by volcanic eruptions, ranging from 0.001 mm (comparable to talcum powder) to 2 mm (coarse grit). Volcanic ash causes breathing problems, alters weather patterns, threatens the health of livestock, and interrupts power generation and telecommunications.

5.5.2 Extent of volcanic ash from the 2010 Eyjafjallajökull eruption

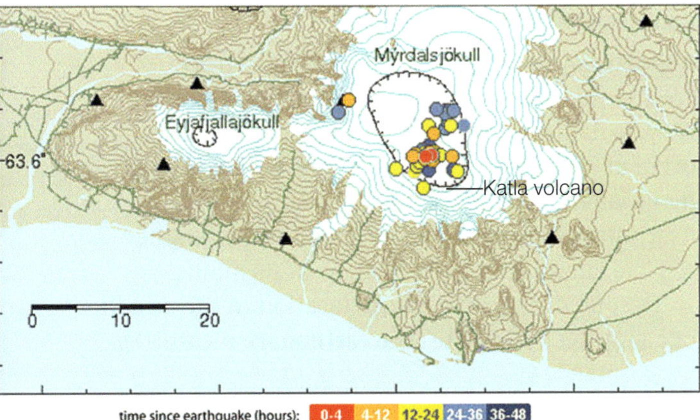

5.5.3 Topographic map showing seismic activity of the Katla volcano 9–10 July 2011. The blue lines show the extent of the ice caps over Eyjafjallajökull and Mýrdalsjökull. The colour spots show the extent and cooling of the eruptions over this period of time.

It may become so dense that the sky turns grey-black during the day, lowering temperatures.

Aircrafts avoid airspace containing volcanic ash as jet engines suck in the ash, jamming aircraft engines and contaminating air inside the cabin.

Katla

Katla is another subglacial volcano in Iceland, 25 km east of Eyjafjallajökull. A Katla eruption is likely to be 10 times more powerful than the 2010 eruption of Eyjafjallajokull. With a caldera of 10 km, it has the potential to flood the Atlantic Ocean with billions of litres of water from its melting snow. In 1755, the volume of water produced by its eruption was equivalent to the combined volume of all the world's rivers.

In July 2011, Mýrdalsjökull—the glacier where Katla is located—experienced several earthquakes. Based on previous patterns, large eruptions from Katla occur every 40 to 80 years. On this basis one is overdue—and the world will hold its breath in the coming years.

Geo**info**

Sulphur dioxide from the 1783–1784 Laki eruption in Iceland caused cooling of Europe and North America.

Geo**activities 5.5**

Knowledge and understanding

1. Why do volcanic eruptions occur in Iceland?
2. Explain the events of 14–20 April 2010 in Iceland.
3. What were the global impacts of the Eyjafjallajökull eruption?

Inquiry and skills

4. Refer to 5.5.2.
   a. What countries were affected by the ash cloud?
   b. Why were aeroplane flights cancelled?
5. Refer to 5.5.3 and the internet.
   a. Explain the intensity of seismic activity around Katla in the period 9–10 July 2010.
   b. The Katla volcano is displaying signs of unrest. What does this mean? What could be its impact on the environment?
6. *It is inevitable that somewhere in the world, a volcano is about to erupt.* Research this statement and identify one volcano experiencing signs of volatility at present. Include a summary risk and impact assessment.

ISBN 978 1 4586 6277 4

5.6 Shaky ground: earthquakes

Every day, 8000 micro-earthquakes occur—most pass unnoticed as they either are too weak or take place in remote locations. On the other hand, a disastrous earthquake occurs approximately once a year, killing people and destroying properties.

Earthquakes are graded on the **Richter scale**, which measures the energy released during an earthquake. Energy released ranges from less than 2 to over 10, or magnitude M2–M10. The largest recorded earthquake was the Great Chilean earthquake in 1960, which registered M9.5. The Indian Ocean earthquake in 2004 registered M9.1–M9.3.

Earthquakes trigger the collapse of buildings, activate tsunamis and cause fires. Damage could be reduced by improved building construction, early warning systems and evacuation plans.

Big vibrations

An **earthquake** is the sudden release of energy in Earth's crust creating vibrations. They are generally caused by slippage along geological faults, but also occur during volcanic activity, mine blasts and nuclear tests. Small earthquakes occur when drilling for hot rocks to supply geothermal energy, and when drilling for gas and oil from shale rocks (fracking). Some vibrations also cause landslides and avalanches.

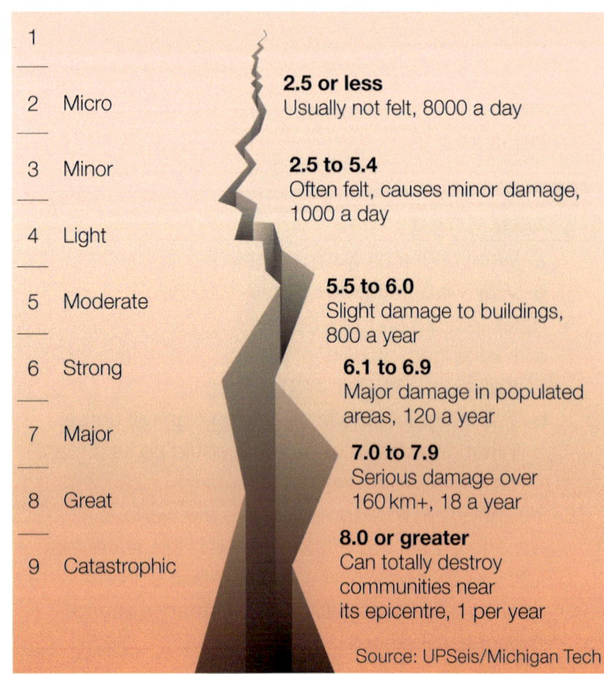

5.6.1 Earthquake magnitude, frequency and impacts

5.6.2 Deadliest earthquakes according to deaths since 115

Rank	Death toll (estimated)	Location	Year
1	830 000	Shaanxi, China	1556
2	230 000–316 000	Port-au-Prince, Haiti	2010
3	273 400	Haiyuan, China	1920
4	260 000	Antioch, Syria	115
5	250 000–300 000	Antioch, Syria	526

The underground point or origin of an earthquake is the **hypocentre**. The point directly above the hypocentre on the ground is the **epicentre**. The damaging effects of earthquakes vary according to:

- *magnitude*—higher magnitude means greater damage
- *location*—distance from the epicentre
- *depth*—shallow or deep focus (shallow means greater damage)
- *rock type* and *soil*—soft rock and soils (e.g. sand) amplify the shaking ground while compacted soil is more resistant to earthquakes
- *foreshocks* and *aftershocks*—foreshocks and aftershocks are smaller earthquakes before and after, respectively, the main earthquake (the *mainshock*)
- *population*—population density, and quality of buildings and infrastructure.

Earthquakes in Australia

Approximately 80% of earthquakes take place around the edge of the Pacific Plate, for instance in New Zealand and Japan. **Inter-plate earthquakes** occur at plate boundaries when plates collide or move laterally along each other, such as in West Asia. **Intra-plate earthquakes** occur in the middle of plate, along fault lines, such as in Australia.

Despite Australia's low-risk location in the middle of a large tectonic plate, one earthquake measuring

Geoinfo

In 1941 the biggest quake in Australia of M7.2 was at Meeberrie in Western Australia.

ISBN 978 1 4586 6277 4

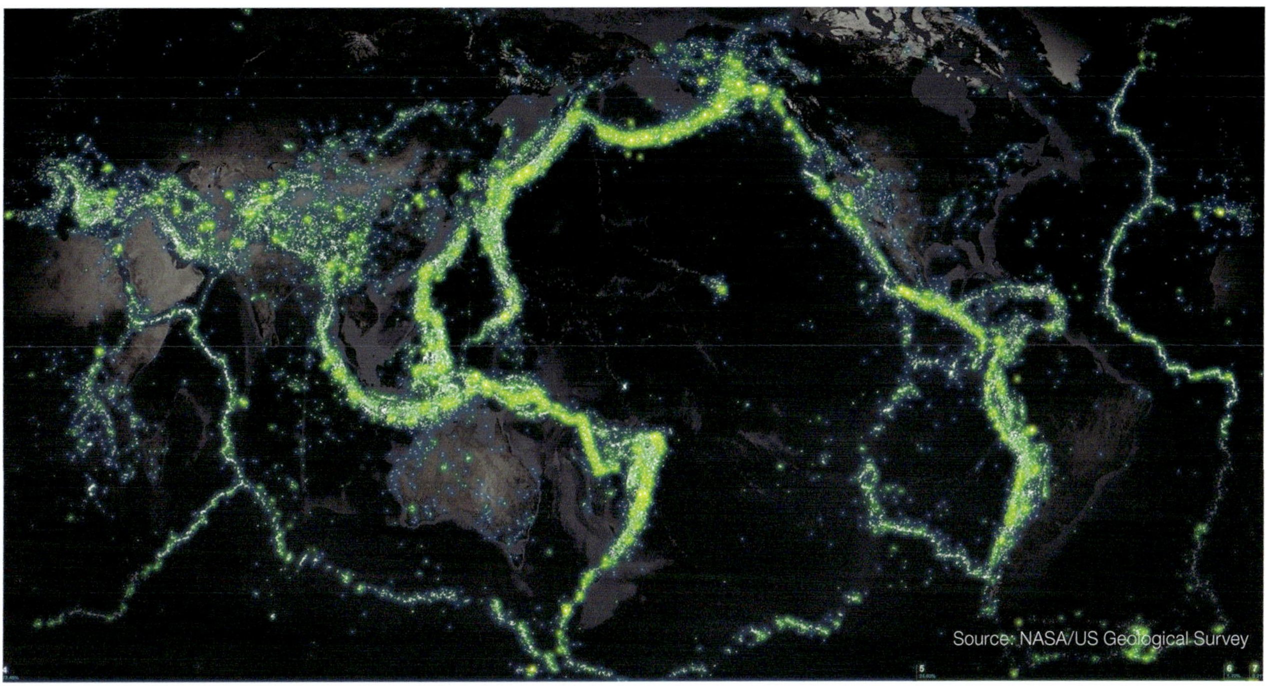

5.6.3 Satellite image showing spatial distribution of earthquakes by magnitude since 1898

M2 or greater occurs every day. Over the last 80 years, 17 earthquakes registered M6 or more. The rate is approximately one earthquake every five years, compared to the world average of 140 per year. High-risk earthquake areas in Australia include the areas surrounding Perth and Adelaide, which is the highest earthquake hazard capital city. In 1989, Newcastle (New South Wales) was partially devastated by an earthquake measuring M5.6. It was the first lethal earthquake in Australia, claiming 13 lives. However, in 1917 one miner died and five were injured in an underground rock fall triggered by an earthquake in Kalgoorlie, Western Australia.

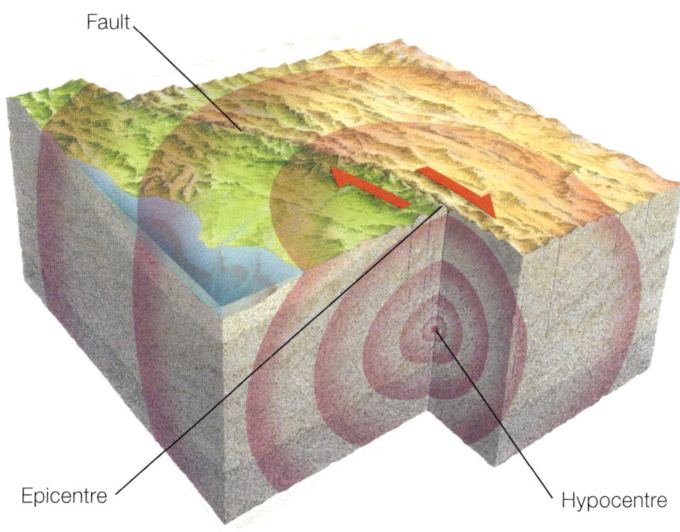

5.6.4 The hypocentre and epicentre of an earthquake

Geoactivities 5.6

Knowledge and understanding

1 What are the causes of earthquakes?
2 Explain how the Richter scale is used to measure the strength of earthquakes.
3 Explain the difference between an inter-plate and intra-plate earthquake.

Inquiry and skills

4 Refer to 5.6.1.
   a Distinguish between these magnitudes of earthquakes: micro, moderate and catastrophic.
   b Research the impacts of two large earthquakes using an annotated photo story.
5 Refer to 5.6.3 and an atlas.
   a Explain the spatial distribution of earthquakes.
   b Why is the map important to future planners of mines and cities?
6 Refer to 5.6.4. Distinguish between the hypocentre and epicentre of an earthquake.
7 Research earthquakes in your state or territory. Investigate where they were located and the impacts on people and places. Present your findings using web 2.0 tools.

ISBN 978 1 4586 6277 4

5.7 Deadly Haitian and Chilean earthquakes

The Caribbean consists of islands surrounded by the Caribbean Sea and Atlantic Ocean. Located on the Caribbean Plate, the area has experienced many major earthquakes. In 1946 a M8.1 earthquake occurred off the north-east coast of the Dominican Republic. It triggered a tsunami killing around 1600 people.

Haitian earthquake

The Republic of Haiti is located in the Caribbean. The island experiences frequent natural disasters, including hurricanes, landslides, tsunamis and earthquakes. On 12 January 2010, the Caribbean Plate moved east in relation to the North American Plate. The movement created a M7 earthquake that struck Haiti and devastated Port-au-Prince, the capital city, which has a population of 2 million people. The epicentre was only 20 km from Port-au-Prince and the hypocentre was shallow. As a shallow earthquake located close to a large urban population, it created greater shaking and damage than if it had been located deeper and further away. It also generated a tsunami with waves over 3 m sweeping boats and debris into the ocean.

Haiti Humanitarian Response

Cap-Hatien
HAITI
Gulf of Gonave
Saint-Marc
Gonave Island
DOMINICAN REPUBLIC
Port-Au-Prince
Petit Goave
Epicentre
Cayes
Damage
Lightest
Heaviest

Mission 4636

Mission 4636 was a free texting service set up for earthquake Haiti victims who needed emergency aid. Crowdsourcing platform CrowdFlower became the switchboard for Mission 4636, translating and geolocating information of missing persons from text messages. This was referred to as micro-tasking.

Social media and SMS

Social media (e.g. Facebook, Twitter and chat rooms) and SMS were used to coordinate humanitarian relief after the earthquake. The main information systems were SAHANA Free and Open Source Disaster Management System, Ushaidi (crowdsourcing platform) and the UN's OneResponse website.

Ushaidi allowed people to collect and share stories (via SMS and email) and UN agencies and aid organisations used Facebook, Twitter and wikis. These technologies illustrated the importance of harnessing the community to improve collaboration and reduce adverse impacts.

Media

As normal communications channels were upset, social media played a key role in providing news. Twitter feeds provided pictures of the earthquakes. The Guardian newspaper's blog used social media and CNN's filmed iReports to help rescuers locate missing people.

Radio from Port-au-Prince covered most of the local news. There were also SMS call-in programs.

Maps

Maps were shared between rescuers and relief workers, including US Geological Survey maps showing real time earthquakes and aftershocks. Google Earth Outreach, NASA's Earth Observatory maps and NOAA provided updates, and Google Earth provided OpenStreetMap from satellite imagery—all of which improved the relief response.

5.7.1 The humanitarian response to Haiti's earthquake. Social media and ICT played an important role

ISBN 978 1 4586 6277 4

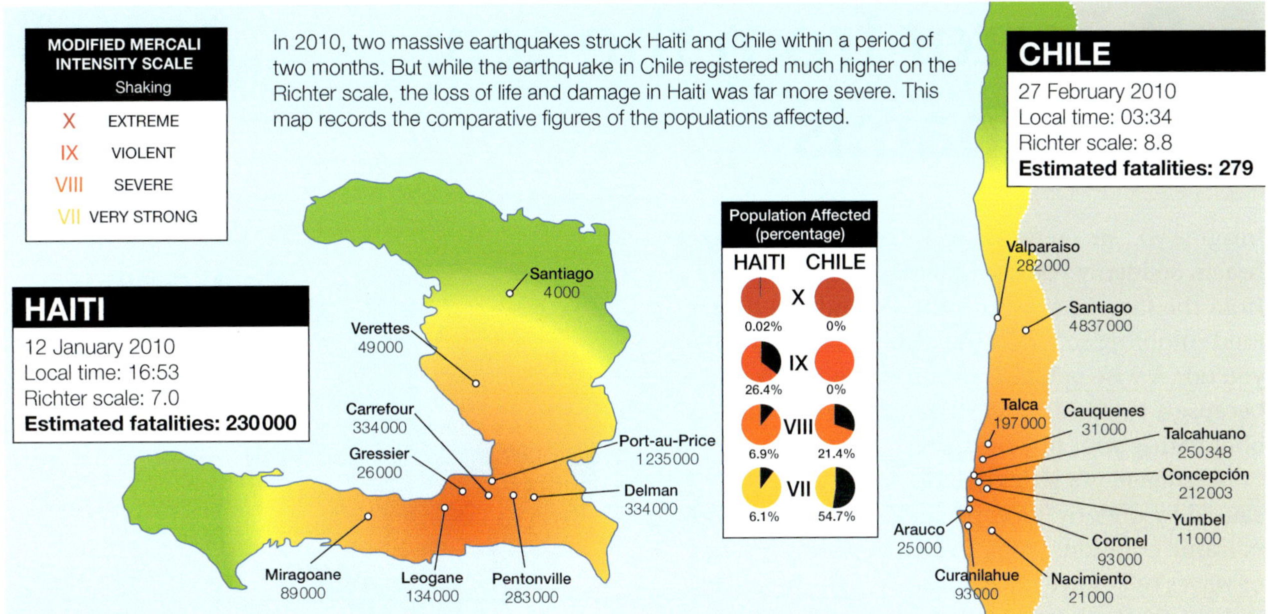

5.7.2 A tale of two earthquakes: Haiti and Chile

By 24 January, there had been at least 52 aftershocks measuring M4.5 or greater. Approximately 230000–316000 people were killed and over one million were left homeless.

Global citizenship

As Haiti is a developing country, the government had inadequate resources to manage a disaster of this magnitude. Governments and non-government organisations (NGOs) around the world provided aid to Haiti. Unfortunately aid was hampered by:

- damaged infrastructure, such as roads and port, and only one single-strip airport
- cuts in power and communications
- a weak government made further ineffective by the collapse of the Parliament House building.

International organisations rescued people, buried the dead, and provided medicine, water, food and shelter. 'Sniffer' dogs located people, field hospitals were established and 500 camps provided shelter for the homeless. A UN 'Food Aid Cluster' fed two million people.

Chilean earthquake

In February 2010, Chile suffered a large earthquake of M8.8 that triggered a tsunami. About 280 people died and buildings were destroyed. While the magnitude of the earthquake was greater than Haiti's the destruction was less severe. Chile's stricter building code and the less dense population where the earthquake hit meant that the impact was not as devastating.

Geoactivities 5.7

Knowledge and understanding

1. What disaster occurred in Haiti in 2010?
2. Describe the impacts of the earthquake in Haiti.
3. Describe the global response to the Haiti disaster.
4. Explain how aid is hampered in a developing country such as Haiti.

Inquiry and skills

5. Refer to 5.7.1.
   a. How did ICT (e.g. SMS, CrowdFlower, Facebook, Twitter and blogs) contribute to the humanitarian response?
   b. Explain how maps and the media became useful tools in a humanitarian crisis.
6. Refer to 5.7.2.
   a. Distinguish between the location, date and Richter scale rating of the Haitian and Chilean earthquakes.
   b. Which country experienced the greatest number of fatalities?
   c. What percentage of the population was affected by violent shaking in Haiti and Chile?
   d. Compare and contrast the cause and impacts of earthquakes in 2010 in Haiti and Chile. Present your findings as a Prezi.
7. Reflect on your learning and research individual and collective actions to reduce the impacts of future earthquakes around the world.

Chapter 5

ISBN 978 1 4586 6277 4

5.8 Terror from the sea: tsunamis

Imagine you are on holidays, walking along a peaceful beach. Suddenly you see water being sucked away from the shoreline towards the sea and you hear a loud rumbling sound in the distance. What will you do? A wise option is to run to higher ground because a tsunami could strike, killing you and obliterating your holiday town.

A **tsunami** is a series of large waves generally caused by underwater earthquakes, landslides and volcanic eruptions. After the Earth moves, a series of waves or a 'wave train' forms. When the ocean is deep, the wave height of a tsunami is not high, but it can travel as fast as a plane at 500 km/h. If you were in a boat at sea, you would probably be unaware that a tsunami was passing underneath you. However, once a tsunami reaches the coast, the shallow water slows the speed of the waves and causes the waves to increase in height. Its destructive power changes the coastal landscape as it floods low-lying areas, where many towns and cities are located.

Tsunamis around the world

Japan has the most recorded tsunamis in the world, Hawaii averages one per year, and the tallest tsunami occurred in 1958 at Lituya Bay, Alaska, reaching 524 m. The deadliest tsunami was the 2004 Indian Ocean tsunami, generated by an undersea earthquake with an epicentre located off shore from Sumatra. It killed between 230 000 and 310 000 people in 11 countries across the Indian Ocean. Indonesia suffered the most damage, followed by Sri Lanka.

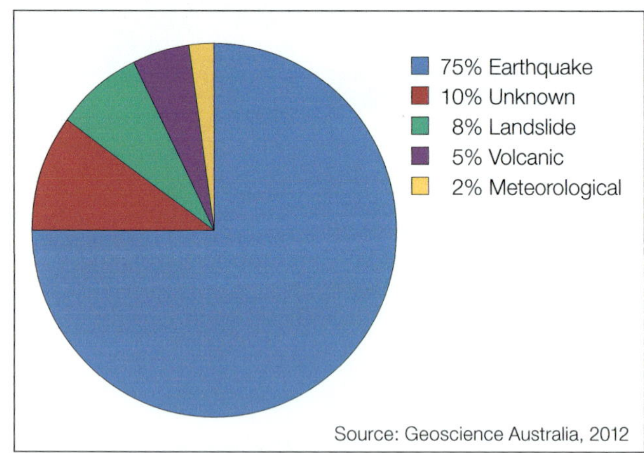

5.8.2 Causes of tsunamis

Approximately 90% of tsunamis occur in the Pacific Ocean, as the Pacific Ring of Fire accounts for 90% of the world's earthquakes and 75% of its volcanoes. When Krakatoa erupted in 1883 (see 4.2) it produced a 35 m tsunami which destroyed coastal communities in Java and Sumatra. The wave was felt in the English Channel (UK) 37 hours later.

There is evidence that over the past few thousand years the Australian coast has experienced tsunamis generated from earthquakes and volcanic eruptions from countries such as New Zealand, Chile, Hawaii, Indonesia and Japan.

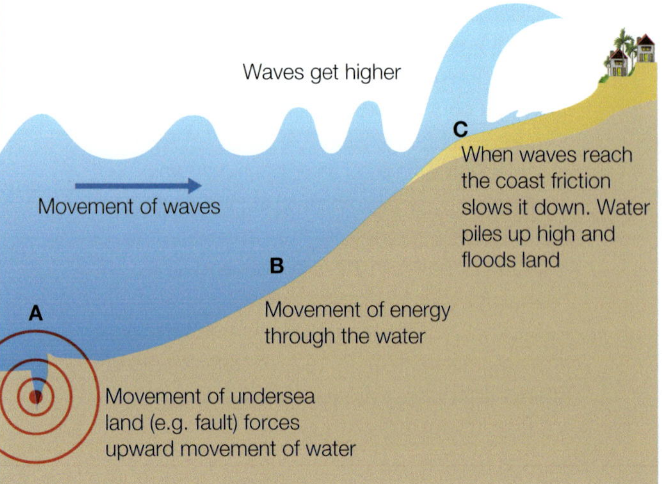

5.8.1 Formation of a tsunami

5.8.3 Worst tsunamis in history

Rank	Estimated death toll	Event
1	240 000	2004 Indian Ocean tsunami
2	18 000	2011 North Pacific Coast, Japan tsunami
3	60 000	1755 tsunami following Lisbon earthquake, Portugal
4	40 000	1883 tsunami following Krakatoa eruption, Indonesia
5	31 000	1498 tsunami following Meio Nankaido earthquake, Japan

ISBN 978 1 4586 6277 4

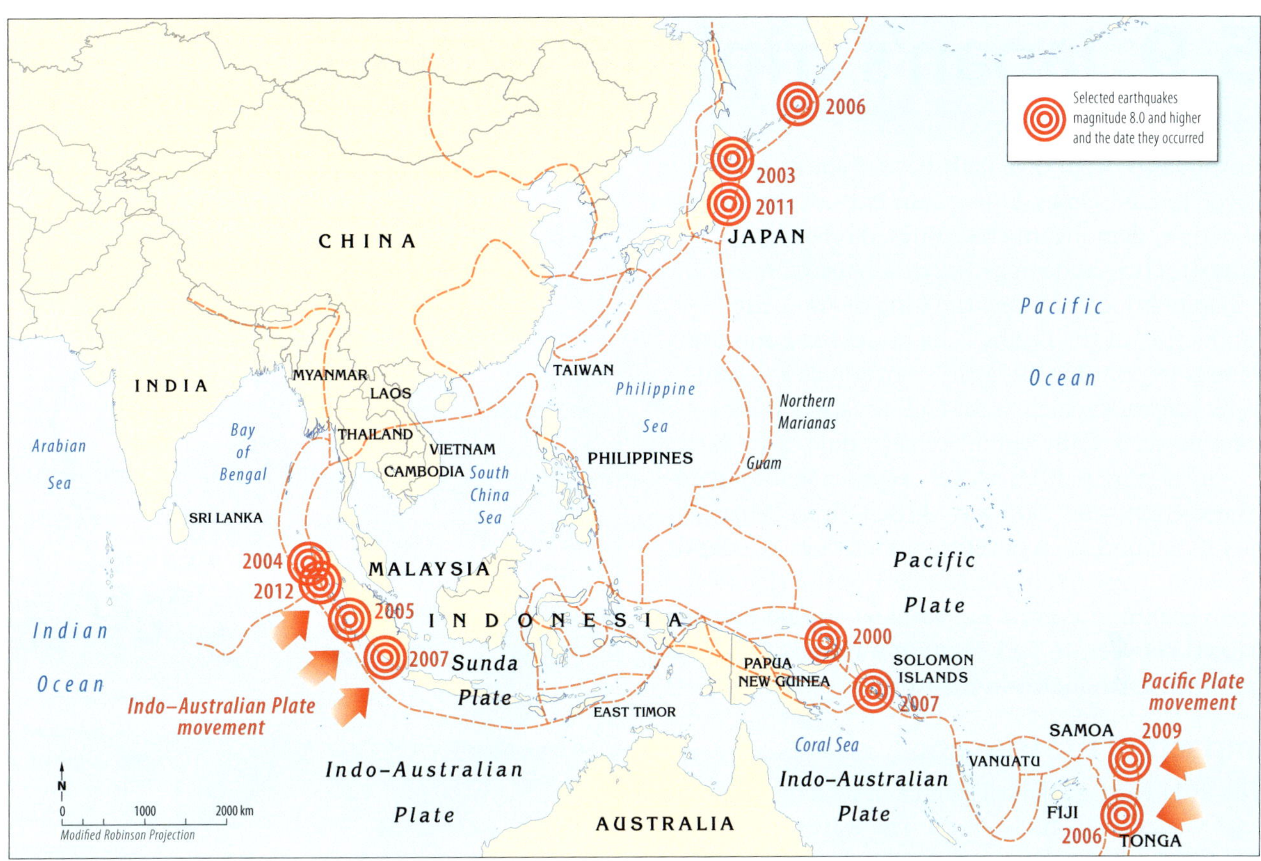

5.8.4 Major earthquakes near Australia since 2000

On 29 September 2009, Samoa experienced an M8.3 earthquake. The earthquake, only 18 km beneath the Earth's surface, triggered a tsunami that battered Apia (the capital of Samoa), Pago Pago (the capital of American Samoa) and Tonga. With waves over 3 m high, the tsunami engulfed cars and homes, flattened villages and killed 190 people.

Tsunami preparation and responses

The National Oceanographic and Atmospheric Administration (NOAA) provides maps illustrating the projected travel times of tsunamis, and deep ocean tsunami buoys detect approaching waves. The Pacific Tsunami Warning System (PTWS) and the International Tsunami Information Centre (ITIC), aim to save lives and minimise damages. Maps and satellite images play an important role after disasters as they provide relief workers with information on access routes to areas in crisis.

Geo**info**

- Most domestic and wild animals detect approaching tsunamis.
- The Ring of Fire runs for 40 000 km around the edge of the Pacific Ocean.

Geo**activities 5.8**

Knowledge and understanding

1 What is a tsunami?
2 Explain the impacts of tsunamis on people and landscapes.

Inquiry and skills

3 Refer to 5.8.1. Explain the changes to waves from A to C.
4 Refer to 5.8.2 and list the causes of tsunamis.
5 Refer to 5.8.4.
   a Countries located above and beside Australia are located on shaky ground. Analyse the map and list five earthquakes over M8 between 2000 and 2012.
   b Research and describe the impacts these earthquakes had on Australian coasts.
6 Suggest strategies you would implement to reduce the impacts of a tsunami on a coastal resort. Present as an oral report.
7 Using ICT, research the 2010 Mentawai Islands (Indonesia) earthquake and tsunami. Where are the islands located? What were the causes and impacts? How did organisations such as SurfAid International make a difference to the lives of people? Present as a Prezi.

ISBN 978 1 4586 6277 4

5.9 Japan's triple disaster

Earthquakes occur on a daily basis, frequently causing havoc to landscapes, killing hundreds of thousands of people, demolishing buildings, damaging infrastructure and triggering deadly tsunamis.

Japan is located along the Ring of Fire—the subduction of the Pacific Plate under the Eurasian Plate is responsible for Japan's earthquakes as well as its volcanoes, such as Mt Fuji. In Japan, minor tremors occur daily but major earthquakes are rare.

On 11 March 2011, an M9 earthquake hit Japan. Its epicentre was 72 km east of the Oshika Peninsula of Tohoku and its hypocentre underwater at a depth of 32 km. The earthquake caused Honshu (Japan's main island) to move 2.4 m east and shifted Earth on its axis between 10 and 25 cm. It was one of the most powerful earthquakes in the world.

5.9.2 Impacts of the earthquake in Japan

Impacts of the tsunami

The 2011 earthquake triggered a destructive tsunami with waves up to 40.5 m high. The waves travelled 10 km inland, destroying everything in their path. Tsunami warnings were issued and evacuations ordered along Japan's Pacific coast and to over 20 countries located around the Pacific Rim. Japan had invested billions of dollars in anti-tsunami seawalls standing 12 m high and lining 40% of its coastline. However, the tsunami washed over the top.

Nuclear power stations were automatically shut down following the earthquake. Cooling is required to remove heat after a nuclear plant has been shut down—a process powered by emergency diesel generators. At the Fukushima Daiichi nuclear power plants I and II, tsunami waves destroyed the diesel backup system, which led to explosions at Fukushima I and leakage of radiation. Over 200 000 people were evacuated from the surrounding area. Evidence was found of radiation in vegetables, water and milk, and as a result many countries blocked imported food from the region.

One disaster impact often overlooked is that 5 million tonnes of rubbish washed into the Pacific Ocean. Approximately 70% sank near Japan and 30% floated to the Pacific north-west coast and hit the west coast of Canada in 2014.

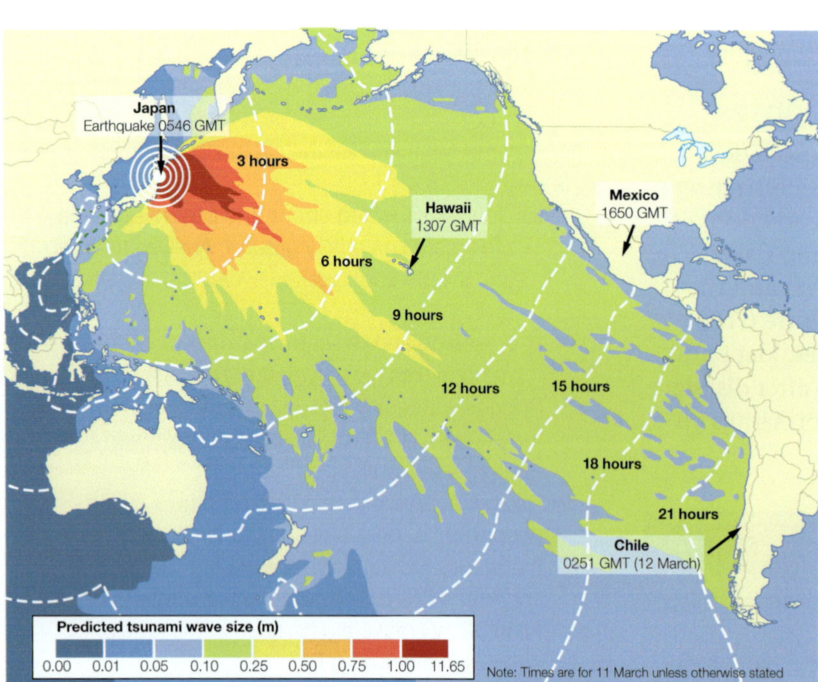

5.9.1 Japan's tsunami wave heights and how long it took to travel across the globe

ISBN 978 1 4586 6277 4

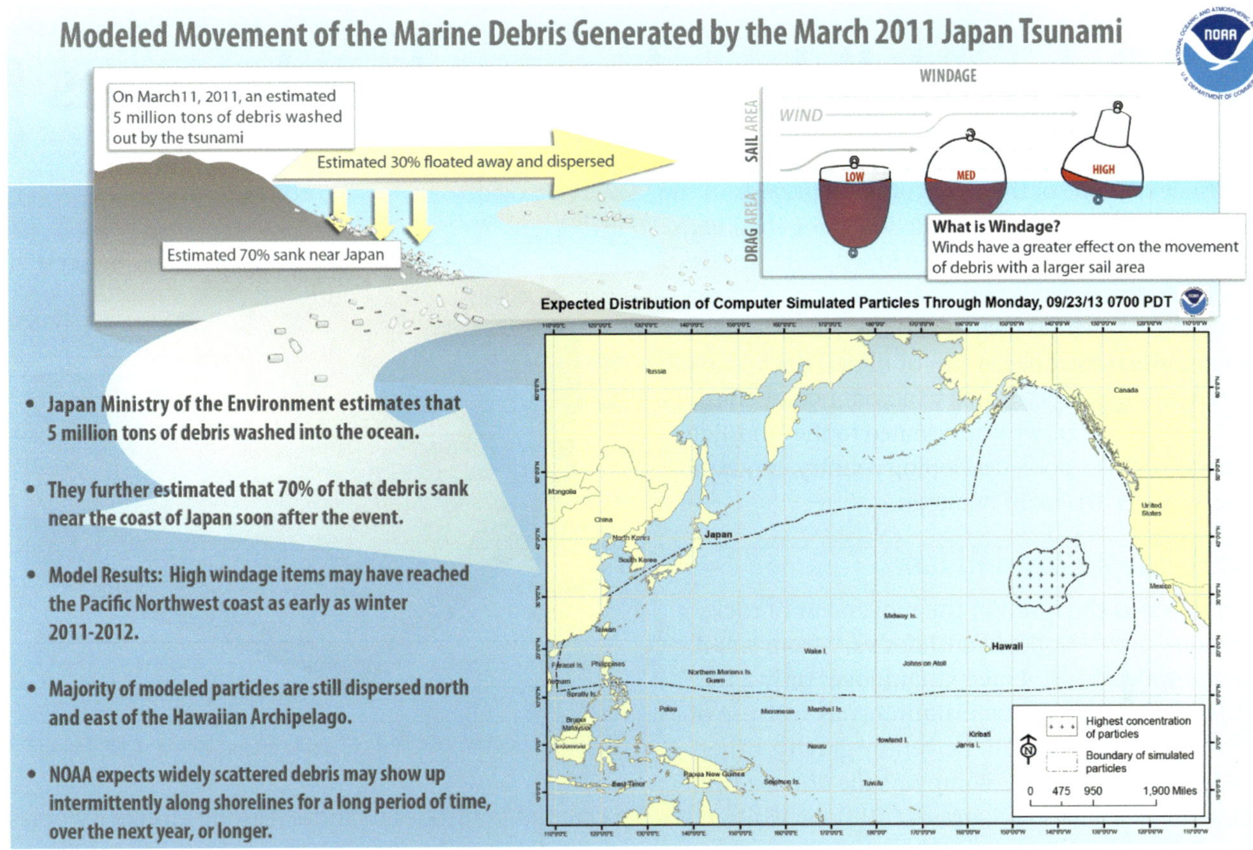

Source: National Oceanic and Atmospheric Administration (NOAA), USA

5.9.3 Movement of marine debris after the Japanese tsunami

Japan's domino effect

The earthquake and its tsunami resulted in around 18 000 deaths, 6000 injured and 3000 missing people. Transport networks, including roads and railways, and over one million buildings were damaged or partially destroyed. Around 4.4 million households were left without electricity and 1.5 million without water. Japan's government predicted the cost could reach $235 billion, making it the world's most expensive natural disaster to date.

In the hours following the earthquake, internet users turned to YouTube and Twitter. Micro-blogging platforms provided real time stories with earthquake updates.

Earthquakes and tsunamis are expected to continue occurring in the region, as the Pacific Plate is smashing against the Japanese Plate boundary at a speed of 8.9 cm per year.

Geoinfo

- It has been 1200 years since an earthquake of this magnitude hit the Japanese plate boundary.
- Sendai residents had 8–10 minutes warning before the tsunami struck.

Geoactivities 5.9

Knowledge and understanding

1 Why is Japan vulnerable to earthquakes and volcanoes?
2 Where were the epicentre and hypocentre of the 2011 earthquake located?
3 Describe the cause and magnitude of the earthquake.
4 Explain the multiple impacts of the earthquake.
5 How did Japan prepare for the earthquake and tsunami?

Inquiry and skills

6 Refer to 5.9.1.
   a How long did the tsunami take to reach north and east Australia and what was the height of the waves?
   b What was the wave height around Hawaii?
7 Refer to 5.9.2.
   a Refer to an atlas or the internet and name the places that experienced 'severe coastline damage'.
   b Calculate the area threatened by radiation.
8 Refer to 5.9.3.
   a Describe the movement of debris after the Japanese tsunami in 2010.
   b Investigate the impact of rubbish on oceans and coastal locations in Canada. Collect geographical data using ethical protocols. Present your findings as a newspaper report.

5.10 Landslides change landscapes

Landslides are one of the most common hazards in the world, causing death and destruction along their path. Approximately 8000 people are killed by landslides each year. In 1999 a landslide in Vargas, Venezuela, killed 10 000–30 000 people and the village of Los Corales was buried under 3 m of mud. Heavy rains, flash floods and land cleared to accommodate the high population growth contributed to the landslide. As Venezuela is a poor, developing country, thousands of survivors still remain homeless.

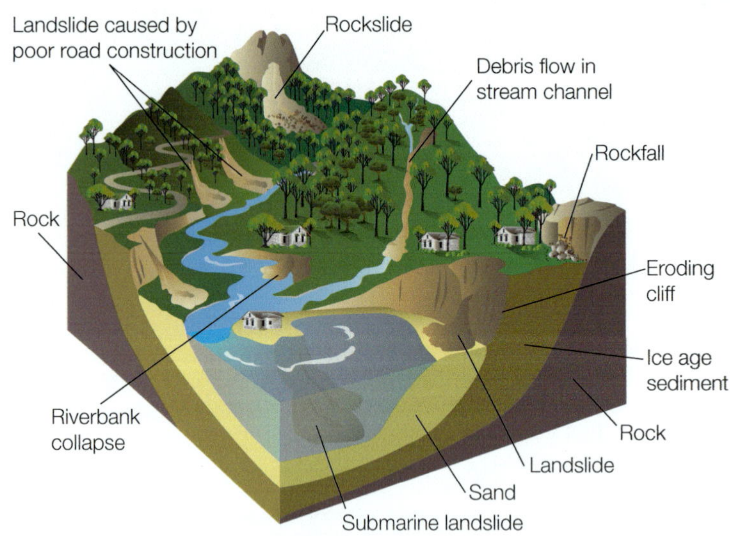

5.10.2 Location of different types of landslides in a landscape

Landslides: slow and fast

A **landslide** is the result of the movement of rocks and soil down a slope. Landslides vary from small, hardly noticeable soil creep that moves only a few millimetres a year to the instantaneous collapse of a cliff causing a rock fall. Driven by gravity, other landslides flow, topple, slump or slide down a hill. Some impact on a small area or hill slope while others travel for kilometres. The most destructive landslides are soil and rock flows containing high quantities of water, which cause the slope to lose cohesion. These are referred to as mudslides. In 2010 a mudslide in Gansu, China, killed 1765 people.

Landslides are caused by natural and human factors. They occur across Australia in locations such as coastal cliffs, the Great Dividing Range, Mt Wellington (Tasmania), Mt Lofty Ranges (South Australia), Strzelecki Ranges (Victoria) and the Alpine region (NSW). In 1997 a landslide in Thredbo, NSW, killed 18 people. Evidence of potential landslides includes leaning electricity poles and trees, broken water pipes, sunken roads and cracked footpaths. To avoid the likelihood of hazards becoming disasters, Geoscience Australia requests all inhabitants become landslide spotters and report any such signs or landslides they observe.

Hazard risk and vulnerability

The majority of fatal landslides occur in remote mountain areas prone to heavy rainfall and earthquakes. Landslide hotspots are along the southern edge of the Himalayan Mountains and in west China, Indonesia, the Philippines and Central America.

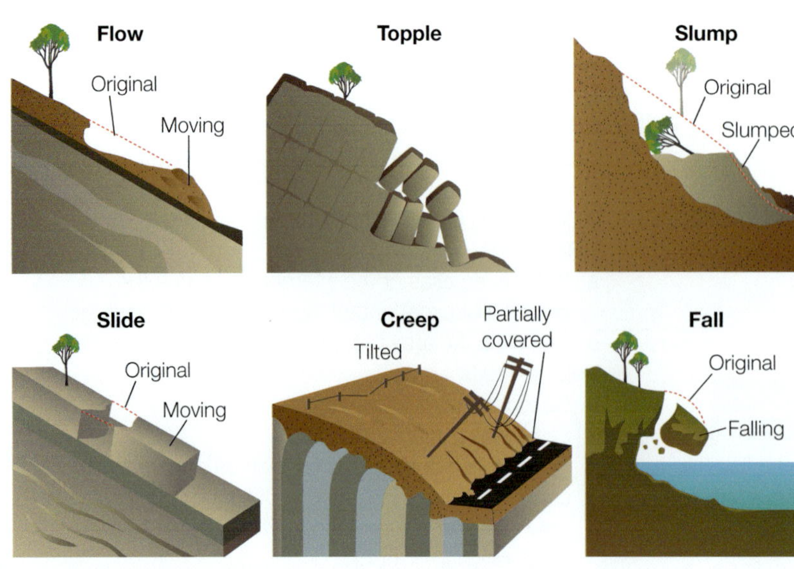

5.10.1 Types of landslides

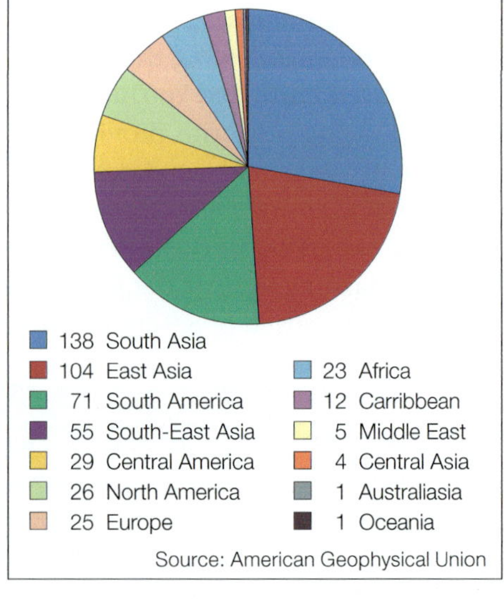

5.10.3 Number of landslides, 2002–2010

ISBN 978 1 4586 6277 4

5.10.4 Oblique aerial view of the landslide in Colonia Las Colinas, El Salvador

In 2001 an M7.6 earthquake caused a landslide in El Salvador which buried parts of Colonia Las Colinas and resulted in at least 944 dead and almost 108 000 homes damaged or destroyed. Countries have implemented programs to reduce landslide risk, such as building retaining walls, relocating people at risk, and implementing early warning systems.

Of global concern is the fear that landslides are increasing in number. As population density increases, more people are forced to live and work on unstable land. Environmental degradation (such as deforestation) makes landslides more likely. A contributing factor is the increase in rainfall intensity in some areas, possibly associated with climate change.

5.10.5 Without funding and research the death toll from landslides will only get worse

Landslide hazard analysis, mapping and GIS assist in the development of sustainable land use planning. Before and after aerial photographs and satellite imagery are used to reveal how the landscape changed, what triggered the landslide, and strategies devised for effective recovery.

Geo**activities 5.10**

Knowledge and understanding

1 What are the causes of landslides?
2 How do steep slopes and rainfall contribute to landslides?
3 Landslides vary in speed and distance. What does this mean?
4 Where are landslide hotspots located?
5 Suggest procedures to reduce the liklihood of a landslide hazard developing into a disaster.
6 Explain why the number of landslides is expected to increase in the future.

Inquiry and skills

7 Refer to 5.10.1 and 5.10.2.
  a List three types of landslides.
  b Describe how soil creep differs from rock fall.
  c Explain how landslides occur in different places. Which landslide would concern you the most? Provide reasons for your answer.
8 Refer to 5.10.3.
  a What region has the greatest number of landslides?
  b State the number of landslides in Australia and explain how it compares with the global total.
  c Using ICT research one landslide disaster in Australia, such as the Thredbo landslide. Include a map and the date, cause, impacts and management.
9 Refer to 5.10.4.
  a Draw and label a line drawing of the photograph.
  b Imagine you lived in Colonia Las Colinas in El Salvador. Describe the impacts on your community.
10 Refer to 5.10.5. Explain the message in the cartoon.
11 View a computer simulation of a slump landslide on YouTube (see Geolinks). Present your thoughts as a newspaper article.
12 Refer to the ABC news website on landslides (see Geolinks).
  a Draw a table listing 10 landslides. List causes in the first column, list impacts in the second column and whether they were natural or human-induced in the third column.
  b What are the impacts of landslides on poor people with poorly constructed buildings?
  c Reflect on your learning and describe strategies to reduce future landslides.

ISBN 978 1 4586 6277 4

5.11 Avalanche alert: white death!

During the Middle Ages in Europe, an avalanche was referred to as 'white death', thought to be sent by witches who rode on snow-white surges. Today, adventure sports in scenic snow-clad mountains can be fraught with danger, with sudden avalanches killing over 150 people per year.

An **avalanche** consists of snow, ice and rock falling rapidly down a slope. When weighing millions of tonnes and travelling 320 km/h down a hill, they kill skiers and destroy mountain resorts. In 1970, an avalanche crushed the town of Yungay in Peru, killing 29 000 people. During World War I, avalanches in the European Alps thought to be triggered by artillery fire killed 60 000 Italian and Austrian troops.

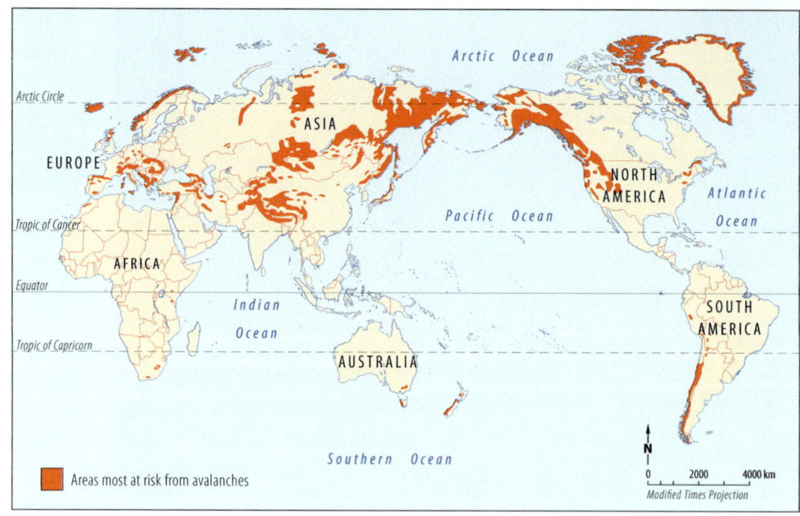

5.11.2 Spatial distribution showing areas at high risk from avalanches

Causes of avalanches

Scientists estimate more than 1 million avalanches occur each year—the European Alps alone experience about 250 000 each year. Landscapes containing steep, high mountains tend to suffer the largest number of deadly avalanches as the speed of falling snow escalates while travelling down the long, steep slope.

Approximately 90% of avalanches are triggered by humans, including snowmobilers, skiers and snowboarders. However, nature also plays an important part in the stability of snow on slopes. Important factors include:

- *gradient*—avalanches generally occur on steep slopes between 25° and 40°
- *temperature*—a rise in temperature weakens the bonds between the layers of snow
- *high winds and thunderstorms*—after a snow storm or high winds the possibility of an avalanche increases (e.g. in 1901 in Wellington, New Zealand,

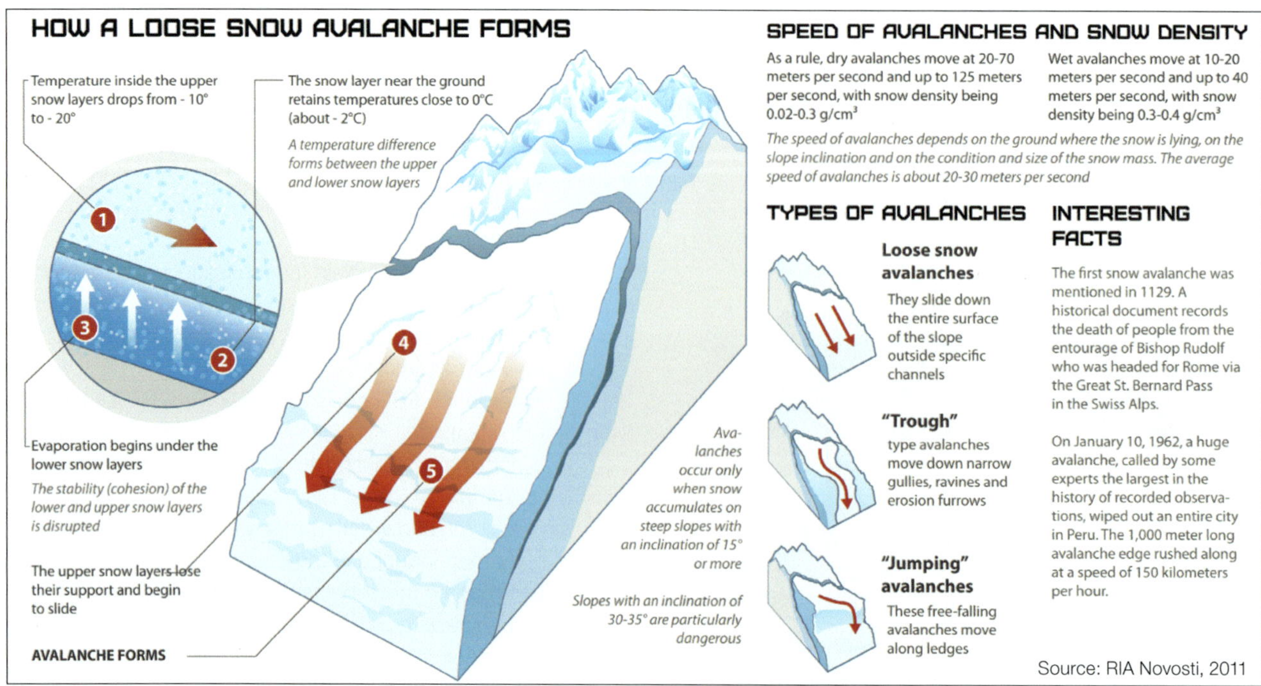

5.11.1 Causes of avalanches

ISBN 978 1 4586 6277 4

a thunderstorm unleashed an avalanche that hit a train depot, sending trains 45 m downhill and killing 96 people)
- *snow build-up*—if snow builds up by more than 2 cm/h it produces unstable conditions on slopes. More than 30 cm of snow build-up within 24 hours triggers a hazard warning
- *loose snow* and *air pockets*—loose snow and air pockets caught under the snow increase the chance of an avalanche
- *overloading*—when the weight of snow on top increases until it detaches from the underneath snowpack and moves down the slope
- *earthquakes*—can jolt the snow and trigger an avalanche. In New Zealand in 2010, an earthquake near Christchurch caused an avalanche in Canterbury, around 46 km away.

In 2000, Joel Roof was snowboarding in the Teton Mountain Range in Wyoming, USA. He triggered a slab avalanche, which buried him under 2 m of snow. The mountain had received 56 cm of snow during the week and 8 cm the previous evening.

5.11.3 Glory Bowl Avalanche Path, Wyoming, USA

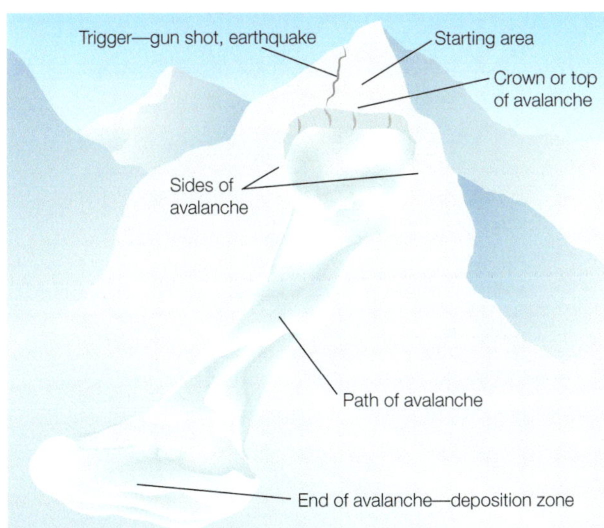

5.11.4 Profile of an avalanche

Classifying and managing landslides

There is no universally accepted classification of avalanches. They can be described by their size, cause or destructive potential. However, some are classified into the following categories:

- *slab avalanche*—a giant block that occurs when harder layers of snow set on top of softer layers of snow. The bottom softer layers are unable to support the harder top layers
- *sluff avalanche*—a small slide of dry, powdery snow. These are common following heavy snowfall of 2.5 cm/h or more
- *wet avalanche*—occurs after warm temperatures or a spring thaw.

Knowledge of human and natural causes of avalanches and how they move has resulted in the revegetation of steep slopes and the construction of snow nets and fences made of steel and wood. Avalanche dams have been built to stop or deflect avalanches above railways lines and roads.

Geo**activities 5.11**

Knowledge and understanding

1. What is an avalanche?
2. Why are avalanches referred to as 'white death'?
3. Where are most avalanches located?
4. Explain the main causes of avalanches.
5. Describe how natural and human forces contribute to avalanches.

Inquiry and skills

6. Refer to 5.11.1.
   a. Explain how changes in temperature can cause an avalanche.
   b. What factors determine the speed of an avalanche?
   c. List three types of avalanches related to their movement.
   d. Explain the relationship between the gradient of the slope and the possibility of an avalanche.
   e. What is a jumping avalanche?
7. Refer to 5.11.2. Refer to an atlas or the internet and name four areas and their mountains at high risk from avalanches.
8. Refer to 5.11.3 and 5.11.4.
   a. Describe the movement of the slab avalanche.
   b. What kind of triggers are typical for this kind of avalanche?
   c. Research recent avalanches and present findings as a newspaper article. Include a map showing the location of the avalanche, and apply appropriate cartographic conventions.
9. Reflect on your learning and suggest actions to reduce adverse impacts of avalanches on people and places.

Chapter 5

ISBN 978 1 4586 6277 4

5.12 Disaster management and technology

Geomorphic hazards frequently evolve into disasters. While little can be done to block huge tsunami waves caused by undersea earthquakes or stop volcanic explosions, the United Nations Office for Disaster Risk Reduction (DRR) aims to reduce damage to people, properties and livelihoods.

There are four main disaster management strategies: *preventing, preparing, responding* and *recovering*. The DRR focuses on preventing and preparing before the disaster, rather than responding and recovering after the disaster. Improvements in preparing for geomorphic disasters include education, early warning systems and evacuation procedures.

Technology prevents and responds

The World Disasters Report focuses on the spread of technologies to prevent and respond to geomorphic disasters. Mobile phones, crowdsourcing, crises mapping, digital data collection, improved seismic networks and tsunami early warning systems led to a decrease in loss of lives over the last 10 years.

Geo-referencing of maps shows the location of the worst affected areas, and the United Nations Office for the Coordination of Humanitarian Affairs (OCHA) network of volunteers enables governments and aid organisations to respond quickly with shelter, food and water.

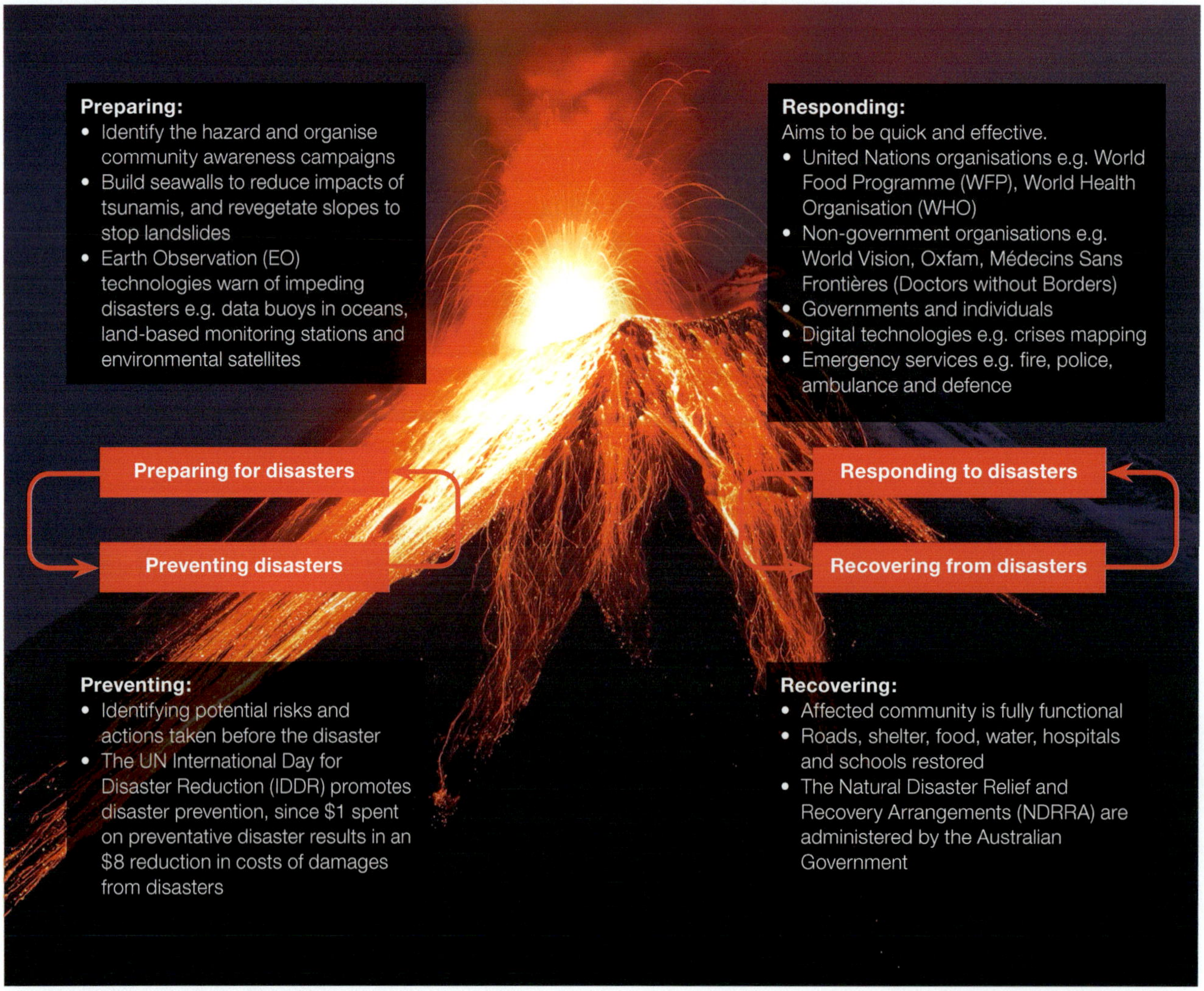

Preparing:

- Identify the hazard and organise community awareness campaigns
- Build seawalls to reduce impacts of tsunamis, and revegetate slopes to stop landslides
- Earth Observation (EO) technologies warn of impeding disasters e.g. data buoys in oceans, land-based monitoring stations and environmental satellites

Responding:

Aims to be quick and effective.

- United Nations organisations e.g. World Food Programme (WFP), World Health Organisation (WHO)
- Non-government organisations e.g. World Vision, Oxfam, Médecins Sans Frontières (Doctors without Borders)
- Governments and individuals
- Digital technologies e.g. crises mapping
- Emergency services e.g. fire, police, ambulance and defence

Preparing for disasters → **Responding to disasters** → **Recovering from disasters** → **Preventing disasters** → **Preparing for disasters**

Preventing:

- Identifying potential risks and actions taken before the disaster
- The UN International Day for Disaster Reduction (IDDR) promotes disaster prevention, since $1 spent on preventative disaster results in an $8 reduction in costs of damages from disasters

Recovering:

- Affected community is fully functional
- Roads, shelter, food, water, hospitals and schools restored
- The Natural Disaster Relief and Recovery Arrangements (NDRRA) are administered by the Australian Government

5.12.1 Disaster management strategies

ISBN 978 1 4586 6277 4

Avalanches	Earthquakes	Volcanoes	Tsunamis
• fences • avalanche dams • radar detection • geophones • snow pits • SiteMonitor 3DLM	• seismometers and seismographs • earthquake early warning systems • emergency earthquake alert	• seismometers and seismographs • videos, infrared cameras, satellite imagers, webcams	• deep-ocean tsunami detection buoys • NOAA's DART real-time monitoring system

5.12.2 Technology used to prevent geomorphic disasters. Photo shows an avalanche warning on Mont Blanc

In the future, technology will help to eliminate earthquake damage to buildings; for instance, building on top of a cushion of air, using sensors on buildings to detect seismic activity, and incorporating shock absorbers to reduce the magnitude of vibrations.

5.12.3 Soccer ball–shaped house resistant to earthquakes and floats on water, Japan

Geo**activities 5.12**

Knowledge and understanding

1. List four main disaster management strategies.
2. Explain how technology has reduced the number of deaths and damaged buildings in disasters.

Inquiry and skills

3. Refer to 5.12.1. Describe what is meant by preventing and preparing for a disaster.
4. Refer to 5.12.2. Explain how different types of technology assist people to prepare for an earthquake and a volcanic eruption.
5. Research the advantages of the ESRI Global Disaster Resilience App (e.g. evacuation routes and safe shelter locations).
6. Design an electronic poster for the International Day for Disaster Reduction (IDDR).
7. Technology is an important weapon against geophysical disasters. What are your thoughts on this statement?

ISBN 978 1 4586 6277 4

5.13 Contemporary hazard: Nepal's earthquakes

Nepal is one of the most geologically hazardous countries on Earth. The country is located on a fault line where the Indian plate collided with the Eurasian plate 40–50 million years ago. At present, the Indian plate is moving north at a rate of 5 cm a year—twice the speed at which fingernails grow. In 2015, the plate moved over 6 metres, generating a huge earthquake on 25 April. This was followed by a large aftershock on 12 May.

The impacts of the earthquake and the aftershock killed over 9000 people and destroyed 36 of the 75 Nepali districts. Reasons for the extent of the damage include:

- the *hypocentre* or depth of the quake (see 5.6) was shallow, which caused more aftershocks and damage than quakes originating deeper in the ground
- the *epicentre* or location of the quake (see 5.6) on 25 April was near Kathmandu, the capital city
- Kathmandu is located on an *ancient sedimentary basin of soft rocks* that amplified the earthquake's movements in the ground
- Nepal has *lax building regulations* that permitted poorly constructed low-cost buildings. Around 75% of deaths were due to collapsing buildings, and not the earthquake or aftershocks.

As a result, densely populated Kathmandu, with a population of over one million people, experienced the greatest damage.

Spatial distribution and impacts

The spatial distribution of the earthquake was widespread. It extended from urban Kathmandu to remote rural villages, and in its path it:

- damaged Kathmandu's Durbar Square, a World Heritage Site
- flattened villages near the epicentre
- destroyed 80% of houses in rural areas.

The movement of the plates also generated avalanches and landslides in distant locations:

- *Avalanches*: on 25 April the earthquake triggered avalanches on Mt Everest, killing 22 people. As it was the prime climbing season, between 700 and 1000 people were ascending Mt Everest or staying at Everest Base Camp.
- *Landslides*: on 4 May, a post-seismic landslide obliterated Langtang village, resulting in 300 people being buried under metres of ice and rock. Most of the trekkers and international search teams working through the earthquake debris from 25 April were killed.

5.13.1 Comparing earthquake and aftershock in Nepal 2015

	Gorkha earthquake: 25 April 2015	Aftershock: 12 May 2015
Richter scale	Magnitude of 7.8	Magnitude of 7.3
Epicentre	80 km north-west of Kathmandu	65 km north-east of Kathmandu
Hypocentre	Depth of 15 km	Depth of 10 km
Impacts	• Shaking triggered avalanches on Mt Everest and landslides in the Langtang Valley • Tremors felt in northern India, Bangladesh, China, Tibet and Bhutan	• Tremors felt in northern India such as Bihar • Fewer deaths and destruction compared to 25 April earthquake

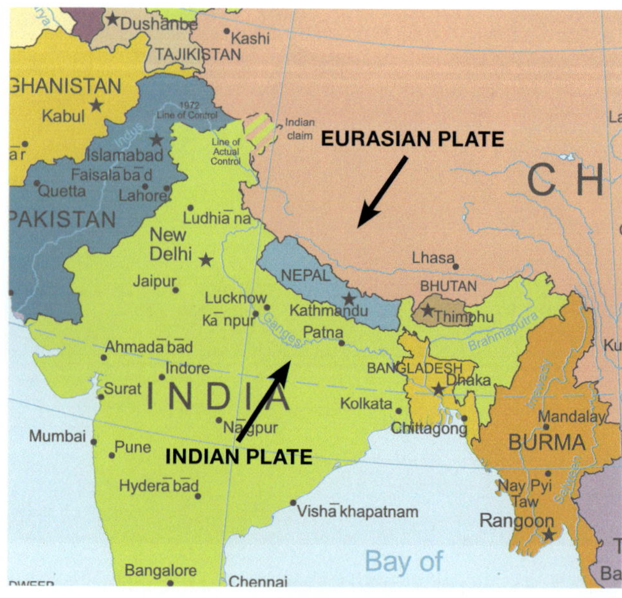

5.13.2 Movement of tectonic plates

ISBN 978 1 4586 6277 4

5.13.3 Impacts of Nepal's earthquake on buildings

5.13.4 A woman sits working beside her damaged home

Impacts: poverty increases

The earthquake pushed one million Nepalese below the poverty line. It damaged properties, and affected agriculture, sanitation, health, water and education. It is estimated to have cost the country 35% of gross domestic product (GDP). Unfortunately, Nepal's poorest are receiving the least aid, especially the lower caste Dalits (or untouchables). Poverty has led to the increase in human traffickers preying on girls and women to work in brothels in South Asia.

Unfortunately the worst affected earthquake areas are also vulnerable to floods, rock falls and landslides from annual monsoon rains.

Geo**activities 5.13**

Knowledge and understanding

1 What was the cause of the earthquake?
2 What were the impacts of the geomorphic hazard?
3 What was the extent of the earthquake?

Inquiry and skills

4 Refer to 5.13.1. Compare the epicentre and hypocentre on 25 April with 12 May.
5 Refer to 5.13.2 and explain how the movement of the plates caused earthquakes.
6 Collect three electronic media articles on the Nepalese disaster. Summarise the articles as an oral report.
7 Investigate another current geomorphic hazard. Include location, causes, impacts, responses and the role of technology in monitoring and predicting the hazard. Present investigation using web 2.0 tools.

ISBN 978 1 4586 6277 4

5.14 Responses and technology: Nepal's earthquake

In 2015, Nepal's earthquake and aftershocks affected 5.6 million people, many of whom lived in poorly constructed shelters. The National Society for Earthquake Technology—Nepal (NSET) aims to improve disaster preparedness and responses. To assist the millions of displaced people, NSET encourages the construction of 'Duinale Ghars'—low-cost temporary shelters built with salvaged materials. They are designed to resist heavy monsoon rains, high wind speeds and earthquakes for four years.

Multiple responses

International Search and Rescue Advisory Group (INSARG) and foreign medical teams were essential in the first days after the earthquake, because thousands of people were trapped under debris and many others were suffering serious injuries. Damaged roads meant the delivery of aid to remote mountain locations was extremely difficult. As a solution, the World Food Programme (WFP) used Nepalese porters and donkeys to assist quake victims.

Many countries donated aid, including the UK, USA, Australia and China. Numerous aid organisations, such as United Nations Children's Fund (UNICEF), worked to support children, as more than one million children were left without classrooms. UNICEF also provided water, sanitation, medical supplies and psychological support.

Nepal
- government set up tents and distributed food
- youth centres and mosques became shelters
- 50 000 soldiers involved in search-and-rescue and assisting medical aid

Global organisations
- Médecins Sans Frontières (MSF) supplied medical teams and clean water
- World Health Organization (WHO) mobilised funds to treat spinal cord injuries, which accounted for one third of the injuries

Overseas countries
- India delivered relief materials and rescued people
- Australia contributed $20 million
- Bangladesh provided tents, dry food, water and blankets

Individuals
- digital volunteers
- contribution of money and expertise (e.g. doctors, engineers, social workers, military)

5.14.1 Responses: individual, groups and governments

ISBN 978 1 4586 6277 4

World Vision provided 100000 people with:
- water purifiers
- hygiene kits
- cooking kits
- mosquito nets
- sleeping bags

In addition:
- 3919 people received tarpaulins
- 111 people were provided with child friendly places where children could play safely

5.14.2 Response by communities and World Vision. Photo shows emergency accommodation provided by Nepalese communities

Technology responds

Mobile phones, Twitter feeds and Facebook posts delivered the first glimpses of earthquake-damaged Nepal, and news about loved ones. In response to the disaster, Google revived Person Finder, and Facebook launched Safety Check, urging people to click a button to let their family know they were safe.

Other technology included:

- *Drones*: showed photos of buckled buildings, broken roadways, and refugees living in open spaces. They delivered medical supplies to remote villages.
- *Crowdsourcing*: DigitalGlobe's crowdsourcing, with over 21 000 volunteers, tagged images of damaged buildings and roads. This provided immediate information to response teams.
- *Satellite imagery*: images from Sentinel-1A supported emergency aid organisations.
- *Crises mapping*: located thousands of people living in remote parts in need of medical assistance and basic supplies.
- *Finding Individuals for Disaster and Emergency Response (FINDER)*: microwave-radar technology penetrated rubble 20–30 m deep, looking for a 1 mm movement from heartbeats.
- *Digital volunteers*: within 48 hours, 2182 digital volunteers trawled high-resolution satellite imagery, identifying 3128 damaged buildings.

Chautara, capital of Sindhupalchowk district, is perched on top of a steep hill 50 km east of Kathmandu. The town suffered the largest number of deaths from the earthquake on 25 April, and over 90% of its homes were destroyed. FINDER was used to save the lives of four men trapped beneath rubble for days in Chautara.

5.14.3 Looking for bodies under a collapsed three-storey building in Chautara

Geo**activities 5.14**

Knowledge and understanding

1 What do the following acronyms mean: NSET, INSARG, UNICEF, MSF, WHO and FINDER?

2 Explain how technology helped aid organisations reduce deaths and improve the population's wellbeing.

Inquiry and skills

3 Refer to 5.14.1 and 5.14.2. In groups, research how local, national and global organisations improved the lives of people affected by the earthquake. Present research using web 2.0 tools.

4 Imagine the Nepalese Prime Minister asked you to design an earthquake emergency kit to be used in the event of another earthquake. In groups, decide what you would include in the kit.

5 Find a survivor story of a person affected by the earthquake. Summarise the story.

6 What is your role as an Australian concerning the earthquake in Nepal? Do we have any obligations or responsibilities? If we do, what are they?

ISBN 978 1 4586 6277 4

5.15 Geographical inquiry: Nepal's landslide

Movements of the Earth causing earthquakes and volcanic explosions leave people in need of aid. Local and international aid organisations investigate geomorphological disasters and ask questions such as: Where did it occur? Why did it happen? Who are the vulnerable people? What were the economic costs? Could the community be better prepared? Will aid provide shelter, food and medicine? How can this disaster be prevented in the future? Finding answers to these questions involves working through three mains steps in a geographical investigation:

- acquire geographical information
- process geographical information
- communicate geographical information.

Landslide in Langtang Valley

Langtang is a small village in Nepal, visited by 20000 international tourists each year. The village is located below a steep ridge that contained a glacier and large snow area. On 4 May 2015, a post-seismic landslide obliterated the village, causing 300 people to be buried under metres of snow, ice and rock.

1. Acquire geographical information
- identify an issue
- develop geographical questions to investigate the issue
- collect primary geographical data (e.g. fieldwork, interviews, questionnaires)
- gather geographical information from secondary sources (e.g. internet, journals, newspapers)
- record information

2. Process geographical information
- evaluate information for bias and reliability
- represent information in appropriate forms such as maps, graphs, statistics, spatial technologies and visual representation
- interpret data and information gathered

3. Communicating geographical information
- communicate results using a variety of strategies
- reflect on the investigation findings
- propose individual or collective actions
- predict expected outcomes
- where appropriate, take action

5.15.1 Three steps in a geographical investigation. Photo shows impacts of Nepal's earthquake in 2015

ISBN 978 1 4586 6277 4

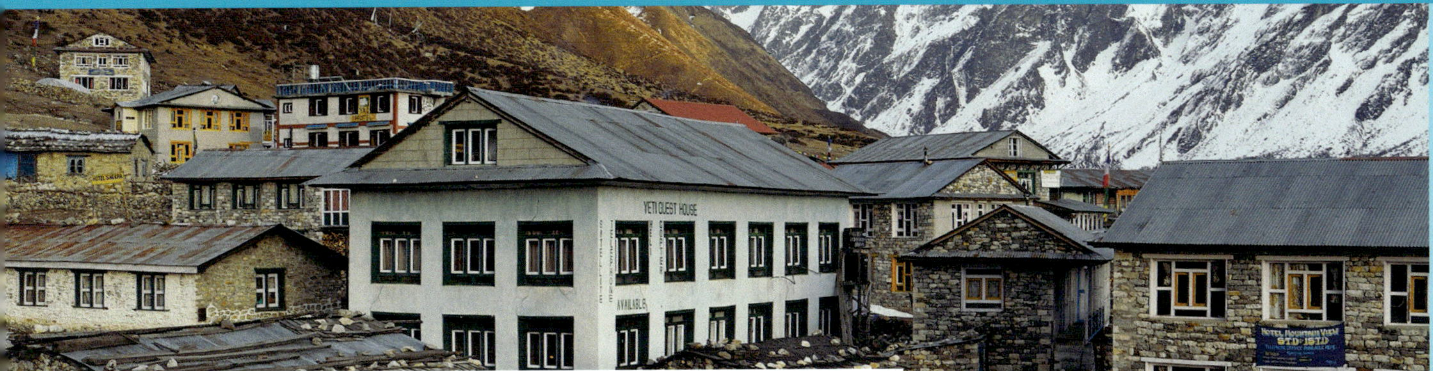

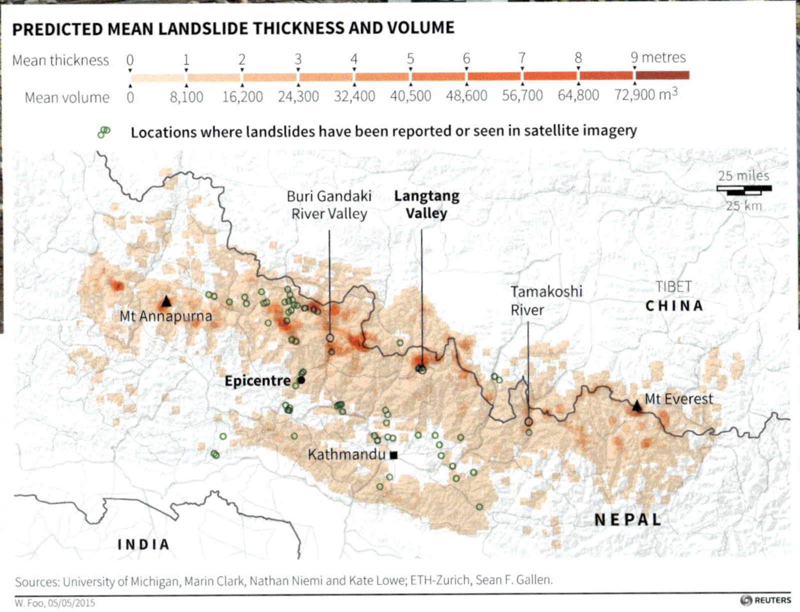

PREDICTED MEAN LANDSLIDE THICKNESS AND VOLUME

Mean thickness 0 1 2 3 4 5 6 7 8 9 metres

Mean volume 0 8,100 16,200 24,300 32,400 40,500 48,600 56,700 64,800 72,900 m^3

Locations where landslides have been reported or seen in satellite imagery

Buri Gandaki River Valley
Langtang Valley
25 miles
25 km
Mt Annapurna
Tamakoshi River
TIBET
CHINA
Epicentre
Mt Everest
Kathmandu
NEPAL
INDIA

Sources: University of Michigan, Marin Clark, Nathan Niemi and Kate Lowe; ETH-Zurich, Sean F. Gallen.

W. Foo, 05/05/2015 REUTERS

5.15.3 The earthquake triggered a large avalanche and mudslides in Lantang Valley

5.15.2 Langtang village before second earthquake hit on 12 May 2015

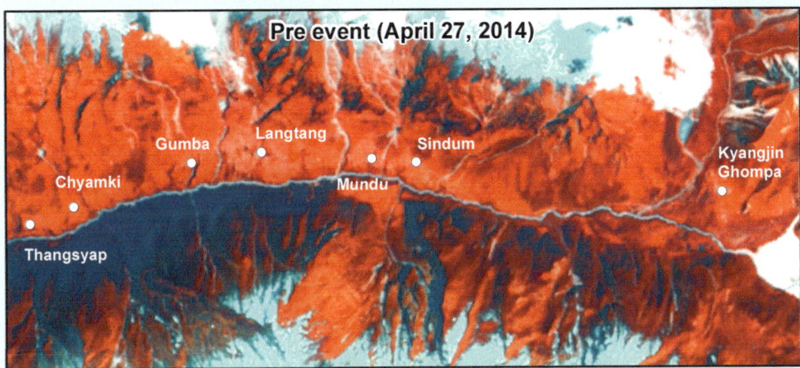

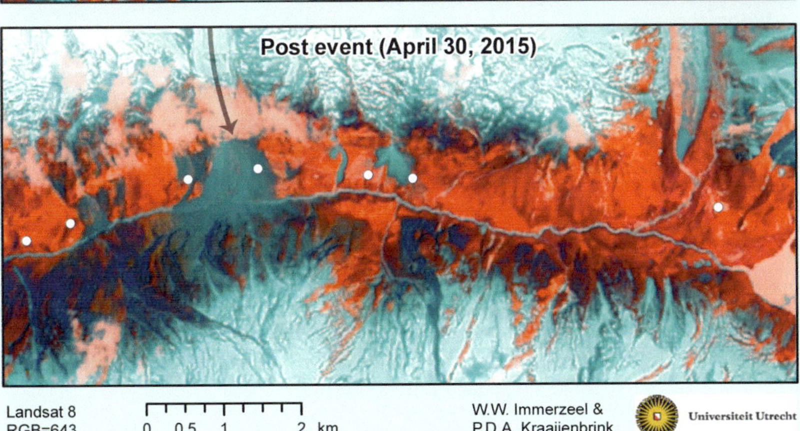

Landsat 8
RGB=643

0 0.5 1 2 km

W.W. Immerzeel &
P.D.A. Kraaijenbrink

Universiteit Utrecht

5.15.4 Satellite image before and after the landslides at Langtang

Geoactivities 5.15

Knowledge and understanding

1 Describe how the village of Langtang was wiped off the map in a few terrifying seconds.

Inquiry and skills

2 Refer to 5.15.1. Using the internet, complete the geographical inquiry process on the geomorphic hazard that became a disaster in Nepal in 2015. Include diagrams, photographs and satellite images.

3 Refer to 5.15.2. Explain the links between earthquakes and avalanches, and earthquakes and landslides.

4 Refer to 5.15.3. List the villages affected by the Langtang landslide and avalanche.

5 Refer to 5.15.4. Explain how satellite imagery provides a different view of the disaster.

6 Although the avalanche damaged the ladders at the Khumbu Icefall, a handful of mountaineers were granted government permission to climb again on 29 April 2015. What are your thoughts on this decision?

ISBN 978 1 4586 6277 4

Geothink

Christchurch earthquake 2011

1 Refer to 5.16.1. Research the 2011 Christchurch earthquake and answer the inquiry questions. Include a map showing the location of the earthquake and photographs to illustrate the impact of the disaster. Present your research as a PowerPoint.

On 22 February 2011 an M6.3 earthquake struck New Zealand's South Island. Caused by the movement of plates, the earthquake was centred 10 km south-east of Christchurch, the second most populous city in the country. The disaster killed 185 people and the cost of rebuilding is anticipated to be NZ$15 billion. The building codes required buildings with a 50-year design life with planning for the possibility of a similar 1-in-500-year event. But should this anticipate a 1-in-2500-year event?

5.16.1 Christchurch earthquake 2011

Japanese earthquake and tsunami 2011

2 Refer to 5.16.2 and 5.16.3 and unit 5.9. Imagine you were a reporter in Minamisanriku in Japan when the tsunami hit the town. Interview a person living in the town covering the following points.

- a Where is the town located?
- b What was the cause of the disaster?
- c What were the impacts of the disaster?
- d Why do you live in a town located on a fault line that has experienced deadly tsunamis in the past?
- e What precautions were taken to reduce the impacts of the tsunami on the town?
- f Should the town be rebuilt or demolished to avoid a hazard becoming another disaster?
- g How does the Tsunami warning system transmit data?
- h Do you think the Tsunami warning system will reduce future disasters?

Minamisanriku is a fishing and tourist town located on the east coast of Japan. In March 2011 a tsunami with waves over 16 m high engulfed homes and wiped out most of the town, most of which is less than 5 masl. One year later, Japan still mourns the deaths of more than half of the town's population.

People were aware they lived in an area where hazards occur as a concrete seawall had been built to protect the town. The town had experienced tsunamis in 1896 and 1993, and it was affected by the 1960 Chile tsunami. The media and schools publicised precautionary measures and how to respond to the hazard. Sirens and loud speakers warned people to move to higher ground. In the wake of the disaster the tsunami-detecting system has been expanded.

5.16.2 Japanese earthquake and tsunami 2011

ISBN 978 1 4586 6277 4

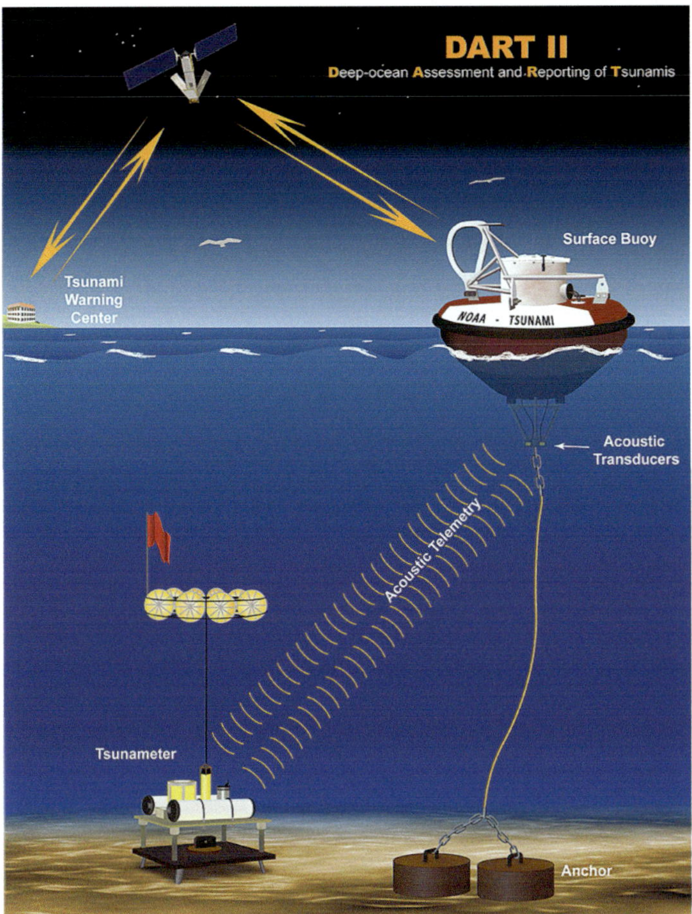

5.16.3 Tsunami warning system

5.16.4 Lack of tap water in Nepal after the earthquake forces woman to wash clothes in the street

Journalist for a day

3 Refer to 5.16.4, the photograph taken after Nepal's earthquake in 2015. Geomorphic disasters make popular news items. If you were employed as a journalist, you would be required to research the validity of the information and determine bias and accuracy before it goes live online or on TV. Answer the following questions before you broadcast your information.

- *a* Would you show photos of dead bodies or grieving parents? Why or why not?
- *b* Would you focus on good stories (survival and aid) or about poor displaced and injured people?
- *c* Should you encourage people to return to Nepal as tourism is critical to the economy or focus on danger?
- *d* Would you promote the use of water for long showers in tourist hotels while thousands of Nepalese lack access to running water to drink, cook and wash clothes?

What did I learn?

4 Answer the following inquiry questions:

- *a* What is a geomorphic hazard?
- *b* What are the causes of geomorphic hazards?
- *c* How does a geomorphic hazard become a disaster?
- *d* What is the spatial distribution of one geomorphic hazard, such as an earthquake?
- *e* What are the impacts of one geomorphic hazard on people and environments?
- *f* How do individuals, non-government organisations, governments and international organisations respond to geomorphic disasters?
- *g* How can technology monitor and predict geomorphic hazards?
- *h* Why do geomorphic disasters have a greater impact on poor people and poor nations?
- *i* Why do people continue to live in areas that experience geomorphic disasters?

Chapter 5

ISBN 978 1 4586 6277 4

unit 2

Place and liveability

Santorini is a Greek island in the southern Aegean Sea. Santorini remained after a volcanic explosion, which destroyed early settlements and created the existing geological caldera.

'Place and liveability' focuses on the factors that influence the decisions people make about where to live, and their perceptions of the liveability of places. This unit examines how the liveability of places is influenced by accessibility to services and facilities, and by environmental quality. People's choices about where to live are influenced by culture, age, income, social connectedness, community identity and perceptions of crime and safety. Strategies used to enhance the liveability of places—especially for young people—are examined. The unit includes local and global examples, with a focus on Australia, Europe and the Asia region.

Key inquiry questions

- What influences the liveability of places?
- Which is the most liveable city in the world?
- What strategies can be taken to improve liveability?

Outcomes

At the end of this unit students should be able to:

- locate and describe the features of places that impact on liveability
- explain how liveability changes over time due to the interactions between people, places and environments
- examine the perceptions of different people on the liveability of places, and the role of people, organisations and governments in enhancing the liveability of places
- acquire, process and communicate geographical information on the liveability of a place.

chapter 6

Liveability: perceptions and influences

'Anywhere is paradise; it's up to you.'

Author unknown

Geovocab

affordability: cost can be met from a person's income

DINK: Double Income No Kids—two-income family with no children and a high disposable income

ecological footprint per capita: amount of land used to provide each person with their needs

exploitation: taking advantage of

gated suburb: enclosed housing estate

ghost town: once-flourishing town wholly or nearly deserted, usually as a result of the exhaustion of some natural resource

grey nomad: travelling retiree

lifestyle segment: section of a population following a particular way of life

nomadic: moving from place to place, usually when seasons change

persecuted: suffered persistent mistreatment by another group

population density: number of people for every square kilometre of land

refugee: In 1951 the United Nations High Commission for Refugees (UNHCR) defined a refugee as someone who, 'owing to a well-founded fear of being persecuted for reasons of race, religion, nationality, membership of a particular social group or political opinion, is outside the country of his nationality, and is unable to, or owing to such fear, is unwilling to avail himself of the protection of that country'

regional: area outside large cities centred on a large town

ISBN 978 1 4586 6277 4

A nomad camp between Chakhcharan and Jam Pal Kotai i Guk Aimaq, Afghanistan.

sea change: movement to the seaside for a change of lifestyle
sea gypsy: nomad, constantly moving around the oceans fishing and collecting seafood, and living most of their lives on boats or homes built in the sea
sense of place: feeling of belonging to a place and of a place being a part of a person's identity
tree change: movement to the country for a change of lifestyle
web 2.0 tools: websites that are not static and allow input from the user
yuppy: Young Urban Professional, aged 20–40 years of age, working in a professional occupation (e.g. law or business) and very career focused

ISBN 978 1 4586 6277 4

While some people have very little choice about where they live, most people make their locational choices for a variety of social, cultural, economic, environmental, political and technological reasons (influences). People live in some extreme and often dangerous locations; for instance, places with freezing cold temperatures or exposure to natural hazards, like volcanoes, and human activities, such as war. To many people, these places are unliveable because they do not offer the conditions they are prepared to live with.

Individuals also perceive places differently, depending on their age, their preferred lifestyle, and the ability of a place to satisfy their needs and wants. For example, Amish communities offer a sustainable lifestyle, while gated suburbs promise safety. Employment and housing affordability are two factors heavily influencing our perceptions of liveability. Whatever our choices, it is our individual perspectives about liveability that will decide where we live.

Think, puzzle, explore

- **Place** What are the factors that affect a place where people live?
- **Space** Where are the world's most naturally dangerous places?
- **Sustainability** How does sustainability influence the liveability of places?
- **Scale** How does liveability vary within and between countries?
- **Environment** Why do people live in extreme places such as deserts?
- **Interconnection** What are the interconnections between the environment and liveability?
- **Change** Why do people change where they live?

Geo**skills** in focus

- **Planning** a geographical inquiry into the liveability of a place
- **Collecting** and **interpreting** geographical data
- **Evaluating** and **analysing** geographical data on some extreme places in which people live
- **Concluding** and **communicating** information using a range of texts
- **Reflecting** on the inquiry process and **responding** to an issue

6.1 Living in extreme places

Spectacular and wild places capture our imagination. They include the highest, driest, hottest, coldest and wettest environments on Earth, where **population densities** are very low. People have lived in these environments for thousands of years. Indigenous communities such as the Bedouin in the Sahara Desert and the Yanomami in the Amazon Rainforest may live in extreme places because:

- *sacred ancestral sites* are located there
- they have developed a common sense of identity in these locations through *culture*, *language* and *lifestyle*
- *traditional ecological knowledge* enables them to live there sustainably
- they possess *legal title* to the land.

6.1.1 An indigenous hunter uses a snow machine

Nunavut: cold and remote

The Inuit are Indigenous people who live in the Arctic and sub-Arctic regions of Greenland and North America. Many Inuit live in the Canadian territory of Nunavut. The Arctic experiences a polar climate with long, cold winters and short, cool summers. Precipitation is low and mostly falls as snow. For thousands of years the Inuit have lived as **nomadic** fishers and hunters using traditional knowledge to survive. Their culture is based on seasonal cycles of the land and sea. Igloos and dog sleds are used in winter and animal skin huts and kayaks in summer. The traditional diet of the Inuit consists of whales, walruses, seals and fish, because most plants cannot grow there.

Today Inuit live and work in permanent communities along the coast. Homes are modern and connected to the internet via satellite. Hunting parties use GPS and motorised vehicles to move between communities where there are no roads or rail lines. Everything arrives or leaves by plane or sea, adding to the cost of food and clothing. The Inuit still rely on hunting and fishing as their 'supermarket'.

6.1.2 Location of Nunavut territory in Canada (green)

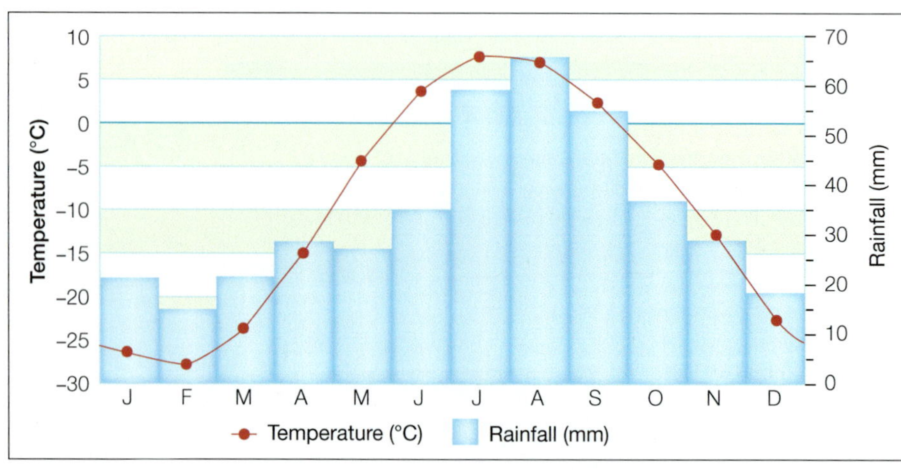

6.1.3 Average rainfall and temperatures in Iqaluit (Nunavut), Canada

ISBN 978 1 4586 6277 4

Geoinfo

- In Inuit language, Nunavut means 'our land'.
- The population of Nunavut in 2015 was 36 886 and the unemployment rate 16.8%.

Sense of place

Animism (the belief that natural things such as animals and plants have a spirit) and stories of supernatural creatures are important to Inuit culture and an important reason for many Inuit to live in Nunavut. The aurora borealis is believed to be images of relatives and friends dancing in the afterlife or the souls of killed animals. These beliefs contribute to the Inuits' **sense of place**, belonging and community identity.

Impact of climate change

Inuit families separated from their villages and hunters unable to reach bears and seals because of melting ice are warning signs that climate change is threatening traditional lifestyles. Ancestral knowledge, scientific research and technology are being used to create liveable and sustainable Inuit communities for the future. Traditional knowledge is taught in schools and tribal elders are honorary teachers. Architects are designing buildings for communities affected by climate change.

Geoactivities 6.1

Knowledge and understanding

1. List three reasons why the Inuit live in the Arctic.
2. Explain the high cost of living in Nunavut.
3. Discuss the advantages and disadvantages of living in Arctic and sub-Arctic regions.
4. Do the Inuit live in Nunavut for economic, social, cultural, political and/or environmental reasons?

Inquiry and skills

5. Refer to 6.1.2.
   a. Estimate the latitude and longitude of Iqaluit.
   b. Calculate the distance from Iqaluit to Cambridge Bay.
   c. Estimate what percentage of Nunavut is above the Arctic Circle.
6. Refer to 6.1.3.
   a. Describe the climate of Iqaluit. What are the hottest and coldest months, annual temperature range and annual precipitation?
   b. What would it be like in Iqaluit in January?
   c. What would happen to the Arctic Ocean around Nunavut in winter?
   d. How do people make cold climates more liveable?
7. Refer to 6.1.4 and explain how technology and the aurora borealis influence the liveability of Nunavut for Inuit people.
8. In groups investigate why people live in other extreme places, using the examples in 6.2, 6.3, 6.4 or 6.5.

6.1.4 The aurora borealis and an inukshuk—a stone marker traditionally used by the Inuit

ISBN 978 1 4586 6277 4

6.2 Floating worlds and culture

Coastal islands and cities such as the Maldives and New York may become unliveable in the future because of climate change. While the Netherlands continues to build structures to hold back the sea, Dutch architects are now designing floating cities, like the Lilypad, to house climate refugees.

The Indian and Pacific oceans are already home to thousands of nomadic fishers and hunters. The Bajau Laut and the Moken are two of Asia's **sea gypsy** communities. They live on small boats with no fixed address and rely on the sea for food and trade.

Moken spend up to nine months at sea, only coming ashore to bury the dead or during the stormy wet season.

Sea gypsies: body and soul

Sea gypsies have developed unique physical adaptations. They can free dive to over 20 m, slow their heart rate to 25 beats per minute to reduce buoyancy, and have extraordinary underwater vision. The Moken and Bajau Laut have a spiritual bond with the sea. Bajau Laut thank the god of the sea for good catches and use mediums to remove bad spirits.

Overfishing and piracy are threatening the safety and food security of the oceans for the sea gypsies, while young people are seeking employment and a modern, sustainable lifestyle on land.

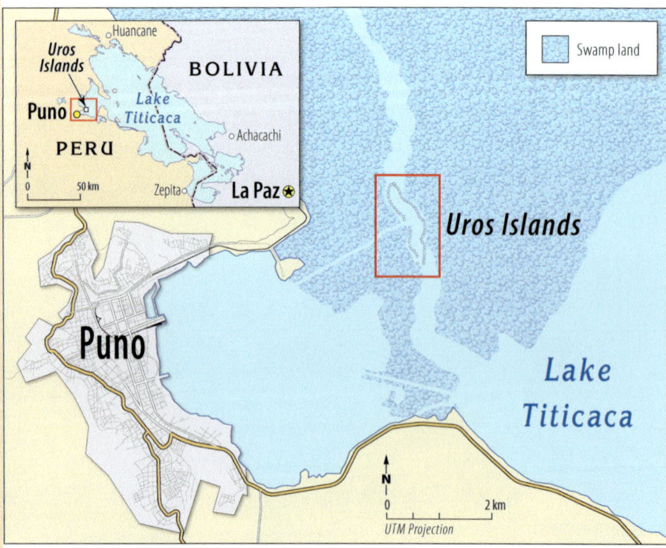

6.2.1 Lake Titicaca

i Geoinfo

- Many sea gypsies are stateless, meaning they do not belong to a country.
- When sea gypsies step on solid ground they often suffer land sickness.

6.2.2 Young sea gypsies in Semporna, Borneo. Today some sea gypsies still live in small communities built on stilts in shallow coastal bays.

Reed islands

Islands made from floating aquatic plants can be found in Lake Titicaca, Peru. The traditional Uros people made permanent islands from layers of totora— a thick reed. The reeds were used to make homes, watchtowers and boats and the Uros survived by fishing, hunting and collecting birds' eggs. Storms were the biggest threat to the liveability of the islands. Today the lake is still home to several hundred Uros. Traditional activities are still important but life has modernised with solar power, television, the internet and motorised boats.

Cultural survival

The Uros moved their islands to safe locations near shore after violent storms in the 1980s. This meant they could work and study in Puno, and tourism expanded. Uros showcase their culture to 200000 tourists a year with 80% of the population now in tourism-related employment. These changes have

ISBN 978 1 4586 6277 4

improved the islands for younger generations who now study hospitality, foreign languages and tourism. While the Uros have been criticised for allowing the 'disneyfication' of their culture, tourism may guarantee the future liveability of the islands and the survival of the Uros culture.

A floating future

Scientists warn of rising sea levels placing several areas of the globe in danger of vanishing from the map, disappearing under water. Society must adapt and perhaps floating houses are an option.

6.2.3 Lilypad islands—possible future floating worlds

Geo**activities 6.2**

Knowledge and understanding

1 Explain why the Dutch are planning floating cities, such as the Lilypad.
2 List three examples of the connection sea gypsies have with the sea.
3 What two issues are threatening the liveability of oceans for sea gypsies?
4 Suggest ways that tourism would guarantee the survival of the Uros culture.
5 Create a definition for the term 'disneyfication'.
6 How important is culture as a location factor for the Uros and sea gypsies?
7 What changes to the reed islands of Lake Titicaca made them more liveable for young indigenous Uros people?

Inquiry and skills

8 Refer to 6.2.1.
 a Refer to an atlas to determine the latitude and longitude of Lake Titicaca.
 b Measure the distance from Puno to the Uros Islands.
 c What is the approximate length and breadth of Lake Titicaca? Calculate the area of the lake.

9 Refer to 6.2.2.
 a Is there any evidence of sanitation in the photograph? What do you think happens to human waste?
 b As a class discuss the basic needs of young children. Create a list of these needs.
 c Watch a YouTube clip or documentary on sea gypsies (see Geolinks). Draw a conclusion about the suitability of the ocean as a place for children to live.

10 Refer to 6.2.3.
 a What features of the Lilypad make it a sustainable option for living in a world facing climate change?
 b Would you consider living on a floating island? What would be the advantages and disadvantages?
 c Where might people living on floating islands get employment?
 d What would the island need to contain to be an attractive place for teenagers?

11 Investigate the floating villages in Halong Bay, Vietnam.

ISBN 978 1 4586 6277 4

6.3 Nature, danger and livelihood

There are over 1500 active volcanoes in the world—an average of 20 erupt daily. Although volcanoes often bring destruction and death, millions of people choose to live near them because:

- volcanic soils are *rich in nutrients* important for agriculture
- they *provide employment* in tourism (e.g. hot mud baths), construction (e.g. providing stones for buildings) and mining (e.g. the sulphur used in medicine)
- they provide *geothermal energy*.

Volcanoes—life and death

Indonesia sits along the Pacific Ring of Fire—where Earth's tectonic plates collide and cause earthquakes and volcanic eruptions. Some 240 million people live in the shadow of Indonesia's active volcanoes, where ash clouds, mudflows, poisonous gases, lava flows and tsunamis threaten lives and livelihoods. In 2011, clouds of ash from Mt Merapi caused the evacuation of over 50000 people.

Volcanoes have cultural and economic significance. Indonesians believe that rumbling volcanoes signal restless gods. People perform rituals in which they offer rice, money and chickens to the gods to appease them, and Balinese sleep with their head towards nearby volcanoes. Hindu priests climb Mt Agung in Bali and collect hot water to sprinkle on rice farms to ensure a profitable harvest. Volcanic eruptions might destroy crops and livestock, but the rich soils allow farmers to harvest three crops of rice a year.

Geoinfo

- Ash clouds cause 28% of volcano-related deaths.
- About 50–60 volcanoes erupt every year.

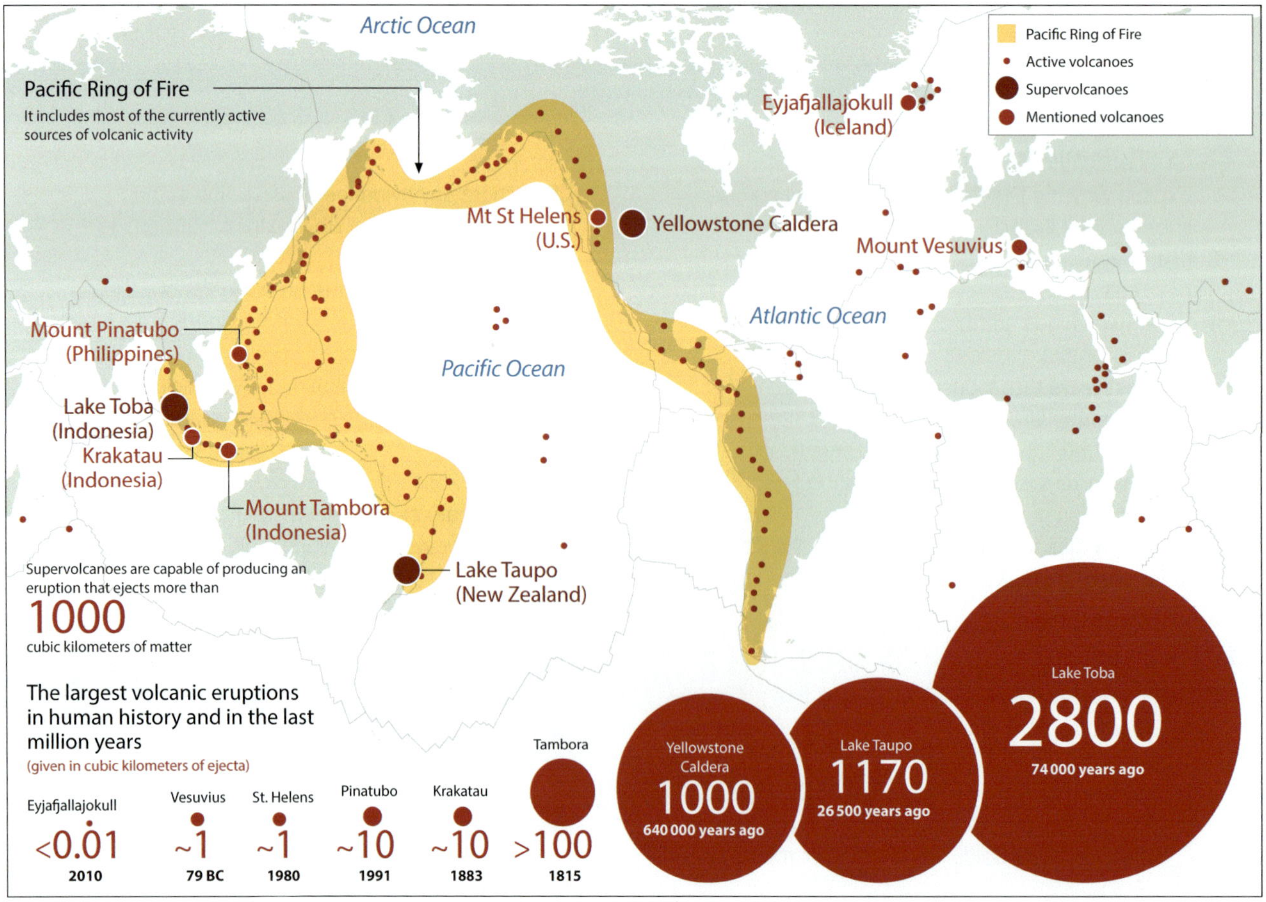

6.3.1 The world's most dangerous volcanoes

ISBN 978 1 4586 6277 4

Why take the risk?

About 500 million people across the planet live on or near volcanoes, but consider themselves to be safe because:

- there are often many years between eruptions
- modern technology monitors volcanic activity
- early warning systems reduce the risk of death.

Cultural beliefs and the benefits of modern technology, as well as the economic benefits of living near volcanoes, make these dangerous locations liveable for millions of people. Not all people are happy living alongside volcanoes, but many can't afford to move.

Is Australia safe?

In 2010 and 2011, volcanic eruptions in Iceland and Chile produced millions of tonnes of ash, which circled the globe and delayed air travel in Europe, South America and Australia. Australia's proximity to the Pacific Ring of Fire puts people at risk. Indonesia sits beneath the aviation corridors linking Australia to Asia and Europe. Darwin's global ash-monitoring centres issued over 1700 advisories to airlines on threats from active volcanoes in 2011. The Bureau of Meteorology stated it was 'virtually impossible to fly in and out of Australia without going over volcanic activity'.

6.3.2 Indonesians live with volcanoes every day

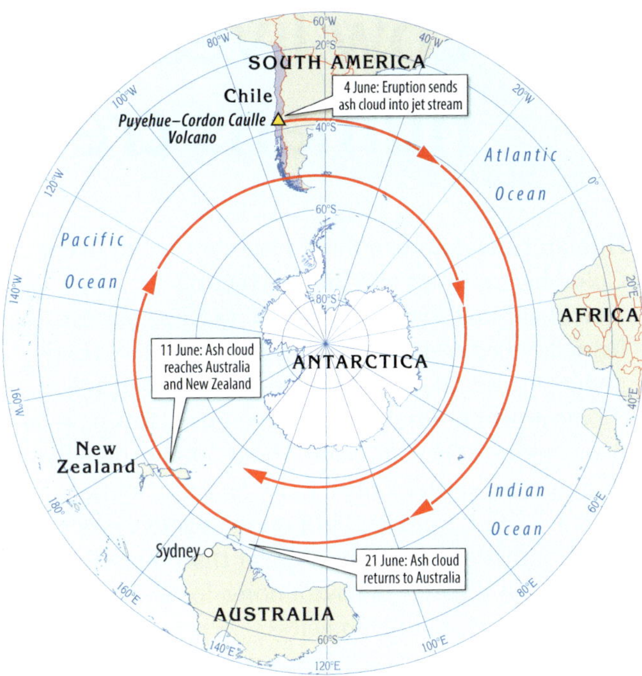

6.3.3 Ash clouds travel the globe in 2011

Geoactivities 6.3

Knowledge and understanding

1 What are the economic benefits of living near volcanoes?
2 Explain the disadvantages of living near volcanoes.
3 Indonesians accept volcanic activity for cultural reasons. What does this mean?
4 How has technology reduced the danger of living near volcanoes?
5 How are Australians affected by volcanic activity?
6 Why do you think the economic benefits of living near volcanoes seem to be more important to Indonesians than the risk to life and property?

Inquiry and skills

7 Refer to 6.3.1.
   a Where is the Ring of Fire located?
   b Discuss the relationship between Indonesia's volcanoes and the Ring of Fire.
   c Name and locate the largest supervolcano.
8 Refer to 6.3.3.
   a Explain how a volcanic eruption affects people living in distant places.
   b How long did it take the Chilean ash cloud to reach Australia each time?
   c Use a world globe or ICT to explain the direction that the ash cloud travelled the world.
9 Illustrate one connection between volcanoes and liveability using a flow diagram.

Chapter 6

ISBN 978 1 4586 6277 4

6.4 Low liveability: where humans create danger

Human behaviour can make places unsafe to live in. Colombia's killings and Rio de Janeiro's 'quicknappings' have contributed to these places being labelled as dangerous. In Rio people have been abducted and taken to ATMs to pay their ransom.

Most people who live in dangerous places have to do so because they are poor. About 5000 people live in a squatter settlement at Tudor Shaft in South Africa. Their homes are built on radioactive ground, where radiation is 15 times higher than normal. One wonders: is it better to have a home located in a dangerous place or be homeless?

Nuclear neighbours

Approximately 1 million people live within a 30 km radius of a nuclear power station. After the nuclear disaster at Fukushima, Japan, people questioned the safety of living adjacent to these plants. In 2011, the Fukushima Daiichi plant was severely damaged by an earthquake and tsunami.Radiation made the area unliveable, forcing thousands to evacuate.

Nuclear power provides clean, cheap electric power as an alternative to coal-fired power stations. In July 2012 there were 435 nuclear power plant units in 31 countries, with another 62 under construction.

Although reactors are built to the highest safety standards, accidents can cause death, sickness, starvation and environmental contamination. Natural hazards, human error, increasing population, war, terrorism and ageing plants create safety risks.

The path of radiation ...

Direct radiation
Exposed fuel rods and leaked radioactive material lead to high radiation doses in the area immediately surrounding a reactor. Particularly dangerous for workers is **gamma radiation**, which even protective suits are unable to shield from.

Radioactive particles
The venting of steam from cooling waters or fire or explosions at reactors can send radioactive elements like **cesium, plutonium or iodine** directly into the atmosphere.

Air
Fine radioactive aerosols or particles can even enter buildings with closed windows, through air vents and air conditioning systems. Residents located near the evacuation zone are advised to remain in closed buildings.

Ocean
The ocean can be polluted through contaminated cooling water and also by rain. Radioactive particles can enter the food chain through fish or other sea animals.

Lakes, soil, groundwater
Radioactive particles spread by rain can gather in lakes or reservoirs that are used to collect drinking water.
Most of the particles are filtered out by the soil, but some can still make their way into the groundwater.

Agriculture
Contaminated soil is another path through which radioactive particles can make their way into the food chain, entering into animal feed, milk, eggs or meat products. Radiation can also enter into plant-based foods like cereals, cabbage, lettuce or fruits through rainfall or the soil.

... Effects on Humans

Gamma radiation
The electromagnetic waves wreak havoc with genes and disrupt the creation of blood cells. It leads to the danger of uncontrolled cell growth and tumors.

Iodine-131
Half life: 8 days
If accumulated, it can trigger cancer of the thyroid. Consumption of iodine tablets can hinder the accumulation of radioactive iodine.

Cesium-137
Half life: 30 years
Can be passed through food into muscle and nerve cells and cause cancer. Can also create deformities in unborn babies or lead to miscarriages.

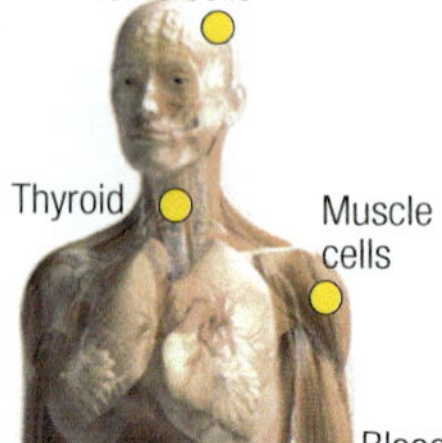

DER SPIEGEL

6.4.1 The path of radiation

ISBN 978 1 4586 6277 4

Geoinfo

Contaminated tuna from Japan's nuclear accident have been caught in California.

After a nuclear accident

The effects of the world's worst nuclear accident in 1984, at Chernobyl in Russia, continue today. Radioactive isotopes spread across 200 000 km^2 of Europe, causing 134 workers to die and thousands of children to develop cancer. The land became too contaminated to farm. Farmers are currently growing radioactive crops to make biofuels—the crops suck up radioactive materials, making the soil suitable to grow food within decades rather than hundreds of years.

For many people, the fear of a nuclear accident would prevent them from living beside a nuclear neighbour. After the clean-up of the Fukushima area, evacuated people will need to decide whether they should return home.

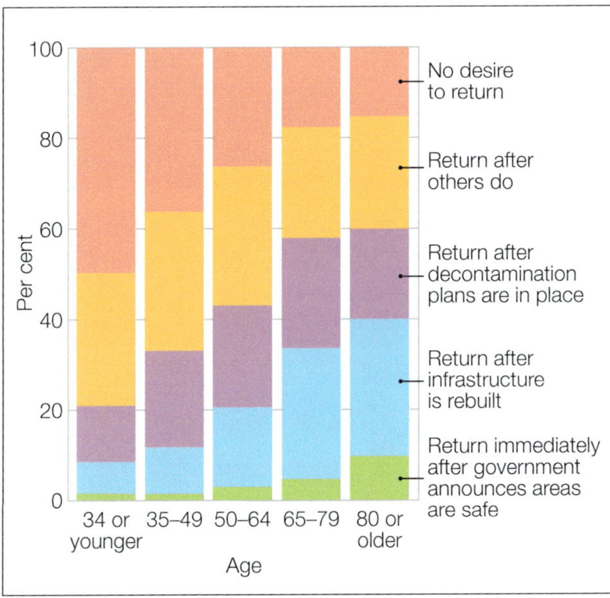

6.4.2 Will Japan's evacuees return home?

War zone: Afghanistan

In every war zone, civilian lives are at risk. The United Nations has declared Afghanistan an extremely dangerous place. Between 2001 and 2014, over 26 000 civilians died in Afghanistan as a result of the war. People in war zones often seek safety in camps for Internally Displaced Persons or refugees where conditions are poor but they are safe.

6.4.3 Effects of war on a family home in Afghanistan

Geoactivities 6.4

Knowledge and understanding

1 Construct a mind map showing potential threats to the safety of nuclear plants.
2 Suggest reasons why people live close to nuclear power plants.
3 What made the area around the Fukushima power plant unliveable?
4 Explain why Afghanistan is a dangerous place for civilians.

Inquiry and skills

5 Refer to 6.4.1.
  a Explain three ways someone could be contaminated by radiation.
  b List three possible impacts on the health of people contaminated by radiation.
6 Refer to 6.4.2.
  a Discuss the influence of age on people's plans to return home. Use statistics.
  b Would you return knowing what the radiation risks are? Explain your answer.
7 Survey your class:
  a Develop a questionnaire that will reveal the attitude of your classmates to living near a nuclear reactor.
  b Graph the results, analyse them and make a concluding statement.
8 Which do you consider to be the most liveable place: a war zone or a place next to a nuclear power plant? Discuss with a classmate and come to a consensus. Present your opinion to the class.

ISBN 978 1 4586 6277 4

6.5 Seeking safety and shelter: Dadaab

Families pack their possessions and walk from Somalia over the border into Kenya. Over 1000 Somali arrived at Dadaab each day during the 2011 famine.

On arrival they join the queue to be processed. The UNHCR, or one of 18 non-government organisations in Kenya, such as Red Cross and Care, run the camps.

Families are provided with food rations and other essentials including tents, kitchen sets, firewood and fuel-efficient stoves.

In a drought it is children who suffer the most. Many require rehydration and inoculation. Most are malnourished.

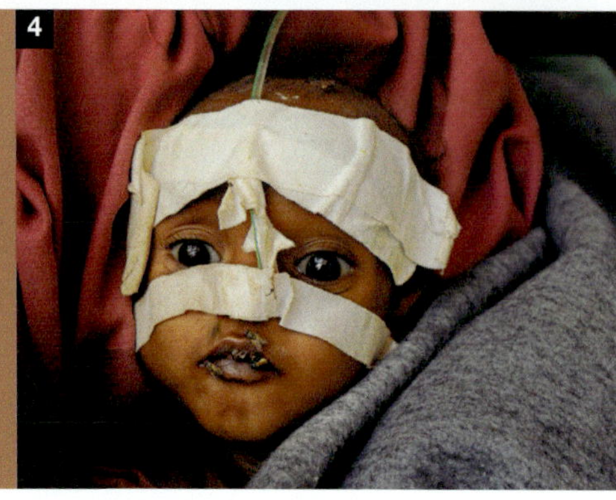

Queuing up for food and water takes many hours each day.

Families wait for news that it is safe to return home. Some who returned home to Somalia returned in the drought crisis of 2011. Some never leave.

When the camps have been declared full, new arrivals construct houses outside the camp boundaries using whatever resources they can find.

In the dry season the camps become hot dustbowls and water is scarce. In the wet season flooding rains mean people have to relocate and water-borne diseases become an issue.

6.5.1 Life for Somalian refugees in a refugee camp

ISBN 978 1 4586 6277 4

Escaping to a **refugee** camp is the decision of desperate people facing conflict, **persecution** or starvation. Leaving your home and crossing the border into another country to seek refuge is a last resort for people whose homes have become unliveable. Refugee camps are built as temporary places to live, where people have their basic needs for shelter, food, water, medical supplies and safety met until they can return home. When going home is impossible, some refugees are resettled in other countries.

The Kenyan government opened its borders to refugees from Somalia fleeing conflict and famine in 1991. Twenty years later, the five Dadaab camps—the world's largest—are home to over 400000 people, mostly from Somalia. Children born in the camps are 'stateless' as they do not belong to any country. Although refugee camps are set up as safety havens, they can still be dangerous places for orphaned children and women.

6.5.2 Food scarcity in East Africa, 2012. Refugees flee Somalia because of food insecurity

Geo**info**

- Women and children make up about 80% of the Somali refugees in Dadaab.
- Many children born in the camps have never been to their 'home' country.

Geo**activities 6.5**

Knowledge and understanding

1. Identify reasons for the Somali refugee crisis.
2. Why do Somali refugees have no choice about where they live?
3. How do officials ensure the basic needs of refugees are met?
4. Why have some Somali refugees been in camps for 20 years?
5. Why do many refugees require medical attention on arrival at Dadaab?
6. Why do you think refugee camps can be dangerous places for orphaned children and women?
7. Do you think safety is a major factor for Australians choosing a place to live?

Inquiry and skills

8. Refer to 6.5.1.
   a. Use the photographs to describe the physical environment and living conditions at Dadaab.
   b. Create a simple flow diagram to show the steps involved for a refugee fleeing from Somalia to Dadaab and their return home or resettlement.
   c. Think about the possessions you own. Create two lists. In one list put possessions you would take with you if you had to flee your home on foot not knowing when you could return. The other list will be what you leave behind.
9. Refer to 6.5.2.
   a. What is 'food security'?
   b. What percentage of Somalia was suffering from some type of food insecurity in 2012?
   c. Using a column graph, graph the populations (by country) that are facing food insecurity in East Africa.
   d. Suggest why most Somali refugees go to the camps at Dadaab.
10. Research the current status of the Dadaab refugee camps, including population statistics.
    a. Investigate current conditions that influence liveability in Somalia and Dadaab.
    b. Propose one strategy that could improve the liveability of Somalia, and allow the refugees in Dadaab to return home.
    c. Propose one strategy that could improve the liveability of Dadaab camps for refugee children.

Chapter 6

ISBN 978 1 4586 6277 4

6.6 Children: poverty, no playground, no choice

Children have no choice about where they live. In places with extremely low liveability, poverty, danger and **exploitation** reduce the quality of life and future prospects of children. Humanitarian workers and journalists believe the most dangerous places to be a child are in Africa and Asia.

The UN Convention on the Rights of the Child states that children have a right to adequate food, health, shelter and education, plus to safety, freedom, play and leisure. The most liveable places (e.g. Australia) must help protect the rights of all children.

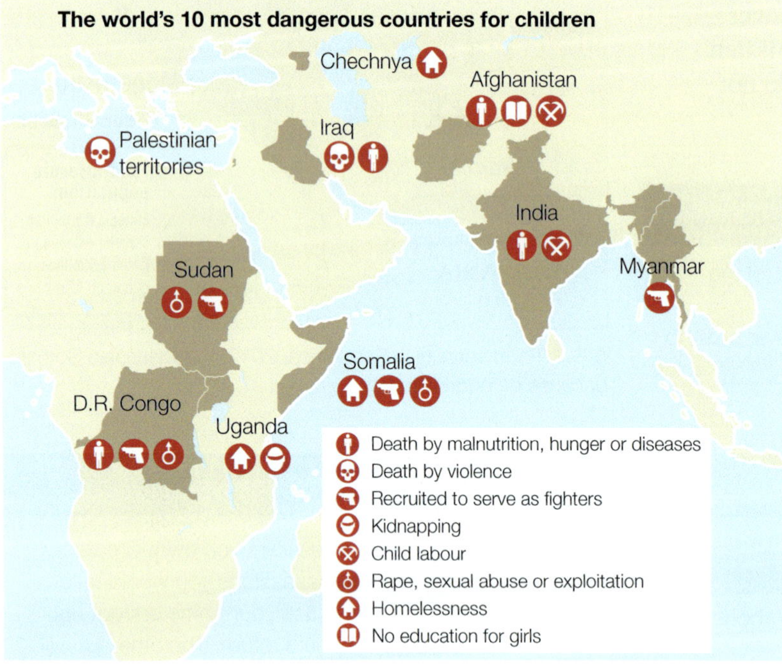

6.6.1 Children in dangerous countries

6.6.2 United Nations campaign to reduce child labour

No choice: poverty

The least liveable places for children are the poorest. Extreme poverty results in stunted physical and mental development. UNICEF estimates 190 million children are stunted by malnutrition and have impaired cognitive growth (development of brain functions). The poorest children are often orphans and street dwellers. For girls, poverty can mean being forced against their will to marry at a very young age, sometimes to pay off family debts.

No choice: exploitation

Poverty leads to exploitation and child labour. An estimated 200000 children work on cocoa farms in West Africa so the world can enjoy chocolate. The US Department of Labor says 71 countries use child labour—often in dangerous situations, such as gold mines in Chad and rock quarries in India. Many children work 80 to 100 hours a week to support their families.

Perhaps the worst form of child labour and violation of human rights is the use of children in armed conflict. Up to 300000 children in 17 countries fight for armies, perform acts of violence and witness the atrocities of war, often against family members. Some are as young as seven years of age, and

ISBN 978 1 4586 6277 4

6.6.3 Children in armed conflict

most are in their teens. Child soldiers suffer physical and psychological damage. They take many years to recover, reconnect with their communities and feel safe enough to imagine a more positive future.

Liveable places for children

The eradication of poverty, child labour and child marriage are essential to improve the liveability of places for children. The following groups act on behalf of children to improve liveability and ensure the basic human rights of children are met:

- *international organisations* (e.g. UNICEF)
- *non-government organisations* (e.g. Care and World Vision)
- *national governments* (e.g. Australian labour laws).

Individual Australian citizens assist organisations working for the rights of children by making donations, volunteering and supporting action to end child exploitation.

Geo**info**

- Children working after school is not child labour.
- A survey in Kolkata in India revealed that 70% of children who lived on the pavements had stunted growth.

Geo**activities 6.6**

Knowledge and understanding

1 Why do children have no choice about where they live?

2 Explain the link between liveability and the rights of children.

Inquiry and skills

3 Refer to 6.6.1.
   a Identify the most dangerous countries for children.
   b What is the underlying cause of the situations described on this map?

4 Refer to 6.6.2.
   a What is the message behind this advertising campaign?
   b Who do you think the campaign is aimed at?
   c Estimate the age of the children in the images.
   d What aspects of childhood would these children be missing?
   e What role could governments play in making their countries more liveable for children?

5 Refer to 6.6.3.
   a Why do places with conflict create low liveability for children?
   b Why would children become soldiers to seek revenge?

6 Create a poster (digital or hard copy) titled 'Let's make places liveable for children'.

Chapter 6

ISBN 978 1 4586 6277 4

6.7 Skill: using and interpreting cartoons

The use of cartoons can be a very valuable tool in getting messages or ideas across about the liveability of places. It is the idea behind the cartoon that is important not the artistic quality of the drawing. Cartoons can be used to:

- identify key geographical concepts (e.g. liveability)
- analyse environmental and social issues (e.g. climate change)
- present different points of view or perspectives.

Cartoons can promote discussion, analysis and critical thinking through the identification of stereotypes and bias, keeping in mind that liveability is a personal perception.

Techniques

Cartoonists use a variety of techniques to express a message:

- *Caricatures* are pictures of drawings that emphasise and exaggerate the peculiarities of people, things or places, usually in a humorous way.
- *Symbols* are easily recognised (e.g. crown = royalty, $$ = money).
- *Colour* suggests the tone or seriousness of an issue, and positive or negative feelings.
- *Headings* and *captions* give clues to the purpose of the cartoon.
- *Irony* uses words to express something different from their usual meaning, or satire, which uses sarcasm or irony to poke fun at or attack something.

Ten key questions for interpreting cartoons

The ten key questions scaffold can assist in analysing cartoons. Create a mind map or annotate the cartoon as you answer the questions, making sure you relate it to your topic; for example, liveability. Summarising your ideas in this way will help you to discuss or write about the message of the cartoon.

Message: the use of child labour reduces the liveability of places for children and robs them of their childhood

Symbolism: 'ball and chain' = slavery

Box of supplies: lots of work for one child

Context: across the world thousands of children are forced to work because of poverty

Facial expressions: happiness

Message: playing is a child's human right

Thought bubble: the child wishes he could play

No mouth: no choice, no say

Irony: children denied a childhood so other children can enjoy theirs

Sitting on the floor = poor workplace conditions

seppo.net

6.7.1 Annotate cartoons with your ideas

ISBN 978 1 4586 6277 4

Ten key questions scaffold

1. Does the cartoon have a *title* or *caption*?
2. What *concept* or *issue* is represented?
3. Are any *symbols* used? If so, what do they represent? What is the relevance of the symbols (e.g. heart or cupid = love; wearing black = villain)? Can the symbols be interpreted in different ways?
4. Is *colour* important to the meaning of the cartoon?
5. Do the *facial expressions* indicate any feelings and emotions?
6. Are *caricatures* used in the cartoon? What is being exaggerated? Why? Are the people in the cartoon real? How exaggerated are they? Are these exaggerations *stereotypes*?
7. Is a particular perspective or point of view presented? If so, is it positive or negative? Is it *biased*? What other points of view are there?
8. What is the *purpose* or *motivation* behind the cartoon?
9. What is the *message* of the cartoon?
10. What is the *context* or *background* of the cartoon? What is its topic and what are the related issues and facts? What else do you know about the topic or issue? Is it a global or local issue? Where is it happening? What are the circumstances? Can you link the cartoon to this current topic?

6.7.2 Cartoons vs photographs

Cartoons versus photographs

Cartoons can have advantages over photographs when studying geographical issues. Cartoons can present perspectives and give a broader context to an issue. However, they can trivialise issues and enforce stereotypes. Photographs put real faces and places to an issue so they may be taken more seriously. The advantages and disadvantages of cartoons must be considered when they are used to investigate geographical issues.

Geo**activities 6.7**

Inquiry and skills

1. Refer to 6.7.2.
   a. Why is the child inside the soccer ball?
   b. What does the size of the foot indicate?
   c. Do you think the player knows a child made the ball?
   d. Why might the child be determined to finish the sewing?
   e. What insight does the cartoon give to the issue that cannot be interpreted in the photo?
   f. Why are cartoons often used in conjunction with stories and photographs?
2. Search the internet for a geographical cartoon relevant to the topic 'Liveability', or choose a cartoon from this textbook. Use key concepts such as environmental quality, safety, human rights, child exploitation or resources to focus your search.
   a. Follow the ten key questions scaffold to analyse the cartoon of your choice.
   b. Draw a mind map of your ideas OR annotate a copy of your cartoon manually or digitally.
   c. Write a paragraph explaining the main message and context of the cartoon. Refer to liveability in your response.

ISBN 978 1 4586 6277 4

6.8 Places, resources and jobs

6.8.1 Abandoned buildings at Bodie, California, USA

Saloons, gambling halls and opium dens were all features of life in Bodie—a thriving Wild West gold rush town of 10 000 people. Settlers came from across the globe, but when the gold ran out the town was abandoned. Bodie soon became the USA's most famous **ghost town**.

Resources attract people

Human migration and settlement are linked to the availability of environmental resources. Populations concentrate around resources that provide employment. New towns emerge with the discovery of minerals, like in the Pilbara, WA. Ghost towns

Rubbish pickers at a dump in Brazil

Camel caravans transport salt to the Ethiopian Highlands

The Afar mining salt at Lake Assal in Djibouti, East Africa

Ship cutters at Alang in India

6.8.2 Local resources provide jobs for poor people in developing countries

ISBN 978 1 4586 6277 4

develop when a resource is depleted (e.g. Big Bell, WA) or when railroads and roads bypass them. In developing countries, garbage dumps become liveable because waste creates employment.

Necessity and choice

At Lake Assal—a desert salt lake in Djibouti (East Africa)—salt provides an income for the Afar people. Temperatures reach 55 °C and there is no shade or fresh water. Work starts at 4 am and finishes by 8 am to avoid the heat, and water is delivered once a week by truck. Camel caravans transport the salt across the desert to towns in the Ethiopian Highlands. In a country with 60% urban unemployment there are few other job opportunities.

In Australia high wages attract thousands of workers to remote places, such as the Kimberley region and Bass Strait. Money compensates workers for difficult environmental conditions and isolation.

In the USA, 'ice road truckers' deliver critical supplies to remote oil communities, such as Deadhorse, Alaska. Treacherous conditions, including blizzards, whiteouts, steep icy slopes and rockfalls, cause serious accidents and deaths. This dangerous job attracts people to Alaska prepared to risk danger to make money.

Trash is treasure

In developing countries millions of city residents live in slums near garbage dumps where they work as waste pickers. An estimated 1.5 million waste pickers in India collect materials such as plastic, metals and e-waste, which they sell for as little as $1 a day. They are exposed to poisons, chemicals and disease, but without education, access to transport and other opportunities there are no other work options.

Young men move to Gaddani (Bangladesh) and Alang (India) where the 800 ships that become obsolete each year are grounded. They work as ship breakers, facing death and disability each day. Rusty machinery, jagged steel and chemicals like mercury create hazardous working conditions. On one occasion a worker was cut in half by a moving cable.

Many Australians would never consider living in some of these places. Would you?

Geoinfo

- Each camel in a 'camel caravan' carries twenty 7 kg slabs of salt.
- Waste picking is known as the industry that is always hiring.

Geoactivities 6.8

Knowledge and understanding

1 Draw and complete the table for five locations mentioned in these pages. An example has been completed for you.

Place	e.g. Big Bell
Country	Australia
Resource being exploited	Gold
Necessity or choice for workers	Choice
Key concepts	Ghost town resource exhausted

2 Name two jobs dependent on Lake Assal's resources.

3 Create a simple flow diagram to illustrate the development of ghost towns.

4 Which would be worse: waste picking or ship breaking? Justify your answer after considering the advantages and disadvantages of each job.

5 Write a statement that links liveability with employment.

Inquiry and skills

6 Refer to 6.8.1 and 6.8.2 and Geoinfo.
  a What resources made Bodie and still make Lake Assal liveable places?
  b Why would waste picking be called the industry that is always hiring?
  c Why would waste picking be considered a health hazard?
  d Why do many of the world's poorest people work in places that Australians would consider unliveable?

7 Use Google Earth or other satellite imagery online to study the beach at Alang.
  a Count the number of grounded ships.
  b Study photographs of different ships on the beach to comment on their size.
  c Examine the surrounding environment to determine other sources of employment around Alang.

Chapter 6

ISBN 978 1 4586 6277 4

6.9 Stages of life and perceptions of liveability

Liveability is perceived differently by each individual, whose views about places will change as life progresses. Generation Y includes the 'boomerang kids' who move away from the family home in their twenties to assert their independence, then return within a few years, usually for financial reasons. In this example, the family home is considered less liveable by young adults at one stage, but more liveable at a later date.

Some people choose to live all of their life in the place they were born, for instance their suburb or the family farm. Others will relocate in response to changes in family status, income, employment, lifestyle choice and stage of life.

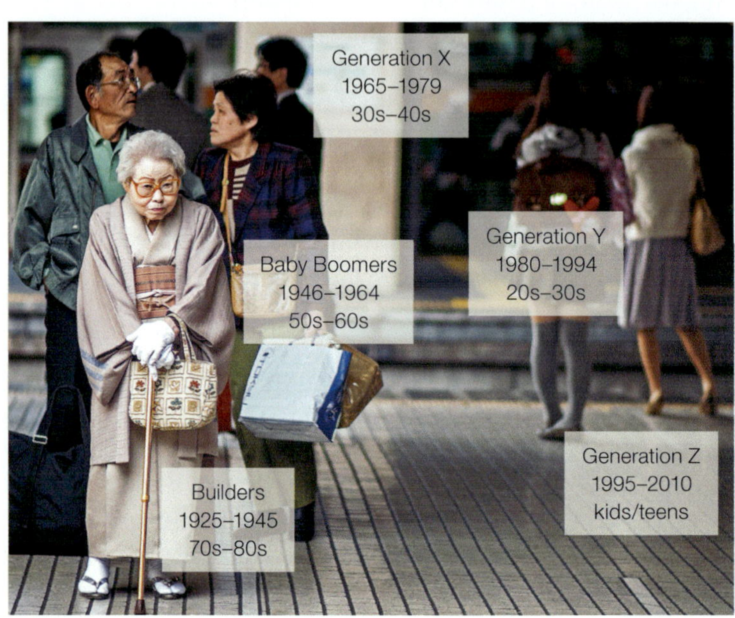

6.9.2 The different generations: where will their train go?

6.9.1 Locational choices affected by life stages

ISBN 978 1 4586 6277 4

Life stages

Based on age, life stages have a big impact on choice of place to live. Parents of young children might prefer places where there are ample playgrounds and childcare centres, whereas young single adults may prefer living close to city jobs and nightlife. Young people leave rural and **regional** places to be closer to education, employment and lifestyle facilities, while older generations may sell their family home and move to aged care facilities. Some retirees sell up and become **grey nomads**; others opt for a **tree change** or a **sea change**.

A survey by *The Economist* found that around 70–80% of students, people with their first job or those developing their careers chose to live in inner-city urban locations. By contrast, 53% of families with children chose to live in the suburbs or city outskirts. Retirees had different preferences, with 42% choosing to live in smaller towns or villages. Rural living was preferred by 15% of retirees compared to just 1% of students.

Lifestyle segments

For older generations, becoming an adult meant settling down, getting married, having a family and buying a home. Society has become more complex in recent years. The declining importance of marriage, delays in starting a family, the cost of living and the ageing of society have led to the development of segments within society. Belonging to a **lifestyle segment** such as **DINKs** (Double Income No Kids) or **yuppies** (Young Urban Professionals) will influence a person's perception of a place's liveability.

6.9.3 Lifestyle segments

Population profiles

Population profiles reflect the liveability of places for different age and lifestyle groups. These 'pyramids of age' (see Geothink) help councils plan facilities to meet the demands of each group in their populations. Planning age-friendly places for the older generations and liveable communities that are inclusive of all age groups are current trends across the globe.

Geo**info**

- Downagers (over 60s who feel and act much younger) are the fastest growing lifestyle segment over 60 years of age in Australia.
- Eight per cent of Australians aged 30–34 still live at home with their parents.

Geo**activities 6.9**

Knowledge and understanding

1 How do 'boomerang kids' illustrate the changing perceptions of liveability in Australia?
2 What is meant by 'life stages'?
3 Do you think the cartoon in 6.9.1 accurately portrays all stages of life? Why?
4 State the preferred location to live in for students, families with children, and retirees.

Inquiry and skills

5 Refer to 6.9.1.
   a For each of the four age groups in the cartoon suggest another option for where they might choose to live.
   b Which living options would you choose for each of the three stages of life when you are no longer a dependant child?
6 Refer to 6.9.2. What type of location (e.g. house in the suburbs) might the people from each generation on this Tokyo railway station be travelling home to? Give reasons for your answers.
7 Refer to 6.9.3.
   a Investigate the different lifestyle segments mentioned in the Wordle (see Geolinks).
   b Create a mind map to define the segments.
   c How would identifying lifestyle segments assist future planning to improve the liveability of places?
8 Examine the influence of stage of life and lifestyle on liveability. Write a 500-word written report.

Chapter 6

ISBN 978 1 4586 6277 4

6.10 The Amish: sustainable lifestyle communities

Some people choose to live an alternative lifestyle, like the Amish who migrated to the USA in the 18th century from Germany, Switzerland and France. They originally settled in the US state Pennsylvania where they worked as farmers and lived in close-knit religious communities. In the USA and Canada there are around 273 000 Amish now living in 427 settlements across 31 states. Two-thirds reside in Ohio, Pennsylvania and Indiana. The Amish population has doubled in the last 20 years—at the present rate the population is anticipated to reach 1 000 000 by 2050.

Unique lifestyle

The Amish speak a German dialect, and the adults do not vote or go to war. Their community promotes humility, thrift and obedience to higher authorities. The Ordnung is an outline of the Amish faith, defining what it means to be Amish—from their dress and hair length to farming techniques. Dress is simple and watches, cosmetics, jewellery and personal photographs are forbidden.

Most Amish cultivate their fields with horse-drawn machinery, ride around in horse-drawn buggies and live without electricity. Conveniences such as televisions, cars, telephones and tractors are generally forbidden as they create temptation and inequality. As the Ordnung varies between geographic locations and communities, some Amish drive cars and use disposable nappies while others are not even permitted battery-powered lights.

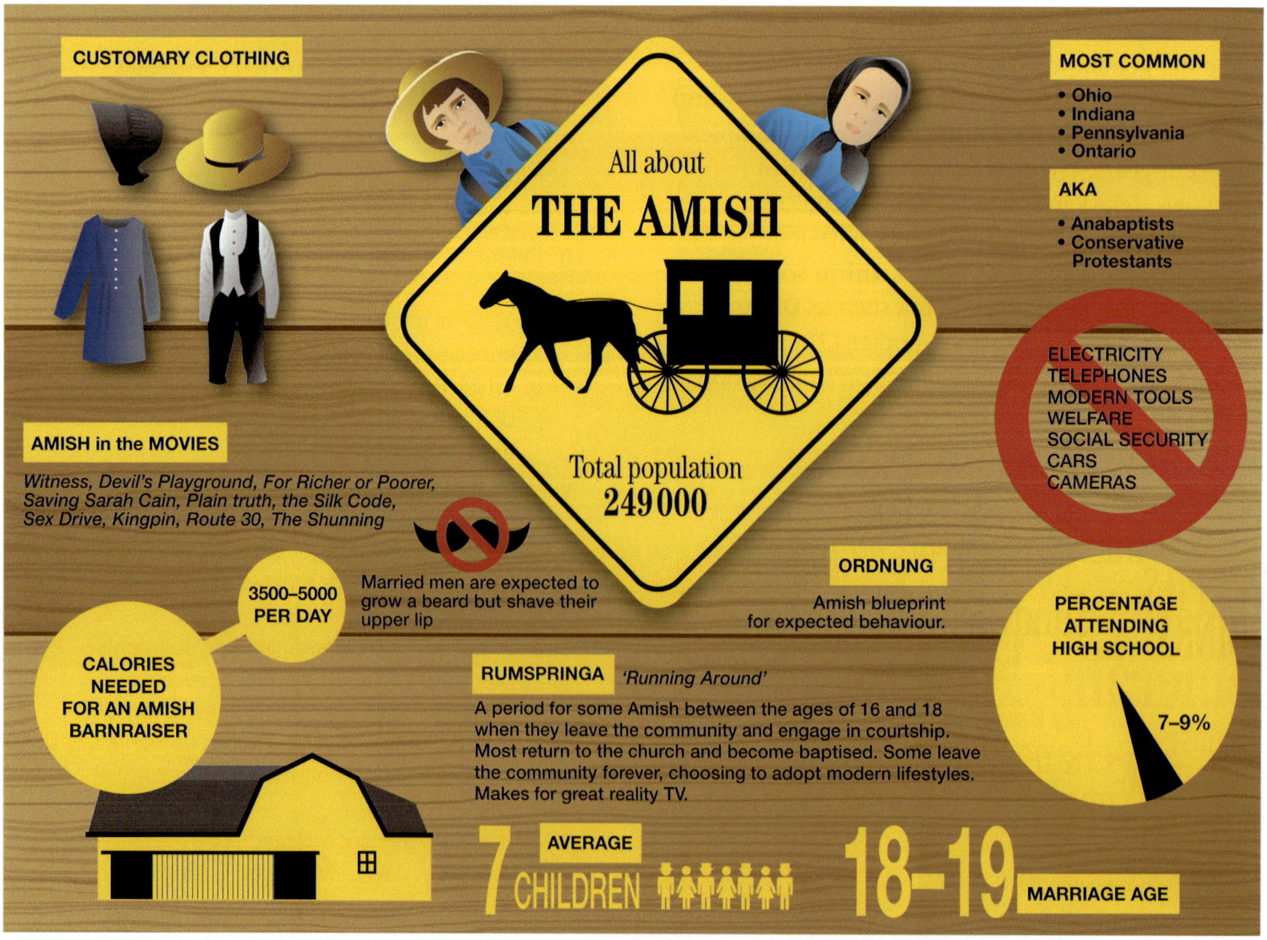

CUSTOMARY CLOTHING

All about
THE AMISH

Total population
249 000

MOST COMMON
- Ohio
- Indiana
- Pennsylvania
- Ontario

AKA
- Anabaptists
- Conservative Protestants

ELECTRICITY
TELEPHONES
MODERN TOOLS
WELFARE
SOCIAL SECURITY
CARS
CAMERAS

AMISH in the MOVIES

Witness, Devil's Playground, For Richer or Poorer, Saving Sarah Cain, Plain truth, the Silk Code, Sex Drive, Kingpin, Route 30, The Shunning

Married men are expected to grow a beard but shave their upper lip

3500–5000 PER DAY

CALORIES NEEDED FOR AN AMISH BARNRAISER

ORDNUNG

Amish blueprint for expected behaviour.

PERCENTAGE ATTENDING HIGH SCHOOL

7–9%

RUMSPRINGA *'Running Around'*

A period for some Amish between the ages of 16 and 18 when they leave the community and engage in courtship. Most return to the church and become baptised. Some leave the community forever, choosing to adopt modern lifestyles. Makes for great reality TV.

7 AVERAGE CHILDREN

18–19 MARRIAGE AGE

6.10.1 All about the Amish

ISBN 978 1 4586 6277 4

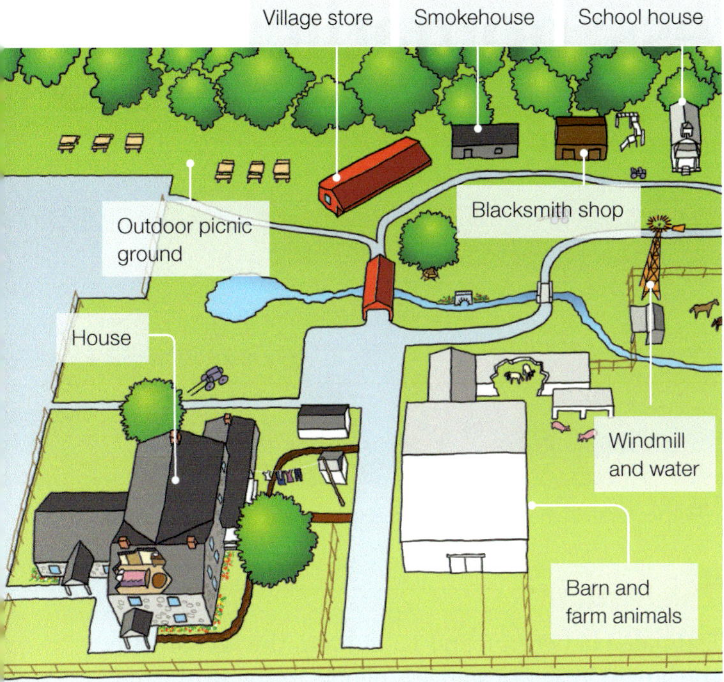

6.10.2 Tour of an Amish farm

Teenagers and changes

'Rumspringa' means to run around. It allows Amish youths to experience life outside the community before joining the church and marrying. About 85% of youths choose to return to the community after Rumspringa, indicating the popularity of the lifestyle. Amish weddings involve no rings, flowers or caterers. Men are clean-shaven prior to marriage and grow beards after marriage.

Amish live a separate life from the modern world, strengthening their community and beliefs. They are expelled from the community for breaching religious guidelines, including marrying outside the faith.

Although farming continues to be important, higher prices for land and equipment have meant some Amish taking jobs off the farm. As a result, more families are involved in small businesses, such as furniture or quilt shops, bakeries or construction. Most Amish are self-sufficient and use Earth's resources sustainably.

Geo**info**

The Amish are well known for the practice of barn raising. The community comes together to build a barn for a community member, contributing labour and organisation, and socialising together.

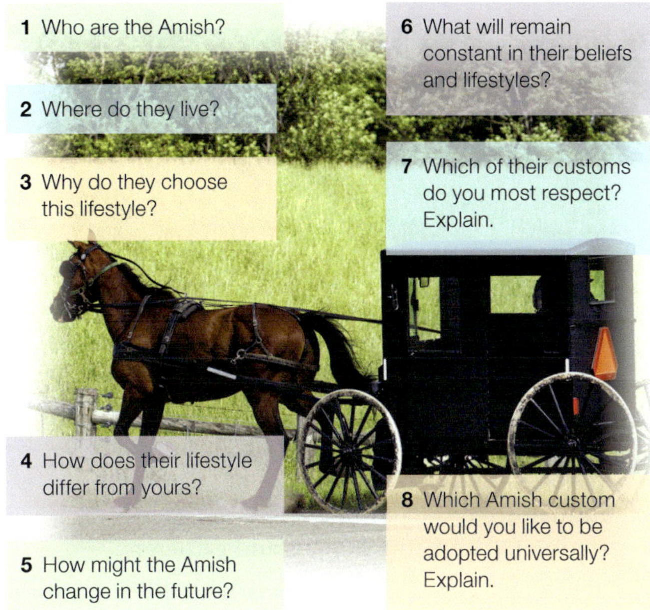

6.10.3 What did I learn?

Geo**activities 6.10**

Knowledge and understanding

1. How does the Ordnung guide the lifestyle of the Amish?
2. Discuss the ways in which the lifestyle of the Amish is sustainable.

Inquiry and skills

3. Refer to 6.10.1.
   - a Describe the clothes worn by the Amish.
   - b What is generally forbidden in an Amish community?
   - c What is the percentage of Amish who attend high school? What is the impact of little education on their future lifestyle?
   - d Explain why the population has doubled in the last 20 years.
4. Refer to 6.10.2.
   - a Explain why this village is an example of a self-sufficient sustainable community.
   - b Amish communities have become tourist attractions. List the advantages and disadvantages of this to the communities.
5. Refer to 6.10.3 and answer the key geographical questions.
6. In groups draw a mental map reflecting the daily lives of the Amish.
7. Explain why teenagers would choose to return to the Amish community after their period of freedom.
8. Examine a map showing the location of Amish communities in the USA. Describe the distribution of the communities.

Chapter 6

ISBN 978 1 4586 6277 4

6.11 No fixed address: nomadic lifestyles

The nomadic lifestyle

Rather than settling permanently in one location, nomads move from place to place. There are about 30 million nomads today, usually living in regions with scarce resources. Nomadic herders, such as the Berbers in Morocco or the nomads in Mongolia, move animals to fertile pastures and water sources. Traditional nomadic hunters and gatherers, such as Aboriginal and Torres Strait Islander Peoples, followed seasonal routes to find native plants and animals.

Peripatetic (travelling) nomads traditionally moved among sedentary populations offering a craft or trade. Some were animal dealers, snake charmers, musicians, dancers and fortune tellers. These nomads include the Romani (or Roma), who originated in India and now live in many countries, including Australia and the USA, but are concentrated in Europe.

6.11.2 Berber tents in the Sahara Desert. Berbers are nomadic herders.

The Romani

Estimated between 6 and 12 million, the Romani are the largest group of stateless people in the world and Europe's biggest minority group. They are sometimes called 'gypsies', but this term is no longer used because it causes offence. Instead, they are called Romani peoples, or they may be known by other names in different European countries (e.g. Tsiganes in Hungary, Gitanes in France or Sinti in German-speaking parts of Europe).

6.11.1 The Romani in Europe

ISBN 978 1 4586 6277 4

Over centuries, the Romani were persecuted—particularly in Nazi Germany. Discrimination still occurs today, as most Roma have poor access to education, health services and employment. Poverty rates are 10 times higher than in the non-Roma population.

Each Roma group has different traditions and cultures. Contrary to popular belief, most Roma live in one place. Many do not choose to move, but are forced to move on by authorities as they have no homeland or migrate in search of a better life.

Irish Travellers

The Romani are referred to as 'Travellers' in Ireland. Their lifestyle was shown worldwide on the television documentary *Big Fat Gypsy Weddings*. The program has been criticised for presenting negative stereotypes about these peoples and for leading to increased bullying of Traveller children at schools in the UK.

Most Travellers are proud of their culture. They follow 'The Travellers Code', which dictates their moral beliefs and actions. Many girls are married at 16 years and those Travellers following the Roman Catholic religion ensure their child's first communion is a large and extravagant celebration.

There are over 27 000 Travellers. Many lack skills and qualifications, so they receive low wages. The majority are dependent on social welfare. Those who work are generally involved in horse trading, breeding greyhounds and recycling scrap metal.

Traveller accommodation includes group housing as well as permanaent and temporary halting sites.

6.11.3 Girls dressed in their Roman Catholic Communion dress

The Irish Traveller Movement is a partnership between Travellers and settled people committed to seeking equality for Travellers.

Geo**info**

- There are 25 000 Romani peoples in Australia.
- Poet Henry Lawson may have been Romani.

Geo**activities 6.11**

Knowledge and understanding

1. Identify groups who live a nomadic lifestyle.
2. Explain why it is difficult to calculate the exact number of nomadic people.
3. Romani have experienced racism, discrimination and human rights abuses. Explain what this means.
4. Discuss the lifestyle of some Romani.

Inquiry and skills

5. Refer to 6.11.1.
   a. List five countries with the largest Romani population.
   b. Name five countries with the largest percentage of Romani.
6. Refer to 6.11.2.
   a. Draw and label a sketch of the photograph.
   b. Imagine this is your lifestyle preference. Write a report on the liveability of the places you camp and the features of your lifestyle.
7. Watch short video clips, such as 'Roma: A New Generation'. (See Geolinks.)
   a. Describe conditions in Roma settlements, such as access to services and facilities, and environmental quality. Comment on the liveability of these settlements.
   b. Discuss why it is difficult for nomads to improve the liveability of their settlements.
   c. Identify the features that contribute to the community identity of Romani settlements.
   d. Suggest why their community identity is important to the Romani people.
8. In groups discuss your opinions on the following comments. Present your thoughts as an oral report.
   a. 'More money should be given to councils to develop official caravan parks for the Romani as they have no place to stop.'
   b. 'Travellers are one of the most disadvantaged, impoverished, under-educated, despised and ostracised groups, who live on the margins of Irish society.'

ISBN 978 1 4586 6277 4

6.12 Safety behind barriers

Is privacy and safety in your own home and neighbourhood a high priority? Are there people living near you that you would prefer not to mix with? Concerns over personal safety and who we want as neighbours helps to explain the global explosion in 'gated' or fortified communities. Living behind a barrier is an option more people now have when choosing a place to live.

A **gated suburb** is a residential community where the entry is controlled by perimeter walls or fences. Some of these communities have security guards and surveillance cameras. Safety is the prime liveability feature.

The demand for safe housing increases the more problems a city has—hence the rapid growth of gated communities in cities of developing nations (e.g. Jakarta in Indonesia).

Protection, prestige, privacy

Michael Jackson's family and Justin Bieber are residents of Estates at the Oaks, a gated community near Los Angeles. It offers the three Ps (protection, prestige and privacy) and is protected by paparazzi-proof gates.

The global appeal or liveability of modern walled communities today lies in promises of:

- safety and privacy
- exclusive access to facilities (e.g. golf courses)
- being with people of similar status (e.g. famous).

Expatriates, people in show business and medium- to high-income earners are the main residents in gated communities.

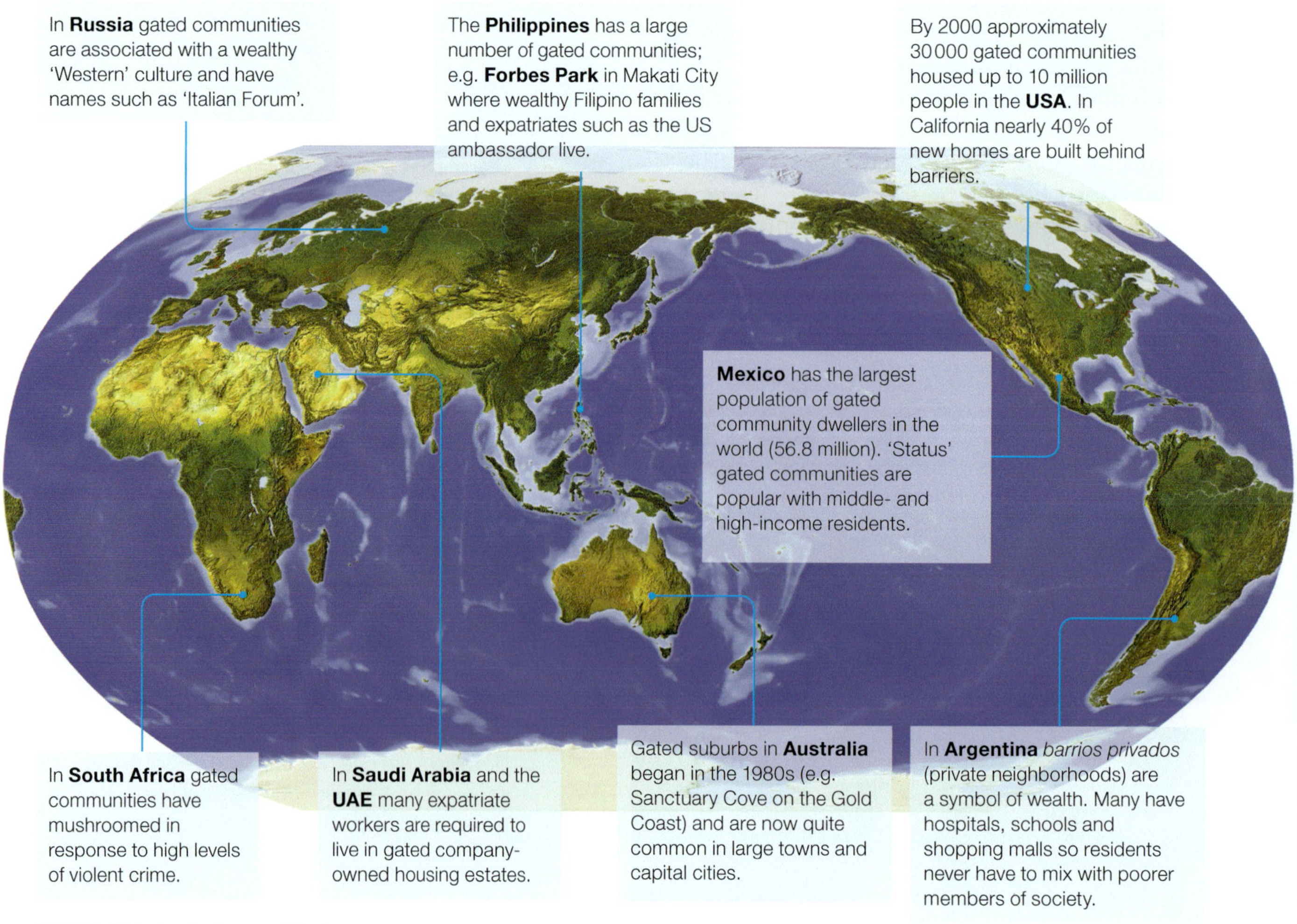

In **Russia** gated communities are associated with a wealthy 'Western' culture and have names such as 'Italian Forum'.

The **Philippines** has a large number of gated communities; e.g. **Forbes Park** in Makati City where wealthy Filipino families and expatriates such as the US ambassador live.

By 2000 approximately 30 000 gated communities housed up to 10 million people in the **USA**. In California nearly 40% of new homes are built behind barriers.

Mexico has the largest population of gated community dwellers in the world (56.8 million). 'Status' gated communities are popular with middle- and high-income residents.

In **South Africa** gated communities have mushroomed in response to high levels of violent crime.

In **Saudi Arabia** and the **UAE** many expatriate workers are required to live in gated company-owned housing estates.

Gated suburbs in **Australia** began in the 1980s (e.g. Sanctuary Cove on the Gold Coast) and are now quite common in large towns and capital cities.

In **Argentina** *barrios privados* (private neighborhoods) are a symbol of wealth. Many have hospitals, schools and shopping malls so residents never have to mix with poorer members of society.

6.12.1 Gated suburbs go global

ISBN 978 1 4586 6277 4

Gated communities prosper as crime rates rise

Rising crime rates and soaring population growth is leading to a boom in popularity for lifestyle villages across Western Australia. Within the Perth metropolitan area, there are now over 15 gated communities, with many others across the state.

The appeal for owners? Safety from home burglary and a resort-inspired lifestyle—most gated communities boast extra amenities, including swimming pools, saunas and entertainment facilities. And they're affordable, too: a selling point for baby boomers looking to downsize in their retirement years.

The growing popularity of gated communities is part of a global phenomenon. Trends in Asia suggest the future of gated communities might be in increased technology, such as licence plate scanning.

While the purported benefits of gated villages are hooking buyers now, some investors aren't convinced, remaining loyal to the diversity offered by traditional suburban living.

Segregation

The benefits of creating gated or closed neighbourhoods and suburbs are debated because they segregate people on the basis of income and status. *Star Wars* film director George Lucas was criticised by wealthy neighbours when he attempted to redevelop his gated property into housing for low-income workers.

6.12.2 Freedom or prison?

Critics claim crime rates are no lower in gated communities. In some US cities criminals target gated communities because of their wealth. Living behind barriers stops people becoming involved in creating better communities. Residents are less likely to be volunteers and involved in community projects. Reducing the interaction of people from different races, cultures and classes creates less liveable, inclusive and vibrant communities.

Geo**info**

China's gated suburbs resemble those in California with artificial lakes, ranch houses and Hummers.

Geo**activities 6.12**

Knowledge and understanding

1 List the reasons people choose to live in gated suburbs.
2 How do gated suburbs provide safe environments for residents?
3 Give examples of the 'exclusivity' of gated communities.
4 What groups of people make up the bulk of residents in these communities? Suggest reasons.
5 What arguments are used against the provision of gated communities?

Inquiry and skills

6 Refer to 6.12.1.
   a Explain why gated suburbs are considered a global phenomenon.
   b Where do expatriates make up a large percentage of gated suburb residents?
   c Give examples of the scale of gated communities in two countries.
7 Refer to 6.12.2.
   a What perspectives are presented in this cartoon?
   b Which perspective do you support? Justify your answer.
8 Refer to the article.
   a Why are gated suburbs referred to as 'lifestyle villages'?
   b What changes promoted the growth of gated suburbs in Perth?
   c What features of Asian estates will be introduced into Perth?
   d List the advantages and disadvantages of gated communities in a table.

ISBN 978 1 4586 6277 4

6.13 Living abroad: opportunity

An increasing number of people choose to live temporarily outside their home country. A connected and globalised world has created overseas work opportunities. People move abroad for lifestyle changes at different phases of their lives, for instance retirees working as volunteers or young people taking a gap year after school. An estimated 3 million students are studying abroad, a number increasing by 12% annually. The liveability of potential destinations affects the choices students make.

Studying overseas

Students are choosing to complete all or part of their education in another country:

- Fifty-two per cent of international students are from Asia, especially China, India and Korea.
- China is the biggest exporter of students (e.g. one in four American foreign students is Chinese).
- There were almost 900000 international students in the USA in 2014.
- Some small countries have more students studying abroad than at home (e.g. São Tomé).

Australia

Australia attracts thousands of foreign students; over 500000 came in 2013. Twenty-eight per cent of all successful student visa applications were from China, 10% from India and 5% from Korea. The top destinations for Australian students are the USA, New Zealand, the UK, Canada and Germany.

Liveability

Students choose overseas study to learn languages, develop career opportunities, experience new cultures and make new friends. Study locations are influenced by liveability factors, such as:

- the language of tuition
- course fees
- the quality of courses and facilities
- living expenses
- standard of living
- geography (e.g. climate, location and culture).

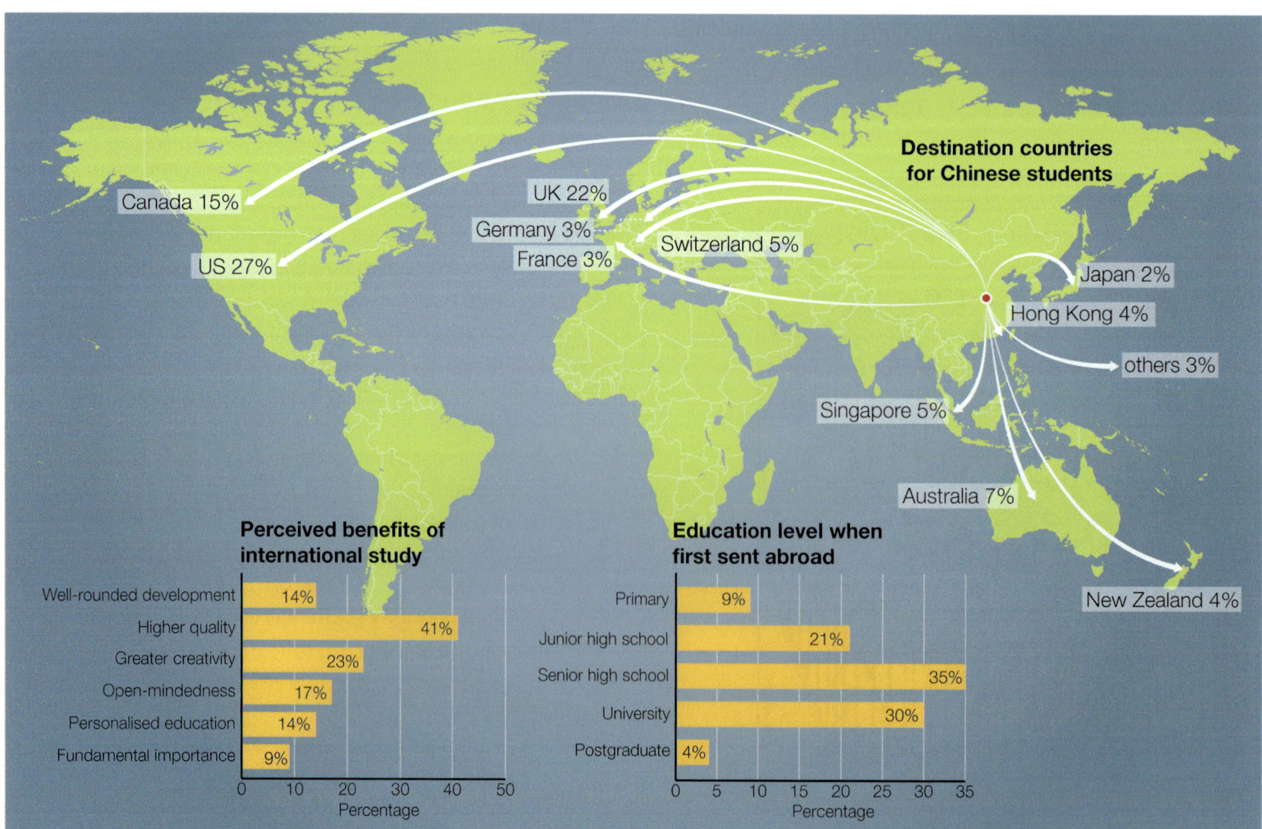

6.13.1 Chinese students abroad

ISBN 978 1 4586 6277 4

6.13.2 QS Best Student Cities in the World 2015

QS Rank	City	Country	Student mix	Desirability	Affordability	Overall
1	Paris	France	83	83	54	412
2	Melbourne	Australia	100	94	40	397
3	London	UK	93	71	28	392
4	Sydney	Australia	95	98	36	388
5	Hong Kong	China	80	88	63	387
6	Boston	USA	87	89	49	386
7	Tokyo	Japan	55	95	62	385
8	Montreal	Canada	96	83	60	380
9	Toronto	Canada	90	100	51	375
10	Seoul	South Korea	71	66	54	372

Note: Not all indicators are shown in this table.
Source: www.topuniversities.com/city-rankings/2015

6.13.3 In Norway a Big Mac costs over US$7.50, in the USA US$4.56 and in China US$2.61

Geo**info**

- Over 2 million Chinese students studied abroad in the last decade.
- The number of Australians studying overseas is increasing by 10% a year.

University rankings, such as the QS World University Rankings, are used to influence student choices. The QS Best Student Cities ranks university cities on several liveability criteria, including the number of foreign students (student mix), the number of universities, the quality of life (using liveability rankings, e.g. Mercer), the availability of student jobs and **affordability**. *The Economist* magazine's Big Mac Index is one measure of affordability students might relate to—it ranks the burger's cost in various countries. The biggest obstacle to overseas study is affordability, making it inaccessible to most of the world's poor.

English-speaking, safe, developed and politically stable countries are considered the most liveable. Canada, New Zealand, the UK and the USA are the main student destinations.

It is estimated that 90% of China's wealthiest families plan to send their children overseas to study. As family incomes of more Chinese increase, the number of international Chinese students will continue to increase.

Geo**activities 6.13**

Knowledge and understanding

1. List groups of people who live temporarily in other countries. Why are the numbers increasing?
2. What are the top source and destination countries for international students?
3. Suggest reasons for the large number of Asian international students.
4. Why is the language of tuition important in choosing a country to study in?
5. What is the relevance of the Big Mac Index?
6. Justify the value of overseas study.

Inquiry and skills

7. Refer to 6.13.1.
   a. Rank the destinations of Chinese students in descending order of importance.
   b. What is the main reason for Chinese parents sending their children to study overseas?
   c. What areas of the world are not popular with Chinese students? Suggest why.
8. Refer to 6.13.2.
   a. Which continents have the most liveable student cities? Suggest a reason for this.
   b. Which city ranks highest on student mix? Why is this an important liveability indicator?
   c. Identify qualities of Paris that account for its first place ranking.
   d. What makes Australia very liveable for overseas students?
   e. What is the least liveable feature of Melbourne for an overseas student?
9. Refer to 6.13.3 and research the Big Mac Index online. Explain the significance of the Big Mac Index to students studying abroad.
10. Given the opportunity, where would you like to study overseas? Explain your choice.

Chapter 6

ISBN 978 1 4586 6277 4

6.14 Reflecting on liveability

Throughout chapter 6 you have been investigating what makes places liveable for different people. The focus has been on places that are considered extreme due to factors such as remoteness (Nunavut), climate (Lake Assal), technology (Dubai), danger (Afghanistan) and poverty (Bangladesh). Many people, however, do not live in extreme places, and they make choices about where they live a number of times during a lifetime. Some people choose to stay in the same place their whole lives, while others move frequently. What factors will make one place more liveable than another for you?

6.14.1 This Wordle contains a comprehensive list of factors for you to consider when completing Geoactivities 7.8.

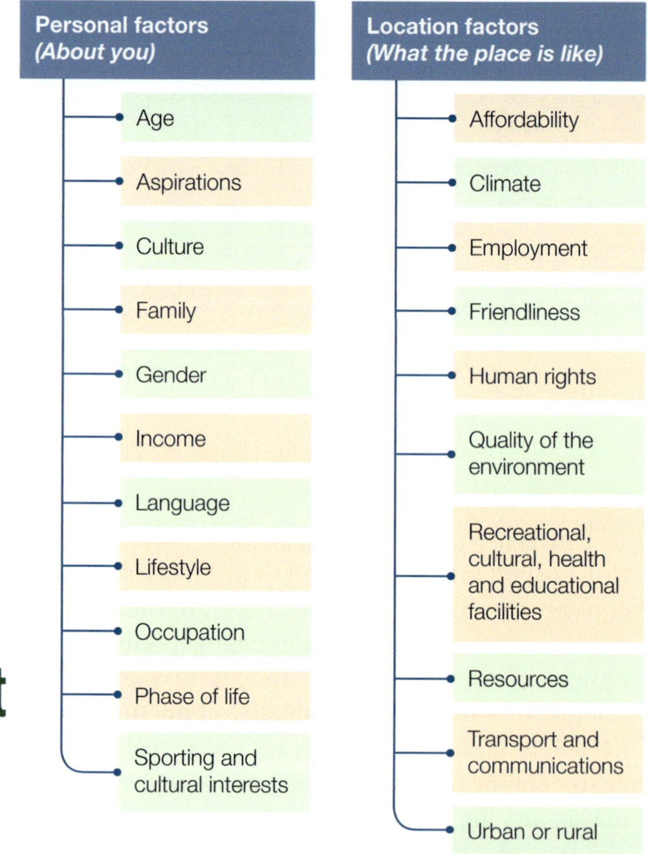

6.14.2 Factors affecting locational choices

Factors affecting where we live

Our decisions are affected by a combination of personal and location factors, such as income or climate. The importance of different factors (e.g. access to schools) will vary between individuals and will change over time. Some factors, such as personal safety, are important in all phases of our lives. In recent years environmental quality has become more important.

Everyone wants to live in a place where their quality of life (QOL) is the best it can be. Global surveys are conducted annually to determine the most liveable cities. In 2013, according to one survey, four of the top 10 liveable cities were in Australia. Each year's results are widely debated due to city rivalry (e.g. Sydney vs Melbourne), and national pride (e.g. Canada vs Australia). Community leaders now use liveability to plan better places for people to live.

Geoinfo

- In 2011 Melbourne and Adelaide were each ranked first in different liveability surveys.
- Vancouver ranked first on a global liveability survey in 2011 but ranked 29th in a national survey.

Criteria for liveability

Human rights such as freedom, the availability of goods and services, personal safety, education, health care, culture, environment, recreation, employment, political and economic stability, public transportation, climate and access to nature can influence an individual's perception of liveability.

In one survey, residents were asked to determine the importance of employment, crime and safety, local hospitals and local schools in their ideal community. There were some surprising results. In London these features were not considered as important as they were in Jakarta (see 6.14.3).

ISBN 978 1 4586 6277 4

Cycling culture in the USA has boomed in recent years, prompting new inquiries into government spending on bike and pedestrian infrastructure.

A recent survey has demonstrated support for increased funding in this area, with almost 50% of respondents calling for a rise in spending. At present, bike paths and footpaths account for just 2 per cent of the annual transportation budget.

Research has shown that interest in bike riding is distributed evenly across genders and political persuasions, with a spike in popularity among the Generation Y age group.

WHEN YOU THINK OF THE COMMUNITY YOU WOULD MOST LIKE TO LIVE IN, HOW IMPORTANT ARE EACH OF THE FOLLOWING?

AVAILABILITY OF EMPLOYMENT
SAFETY AND CRIME
LOCAL HOSPITALS
LOCAL SCHOOLS
LESS IMPORTANT
VERY IMPORTANT

With the exception of London, global city dwellers say safety/crime and local hospitals are most important when choosing an ideal community in which to live. Local schools are more important to residents of Jakarta, New York, Abu Dhabi and Delhi, as is the availability of employment in those cities.

TOKYO: WORK 57% · POLICE 88% · SCHOOL 42% · HOSPITAL 81%

JAKARTA: WORK 85% · POLICE 94% · SCHOOL 95% · HOSPITAL 90%

LONDON: WORK 45% · POLICE 50% · SCHOOL 38% · HOSPITAL 44%

NEW YORK: WORK 79% · POLICE 96% · SCHOOL 70% · HOSPITAL 95%

AMSTERDAM: WORK 53% · POLICE 93% · SCHOOL 35% · HOSPITAL 55%

ABU DHABI: WORK 99% · POLICE 99% · SCHOOL 93% · HOSPITAL 99%

BEIJING: N/A · POLICE 88% · SCHOOL 68% · HOSPITAL 85%

DELHI: WORK 87% · POLICE 88% · SCHOOL 93% · HOSPITAL 99%

6.14.3 An ideal community—perceptions on the importance of access to services and facilities in different cities

6.14.4 Quality of life is an important part of a city's liveability

Geo**activities 6.14**

Knowledge and understanding

1 Why do the factors affecting people's decisions change over time?

2 Do you think 'liveability' and 'quality of life' mean the same thing? Explain.

3 List liveability factors that would be important to you but are missing from the Wordle on page 228.

Inquiry and skills

4 Refer to 6.14.2. Make a list of the personal and location factors that affect where you and your family currently live.

5 Refer to 6.14.3.

a Write a paragraph comparing the differences between Tokyo and Jakarta. Suggest reasons for the differences.

b Create a pictogram for yourself using the same four criteria.

6 Suggest how increased funding for footpaths and bicycle paths would improve the liveability of all places for people of all ages.

Chapter 6

ISBN 978 1 4586 6277 4

6.15 Skill: communicating geographical information

Online presentation tools are known as **web 2.0 tools**. These tools provide exciting ways for geographers to communicate geographical inquiry or research findings. Web 2.0 tools allow you to create:

- graphs using your own statistics (e.g. Create-a-graph)
- surveys that can be collated for analysis and converted to graphs (e.g. Survey Monkey and Edmodo)
- movies, photo stories, animations (e.g. Animoto and Tubechop)
- presentations with links to visual material you have created or found (e.g. Prezi, Glogster and infographics)
- summaries and evaluations in an interesting format (e.g. Wordle and Tagxedo)
- websites that contain qualitative and quantitative data, and include live links to other websites and visual resources
- maps that include fieldwork or geographical inquiry findings placed on a base map using spatial technologies.

Online tutorials make the use of these tools very simple. Some useful and fun tools are listed below.

Communicating geographical information

- Use appropriate methods, including ICT
- Use geographical vocabulary, concepts and conventions
- Display data in graphs, tables, maps or as statistics
- Use written, oral, visual and/or graphic forms

6.15.1 Geographical inquiry involves presenting data

Spatial technologies

Google Maps

Google Maps can be used to create a local (large scale) map, for instance showing changes in water quality along a creek.

Google Earth

Google Earth can be used to create maps with several layers of information that show relationships between geographical features, such as land use and topography.

Google Tour Builder

Tour Builder is used to create interactive tours of sites related to a topic, by adding locations, photos, text and video to base maps in Google Earth. It can be used to show features of an environment or community (see 7.14)

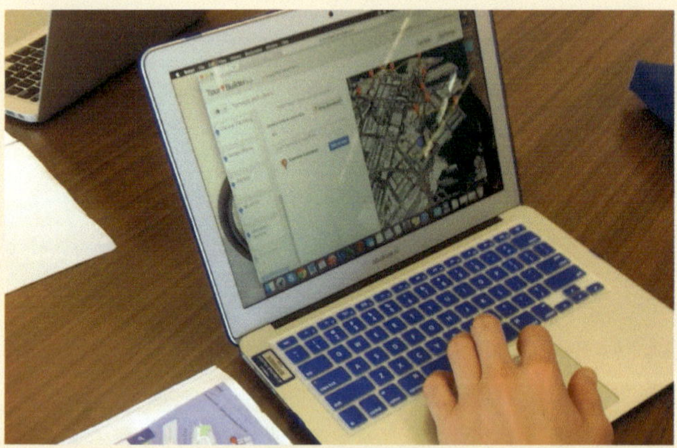

6.15.2 Tour Builder lets you add information to Google Earth maps

National Geographic MapMaker Interactive

National Geographic MapMaker Interactive can be used to create a national or global (small scale) map showing settlement patterns or the location of different types of food production.

Scribble Maps

Add travel paths and outline areas or places on Scribble Maps using moveable boundaries. Photos and labels can be added to create tours similar to those in Google Earth.

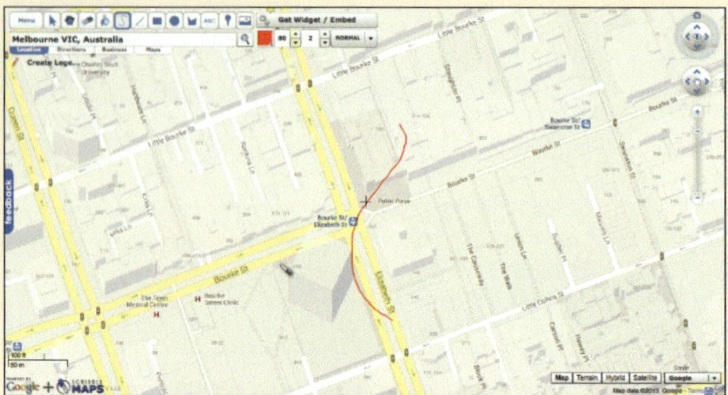

6.15.3 You can export, send and print your Scribble Maps

ISBN 978 1 4586 6277 4

StatPlanet

StatPlanet creates interactive maps using existing data that is imported into the program. It produces maps as well as non-spatial presentations such as graphs, charts and infographics.

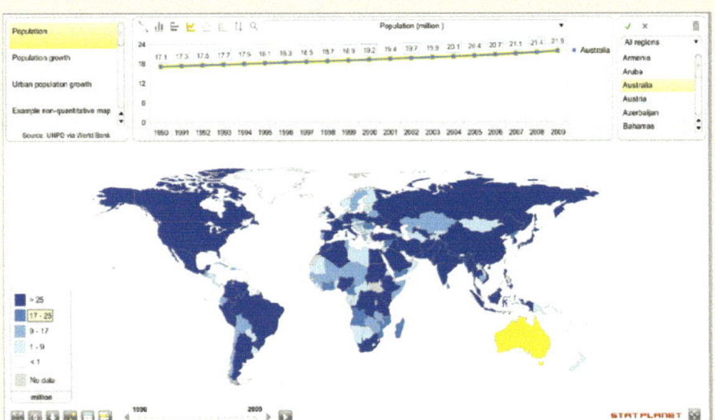

6.15.4 Create interactive maps and other presentations with StatPlanet

Animaps

Animaps allows you to create maps featuring moving markers, pop-up images and text, and changeable lines and shapes. Animaps play like a video that can be paused, slowed or fast-forwarded.

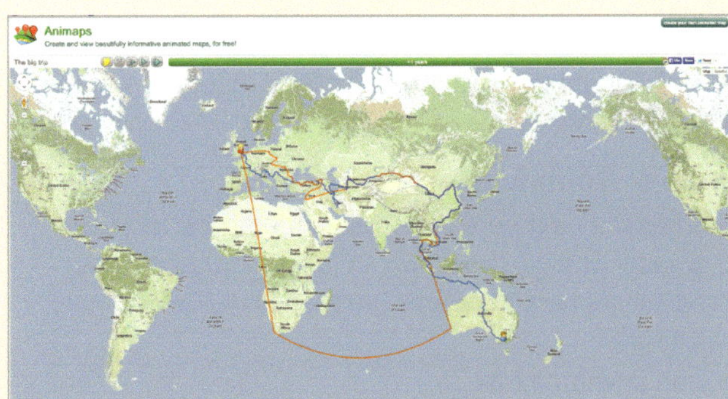

6.15.5 Create an animated map with Animaps

Other web tools

Prezi

Prezi is a zooming presentation tool that allows photographs, Word documents, PDF files, video clips, graphs, tables and other tools to be integrated.

Weebly for Education

This tool can be used to create a free website or blog as a class or an individual. It uses a drag-and-drop feature to add videos, pictures, maps and text, making it easy for students of all capabilities to use.

Tubechop

Allows you to select parts of video clips and other movies to put into your presentations. It also allows you to add together sections from different clips to create a new presentation.

Glogster EDU

A 'glog' is an interactive online poster used to present a variety of information using maps, graphs, posts, and active links to information in video clips (e.g. YouTube) and to websites.

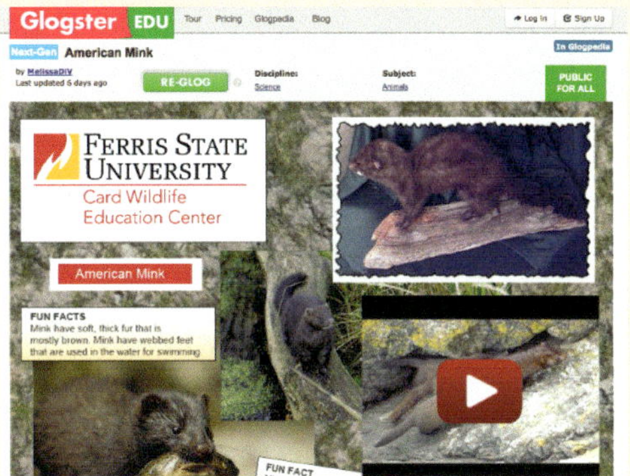

6.15.6 Glogster is used to build online posters

Infographics

Infographics present information visually and graphically using geographical tools such as maps, graphs and diagrams as well as short pieces of text. A lot of information is presented in a small space and in a visually attractive format. You can create your own infographics on sites such as Visual.ly and Infogr.am, while Wordle and Tagxedo can be used for simple infographics like word clouds. Word clouds are useful for summarising topics, identifying key concepts, consolidating ideas and presenting information creatively.

SurveyMonkey

You can create surveys, email the link to your participants and receive completed surveys online. Results can be presented as graphs or tables.

Geo**activities 6.15**

Knowledge and Understanding

1 What are web 2.0 tools?
2 Explain how spatial technologies can be used to communicate geographical research findings. Refer to examples.
3 Explain the advantages of SurveyMonkey over traditional survey methods.
4 Explain why infographics are an interesting way to display information.
5 How is an infographic different from a Glogster?

Inquiry and skills

6 Visit one web 2.0 tool that interests you. Look at samples from the gallery if one exists. Follow a tutorial on creating your own presentation. Create a simple presentation about you and the place where you live. Add features that make your place liveable.

Chapter 6

ISBN 978 1 4586 6277 4

Geo**think**

Pack your bags—criteria-based decision-making

1. Read the scenario. It gives you some information about who you are and a question to answer.
2. Use the geographical inquiry process to complete steps 2 to 4.

Step 1: Your question—Which city is the most liveable for you, Berlin or New York?

Scenario:

You cannot believe your luck. You work for a company with overseas branches in New York and Berlin. They are offering you a two-year contract overseas. You get to choose which of these cities you want to live in.

About you:

Single and 25 years old, you have never travelled overseas. You like open spaces, outdoor activities and sport. In Australia you do a lot of water sports, soccer and cycling. You have always wanted to visit the 'Big Apple' but love the idea of skiing in Europe, and you like cold climates. You want to do a lot of travelling to visit as many countries as you can. The company pays you well for your age but you will have to be careful with your spending. You love history and geography, so visiting museums, historical sites and interesting places is important to you. You only speak English but learnt a bit of German at school. The company will fly you to your destination and home again at the end of your contract.

You have seven days to make a decision.

Step 2: Collecting, recording, evaluating and representing data

Investigate New York and Berlin using the information provided on these pages and research on the internet (see Geolinks). You could look up official tourist sites, infographics and blogs from Australians living there. You need information on location (including neighbouring countries and distance from Australia), climate, transportation, cost of living/affordability, language, entertainment/recreation, places of interest to visit (cultural and natural environments).

Put your information into a table as you collect it. Think about which liveability criteria are the most important for you.

Step 3: Analyse

You should do a SWOT analysis for each location before making your decision. Divide a page into four sections. Label them Strengths, Weaknesses, Opportunities and Threats.

Draw a conclusion about which place best suits your needs. Which place is the most liveable for you? Make your decision.

Step 4: Communicate

To help the next person from your company facing the same choice, present your judgement about the liveability of New York and Berlin using ICT tools, such as:

- an animated film
- a Glogster page
- a Prezi
- a report for a travel magazine with text, graphics and pictures using a publishing template.

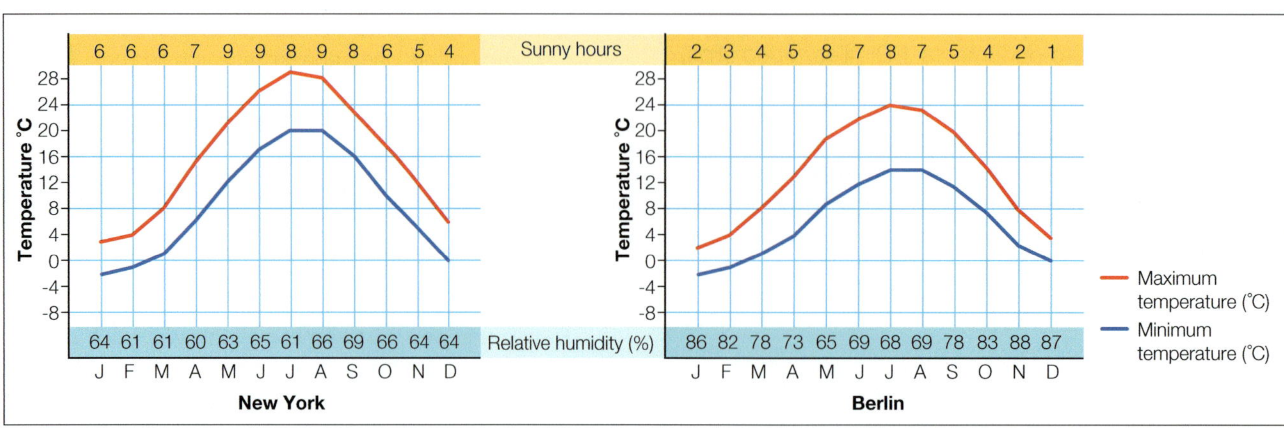

6.16.1 Climate of New York and Berlin

ISBN 978 1 4586 6277 4

Central Park's world-famous Wollman Rink offers ice-skating in winter and inline skating in summer. There's always something to do in Central Park—everything from fun family activities and free public programming to hobbies and sports.

Tropical Islands, the largest indoor water park in the world, is located in a gigantic dome near Berlin. Here you'll find the world's biggest indoor rainforest, Europe's largest tropical spa and sauna complex and a tropical sea with 200 m of sandy beach.

6.16.2 Recreation in New York and Berlin

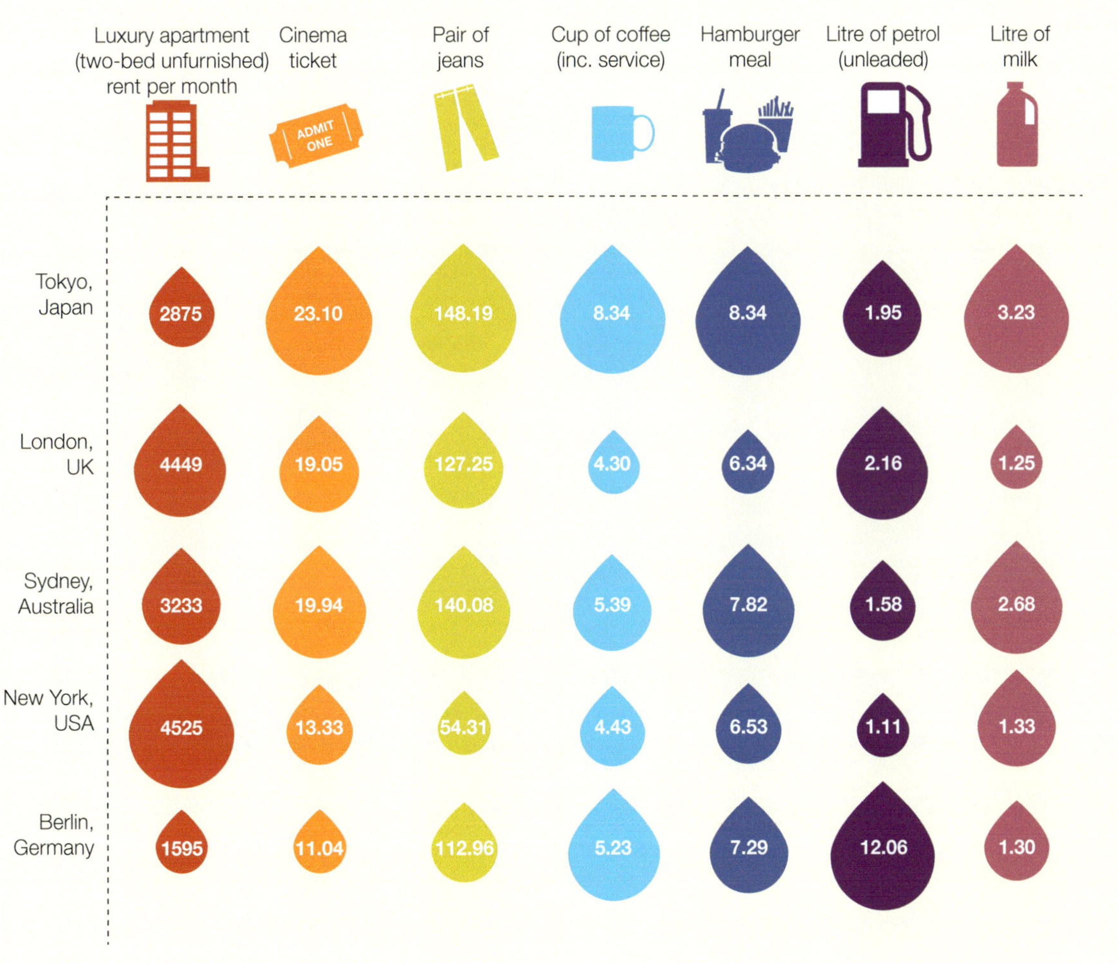

6.16.3 Cost of living in different countries (US$)

ISBN 978 1 4586 6277 4

chapter 7 Liveability: measurement and environmental factors

'The quality of our neighbourhoods, towns and cities have a significant impact on our daily lived experience'

www.urbandesign.gov.au

Aa Geovocab

biomass: material from living or recently living organisms

cultural enclave: place with a concentration of one culture (e.g. Chinatown)

culture of place: character of a place

famine: lack of food

Gross Domestic Product per capita: total income or goods and services produced in a country and divided by the number of people in the country

Human Development Index (HDI): measures life expectancy, education and income

human right: agreed right belonging to all human beings

infrastructure: basic services and structures (e.g. water supply, electricity, transport, schools and hospitals) needed for a society to operate

liveability indexes: variety of indexes that allow places to be ranked (e.g. the Economist Intelligence Unit and the Mercer Quality of Living Survey)

ISBN 978 1 4586 6277 4

Crowds watch skaters compete in the Beach Bowl at Manly Beach, Sydney

Liveability is perceived, structured, organised and managed by people. People make decisions about where they will live using criteria they see as important, such as employment, safety, physical environment, affordability and access to goods and services. The perceptions of people about the liveability of places will depend on their age, gender, lifestyle preference, family status, income, location, culture and ambitions. Various measures are used to compare and rank the liveability of places, including crime rates and environmental sustainability.

Think, puzzle, explore

- **Place** What do you consider to be the essential features of places that make them liveable?
- **Space** Describe some spatial variations in liveability.
- **Environment** What features of the environment enhance liveability?
- **Interconnection** What is the connection between liveability and quality of life?
- **Sustainability** What liveability factors are important for indigenous populations?
- **Scale** How important is culture in the liveability of places for migrants and indigenous peoples?
- **Change** How can liveability be changed and improved?

Geo**skills** in focus

- **Observing** and **questioning** the liveability of their places using the inquiry process
- **Collecting** and **evaluating** geographical tools to assess the liveability of different places
- **Analysing** the liveability ranking of places using different criteria
- **Communicating** information using various text types, digital media and web 2.0 tools

liveability indicators: individual features of society that affect liveability

Millennium Development Goals: eight goals to be achieved by 2015 (e.g. reduce poverty and hunger)

ISBN 978 1 4586 6277 4

7.1 Liveable cities: where living is easiest

Liveability encompasses numerous characteristics that make a place desirable for people to live in (such as access to clean water, education, shelter and health services, as well as freedom from conflict and natural and human disasters).

While opinions vary on the precise characteristics of liveable places, they are generally perceived as healthy, attractive and enjoyable for all people of different backgrounds and ages. Everyone has their own favourite place, as liveability is in the eye of the beholder.

Liveable cities: measuring the immeasurable

A liveable city benefits the people who live and work in it along with those who visit. Different organisations—such as the Economist Intelligence Unit (EIU) and Mercer Quality of Living Survey—rank the world's cities according to a variety of criteria. The EIU ranks Melbourne and Vancouver as top liveable cities but when indicators like green space, lack of sprawl, connectivity and cultural assets were added, Hong Kong comes top. Crime in Cape Town, conflicts in Middle Eastern cities and hunger in African cities reduce their liveability ranking.

The majority of indicators found:

- the highest ranking cities tend to be medium sized as they have less overcrowding and traffic congestion. Consequently, global cities (e.g. New York and Tokyo) are absent from the top of most ranks, as are cities in developing countries (e.g. Mumbai).
- most liveable cities are in wealthier developed countries.

Some **liveability indicators** include air pollution, waste removal, water drinkability, sewerage systems and ecological footprints. Cities with the highest eco-ranking include Calgary in Canada, followed by Honolulu in Hawaii. Port-au-Prince in Haiti ranked at the bottom with only 56% of the population having access to clean water and 22% to sanitation.

Inequalities within Australian cities

Most Australian cities are ranked high on all **liveability indexes**. These cities are socially inclusive, healthy and safe. They have attractive buildings and

7.1.1 Measuring liveability—Sydney scores highly on many liveability measures

The Economist Intelligence Unit

Criteria: 127 cities covering stability, health care, culture, environment, education and infrastructure

Rank	City	Country	Score
1	Melbourne	Australia	97.5
2	Vienna	Austria	97.4
3	Vancouver	Canada	97.3
4	Toronto	Canada	97.2
5	Calgary	Canada	96.6
6	Adelaide	Australia	96.6
7	Sydney	Australia	96.1
8	Helsinki	Finland	96.0
9	Perth	Australia	95.9
10	Auckland	New Zealand	95.7

Mercer Quality of Living Survey

Criteria: 221 cities based on 39 criteria such as housing, recreation, schools and natural environment

Rank	City	Country	Score
1	Vienna	Austria	(01)
2	Zurich	Switzerland	(02)
3	Auckland	NZ	(04)
4	Munich	Germany	(07)
5	Düsseldorf	Germany	(06)
6	Vancouver	Canada	(04)
7	Frankfurt	Germany	(07)
8	Geneva	Switzerland	(03)
9	Bern	Switzerland	(09)
10	Copenhagen	Denmark	(10)

ISBN 978 1 4586 6277 4

7.1.2 What makes Hong Kong a liveable city?

natural environments, such as parks, rivers and harbours. A high liveability ranking attracts new residents and tourists, fostering economic growth. To maintain high liveability, improvements in transport, **infrastructure**, waste management and pollution need to be sustainably managed.

There are inequalities within Australian cities and within groups living in cities. Some people live on the streets while others live in waterfront mansions. When considering housing affordability and traffic congestion, Adelaide was ranked the most liveable city and Sydney the least liveable city.

Geo**info**

- Vienna ranks high for quality of living and Baghdad is low.
- Luxembourg ranks high for personal safety and Baghdad is lowest.

Geo**activities 7.1**

Knowledge and understanding

1 Explain the term 'liveability'.
2 Name four different methods of measuring liveability.
3 Discuss the differences in the liveability of cities across Australia.
4 Explain how liveability varies within a city.

Inquiry and skills

5 Refer to 7.1.1.
 a Use Google Earth and locate the cities on a world map using the Placemark tool.
 b Discuss the spatial distribution of the cities.
 c Explain the dominance of middle-sized cities in most measurements compared to the large global cities, such as New York.
 d Use the internet to research the latest liveability indexes.
  i Compare these indexes with the indexes in the textbook.
  ii Explain the reasons for the changes.
  iii Discuss how the ranking of cities differs from past rankings.
 e Identify features that contribute to Sydney's liveability.
6 Refer to 7.1.2. In groups, develop your own liveability indicators. Rank them in order of importance. Present your work in the form of a poster or Prezi.
7 Take photographs of what you consider important to the liveability of your local area. Present your photographs as a collage.

Chapter 7

7.2 Low-ranking liveable cities

African, Middle Eastern and Asian cities (such as Harare, Baghdad, Kathmandu and Dhaka) lie at the bottom of the liveability scales, scoring low on infrastructure, health care, education and safety. The impact of the civil war in Libya saw its capital city, Tripoli, deteriorate in the liveability rating, placing it in the bottom 10 locations for the first time.

Harare bottom of list

> In Harare, nine-year-old Dakarai squats and pushes both hands into the pile of rotting food scraps found on the street. He gobbles it down.

Zimbabwe's capital city, Harare, was ranked in the ten least liveable cities in the world in 2014, based on stability, health care, culture, environment, education and infrastructure. The country also ranks low on the **Human Development Index (HDI)** at 156 out of 187 countries, with life expectancy at 60 years due to HIV/AIDS. **Gross Domestic Product (GDP) per capita** is US$1307 with 87% of the working population living on less than US$2 a day. Once regarded as the 'bread basket' of Southern Africa, the country fails to produce sufficient food to feed the population, resulting in food shortages and civil unrest. Human rights abuses and poor infrastructure further contributed to the low liveability of the city. However, greater political stability since 2014 has resulted in some improvements to Harare's liveability.

Popular tourist attractions, such as game parks and the famous Victoria Falls, once brought revenue to Zimbabwe, but many countries now advise their citizens to reconsider plans to visit the country.

Geoinfo

- In Harare, 1300 people share one communal toilet with just six squatting holes.
- Poor sanitation—referred to as the 'silent tsunami'—results in a large number of deaths in parts of Harare.

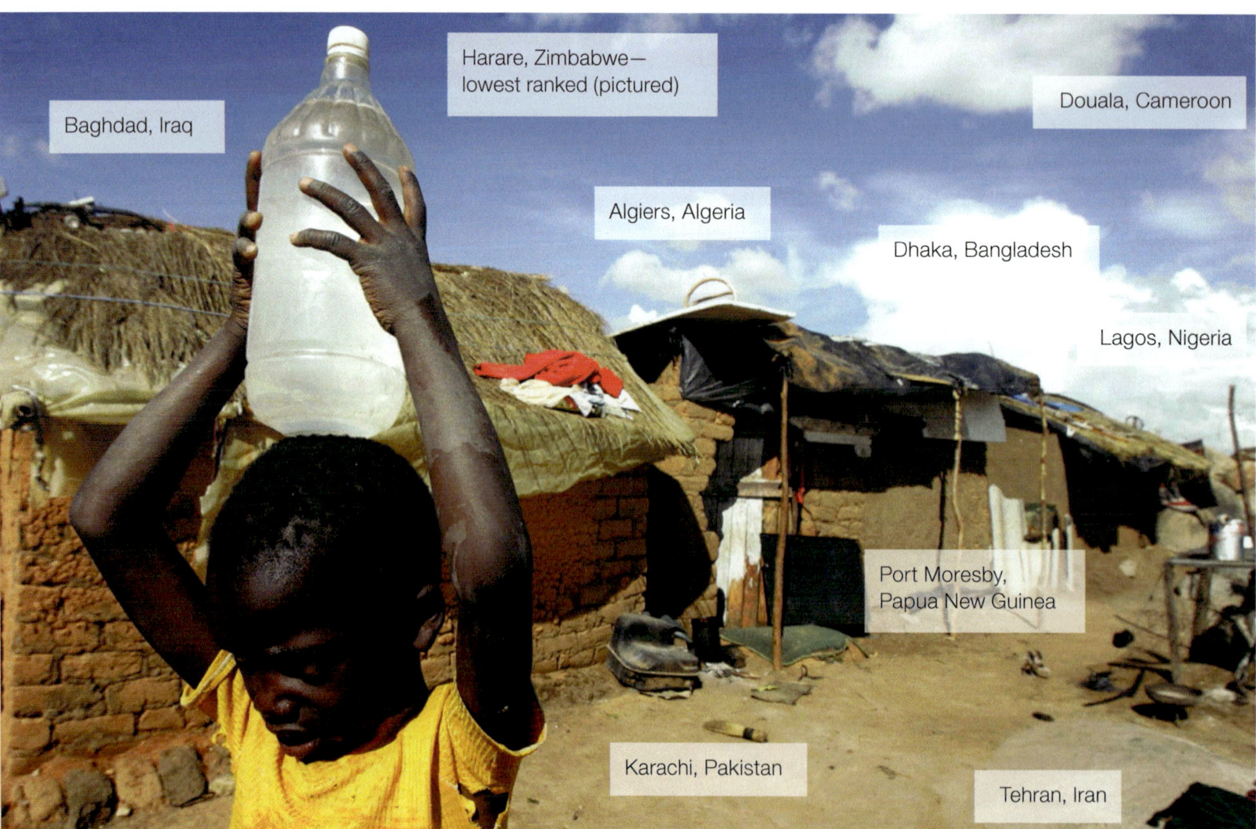

7.2.1 Low-ranking cities—Harare is the world's least liveable city

ISBN 978 1 4586 6277 4

7.2.2 Sadr City

7.2.3 Making Sadr City more liveable will be a formidable challenge

Changing liveability in Baghdad

The city of Baghdad, located on the Tigris River in Iraq, is one of the least liveable cities in the world. The country was ranked 50 out of 130 countries on the HDI before the 1980s war. Baghdad now ranks 120 out of 187 countries, an improvement from 132 a few years ago. Life expectancy is 69 years and average income $US14 000, but only $US4 246 for women.

In the 1970s, oil turned Baghdad into a wealthy city containing beautiful buildings, palaces and mosques. Conflicts, including the 1980s Iran–Iraq War, the 1991 Gulf War and the 2003 invasion of the city by US-led forces caused widespread damage to infrastructure, such as transport, power and water supply. Despite withdrawal of most foreign troops, the city has continued to suffer from infrastructure damage from terrorist attacks.

Rebuilding liveability

Iraq has been slowly rebuilding its infrastructure with the assistance of the international community. Sadr City is a suburb of Baghdad, containing 3 million people. As it is adjacent to an oilfield it became rubble during US air strikes. Water is brought in by truck, electric power is sporadic and residents live in shelters from scavenged materials. Turkish companies are building 75 000 apartments as well as mosques, schools, medical centre, shops, parks and theatres. The challenge is formidable to build this city into a more liveable place.

Geoactivities 7.2

Knowledge and understanding

1 In which regions of the world are cities with low liveability generally located? Suggest the reasons for this.

2 What historical events changed the liveability of Baghdad?

3 Explain why Harare is ranked low on the liveability index.

Inquiry and skills

4 Refer to 7.2.1. In groups, research one city with a low liveability index. Write a brief report indicating why liveability is low. Present your findings as an oral report.

5 Refer to 7.2.2.
   - a What is the distance from Baghdad University to Sadr City?
   - b Name the river running through Baghdad.
   - c Calculate the area of Sadr City.
   - d Calculate the density of the population.
   - e Using Google Earth find the location of Sadr City and draw an outline map of the surrounding features.

6 Refer to 7.2.3. Draw and label a line diagram of the photograph indicating poor liveability indicators.

7 'Liveability refers to the physical, social and mental wellbeing of inhabitants.' Discuss what this statement means.

8 Map the 20 least liveable cities using current rankings and describe their spatial distribution.

Chapter 7

ISBN 978 1 4586 6277 4

7.3 Energy connected to liveability

A liveable place provides access to basic education, adequate shelter and health services, clean water, sewerage systems and power. Access to affordable clean power is important to the liveability of a place and is fundamental to modern life and economic development. Consider the appliances you use daily that rely on electricity—what would happen if the power disappeared?

In 2012, after a massive power breakdown, hundreds of millions of people were left without electricity in northern and eastern India. The internet stopped, shops closed, trains came to a standstill, 200 miners were trapped underground, lifts in buildings didn't work, and hospitals ran on backup generators. The liveability of many places momentarily declined.

7.3.2 Fuelwood is still a source of energy in many countries

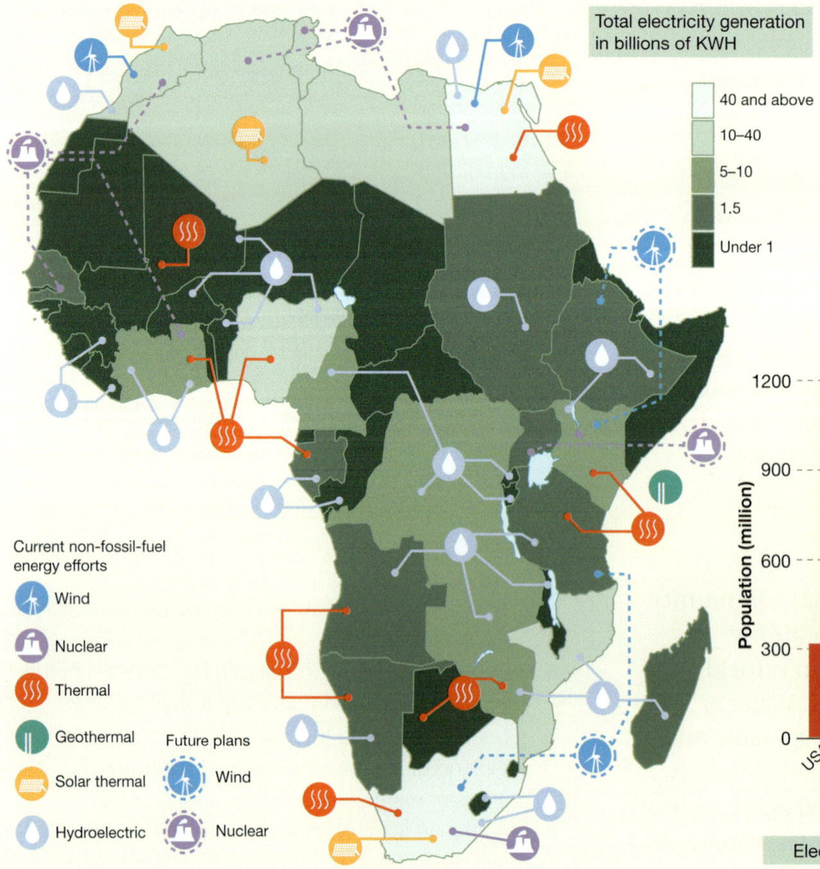

Total electricity generation in billions of KWH

40 and above
10–40
5–10
1.5
Under 1

Current non-fossil-fuel energy efforts
Wind
Nuclear
Thermal
Geothermal
Solar thermal
Hydroelectric

Future plans
Wind
Nuclear

79% of people living in African nations (excluding South Africa) live without access to electricity

25% of the global population live without access to electricity

Every day, New York City alone consumes the same amount of electricity as all of sub-Saharan Africa (excluding South Africa) combined

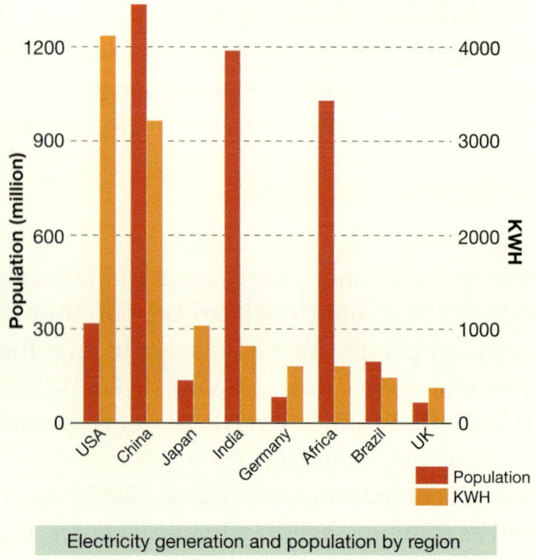

Population (million)
1200
900
600
300
0

KWH
4000
3000
2000
1000
0

USA
China
Japan
India
Germany
Africa
Brazil
UK

Population
KWH

Electricity generation and population by region

7.3.1 Power in Africa: improving the liveability of places

ISBN 978 1 4586 6277 4

Human development and liveability

The Human Development Index (HDI) is a combined measurement of life expectancy, education and standard of living. There is generally a high correlation between the HDI of a country and the liveability of places within the country. This is evident in Africa where countries such as Chad and Niger score low on HDI and their cities and villages score low on liveability indicators. A major contributing factor to their low liveability is lack of power or energy.

Currently 1.5 billion people do not have access to electricity—80% of them live in rural areas of developing countries in Asia and Africa. Those who have access to electricity often pay high prices and the services are unreliable.

'Energy poverty' occurs in Africa, where children are unable to complete their homework at night, sick people are unable to take medicines requiring refrigeration and over 2.4 billion people rely on traditional **biomass** for cooking and heating. Lack of other energy sources forces women and children living in poor rural areas to walk kilometres in search of fuelwood.

Enhancing liveability: clean energy

The liveability of places in developing countries is being transformed by clean, affordable energy, including:

- *Solar photovoltaic technology*—Lanterns, computers and refrigerators are charged using the sun's rays each day.
- *Micro-hydroelectric systems*—These provide electricity for machinery.
- *'Poo power'*—Cow manure and biogas fuel provide a sustainable source of power and fertiliser.

7.3.3 A dangerous practice: people desperate for electricity link into whatever supplies are available

Geo**info**

- China established a Brightness Campaign to provide electricity to 30 million people using solar energy.
- By 2030, a predicted 1.5 million people will die prematurely from using fuelwood in homes with poor ventilation.

Chapter 7

Geo**activities 7.3**

Knowledge and understanding

1 Explain the acronym HDI.
2 Describe the links between energy and liveability.
3 Discuss how reliable and affordable energy has transformed the liveability of places in developing countries.

Inquiry and skills

4 Refer to 7.3.1.
   a Use the world map in the back of the book to list the African countries where electricity generation is lowest.
   b Refer to the graph. The proportion of population to electricity generation varies between countries. Compare the USA, India, China and Africa.
   c The International Year of Sustainable Energy for All aimed to ensure access to energy becomes a right enjoyed by every person. How do you think this could be achieved in Africa? How could it make a difference to the liveability of places in Africa?
5 Imagine your life without electricity. Explain how you would spend your day.
6 Draw a mind map on how clean energy (using continuous resources) can improve the liveability of rural places in developing countries.
7 What makes the situations in 7.3.2 and 7.3.3 dangerous? How can safety be improved?

ISBN 978 1 4586 6277 4

7.4 Safety and crime impact on liveability

Are you safe walking around your local area at night? Why do refugees flee from their home countries? No place is 100% crime free but some are worse than others. The liveability of a place deteriorates if people fear death or damage to property, such as a terrorist attack. The Terrorism Risk Index ranks Somalia at the top, followed by Pakistan.

Peace and liveability

Canadian and Australian cities rank high on the global liveability index with their low crime rates and little threat from political instability or terrorism. However, millions of people live in dangerous countries where kidnapping, murder, disease, terrorism, accidents and theft affect a person's wellbeing.

The Global Peace Index (GPI) uses 23 indicators to rank 162 countries from most to least peaceful. In 2015 Syria was ranked the least peaceful country due to civil war, followed by Iraq, Afghanistan, South Sudan, Central African Republic and Somalia. Iceland ranked the most peaceful country, followed by Denmark and Austria. Australia ranked 9th.

Travelling to cities such as Bogota (Colombia), Port-au-Prince (Haiti), Mogadishu (Somalia) and Port Moresby (Papua New Guinea) is not recommended by most governments for fear of life and property.

Danger reduces liveability

The presence of violent drug cartels—and their wars—impacts adversely on the liveability of places like the slums in El Salvador and Colombia. Ciudad Juárez in Mexico was once a vibrant liveable city of 2 million people. However, over the past five years the government's failure to halt drug cartels has led to over 8000 murders. About 130000 people have fled, leaving businesses and empty houses, and the liveability of the city has declined.

Geo**info**

- Mexodus: violence and lawlessness forcing thousands of middle-class Mexicans to relocate to the USA.
- In 2015, the GPI indicated that 81 countries became more peaceful, and 78 less peaceful.

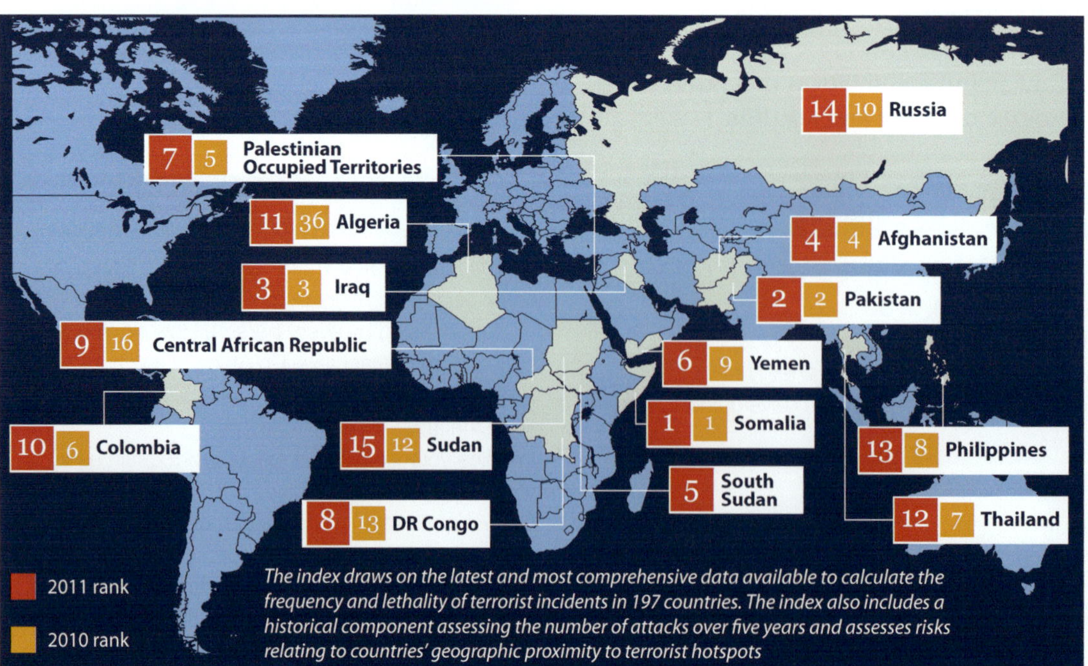

7.4.1 Countries with greatest risk of terrorism using the Terrorism Risk Index

ISBN 978 1 4586 6277 4

City	State	Drug War deaths Dec. 2000–2010	Unoccupied dwellings
Ciudad Juárez	Chihuahua	6437 deaths 450 per 100 000	488 785 dwellings, 111 103 unoccupied 23% empty dwellings
Chihuahua	Chihuahua	1415 deaths 172 per 100 000	293 924 dwellings 44 924 unoccupied 15% empty dwellings
Tijuana	Baja California	1667 deaths 106 per 100 000	553 115 dwellings 111 482 unoccupied 20% empty dwellings

7.4.2 Drug war impacts on the number of unoccupied dwellings and the liveability of communities in Mexico

7.4.3 Mejicanos, El Salvador: an alleged gang member during a drug raid

Security can improve the liveability of a place by protecting people and property. Some societies improve security by imposing curfews, providing alcohol-free areas and implementing street police.

Liveability and kidnapping

Places where kidnapping is rife are not good places to live or visit. Over the past decade, kidnapping rates in Venezuela have soared, with business owners, tourists, athletes and diplomats targeted. Some terrorists use kidnappings to fund their activities and make political statements. Kidnappers in Iraq, Yemen and Iran target civilians, journalists and foreigners, while Somali pirates kidnap people travelling around the Indian Ocean. Despite a major crackdown in Colombia—the so-called 'kidnap capital of the world'—abductions remain a threat.

Geo**activities 7.4**

Knowledge and understanding

1 Explain the Global Peace Index.
2 Name five countries and cities considered unsuitable to visit as a tourist.
3 Explain the relationship between safety and liveability.
4 'The threat of terrorism, conflict and kidnapping plays a large part in liveability rankings.' Explain this statement.

Inquiry and skills

5 Refer to 7.4.1.
  a Rank the countries shown in order from the largest to smallest risk of terrorism in 2011.
  b Research the ranks for the present year and the reasons for the changing trend.
  c Discuss how the risk of terrorism impacts on the liveability of a place.
6 Refer to 7.4.2.
  a Name the city with the highest proportion of drug war deaths.
  b Describe the impact of drug war deaths on the number of unoccupied dwellings.
7 In groups answer these questions, collate the answers and present your findings as an oral report.
  a How safe is your community?
  b Do you feel safe or unsafe at different times of the day?
  c How safe is your community for young boys and girls?
  d Discuss who has responsibility for making places safe.
  e Propose strategies to enhance the safety of your community.

Chapter 7

ISBN 978 1 4586 6277 4

7.5 Transport: mobile and socially connected

People need access to family and friends, goods and services, jobs and opportunities. Places that rank highly in liveability rankings, including Denmark's capital Copenhagen, have transport systems that support people of all ages with safe, clean, efficient and affordable transport. Walking, cycling, driving and mass transit systems (e.g. buses, trains and planes) in the right combination increase connections between people and places.

Vehicle emissions, traffic congestion, dangerous roads and excess noise reduce liveability by affecting human and environmental health. People without access to transport may suffer unemployment and isolation from their communities. Transport systems impact on all aspects of people's quality of life.

Transport and health

The World Health Organization (WHO) blames air pollution for 2 million premature deaths annually. Air pollution in Australia is held responsible for up to 4500 cases of bronchitis, cardiovascular and respiratory disease each year.

Residents of polluted Asian cities (e.g. Beijing and Hong Kong) are so worried about the long-term health impact of pollution that they take their children on holidays to get clean air into their lungs.

Environmental quality

To make Beijing more liveable for the Olympics, the Chinese government introduced strict restrictions on traffic and industry. Pollution, safety and congestion all improved. Traffic volumes fell by 25%, traffic jams dropped 74% and there were 46% fewer accidents.

To reduce pollution levels, and improve environmental quality and people's wellbeing, Berlin's city centre is zoned an 'ecological area', where only vehicles with a green sticker are allowed, and in Paris free public transport is provided one day a week.

7.5.1 Transport and quality of life

ISBN 978 1 4586 6277 4

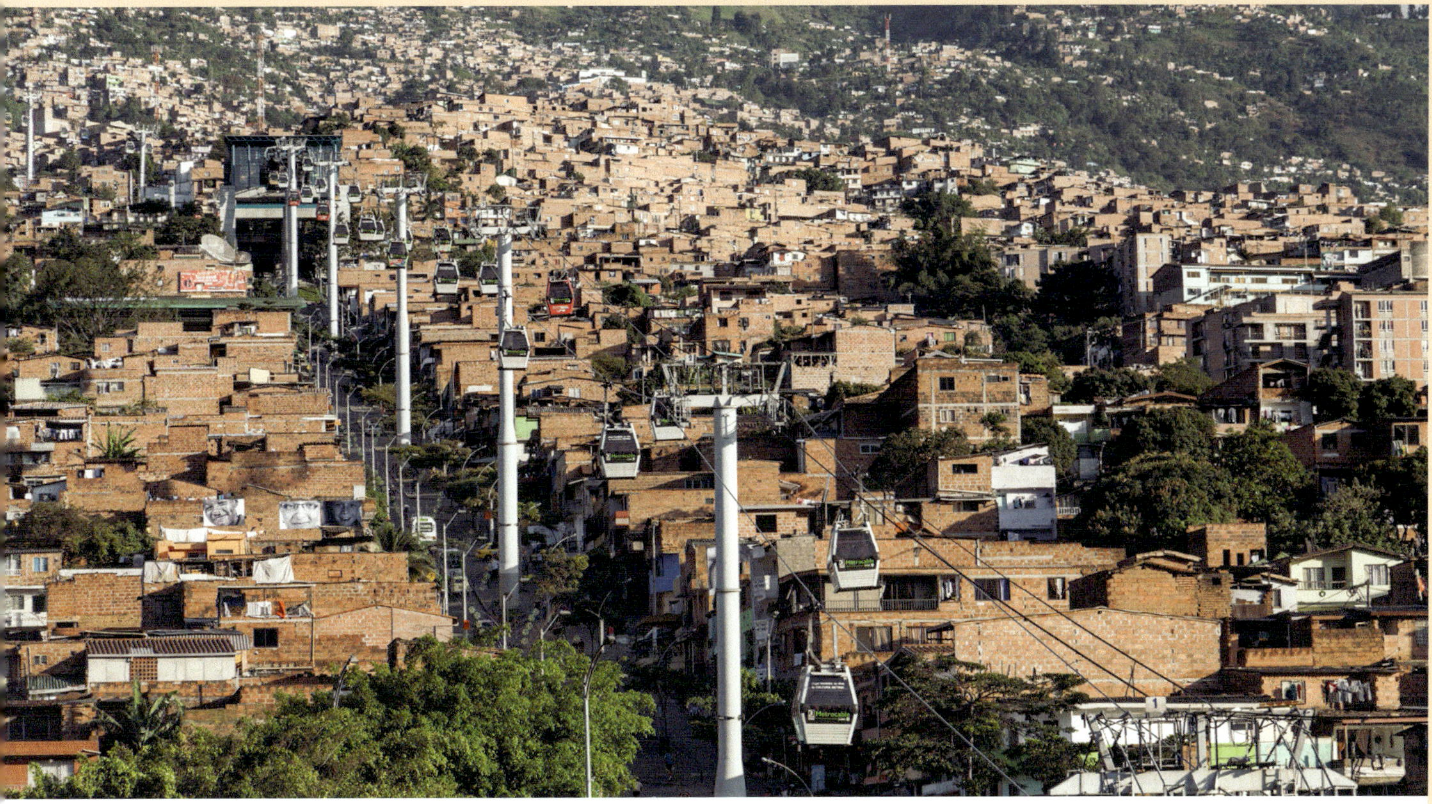

7.5.2 Medellin's cable cars connect slums and the city centre

Connected communities

A reduction in traffic increases interactions between people, because they feel safer, stay on footpaths to talk, walk instead of drive and stay outside for longer.

In the city of Medellin, Colombia, elevated cable cars were built to link hillside slums with the central city and parks. Public facilities were built around the stations. Better access to jobs, community events, entertainment, and health and educational facilities created a more liveable city. A journey that once took 2 hours now takes about 7 minutes. Crime rates have fallen and residents have a new sense of connection with the rest of the city.

Planning for better transport

Transport planning is high on the agenda of city authorities around the world. Giving streets back to the people is a common theme. Improving liveability is being addressed through a variety of transport strategies, including:

- building clean-energy *mass-transit* systems
- constructing *cycleways* and *footpaths*
- creating *walkable neighbourhoods*.

Geo**info**

- Hong Kong's air pollution frequently exceeds the limits set by the WHO.
- Medellin was the first city in the world to use aerial cable cars for a public transport system.

Geo**activities 7.5**

Knowledge and understanding

1. What is the key role of transport systems?
2. Why did China need to improve the liveability of Beijing for the Olympics? How important was transport in achieving this goal?
3. Discuss the connection between traffic congestion and people's wellbeing and connectedness.
4. Explain how transport improvements in Medellin resulted in improved social connectedness and community identity for slum dwellers.

Inquiry and skills

5. Refer to 7.5.1.
   - a Why is transport so important for quality of life?
   - b Choose three ways that transport can have a positive effect on people's lives. Rank them from most important to least important. Justify your answers.
   - c Repeat 6b but with three negative impacts.
6. Refer to 7.5.2. Investigate other cities where cable cars are being introduced for public transport. Present findings using Google Earth Tour Builder.
7. Discuss other factors that influence our social connections to create a concept map.

Chapter 7

ISBN 978 1 4586 6277 4

7.6 Access to shelter

Places are judged on their ability to satisfy basic human needs. Shelter is one of these. From palaces to slums, there are variations in the quality of and access to shelter throughout the world. Film stars living in Hollywood have high security walls enclosing huge mansions with swimming pools. People living in favelas in Rio de Janeiro, Brazil, lack water, a sewerage system and electricity. In India less fortunate people sleep on the streets, under bridges and in drainage pipes.

Many people living on less than US $1 a day reside in overcrowded and poorly constructed dwellings located in areas subject to flooding, landslides or volcanic eruptions. Then there are the millions of people living in refugee camps who require basic shelter provided by the UNHCR. In developed countries, particularly in large cities such as Sydney, homelessness and housing affordability are liveability issues associated with access to shelter.

Right to shelter

Shelter refers to a structure or building providing protection and includes caves, apartments, boats, caravans and houses. Lack of safe, affordable

Everyone has the right to a standard of living adequate for the health and well-being of himself and of his family, including food, clothing, housing and medical care and necessary social services, and the right to security in the event of unemployment, sickness, disability, widowhood, old age or other lack of livelihood in circumstances beyond his control.

Universal Declaration of Human Rights (Article 25.1)

7.6.2 Sleeping rough

A returnee and his son rebuild their home in Afghanistan, which was destroyed by the Taliban. The UNHCR has been helping Afghans settle back home with repatriation packages, shelter kits, mine-awareness training and vaccination programs.

A camp for internally displaced persons (IDPs) in Masisi, Democratic Republic of Congo, where 2300 families sought shelter after fleeing violent attacks throughout North Kivu province.

7.6.1 The UNHCR makes places more liveable by providing shelter

ISBN 978 1 4586 6277 4

shelter is a major contributor to poverty, which affects the liveability of places. Globally, 1 billion people live in inadequate housing, with over 100 million people classified as homeless. On any night in Australia, one in 200 people are homeless. In NSW homelessness has increased 20% since 2006, partly due to lack of affordable housing and crisis shelters. This situation is repeated in cities across the world, including those ranked high on liveability such as Sydney, Melbourne and Vancouver.

Adequate shelter is a **human right**. The **Millennium Development Goals** (MDG) include a goal to improve the lives of 100 million slum dwellers by the year 2020. The figure appears huge but it is only 10% of the global slum population. If nothing is done there will be 3 billion slum dwellers by 2050. Currently the Asia–Pacific region accounts for 60% of the slum dwellers in the world.

Improving liveability

Many organisations work towards improving the liveability of places:

- The *United Nations Human Settlements Program (UN-HABITAT)* promotes socially and environmentally sustainable towns and cities, by providing shelter, water and health facilities. The aim is to create 'cities without slums' and improve liveability. Through 2650 projects in 140 countries UN-HABITAT has helped millions of urban dwellers access adequate shelter.
- *Australia's aid program* improves the liveability of millions of people living in developing countries who lack access to clean water and shelter. It recently provided shelter to refugees in Ethiopia and Somalia and built earthquake-proof schools in Nepal.
- *Community organisations* in many cities provide access to services and facilities for homeless persons and those living in inadequate shelter. In India, slum and pavement dweller organisations built toilets and washing facilities to improve quality of life, while in Melbourne a mobile bathroom facility in a bus provides a safe, clean place to shower, and get clean clothing and hygiene products.
- *Governments* are increasing stocks of affordable housing in large cities, but demand continually exceeds supply as urban populations grow.

North West Frontier Province, Pakistan. The race to provide shelter and supplies to earthquake survivors before the freezing winter set in was intense. By mid-November, the UNHCR had provided 19 356 tents, 152 325 blankets, 71 395 plastic sheets and tens of thousands of jerry cans and kitchen sets.

Geo**info**

Thailand's low-income housing program has cut slum growth rate by 20% in the last 15 years.

Geo**activities 7.6**

Knowledge and understanding

1. What is shelter?
2. What is the relationship between shelter and liveablility?
3. Explain how organisations help make places more liveable by providing access to shelter.

Inquiry and skills

4. Refer to 7.6.1.
   a. Explain how the UNHCR improves liveability for refugees without basic shelter.
   b. Compare your home with these shelters.
5. Refer to 7.6.2 and explain the right to housing.
6. Draw a labelled diagram of a house with all the items that would make it liveable for you.
7. Fieldwork: Photograph the different types of shelters within your local area and list their advantages and disadvantages. Suggest different groups of people who might find each type of shelter most liveable, and explain why.

ISBN 978 1 4586 6277 4

7.7 People need jobs and a place to live

The liveability of places is influenced by access to employment and appropriate, affordable housing. It is not often that both occur in the same location, so people have to make choices. They ask questions such as:

- Is work available where I live?
- Do I have to relocate to find employment?
- If I move can I find suitable housing?
- Where can I afford to live?
- What type of housing do I need and where can I find it?

The most liveable place will be the one with the right balance between work, housing and community features, such as facilities, transport, open spaces, safety and lifestyle.

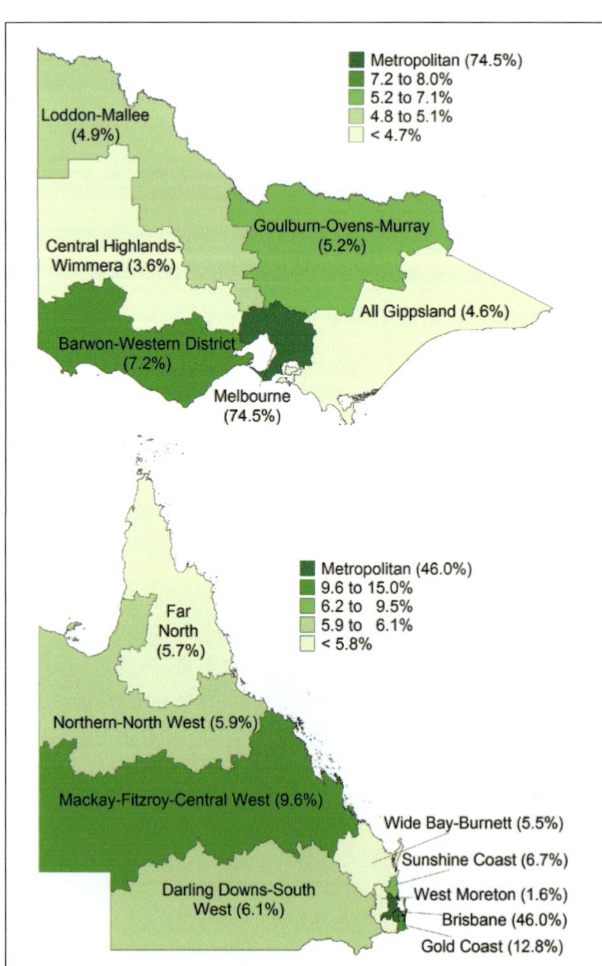

7.7.1 Concentration of jobs in Victoria and Queensland

7.7.2 Housing affordability and employment

Urban employment

Most employment is found in large cities. In Victoria, South Australia and Western Australia 74% of employment is in the capital city. This accounts for the migration of people from rural and regional places to Australian cities.

Within cities the CBD and suburbs share the distribution of jobs. Melbourne's CBD contains 28% of its jobs, with the rest spread throughout the suburbs. Outer suburbs generally have fewer employment opportunities. People in Western Sydney fill 28% of the city's jobs but only a quarter of these are located within the region. Governments are creating employment opportunities in outer city suburbs to improve their liveability.

Employment migration

Employment opportunities create migration. The 2011 Australian census revealed the movement of Australians to Western Australian and Queensland, where a resources boom has created employment in mining and other industries. Mining companies offer high wages as compensation for the low liveability caused by isolation and lack of facilities. Australians are also attracted to employment opportunities in overseas cities where wages and standards of living are high (e.g. Singapore).

ISBN 978 1 4586 6277 4

7.7.3 Increasing concerns about housing affordability: Australians are finding it hard to get a place they can afford in the cities, whether they're renting or buying

7.7.4 Population and employment in NSW

Region	Population 2013	Employment 2014
Greater Sydney	4 757 000	2 427 000
Rest of NSW	2 653 000	1 198 000
TOTAL NSW	7 410 000	3 625 000

Source: https://docs.employment.gov.au/system/files/doc/other/australian_jobs_2015.pdf

Geo**info**

Sydney CBD's population increases from about 50 000 permanent residents to half a million during daytime.
If housing costs more than 30% of household income it is considered unaffordable.

Housing choices and liveability

Most people will live where they can afford to buy or rent and where the appropriate type of housing is available. Generally, housing affordability, which is linked to income, increases with distance from a city centre. Australian cities are among the most expensive in the world for housing affordability.

Where we choose to live is partly determined by where we work and what we can afford. Governments are working to create more liveable communities, offering a variety of affordable housing types (e.g. single dwellings and apartments) with access to employment and facilities.

Geo**activities 7.7**

Knowledge and understanding

1 How is liveability linked to employment?
2 Why will increasing employment opportunities in outer city suburbs increase the liveability of these suburbs?
3 What do you understand by the terms 'housing types', 'housing availability' and 'housing affordability'?
4 How is liveability linked to housing availability and affordability?
5 Why is there a link between liveability and migration?

Inquiry and skills

6 Refer to 7.7.1.
   a Describe the distribution of employment in Queensland. Suggest a reason for the pattern.
   b Suggest a reason for the higher percentage of employment in the Mackay–Fitzroy–Central West region in Queensland.
   c Rank the percentage of employment by region for Victoria.
7 Refer to 7.7.2.
   a What is the implied location in the cartoon? What evidence supports this?
   b Explain the main message behind the cartoon.
8 Refer to 7.7.3. Analyse the liveability of the suburb in the photograph. Consider the advantages and disadvantages, and the groups who may choose to live there.
9 Refer to 7.7.4.
   a Calculate the percentage of NSW employment located in the Greater Sydney region.
   b Compare the distribution of employment in NSW with Victoria and Queensland.

ISBN 978 1 4586 6277 4

7.8 Environmental quality and resources

Vancouver, Canada, is renowned for its extreme natural beauty. The Fraser River, Pacific coast, snow-capped mountains and mild climate contribute to the city's high liveability ranking.

People are attracted by the forests, mountains and water bodies (including oceans, lakes and rivers) for their physical amenity (attractiveness), as well as by resources, such as timber. Popular recreational activities include sailing and kayaking, or bushwalking, camping, skiing and sledding on Vancouver's nearby Mount Grouse.

Climate, landforms and natural resources (such as water, minerals, forests, wildlife and soils) influence the suitability of places for farming, mining and forestry. People live where natural assets satisfy their basic needs or where resources can be exploited and enjoyed. People make adaptations to improve the liveability of places.

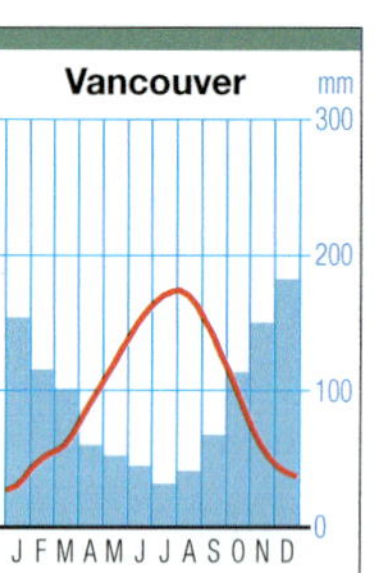

7.8.1 Climate graph for Vancouver: latitude 49°N, longitude 123°W

Natural resources

The first Australians settled the entire continent living sustainably on natural resources from the land and coastal waters. Aboriginal and Torres Strait Islander Peoples living in Northern Australia (e.g. the Yolngu peoples of East Arnhem Land and residents of Mer in Torres Strait) still rely on seasonally available resources, such as wild honey and turtle eggs. Likewise, the resources of rainforests support tribes such as the Penan in Malaysia and the Yanomami in Brazil.

River valleys and coastal plains are the most densely settled places on the planet, providing water, flat land, pasture and rich alluvial soils for agriculture. The Ganges Delta is one of the most fertile places in the world. Growing crops like rice, raising cattle and fishing supports over 140 million people. Population densities reach up to 200 people per km^2. However, the future liveability of the delta is threatened by climate change.

7.8.2 Vancouver is consistently ranked among the world's most liveable cities

ISBN 978 1 4586 6277 4

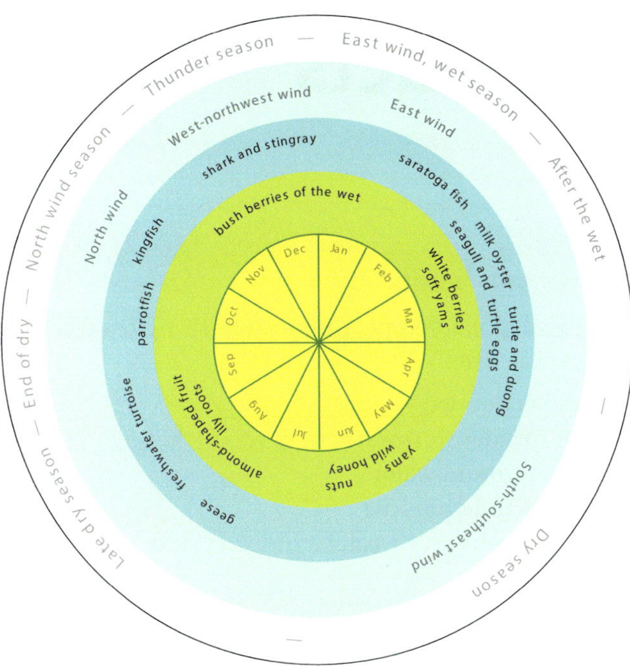

7.8.3 Seasonal calendar of land and sea resources in northern Australia

Adaptation to extreme conditions

In the opal-mining town of Coober Pedy, where summers can reach 45 °C, most of the 4000 inhabitants live underground. The famous underground houses maintain temperatures between 21 and 26 °C. The Berbers who live in the desert at Matmata in Tunisia also carved out cooler underground settlements, which became the site of the famous Star Wars movies. Tourists can stay in Luke Skywalker's home, now used as a hotel.

Natural hazards and liveability

The tornado belt in the USA and the cyclone-prone areas of Australia contain productive agricultural land and rural towns. People have adapted to the threats created by natural hazards with warning systems, evacuation procedures, building regulations and underground shelters. Natural hazards kill around 250 000 people each year yet most survivors return to their properties to rebuild, despite future risks. When the frequency of hazards (such as bushfires, drought and flood) increases, the liveability of rural areas declines to the extent that some people will leave permanently.

Geo**info**

Coober Pedy or 'Kupa Piti' means 'White Man in a Hole' in the language of the traditional landowners.

7.8.4 Underground house in Coober Pedy

Geo**activities 7.8**

Knowledge and understanding

1. Why is the physical environment important for liveability?
2. What natural resources attract people to live in different places?
3. Why do people adapt to extreme environmental conditions rather than live somewhere else?
4. Explain how natural hazards impact on the liveability of places.

Inquiry and skills

5. Refer to 7.8.1 and 7.8.2.
   - a Describe the location of Vancouver.
   - b Draw a sketch of the Vancouver region. Label features of the natural environment that contribute to the city's liveability.
   - c Calculate the yearly precipitation of Vancouver.
   - d Describe the distribution of precipitation.
   - e How would precipitation and its distribution affect Vancouver's liveability?
   - f Describe the landforms on which the city of Vancouver is located.
   - g Write a paragraph explaining how landforms, resources and climate make Vancouver a very liveable city.
   - h Suggest possible threats to the future liveability of Vancouver.
6. Use Google Earth or Google Maps to locate the following places.
   - a Vancouver: 49°N, 123°W
   - b Coober Pedy: 29°S, 134°45'E
   - c Matmata: 33°33'N, 9°58'E
   - d Arnhem Land Yolngu Community at Yirrkala: 12°13'S, 136°55'E.

Chapter 7

ISBN 978 1 4586 6277 4

7.9 Climate change impacts on liveability

Where does a fisher go when fish stocks have disappeared? If there is no fresh water, how can farmers grow crops? If your land or nation is lost to erosion and flooding where do you go? These situations are the real and potential impacts of climate change on liveability.

The Intergovernmental Panel on Climate Change (IPCC) suggests the greatest impact could be on where people live, because environmental changes (such as flooding and freshwater shortages) will make places unliveable. Climate migrants will increase in number, with estimates of 200 million by 2050. The government of Kiribati recently supported a plan to buy over 2000 ha of land in Fiji in the belief that 'moving won't be a matter of choice. It's basically going to be a matter of survival'. Other nations at risk of losing a significant part—or all—of their territory include Tuvalu, Fiji, the Solomon islands, Papua New Guinea, the Maldives and some of the lesser Antilles.

Rising sea levels

Increased global temperatures cause rising sea levels. When seas rise:

- land is flooded
- salt invades freshwater supplies, including underground water
- plants and crops die
- coastal erosion threatens property and livelihoods.

In northern Alaska, a loss of sea ice due to warmer seawater has left the land unprotected from storm waves. Erosion has caused the coastline to retreat, taking villages with it.

Coastal populations

An estimated 10% of the world's population live on coastal or riparian (on a river) land less than 10 m above sea level. The largest cities are in low-lying coastal zones with major ports, tourist facilities, agriculture and fishing industries.

Technology exists to alter coastlines to protect people, cities and farmland. The Netherlands is famous for its dykes, which keep the North Sea from inundating low-lying land. This technology is expensive, and poorer countries may need to use simpler—but less effective—strategies, such as planting mangroves and building seawalls from local materials like shells.

Natural hazards and conflict zones

Due to climate change, some places are predicted to become hotter and drier, others cooler and wetter. Natural hazards (including bushfires, cyclones, insect plagues, droughts and floods) will become more intense and more frequent.

'Being small in area and low-lying, inhabitants will have nowhere to retreat to as the seas inundate their coastlines.'

'Climate change threatens to cause the largest refugee crisis in human history.'

'For the first time in history, you could actually lose countries off the face of the globe.'

'Unlike some people displaced by conflict or persecution who may one day return home, those displaced by the chronic impacts of climate change will require permanent resettlement.'

Source: quotes from Environmental Justice Foundation

7.9.1 Climate change makes places unliveable and causes migration. On Tuvalu, land has been inundated with seawater, destroying food crops and making the soil too salty to grow new crops.

ISBN 978 1 4586 6277 4

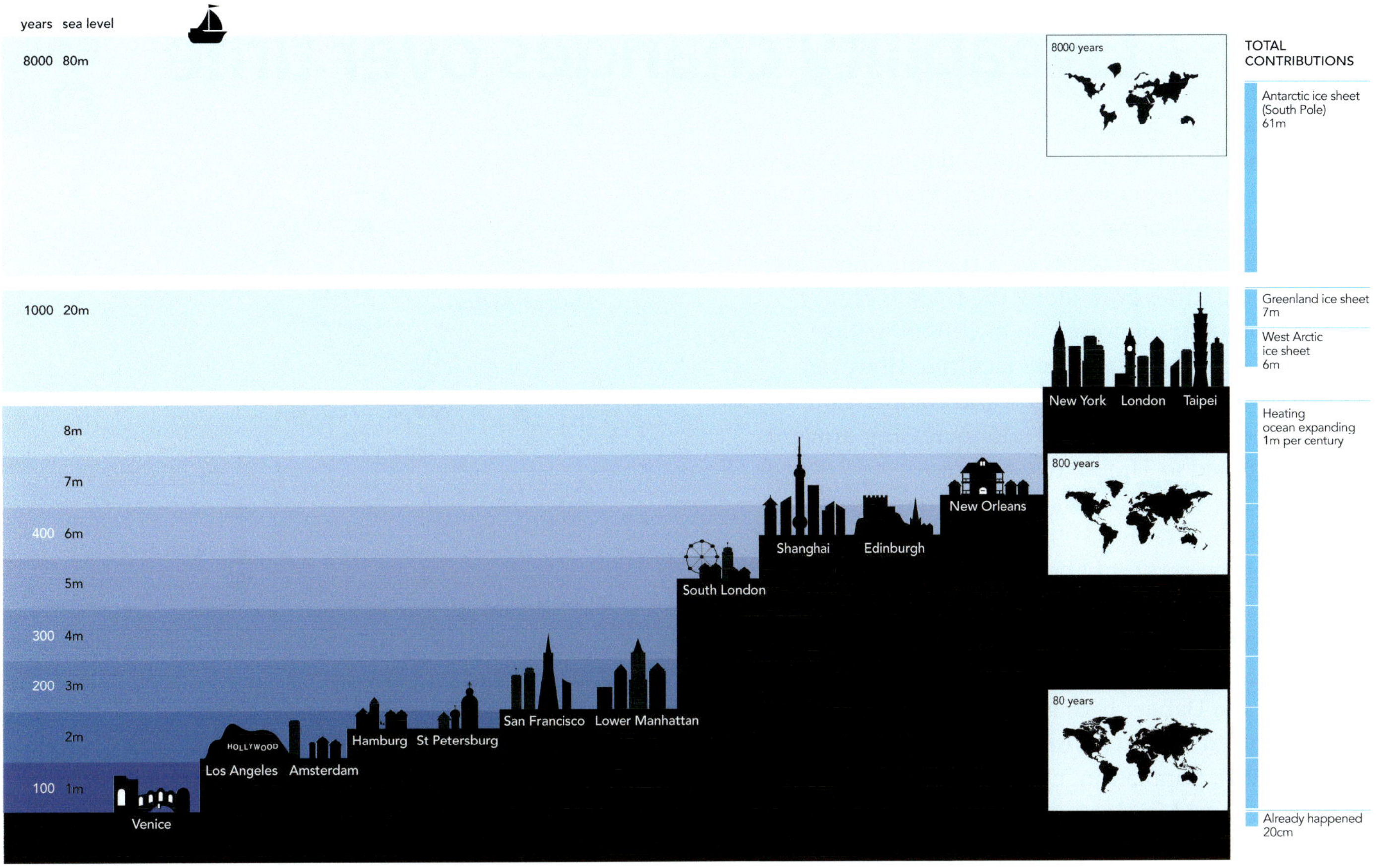

years sea level

8000 80m

1000 20m

8m

7m

400 6m

5m

300 4m

200 3m

2m

100 1m

Venice

Los Angeles Amsterdam

HOLLYWOOD

Hamburg St Petersburg

San Francisco Lower Manhattan

South London

Shanghai Edinburgh

New Orleans

New York London Taipei

8000 years

800 years

80 years

TOTAL CONTRIBUTIONS

Antarctic ice sheet (South Pole) 61m

Greenland ice sheet 7m

West Arctic ice sheet 6m

Heating ocean expanding 1m per century

Already happened 20cm

7.9.2 Coastal cities under threat

Shortages of food, fresh water and land due to climate change will be a source of future conflict (even war), threatening people's safety, homes and livelihoods—making places less liveable. Forty-six countries, home to 2.7 million people, are considered to be at high risk of future conflict over resources. Conflict zones are potential breeding grounds for terrorism—another threat to liveability.

Geo**activities 7.9**

Knowledge and understanding

1 List all of the potential impacts of climate change on the liveability of places.
2 How has the government of Kiribati responded to the threat of rising sea levels? Do you think this is an overreaction? Explain your response.
3 Explain the loss of sea ice in Alaska. How has this impacted on coastal communities?
4 Why might climate change increase future conflicts?
5 Discuss the use of technology to protect the liveability of coastal places.

Inquiry and skills

6 Refer to 7.9.1 and the text.
   a Define the following terms: inundate, displaced, resettlement and chronic.
   b Use these terms to explain why climate change could create a refugee crisis.
   c How will people displaced by climate change be different from other refugees?
7 Refer to 7.9.2.
   a What is the predicted sea level rise over the next 8000 years?
   b What event will contribute the most to future sea level rises?
   c Show the location of the cities on a world map (paper or digital).
   d Which three cities are the most vulnerable to rising sea levels?
   e Suggest three other ways climate change could impact on coastal cities.
   f Use an atlas to list the cities that are also located on a river. Do you think this increases their vulnerability? Explain your answer.
   g Predict three ways these cities might respond to future threats of sea level rise to maintain liveability.

Chapter 7

ISBN 978 1 4586 6277 4

7.10 Liveability changes over time

Timbuktu, in the West African country of Mali, was once a wealthy, powerful trading town and centre for religion, learning and commerce. Salt, gold, cotton, millet and slaves were traded using camel caravans. Cities grew along the trade routes.

Timbuktu supported up to 100 000 people attracted by its wealth and facilities. However, successive waves of invasion and colonisation, changed trading patterns and increasing droughts contributed to Timbuktu's decline.

Today, Timbuktu is a poor and dangerous place to live. Only 15 000 people now live there. One of the poorest nations in the world, Mali ranks near the bottom of the United Nations Human Development Index. In 2012 UNESCO listed Timbuktu's World Heritage mosques and buildings as Sites in Danger because of political conflict and unrest. Poverty, drought, **famine**, conflict, terrorism, population, climate change and slavery make many African countries, such as Mali, the poorest and least liveable in the world.

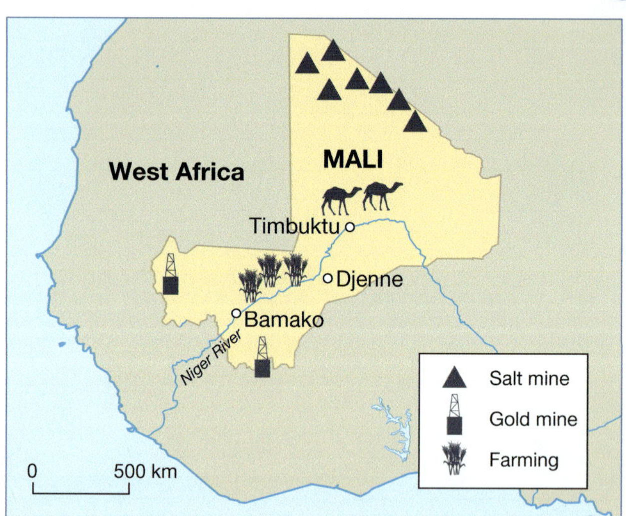

7.10.2 Camel caravans to Timbuktu

The Sahel

In 2012 the Sahel made news headlines and became the focus of famine relief campaigns. It is a zone between the Sahara Desert in Northern Africa and

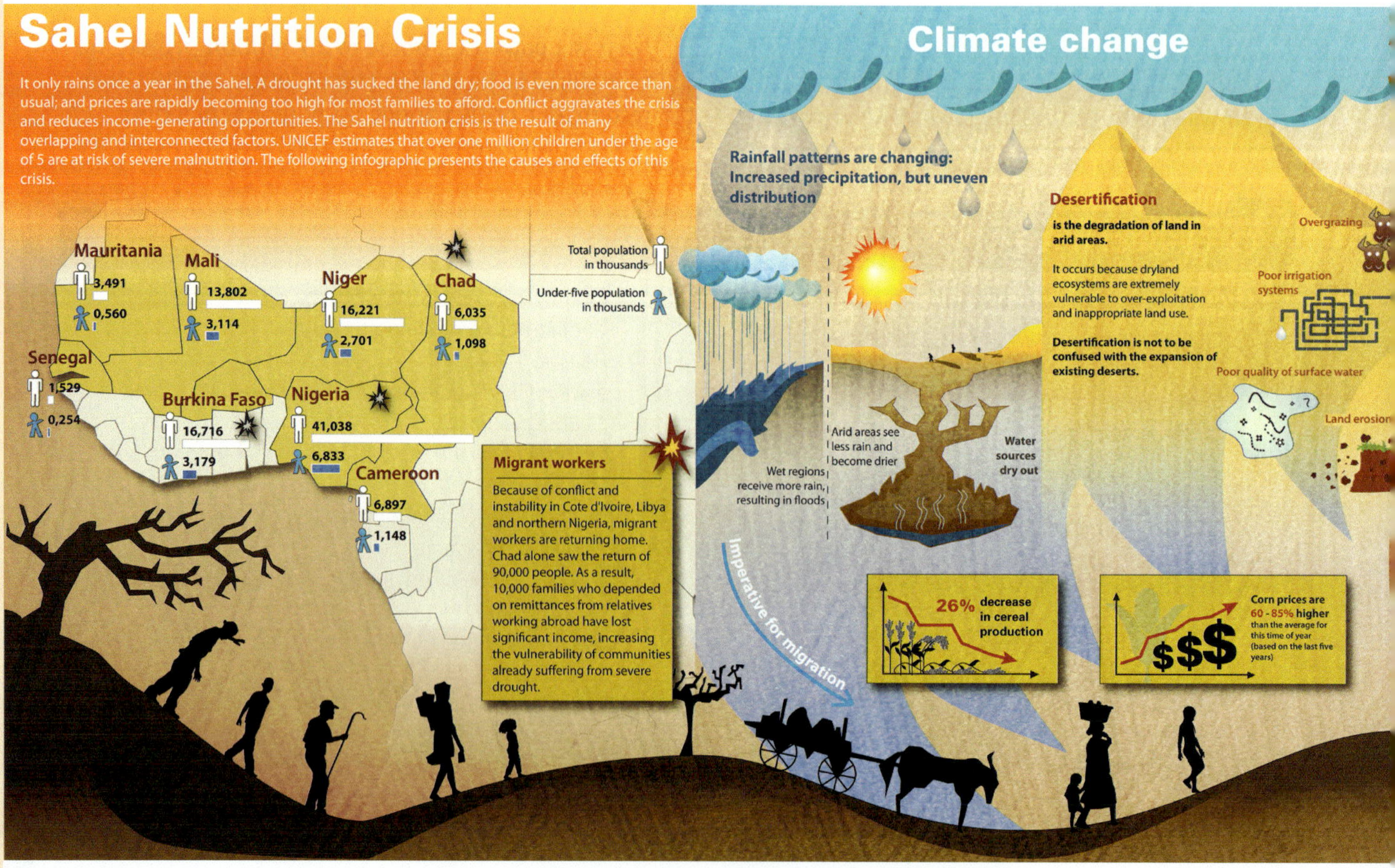

Sahel Nutrition Crisis

It only rains once a year in the Sahel. A drought has sucked the land dry; food is even more scarce than usual; and prices are rapidly becoming too high for most families to afford. Conflict aggravates the crisis and reduces income-generating opportunities. The Sahel nutrition crisis is the result of many overlapping and interconnected factors. UNICEF estimates that over one million children under the age of 5 are at risk of severe malnutrition. The following infographic presents the causes and effects of this crisis.

Total population in thousands
Under-five population in thousands

Mauritania 3,491 0,560
Mali 13,802 3,114
Niger 16,221 2,701
Chad 6,035 1,098
Senegal 1,529 0,254
Burkina Faso 16,716 3,179
Nigeria 41,038 6,833
Cameroon 6,897 1,148

Migrant workers

Because of conflict and instability in Cote d'Ivoire, Libya and northern Nigeria, migrant workers are returning home. Chad alone saw the return of 90,000 people. As a result, 10,000 families who depended on remittances from relatives working abroad have lost significant income, increasing the vulnerability of communities already suffering from severe drought.

Climate change

Rainfall patterns are changing: Increased precipitation, but uneven distribution

Wet regions receive more rain, resulting in floods

Arid areas see less rain and become drier

Water sources dry out

Imperative for migration

Desertification

is the degradation of land in arid areas.

It occurs because dryland ecosystems are extremely vulnerable to over-exploitation and inappropriate land use.

Desertification is not to be confused with the expansion of existing deserts.

Overgrazing

Poor irrigation systems

Poor quality of surface water

Land erosion

26% decrease in cereal production

Corn prices are 60 - 85% higher than the average for this time of year (based on the last five years)

$$$

7.10.1 Events impacting on the liveability of the Sahel

ISBN 978 1 4586 6277 4

the wetter tropical grasslands and forests nearer the equator. Its climate is hot and dry: temperatures often exceed 40 °C. Rainfall is low, varying from 200 mm to 600 mm, and dust storms reduce visibility and degrade the land. Cycles of drought and flood are common.

Sahelian people

Sahelian people, such as the Tuareg, live by growing crops, such as corn, or herding animals. Cropping is possible in wetter places to the south and along rivers (e.g. the Niger). Many herders are nomadic, travelling to find water and pasture. For decades, serious droughts have caused ongoing food crises and famines because food production has fallen and prices have risen. Population growth, conflict and climate change are contributing to the crisis.

Future liveability

The Sahel was once a better place to live. Rainfall was more reliable, farming was adapted to suit seasonal changes in water availability and towns thrived on trade. The future of the Sahel will be determined by climate change and conflict.

Geoactivities 7.10

Knowledge and understanding

1 Outline how the liveability of Timbuktu has changed.

2 List the factors reducing the liveability of places in the Sahel.

3 Suggest why people continue to live in the Sahel.

Inquiry and skills

4 Refer to 7.10.1.
  - a Find evidence of poor liveability in the Sahel.
  - b Create a cross-section diagram to represent the shaded people, land and buildings in the bottom of the image. Label the diagram to show what each image represents.

5 Refer to 7.10.2.
  - a What natural resources led to the growth of Timbuktu as a trading town?
  - b Use ICT to locate the Sahel. Draw a map showing countries, latitude, longitude and the location of Timbuktu.

6 Investigate actions taken by governments and organisations to reduce poverty and improve liveability in the Sahel.

Conflict

Since the start of decolonization, the region has experienced a significant number of conflicts, including large-scale interstate wars, civil wars and localized fighting. In addition, the proliferation of small weapons and drug trafficking have exacerbated violence in this region.

Bush fires

Deforestation

Conflict destroys resources

Conflict over resources

Northern pastoralists have pushed farther south, while southern farmers have expanded cultivation into lands used primarily by pastoralists. Such changes have led to greater competition, tension and violent conflict between livelihood groups.

The insecurity is **hampering the delivery of aid** to those who remain.

Women and girls are being kidnapped and **children are being recruited into armed groups.**

Landmines on the ground have already killed several children.

Southern farmer

Northern pastoralist

Lack of resources generates more conflict

Impact on population

Children under 5 with severe acute malnutrition

Country	Children
Burkina Faso	99,178
Cameroon *(northern regions)*	55,119
Chad *(Sahel belt)*	127,300
Mali	175,000
Mauritania	12,600
Niger	394,000
Nigeria *(northern regions)*	207,718
Senegal *(northern regions)*	20,000

1 million children at risk

Population displaced due to the conflict in Mali

100,000
80,000
60,000
40,000
20,000
0
8 Feb
13 Feb
15 Feb
21 Feb
23 Feb
28 Feb
1 Mar
6 Mar
15 Mar
Mali *Internally displaced persons (IDPs)*
Refugees (displaced across country borders)
Mauritania
Nigeria
Burkina Faso

321,930 people displaced in the Sahel region

Displaced people need:
- Food
- Water and sanitation facilities
- Shelter
- Health care
- Protection

ISBN 978 1 4586 6277 4

7.11 Culture and community identity

Cultural assets, such as art galleries and performance spaces, and cultural connections to people and places through language, beliefs and customs are components of a community's identity. A strong community identity can enhance the liveability of places for different groups of people. Places reflect past and present cultural influences. **Culture of place** can include

- *customs, traditions and beliefs* (e.g. language, dress, dance and festivals)
- *national and international events* (e.g. Mardi Gras and film festivals)
- *art, music and theatre facilities* and *events* (e.g. galleries)
- *historical structures* (e.g. churches)
- *activities allowing creativity and expression* (e.g. street art)
- *lifestyle* (e.g. beach culture).

When people choose a place to live they consider whether the culture of place supports their interests and cultural background.

Cultural facilities and events create places where people can interact. In indigenous communities, culture of place is linked to relationships with the land, concepts of family and spiritual beliefs.

Multicultural places

Migrants take their culture with them to new places, and are likely to settle near people with the same background. This creates **cultural enclaves** where

Number of art galleries	
Paris	1046
London	857
New York	721
Tokyo	688
Berlin	421
Istanbul	267

Percentage of public green space	
Singapore	47
Sydney	46
London	38
Johannesburg	24
New York	14

Public libraries per 100 000 people	
Paris	7
London	5
New York	3
Sydney	3
Tokyo	3

Annual music performances	
Paris	33 020
New York	22 204
London	17 108
Tokyo	15 617
Johannesburg	7400

Bookshops per 100 000 people	
Shanghai	15
Tokyo	30
London	10
Sydney	10
New York	9

Number of cinemas	
Paris	302
Shanghai	230
Istanbul	118
New York	117
London	108
Mumbai	105

Annual dance performances	
New York	6292
Paris	3172
London	2756
Shanghai	1686
Tokyo	1598

Annual theatre performances	
New York	43 004
London	32 448
Paris	26 676
Tokyo	24 575
Shanghai	15 618

7.11.1 Cultural comparison of large cities

ISBN 978 1 4586 6277 4

people feel safe and comfortable. For example, Indian people in Harris Park (Sydney) comprise 43% of the local population. Australians form their own cultural precincts in places like London.

Multiculturalism creates vibrant, diverse places and new experiences to enjoy, for instance Chinese New Year. In New York City mobile food trucks provide a variety of ethnic foods. Vendors—such as Taim Mobile (Middle Eastern) and Rickshaw Dumpling Truck (Chinese)—use Twitter to notify people of their location.

While cultural diversity can increase liveability, ethnic clashes and violence reduce safety, forcing people to leave. For example, Tamils escaping violence have settled in safer cities such as Melbourne where Sri Lankan restaurants are becoming increasingly popular and festivals such as Hindu New Year are celebrated.

Cultural amenities

People interested in history and the arts (music, theatre and art) will be attracted to places that support these activities. The world's cities can be compared based on their cultural facilities. Street art and theatre are used to make places more liveable and to engage people in their communities. Graffiti art is used in the slums of São Paulo in Brazil to improve lives by drawing attention to community issues, create attractive surroundings and build community identity.

Places can be very protective of their culture. In France restaurants are required to serve local produce—local duck appears in shops and on menus throughout Provence. McDonald's restaurants are subtly signposted and serve McBaguettes, minimising the influence of American culture in France.

7.11.2 Street art in São Paulo

Geo**info**

In 2012, the largest group of migrants living in Australia were those born in the UK; the second largest group were New Zealanders.

Geo**activities 7.11**

Knowledge and understanding

1 What is meant by the term 'culture of place'?
2 How does migration influence the culture of places?
3 Why are theatres and museums considered important facilities?
4 How does street culture help build social connections and community identity?
5 What is the benefit to communities of restricting the influence of other cultures?

Inquiry and skills

6 Refer to 7.11.1.
   a Allocate points for each city according to its rank in each category (e.g. rank 1 = 5 marks, rank 5 = 1 mark). Add the points for each city to decide which one offers the best access to cultural facilities.
   b Do you think this system of ranking is a true reflection of the culture of a city? Explain your answer.
   c Why is public green space included as a cultural feature?
7 Refer to 7.11.2.
   a Find images of street art and theatre in Australia. Choose one image to contribute to a class collage.
   b Discuss the choice of images. Would all images contribute positively to the character (identity) of their communities?
8 My ideal cultural environment.
   a Describe the cultural features you would like to have in your neighbourhood.
   b Explain why they are important to you.
   c Would they be appealing to all age groups? Why or why not?
   d Compare the liveability of your current community with your ideal based on the availability of cultural facilities.
   e Write a letter to your local council recommending they take action to provide cultural facilities, predicting the benefits for young people.

ISBN 978 1 4586 6277 4

7.12 Heritage status linked to liveability

The Economist Intelligence Unit (EIU) ranks the liveability of cities using 30 indicators. In 2012 the EIU aimed to improve its assessments using the Spatially Adjusted Liveability Index (SALI), which encompasses seven new indicators: green space, urban sprawl, access to nature, connectivity to the rest of the world, isolation from nearby large cities, lack of pollution, and availability of world class cultural assets. The latter was measured by counting the number of United Nations World Heritage sites located nearby. Hong Kong was ranked first on the SALI, assisted by China's 43 heritage sites. Some argued Hong Kong's dense population was an undesirable liveability quality and in 2015 pollution and unrest reduced its liveability.

World Heritage sites

The World Heritage List includes 962 cultural and natural properties possessing outstanding universal values, such as the Red Fort in New Delhi and Salzburg in Austria. The United Nations Educational, Scientific and Cultural Organisation (UNESCO) aims to protect and preserve these sites, since 'Heritage is our legacy from the past, what we live with today, and what we pass on to future generations'.

While heritage tourism causes environmental degradation from traffic, garbage and sewage disposal, it also enhances the liveability of places. Entry fees contribute to restoring and protecting sites, and jobs revitalise local economies.

In New York and Paris governments provide infrastructure (such as roads, clean water and

Salzburg, Austria—European ecclesiastical city, also associated with Mozart

Red Fort in New Delhi, India, built in 1546. The design is a fusion of traditions— Islamic, Persian, Timurid and Hindu

Ohrid, the Former Yugoslav Republic of Macedonia—situated on the shores of Lake Ohrid, one of the oldest human settlements in Europe

Old Bridge in Mostar, Bosnia and Herzegovina—coexistence of communities from diverse cultural, ethnic and religious backgrounds

7.12.1 World Heritage sites contribute to improved liveability of surrounding places from tourism

ISBN 978 1 4586 6277 4

sanitation), while businesses provide 5-star accommodation near heritage sites, improving liveability and attracting tourism. In Zanzibar, conserving the World Heritage Stone Town's traditional seafront increased the value of surrounding properties and the city's liveability.

Whatever the criteria used to determine the liveability of a place, sustainable practices to protect, maintain and reuse heritage sites is important. The World Bank emphasises the links between conserving natural heritage and sustainable tourism.

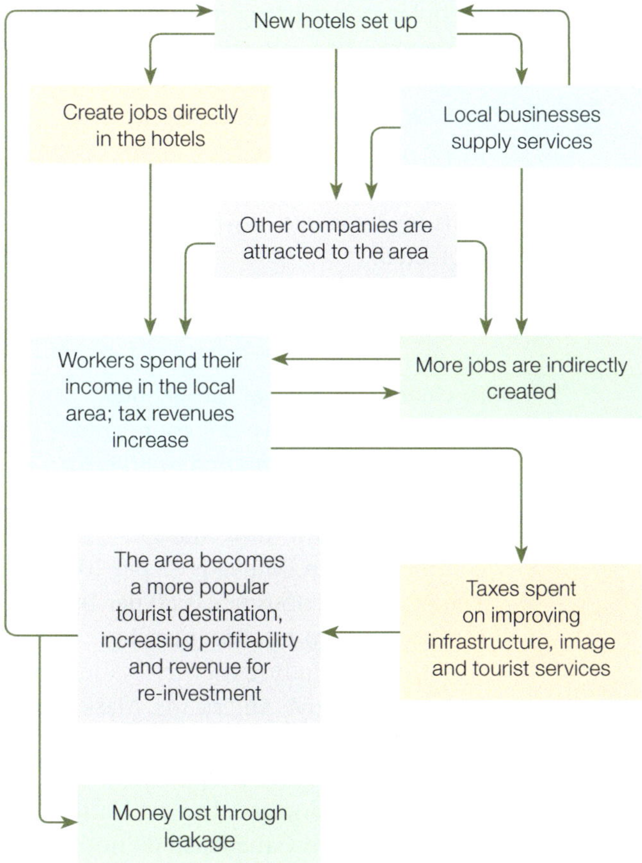

7.12.2 World Heritage tourism has a multiplier effect on the liveability of places

Dracula: World Heritage site

In 1999 Sighisoara was inscribed on the World Heritage List as one of the best preserved medieval fortified towns in Europe. The city attracts tourists for its historic significance—and for being the birthplace of Dracula, or Vlad the Impaler, who was depicted incorrectly in movies as a vampire who fed on blood. A Dracula theme park was planned to lure tourists to Sighisoara, complete with vampire roller coaster rides and red blood food. The aim was to generate income and improve the liveability of the city. After years of protests by groups opposed to the park for environmental or cultural reasons, the plan was abandoned. Governments have safeguarded and restored this UNESCO heritage site and developed sustainable tourism in a liveable city.

7.12.3 Sighisoara, Romania: a popular perception of Count Dracula

Geo**info**

In 2015 there were 1031 World Heritage sites, of which 802 were designated cultural sites. Australia has 19 World Heritage sites.

Geo**activities 7.12**

Knowledge and understanding

1 Explain why the ranking of the world's liveable cities is subjective.
2 List seven new indicators in the Spatially Adjusted Liveability Index.
3 What does the acronym UNESCO mean?

Inquiry and skills

4 Refer to 7.12.1. Explain how World Heritage sites can improve the liveability of surrounding places.
5 Refer to 7.12.2 and explain how heritage tourism impacts on the liveability of a place.
6 Refer to 7.12.3 and design a Dracula Entertainment Park to attract tourists to Sighisoara. List arguments for and against the park to improve the city's liveability.
7 The Australian Government has a responsibility to protect World Heritage sites. How will this improve the liveability of places where World Heritage sites are located? Answer this question in relation to one World Heritage site in Australia (see Geolinks).

ISBN 978 1 4586 6277 4

7.13 Remote indigenous communities

The First Nation Quality of Life Tree provides a snapshot of features that make places liveable for indigenous people. If the tree is nurtured with the physical and cultural necessities for a good life, the resulting community will be liveable and sustainable. Of critical importance is the connection to cultural traditions, language and land that give indigenous people their community identity, as well as the ability to participate in decision-making.

Indigenous rights

The 2007 United Nations Declaration on the Rights of Indigenous Peoples states the right to equality without discrimination. Adequate housing, clean water, sanitation, education, employment and health care are rights that influence liveability. Indigenous people also have a right to have their unique cultures respected, nurtured and protected.

Remote indigenous communities have poor liveability due to lack of access to basic facilities and services. In Guatemala 53% of indigenous 15–19-year-olds have not completed primary school compared to 30% for non-indigenous youth. The infant mortality rate for indigenous communities in Bolivia is 75 per 1000. The non-indigenous rate is 50 per 1000. Indigenous communities in richer countries—including the First Nations in Canada, Native Americans, Alaska Natives and Australian Aboriginal and Torres Strait Islander Peoples—are equally disadvantaged. In First Nations communities in Canada, 53 do not have schools and only three in ten children on reserves graduate from high school.

7.13.1 Quality of Life Tree for First Nation communities in Canada. This tree was developed as part of a project involving consultation with six First Nation communities in Canada as a way of expressing their definition of quality of life and wellbeing. The elements in the tree can be linked to Articles in the United Nations Declaration on the Rights of Indigenous Peoples.

Housing, health and safety

The WHO states that safe, clean housing is essential for good health and wellbeing. Overcrowding and lack of sanitation pose serious health risks. Tuberculosis rates are high among indigenous communities due to overcrowding. In Canada, the tuberculosis rate is around 5 per 100 000 residents compared to 300 per 100 000 for Newfoundland's Inuit communities.

ISBN 978 1 4586 6277 4

Overcrowding is connected to higher rates of domestic violence in the Northern Territory, where violence against women in Aboriginal communities is more than twice the national average. Homes often lack basic infrastructure and require repairs, for instance to plumbing. A high proportion of homes in Native American reservations lack kitchens and sewerage systems.

Culturally appropriate housing

Liveable communities for indigenous people must fulfil cultural as well as physical needs. There is a shortage of accommodation, and many homes are too small and inappropriately designed—an extended family of 12–20 Aboriginal people in Australia may live in a three-bedroom home designed for a nuclear family. The result is overcrowding, homelessness or the addition of lean-to shelters.

In Canada the Inuit still rely on hunting animals like seals. If appropriate rooms are not provided in homes the slaughtering of such animals occurs in living rooms and bathrooms.

A failure to consider indigenous culture and lifestyles in the provision of housing makes places less liveable, while residents are accused of not using their homes appropriately. Building and maintaining adequate facilities in remote communities is very expensive, so despite injections of funding little progress is being made regarding housing quantity, quality and liveability.

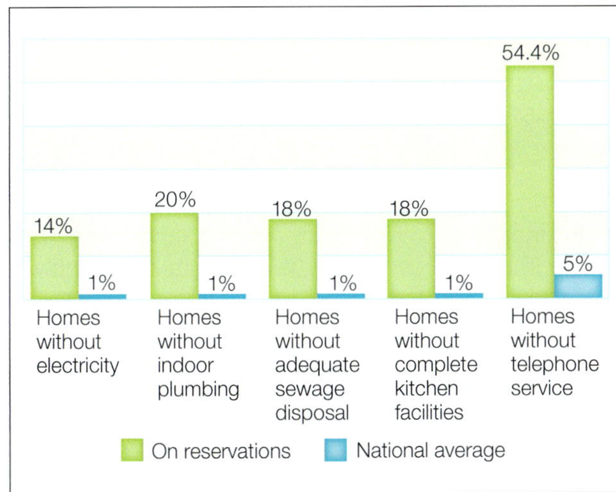

7.13.2 Poor-quality housing on Native American reservations in the USA

7.13.3 Inadequate and overcrowded housing in the Northern Territory

Geo**info**

There are around 350 million indigenous people in 70 countries worldwide.

Geo**activities 7.13**

Knowledge and understanding

1. Why was the Quality of Life Tree created?
2. What makes the Quality of Life Tree applicable to all indigenous communities?
3. What is the main aim of the UN Declaration on the Rights of Indigenous Peoples?
4. Use evidence to describe the poor liveability rating of remote indigenous communities.
5. What are the impacts of overcrowded housing?
6. Explain how housing for indigenous people could be made more culturally appropriate.

Inquiry and skills

7. Refer to 7.13.1.
   a. Create your own personal Quality of Life Tree. Start with the leaves. Decide on the features of the community that would provide you with a satisfactory quality of life. Then add the water droplets. These are the requirements to create the community you want to live in.
   b. How do you think your tree would differ in 30 years time? Explain why it would change.
8. Refer to 7.13.2 and write a paragraph describing the differences in housing quality between Native American reservations and the US national average. Use statistics in your answer.
9. Discuss the features of indigenous communities that make them liveable, despite lack of access to services and facilities.

ISBN 978 1 4586 6277 4

7.14 Geographical inquiry: assessing the liveability of a place

Steps in geographical inquiry

As a Geography student you will complete a geographical investigation to assess the liveability of one place.

The process of inquiry begins with asking questions such as:

- Why does the liveability of places differ?
- What influences my perceptions of liveability?
- What criteria will I use to assess or measure the liveability of a selected place?

Answering these questions involves acquiring, processing and communicating geographical information. Geographical investigation also involves using and interpreting geographical tools and using appropriate tools to communicate research findings. Tools relevant to assessing the liveability of a place could include population profiles, maps, graphs, photographs and spatial technologies such as Google Tour Builder.

When you analyse maps, graphs, photographs, illustrations and statistics, you are undertaking geographical inquiry.

To assess the liveability of a place, you can analyse one data source to make a judgement or you can undertake an inquiry that uses all of the steps shown in 7.14.1.

Assessing the liveability of a place using photographs

Focus questions: How liveable is Nice? What evidence do the images in 7.14.2 provide about access to services and facilities, environmental quality, community identity and social connectedness?

1 Acquire information	2 Process information	3 Communicate information
• identify a focus for inquiry • develop geographical questions • collect primary geographical data (e.g. fieldwork, interviews, questionnaires) • gather geographical information from secondary sources (e.g. internet, journals, newspapers) • record information	• evaluate data and information for bias and reliability • interpret data and information gathered • represent information in appropriate forms such as maps, graphs, statistics, spatial technologies and visual representation • analyse findings and results • draw conclusions	• communicate results using a variety of strategies and tools • reflect on findings • propose individual or collective actions • predict expected outcomes • where appropriate, take action

7.14.1 Steps in a geographical investigation

Mediterranean waterfront

Markets and cafes

Transport and plazas

7.14.2 Nice is a place where environmental quality, community identity and global connectedness combine to create high liveability. Would you want to live there?

ISBN 978 1 4586 6277 4

Fieldwork to measure and assess liveability

Use a variety of strategies and tools to acquire data and information to judge the liveability of a place. Fieldwork involves observing and recording, and could include:

- observing and recording evidence of environmental quality, such as natural surroundings, buildings and visible pollution, through sketching, photography and water quality testing
- recording the availability of services and facilities, such as shops, transport, appropriate shelter for different groups of people and wheelchair access, using photographs, maps, tables and tally sheets
- identifying and photographing evidence of community identity and social connectedness, such as cultural diversity, transport options, walkability, community facilities and meeting places
- assessing the affordability of housing displayed in real estate agency windows.

7.14.3 Assessing the liveability of Pyrmont using fieldwork

Observing and recording	Assessment of liveability
	Close proximity to the city for employment, services and facilities Environmental quality—climate, harbour, open spaces, greenery Social connectedness—open spaces, parks and Darling Harbour nearby
	Recording real estate prices to assess the affordability of Jacksons Landing where waterfront location, expensive high-rise housing and city access impact on housing prices
	Walkability, location, public places and recreational facilities enhance social connectedness and environmental quality Pyrmont's waterfront includes paths, boardwalks, parks and gardens, and the suburb contains facilities such as swimming pool, theatre, cafes and restaurants, schools and supermarkets
	The retail, recreational and entertainment facilities at Darling Harbour contribute to the liveability of Pyrmont Places to meet and relax and access to the city via Pyrmont Bridge for pedestrians and cyclists as well as light rail contribute to social connectedness
	The preservation and adaptive use of historical buildings in Pyrmont adds to its community identity 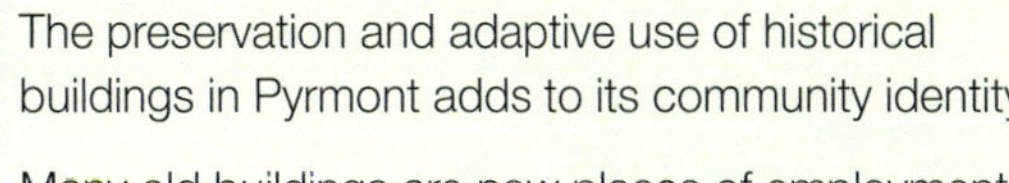 Many old buildings are now places of employment, such as shops and offices, as well as residential apartments, existing alongside modern residential buildings and facilities e.g. the Star

Chapter 7

ISBN 978 1 4586 6277 4

7.14.4 The suburb of Pyrmont in Sydney is liveable for people from many socioeconomic backgrounds, with a mix of affordable housing and expensive apartments. An attractive waterfront location, plus access to employment and facilities such as schools, transport, places for recreation, entertainment and leisure, make the suburb highly liveable for people of all ages. But is it accessible and affordable for everyone?

Using surveys to understand perceptions of liveability

Simple surveys and questionnaires can be useful for assessing liveability for people of different cultural backgrounds, age, life-stage and economic status. These can be as simple as an open response survey (see 7.14.5) or a detailed set of questions about specific influences on liveability such as access to affordable housing or facilities.

Using secondary sources to assess liveability

Observations made during fieldwork can be supported by secondary sources such as graphs, statistics, historical photographs, maps and spatial technologies such as Google Street View. Data from the Australian Bureau of Statistics (ABS) can assist with identifying characteristics of a community that determine community identity; for example, multiculturalism, employment and population structure.

The liveability of places for different age and lifestyle groups is reflected in population pyramids (population profiles). While coastal towns such as Forster–Tuncurry might be very liveable for retirees, they lack facilities for young people. Inner-city suburbs such as Alexandria in Sydney attract young professionals, compared to outer suburbs such as Point Cook in Melbourne that attract families.

<table>
<tr><td colspan="2">Location:</td></tr>
<tr><td>Cultural background:</td><td>Age: Gender:</td></tr>
<tr><td>1. What are the best features of this place?</td><td>2. What features or lack of features make living here difficult?</td></tr>
<tr><td>3. What features or activities contribute to community identity?</td><td>4. What might improve the liveability of this place?</td></tr>
</table>

7.14.5 Sample open response survey

ISBN 978 1 4586 6277 4

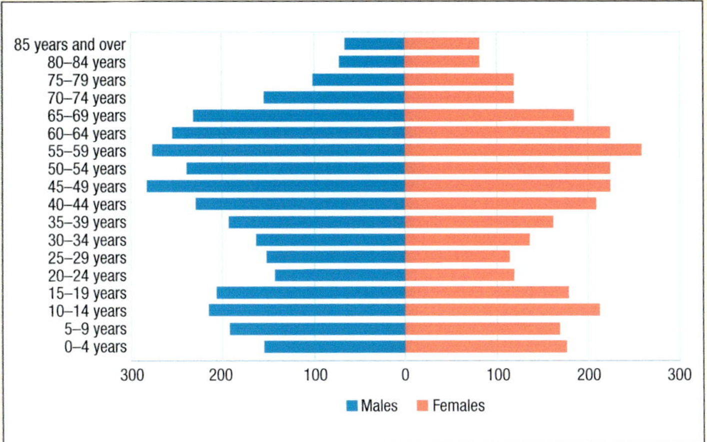

Small rural community Waikerie, SA, 2011
Low liveability for young people due to lack of services, facilities and lifestyle

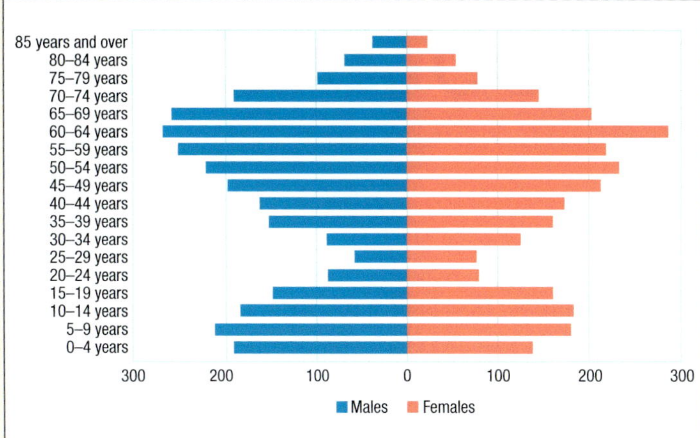

Coastal retirement town Forster–Tuncurry, NSW, 2011
Low liveability for young people due to lack of higher education facilities and employment. High liveability for older people—'sea changers'

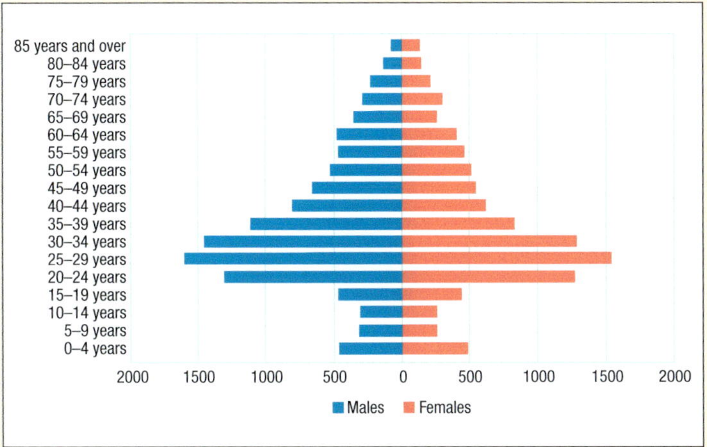

Inner-city suburb Alexandria–Beaconsfield, Sydney, NSW, 2011
High liveability for young professionals and students: close to CBD, restaurants, entertainment and employment

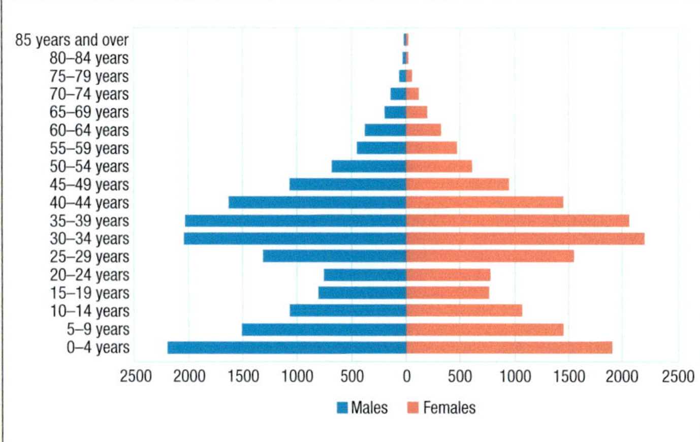

Outer-city suburb Point Cook, Melbourne, Vic., 2011
High liveability for young families evidenced by 20–39s and under 4s.

7.14.6 Population profiles reflect the liveability of places for different people

Communicating research findings using spatial technologies

Spatial technologies are interactive sites that allow data and information from geographical inquiry, including fieldwork, to be added to base maps to create a presentation or new map.

Using Google's Tour Builder, you can create interactive tours of sites visited during your fieldwork. Mapping locations and then inserting photographs, descriptions and data and sharing the tour with other students and teachers makes this application an appropriate tool for communicating your inquiry findings.

Alternative presentations could include creating a web page using Weebly or digital posters using Glogster (see 6.15).

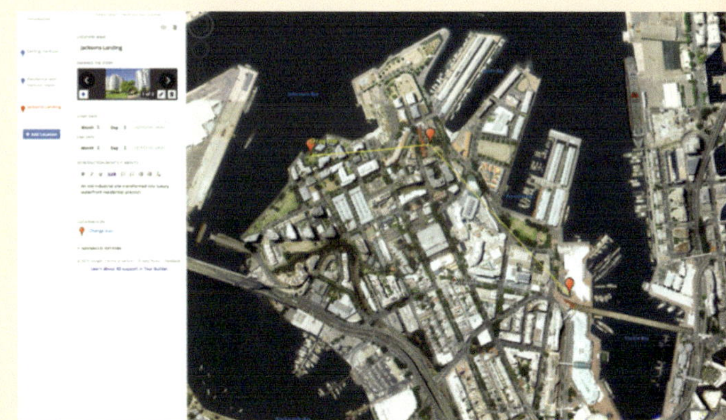

7.14.7 Creating a liveability assessment tour using Google's Tour Builder

ISBN 978 1 4586 6277 4

Geothink

Liveability of Cairns

Found on the coast of Far North Queensland, the city of Cairns is located between the Great Barrier Reef and the lush Atherton Tableland. It has a population of over 130 000 people and is a highly popular tourist spot.

The liveability of Cairns can be assessed using the geographical inquiry process. Answer the following questions using the sources given and your own research.

7.15.1 Satellite image of the Cairns region

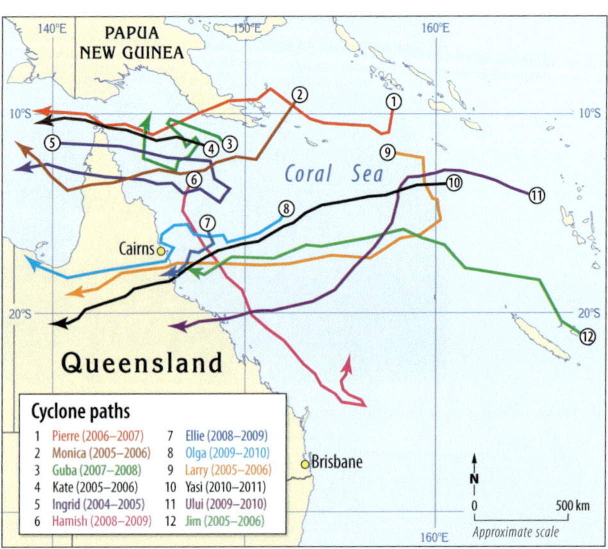

7.15.2 Cyclone paths, 2004–11

7.15.3 Physical environment and location

7.15.4 Panoramic view of Cairns showing environment and facilities

ISBN 978 1 4586 6277 4

7.15.5 Cairns: 2011 census information

Features of society		Cairns	Queensland	Australia
Cultural diversity	Australian	27.3%	27.4%	25.4%
	English	25.3%	28.1%	25.9%
	Irish	8.0%	8.4%	7.5%
	Scottish	6.5%	7.3%	6.4%
	Italian	4.6%	2.0%	3.3%
Employment	Unemployed	7.0%	6.1%	5.6%
	Full time	58.6%	60.0%	59.7%
	Part time	28.1%	28.2%	28.7%
Occupation	Professionals	16.0%	18.9%	21.3%
	Technicians and trades workers	15.9%	14.9%	14.2%
	Clerical/administrative	13.3%	14.7%	14.7%
	Managers	12.5%	12.0%	12.9%
	Labourers	11.8%	10.6%	9.4%
	Community/personal service workers	11.7%	10.0%	9.7%
	Sales workers	10.3%	9.8%	9.4%
	Machinery operators and drivers	6.6%	7.3%	6.6%
Housing costs	Median rent	$240/week	$300/week	$285/week
	Median mortgage repayments	$1647/month	$1850/month	$1800/month

Source: Australian Bureau of Statistics

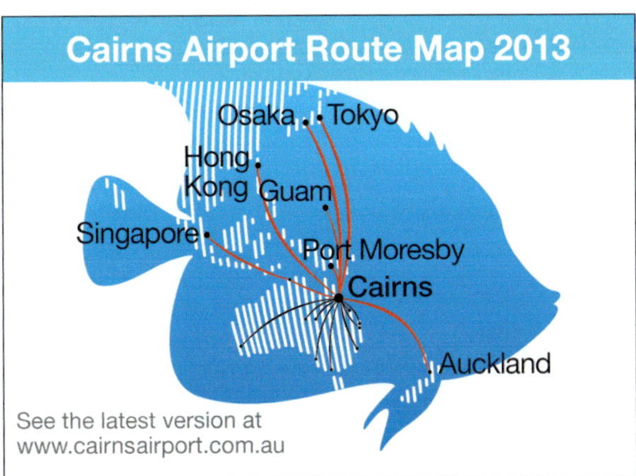

7.15.6 Mobility and accessibility

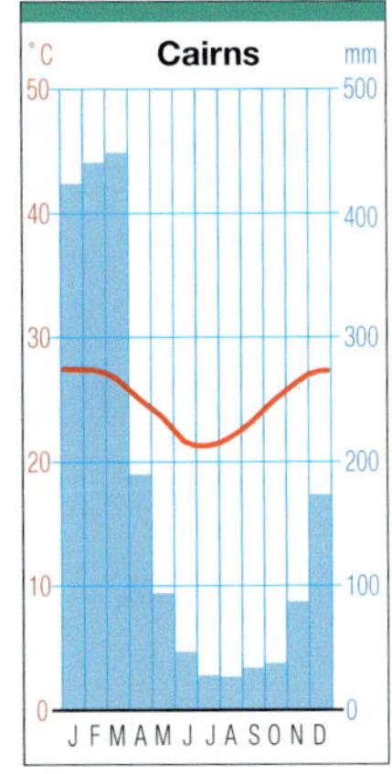

7.15.7 Cairns climate graph

1 Refer to 7.15.1–7.15.7.

a Describe the location of Cairns.

b How has topography influenced the growth of Cairns and other towns in this part of Queensland?

c Describe the climate of Cairns. Refer to annual rainfall, rainfall distribution and temperature.

d Assess the vulnerability of Cairns to tropical cyclones.

e How does Cairns compare to the rest of Queensland and Australia in terms of affordability and employment? Is it cheaper or more expensive? Consider rent and mortgage costs, types of employment and unemployment rates.

f What recreational and lifestyle opportunities are available in the Cairns region?

g Consider the economic, social, cultural and environmental qualities of Cairns. Complete a table listing features that increase and decrease the city's liveability. Consider the impact of future climate change in your answer.

h Is Cairns a place you would like to live now or in the future? Your response to this question will be a postcard to a friend either recommending or not recommending Cairns as a place to live. Choose an image (or collage of images) and a title for the front of your postcard. Write a message (maximum 100 words) that you would put on the back stating your position and briefly justifying it.

Chapter 7

ISBN 978 1 4586 6277 4

chapter 8

Urban, rural, remote places: access, identity and connections

'All settlements require adequate services to make them liveable.'

Susan Bliss

Aa Geovocab

Accessibility/Remoteness Index of Australia (ARIA): measure of the accessibility of places to service centres or, conversely, of the remoteness of places

central business district (CBD): commercial centre of a city, where a large percentage of shops, offices, cultural facilities and activities are found

cosmopolitan: mixture of cultures

ecological footprint per capita: amount of land used to provide each person with their needs

housing density: number of dwellings per square kilometre of land

manufacturing: process whereby raw materials are converted into products

rural–urban fringe: zone separating the edges of large towns and cities and the rural areas surrounding them

ISBN 978 1 4586 6277 4

The liveability of urban, rural and remote places is influenced by environmental, cultural, social and economic forces. People change where they live when they perceive better opportunities in other places. Urbanisation is one result of this change. Many settlements, especially large cities, are highly liveable but are not ecologically sustainable as they possess large ecological footprints. Liveability and sustainability are important when deciding on where to live in the future.

Seoul's growth has turned its cityscape into a dense grid of housing and office towers

Think, puzzle, explore

- **Place** Why do people choose to live in different types of places?
- **Environment** How do the environments of settlements differ?
- **Interconnection** How are different settlements connected?
- **Sustainability** How can cities become both sustainable and liveable?
- **Scale** What action is taken at different scales to improve sustainability?
- **Change** Why do people change where they live?
- **Space** What is the spatial distribution of different types of settlements in Australia?

Geo**skills** in focus

- **Planning** a simple geographical inquiry on urban, rural and remote places
- **Observing** information relevant to urban, rural and remote places
- **Evaluating** different perspectives on the liveability of different types of settlements
- **Communicating** data in a range of appropriate forms, including ICT
- **Reflecting** on personal values and attitudes and how they influence responses to the liveability of places

rural–urban migration: migration of people from rural areas to towns and cities
tree change: movement of people from cities to rural places looking for a new lifestyle
urbanisation: increase in the proportion of people living in urban areas
urban sprawl: spread of suburbs into surrounding rural areas
urban consolidation: increase in housing within a city's boundaries

ISBN 978 1 4586 6277 4

8.1 Urban, rural and remote places

Think about the place where you live. Is it rural, urban or remote? Why do you live there? What are its characteristics? What attracts people to live there? Farms, hamlets, villages, towns, cities, conurbations and megacities are all types of settlements.

Settlements vary in size and the services and facilities they offer. They will change in character over time. The function of a place is the reason it exists. For instance, Dampier, Western Australia, is a port and Queensland's Gold Coast is a tourist destination. People are attracted to places by the opportunities they offer. Karratha grew in response to a mining boom in the Kimberley region of Western Australia. Beechworth in Victoria has transformed from a mining to a tourism and winery district, attracting new businesses and people.

Hierarchy of settlements

Ranking settlements from largest to smallest creates a hierarchy, but a lack of common population cut-off sizes makes comparing settlements across countries difficult. For instance, in Chile a town is defined as having 2000–4999 people, whereas in Greece it is 2000–9999.

Urban places are densely populated towns and cities providing a wide range of educational, medical, entertainment, retail, manufacturing and business functions. Rural places are associated with agriculture and the countryside and include small towns, villages and farms with low population densities, open spaces and fewer services. The zone of change between rural and urban areas is known as the **rural–urban fringe**.

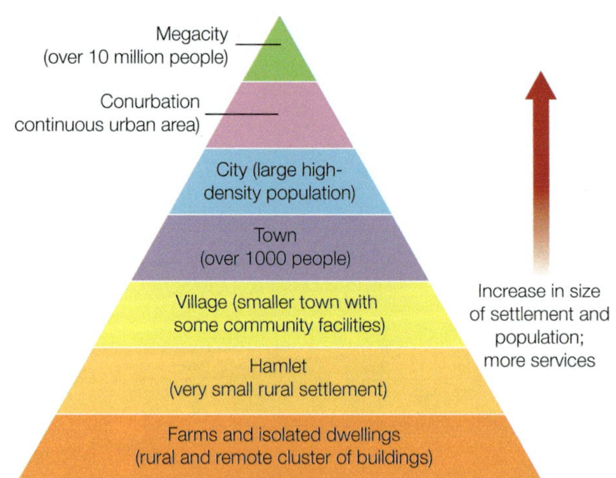

8.1.2 A hierarchy of settlements

There is a global decline in rural populations and an increase in the proportion of populations living in urban places. This process, known as **urbanisation**, impacts on the character and liveability of cities and the places people leave behind. The rapid growth of Sydney is making housing less affordable there, while country towns like Trundle, New South Wales, are offering very cheap accommodation to increase liveability and attract new residents. Liveability declines when services and businesses close down. When this happens, people often leave.

i Geoinfo

- Tokyo is the world's largest settlement with a population of over 35 million.
- Sydney, Australia's largest city, reached a population of 4.5 million in 2012.

8.1.1 Types of settlements: urban, regional and rural

ISBN 978 1 4586 6277 4

A Yayladagi, Turkey

D Guyra, NSW, Australia

B Melbourne, Australia

E Mumbai, India

C village in India

F Shanghai, China

8.1.3 Some settlements across the world

Remote places

Remoteness refers to distance from other places and access to goods and services. People live in remote places for cultural, economic and lifestyle reasons. For example, Arnhem Land is a remote Aboriginal Reserve in the Northern Territory, where traditional Aboriginal cultures thrive. Economic opportunities have increased since tour operators began bringing small tourist groups to the area. Some communities, such as Yolngu, have developed successful economic enterprises, including art, ecotourism and cattle farming. The Bawaka community shares stories, ceremonies, hunting and fishing methods, medicine, art and crafts and promotes its businesses on a website called 'Bawaka Cultural Experiences'.

Settlements vary globally in their characteristics and liveability.

Geo**activities 8.1**

Knowledge and understanding

1 Write out the following sentences and fill in the missing words: *population density*, *cities*, *farms*, *liveability*, *small*, *urbanisation*, *settlements*, *remote*, *conurbations*, *proportion*, *characteristics*, *function*, *megacities*.

People live in a variety of ________ ranging in size from ________ (smallest) to ________, ________ and ________ (largest). In a hierarchy of settlements there are a greater number of ________ places, such as hamlets and villages. ________ settlements are isolated and have limited access to goods and services. The two main differences between rural and urban places are the ________ ________ and the ________ or services they provide. ________ means that a greater ________ of people live in urban places. The ________ and ________ of places is affected by this process.

2 Answer the following questions about the place where you live.

a What type of settlement do you live in? Justify your answer.

b What is the main function (or functions) of that settlement?

c List three advantages and three disadvantages of living where you do.

d What type of settlement would you like to live in when you leave school? Give a reason.

Inquiry and skills

3 Refer to 8.1.3.

a Match the letter representing each image with one of the following types of settlements.

i large megacity

ii large city

iii rural village

iv small country town

v temporary settlement

vi urban slum

b Beside each answer write the words 'developing' or 'developed' to indicate which part of the world each image represents. What characteristics of each place influenced your answer?

c Choose the settlement you would *least* like to live in. Explain why. What could be done to improve its liveability?

d What features of the liveability of places are difficult to see in photographs?

Chapter 8

ISBN 978 1 4586 6277 4

8.2 Rural diversity and identity

Images of the Tour de France show farmlands, villages and historic towns in the French countryside, highlighting rural life in a developed country. The majority of the world's rural population, however, is in the developing world where millions of farmers raise livestock (e.g. sheep) or cultivate crops (e.g. rice). Economic, environmental, social and cultural factors make rural places liveable for almost half of the world's population.

Rural diversity

There is a large diversity of rural environments and lifestyles. In rural areas agriculture can be:

- arable or pastoral
- irrigated or rain fed
- commercial or subsistence
- labour-intensive or mechanised.

Arable farming occurs in better-watered locations, unless irrigation is available, while pastoral or livestock farming occupies drier places. Population densities are higher in arable farming communities, especially where farm sizes are very small (e.g. in Indonesia and China).

8.2.2 Venn diagram showing impacts of drought on communities

Rural settlements

Villages provide basic frequently needed services, which in developing countries may include mobile phone and internet access. Regional centres (e.g. Katherine, NT, and Tamworth, NSW) provide goods and services that are required less frequently.

1 Mechanised cattle mustering in remote northern Australia

3 Reindeer herding in remote Mongolia where population densities are very low

2 Irrigated cropping in the USA

4 Labour-intensive cropping in Africa

8.2.1 Global variety of rural places

ISBN 978 1 4586 6277 4

Common characteristics of rural places

Which of these features of rural places make them more or less liveable?

- life is slower paced and friendly
- the environment is less polluted but the land may be degraded
- populations are ageing as young people move to cities
- community spirit and identity are strong
- there is less cultural diversity
- incomes are lower and employment opportunities more limited
- there is less access to health, education, sporting, entertainment and social facilities
- rural people suffer most when natural hazards occur

Recent trends in rural decline and urbanisation raise questions about the liveability of rural places in the future. A recent media report, 'The slowly dying towns of NSW', discusses these issues.

Geoinfo

Of the 200 US counties with the highest poverty rates, 189 are rural.

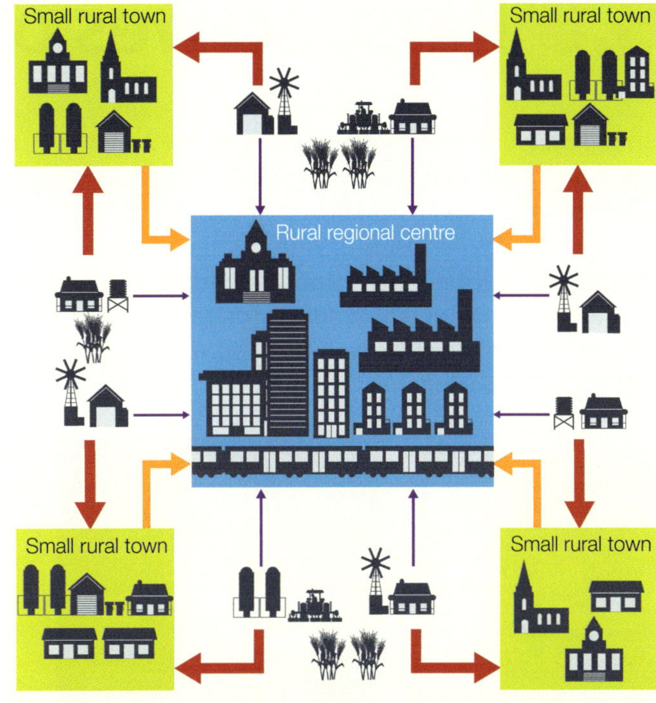

8.2.3 Interactions between farms, small rural towns and rural regional centres

Geoactivities 8.2

Knowledge and understanding

1. Define the term 'rural'.
2. Suggest two reasons why the majority of rural people live in developing countries.
3. Why is there a huge diversity of rural environments and lifestyles across the world?
4. Construct a table to show the advantages and disadvantages of living in rural communities.

Inquiry and skills

5. As a class discuss the economic, environmental, social and cultural factors that would make rural places liveable. Do factors vary between developed and developing countries?
6. Refer to 8.2.1 and study the four rural landscapes.
   a. Which image would represent the highest population density?
   b. Which images would represent commercial farming? Give a reason for your choice.
   c. Rank the images by your preferred choice as a place to live. Justify your first choice.
   d. Create a 'must have' list of the things a rural community would need to provide for it to be deemed liveable for you.
7. Refer to 8.2.2. Divide into class groups to discuss the statement: 'Rural places are less liveable than cities because natural hazards have a bigger impact on them'.
8. Refer to 8.2.3.
   a. Use the diagram to explain the role of large regional towns and cities in rural areas.
   b. Do you think regional centres would be better to live in than small towns? Give a reason.
   c. Do you think regional centres would be better to live in than large cities? Give a reason.
   d. How would the internet and mobile phones have impacted on people living in rural areas in the last 20 years?

Chapter 8

ISBN 978 1 4586 6277 4

8.3 Perspectives on rural life

Rural places: are they dull and boring or the happiest places on Earth? Individuals have different opinions about living in rural places based on their ideas about liveability. Residents of Dull (Scotland) and Boring (USA) became sister communities hoping their amusing names would attract interest in rural life. Now the small NSW town of Bland wants to join with Dull and Boring.

8.3.1 Dull and Boring: the two remote and oddly named places have been linked to attract visitors

Urbanites rarely think about rural and remote places until hazards, such as drought, push up food prices, dust storms carry topsoil to coastal cities, or famine causes death and starvation. Media publicity about natural hazards, declining populations, climate change, lack of facilities, high costs of living, isolation and unemployment presents an unattractive and stereotyped image of life outside large towns and cities. While these are genuine disadvantages of rural life that decrease liveability, large urban places are not necessarily better.

Happiest places

On the Australian Wellbeing Index the happiest places to live are country towns, while large cities and very remote places have the lowest happiness levels. People like the sense of community, safety, trust and belonging in rural communities. Brain scans on people in Britain revealed that urban residents suffer more anxiety, depression and schizophrenia, while exposure to green space reduces stress.

Young people

Young rural residents can be disadvantaged by limited access to educational, employment, transport and entertainment facilities. Children may be sent to boarding school, living away from their families. Rural youth, however, often experience greater freedom, independence and opportunities and take on responsibilities in community life, for instance joining rural fire brigades, which provides valuable life and workplace skills.

In China, poverty, lack of development, environmental degradation and natural hazards are responsible for rural dissatisfaction, resulting in high rates of urbanisation.

8.3.2 Big is not always better

ISBN 978 1 4586 6277 4

Chinese residents were asked: 'Please imagine a ladder with steps numbered from 0 at the bottom to 10 at the top. Suppose we say that the top of the ladder represents the best possible life for you, and the bottom of the ladder represents the worst possible life for you. On which step of the ladder would you say you personally feel you stand at this time?'

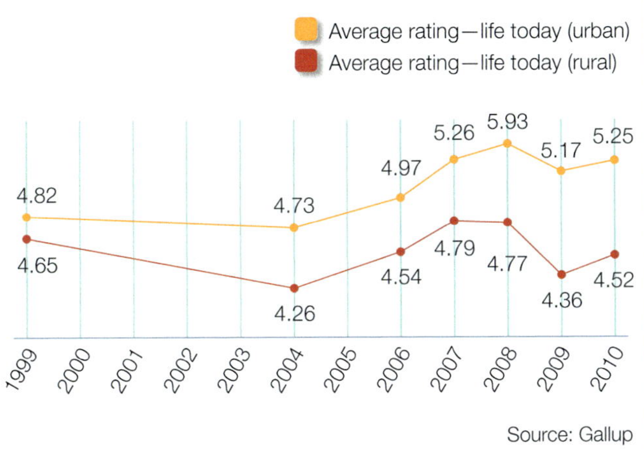

8.3.3 Dissatisfaction with rural life in China

Geo**info**

- A century ago three in five Australians lived outside capital cities, now it's one in three.
- Communities with under 10 000 people across Australia are shrinking, while regional cities grow in size.

Breaking down barriers

Governments, businesses and organisations are promoting rural liveability by:

- *attracting people to events* (e.g. the Elvis Festival and food festivals)
- *recognising the contributions of rural people* (e.g. agricultural shows)
- *increasing people's understanding of rural life* (e.g. the Great Café Challenge, which encourages café-goers to read rural newspapers, and television shows set in rural Australia)
- *improving communications networks*, such as the National Broadband Network.

A trend for people to move from urban to rural places as a lifestyle choice is termed **tree change**. It includes retirees and young families looking to live in cheaper, safer and less stressful places, with strong communities.

Geo**activities 8.3**

Knowledge and understanding

1 Identify the main reasons for the dissatisfaction with rural life in China.

2 The saying 'When the bush comes to town' is used sometimes. What does the term refer to?

3 Why do you think people living in larger rural towns are happy with their lives?

4 List the advantages and disadvantages of rural life for young people.

5 List two ways rural–urban understanding is being improved.

6 Explain the meaning of 'tree change' and why a tree change is attractive to certain age groups. What is the connection between tree changes and liveability?

Inquiry and skills

7 Refer to 8.3.1.
- a What might attract people to live in Dull or Boring?
- b Locate Dull and Boring using Google Earth. Explain why both places are considered rural and remote.

8 Refer to 8.3.2.
- a How do these cartoons represent the extremes of rural and urban life?
- b What is the message of each cartoon?
- c Locate other cartoons contrasting rural and urban living, or create your own.

9 Refer to 8.3.3.
- a How were the scores for this survey developed?
- b Use the same process to score your satisfaction with the place you live.
- c What event might explain the spike in satisfaction in China in 2008?
- d What has been the general trend in satisfaction since 2004?
- e When was the gap between rural and urban satisfaction the greatest? What was the difference?

10 Refer to 'Breaking down barriers'.
- a As a class discuss ways to increase people's understanding about life in rural areas. Draw a mind map summarising the key ideas.
- b Create a television show that could promote rural living for young people. Present a pitch to a television production company to produce your show. Consider who you would cast, the location you would use and the age of the audience you would be targeting.
- c Use a web 2.0 animation tool to create a debate on the topic 'Rural life is dull and boring' or conduct a class debate.

Chapter 8

ISBN 978 1 4586 6277 4

8.4 Remote places

In 2012 a British climber made the first phone call from Mt Everest. At 8840 m altitude and –30 °C, the climber strapped phone batteries to his body to keep them warm and wore an oxygen mask. Communication technology has made many remote places more liveable than in the past, but population densities remain very low because there are fewer opportunities to socialise with other people and limited access to goods and services.

On the Tibetan Plateau (34.7°N, 85.7°E) it can take up to 21 days to travel to the nearest city due to the rugged terrain and 5200 m altitude. Only 15% of places in developed countries are more than one hour from a city compared to 65% in developing countries.

8.4.2 At Zanskar in the Himalayan Mountains children walk to their boarding school for the winter term down a frozen river, because the only road in and out is impassable for six months. The journey takes several days so they sleep in caves each night.

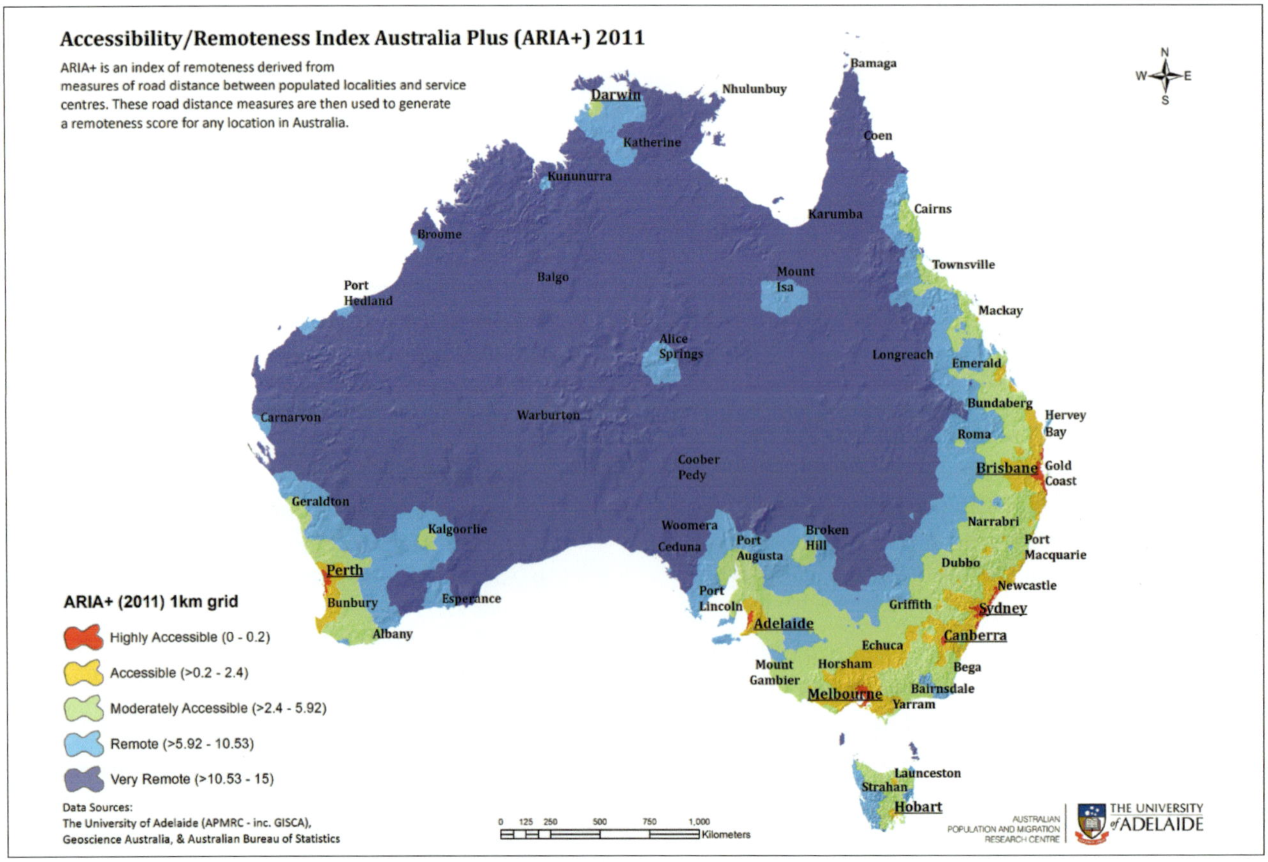

8.4.1 Accessibility/Remoteness Index of Australia (ARIA)

ISBN 978 1 4586 6277 4

Remoteness in Australia

The **Accessibility/Remoteness Index of Australia (ARIA)** defines remoteness by the ease of access to goods and services and opportunities for social interaction. Values range from 0 (high accessibility) to 15 (high remoteness) and are based on road distances to the nearest town.

Cities, where people have easy access to goods and services as well as opportunities to socialise, have the lowest scores and are considered more liveable. Distance increases the cost of goods and services, making remote places more expensive.

Relative remoteness has replaced the use of terms 'urban' and 'rural' for comparing the advantages and disadvantages of living in different places in Australia.

Remote populations

Remote populations contain a high percentage of indigenous people. For instance, 49% of the population in very remote Australian communities are Aboriginal people. Modern conveniences, such as satellite television, make life easier for remote communities but lack of access to services causes inequalities. The health status of remote indigenous populations in Canada, the USA and Australia is well below average.

Resource-rich remote locations, such as Siberia, suddenly become more liveable when natural resources are discovered.

Sixty Minutes: Frozen fortunes

Around the clock, rain, ice or snow, there's a race for resources at the top of the world. Siberia was once the world headquarters of suffering—now this blood-soaked Earth is yielding a staggering cache of oil, gas and gold. And today Siberia's biggest gold mine is about to be made a whole lot bigger. The gold being unearthed in the Olympiada Mine is of such high grade, it's well worth enduring the blizzards and the isolation to exploit it. This eastern slab of Russia is so vast it's almost twice the size of Australia. Despite the treacherous weather, often down to 40 below zero, this operation goes on 24-7, because there's a fortune to be made.

Source: extract from *Sixty Minutes* report, June 2011, http://sixtyminutes.ninemsn.com.au/stories/8219565/frozen-fortunes

Geo**info**

Alert is a small village of only five permanent residents 817 km from the North Pole. The nearest town is a small fishing village over 2000 km away.

Geo**activities 8.4**

Knowledge and understanding

1 Define 'remoteness'.
2 How has technology changed the liveability of remote communities?
3 'Most people do not want to live in remote places.' Provide reasons for this statement.
4 Explain why people live in remote places.
5 Explain how remoteness disadvantages indigenous communities.
6 List goods, facilities and services you would not expect to find in remote places.

Inquiry and skills

7 Refer to 8.4.1.
   a How is ARIA calculated?
   b Estimate the percentage of Australia that is 'very remote'.
   c Which two states have the largest proportion of 'remote' and 'very remote' areas?
   d Measure the distances from Coober Pedy and Bamaga to their state capitals. Show your results in a table.
   e Which category does your home belong to?
8 Refer to 8.4.2. Use a Venn diagram to compare and contrast your trip to school to that of the children from Zanskar. You will need to think of similarities and differences.
9 Create a concept map or Wordle including the facilities and services essential to attract you to a remote place to live.
10 Geographical inquiry: Refer to the article, 8.4.1 and internet sources.
   a List the advantages and disadvantages of living in Siberia.
   b Research remote Australian communities. Use Google Street View for a visual impression of the places, and refer to environment, facilities and services.
   c Create a digital tour of the selected communities using Google Tour Builder or Google Earth. See 6.15.
   d Ask a classmate to evaluate your presentation by identifying positive features and things that could be improved.
   e Adjust your presentation based on the feedback and share with others.

Chapter 8

ISBN 978 1 4586 6277 4

8.5 Skill: graphs and infographics

Tables and graphs are visual tools used by geographers to present data. During the geographical inquiry process, data (information) can be collected from two sources:

- primary data, which you collect yourself (e.g. fieldwork measurements and surveys)
- secondary sources (e.g. the Australian Bureau of Statistics, newspapers and the internet).

Data in the form of numbers is called *quantitative data* (e.g. the number of people living in your suburb). *Qualitative data* is in words or sentences (e.g. the country you were born in or the reason a person migrated). Before any qualitative data can be graphed it needs to be converted into numbers.

8.5.1 Geographical research: students collecting data on a fieldtrip

Organising and presenting data

When data is collected it needs to be processed and organised before a report is written because it helps to analyse and understand the information collected. Often statistics are organised into a table before being graphed. Geographers learn to:

- construct graphs—drawing them by hand or using computer programs (e.g. Excel)
- interpret graphs—the internet makes it easy to collect data already in graph form but you need to decide whether the data is relevant, valid and reliable before using it
- use graphs to present research findings
- acknowledge sources of information from secondary sources.

Every good graph should have the following information (SALTS):

- *Scale*—units of measurement (e.g. millions or tonnes)
- *Axis*—labels on each axis to show the information being presented (e.g. years, countries or percentages)
- *Legend*—the key that gives the meaning of each colour, pattern or symbol
- *Title*—what the graph is about
- *Source*—where the data came from.

Types of graphs

There are many different types of graphs used by geographers.

8.5.2 Types of graphs

Line graphs	A line used to show trends or change over time (e.g. in population and literacy)
Bar graphs	Horizontal bars show change in one variable over time or differences between many variables at one point in time
Column graphs	Vertical bars used in a similar way to bar graphs
Pie graphs	Circles divided into segments, which together make up 100% of a total
Proportional circles	The diameter of a circle used to represent different quantities
Scatter graphs	Points that show the relationship between two variables

Infographics or mashables have developed in recent years to combine sources of data into one visual tool. Infographics have become powerful tools that can be a source of information as well as a medium for communicating research findings. They can be simple or complex, and may contain several types of graphs plus maps, tables, diagrams and other statistics.

Technology and liveability

Wherever people live in the world they are connected to other people and places with communications technologies such as the internet and mobile phones. Of the 7 billion people in the world, one-third use the internet and 1.5 billion use social networking sites such as Facebook. In choosing a place to live, most young people in the developed world would insist on access to these technologies to maintain their connected lifestyle.

ISBN 978 1 4202 3263 9

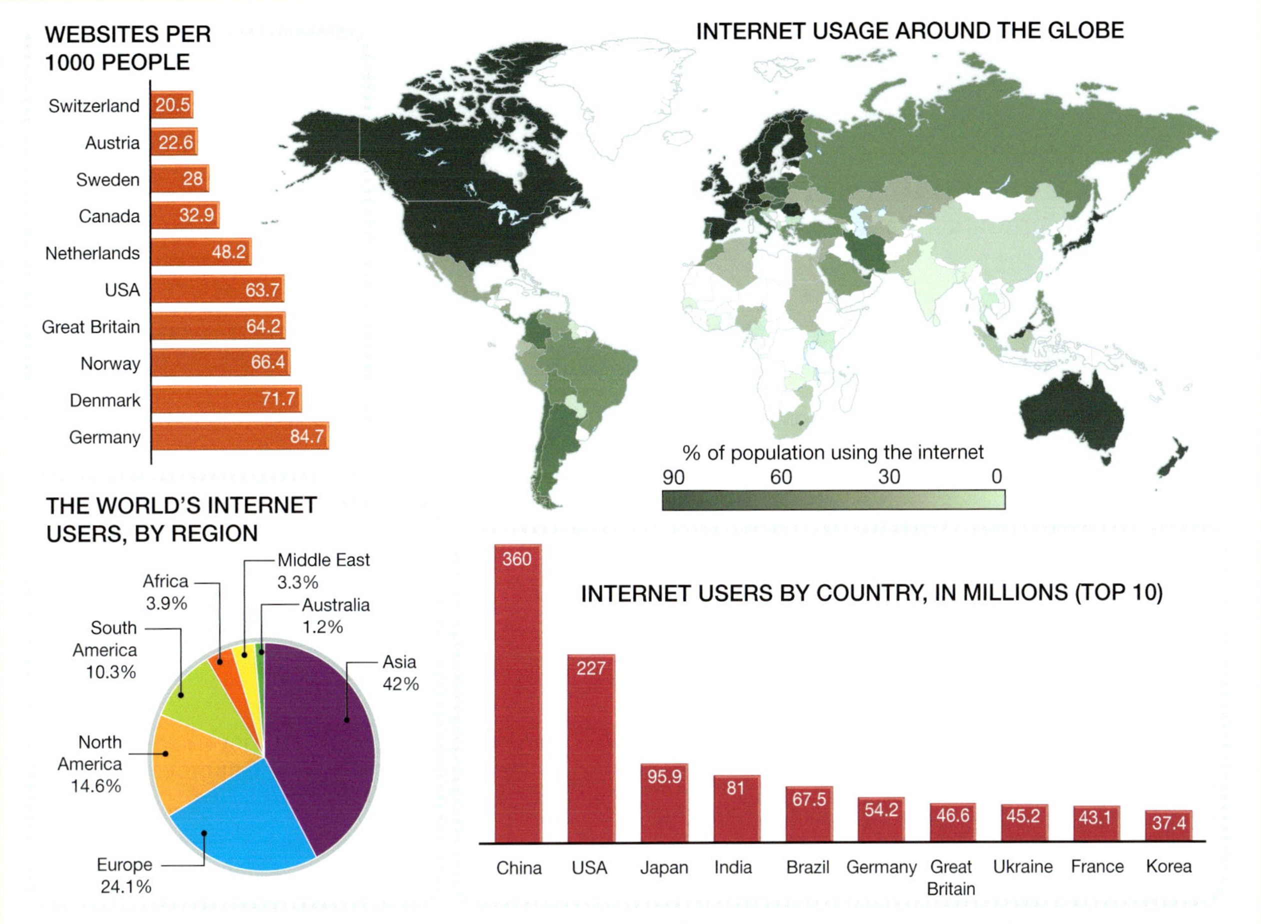

WEBSITES PER 1000 PEOPLE

Country	Websites per 1000 people
Switzerland	20.5
Austria	22.6
Sweden	28
Canada	32.9
Netherlands	48.2
USA	63.7
Great Britain	64.2
Norway	66.4
Denmark	71.7
Germany	84.7

INTERNET USAGE AROUND THE GLOBE

% of population using the internet
90 60 30 0

THE WORLD'S INTERNET USERS, BY REGION

Region	Share
Asia	42%
Europe	24.1%
North America	14.6%
South America	10.3%
Africa	3.9%
Middle East	3.3%
Australia	1.2%

INTERNET USERS BY COUNTRY, IN MILLIONS (TOP 10)

China	USA	Japan	India	Brazil	Germany	Great Britain	Ukraine	France	Korea
360	227	95.9	81	67.5	54.2	46.6	45.2	43.1	37.4

8.5.3 Infographic: global internet usage

Geoinfo

With over 800 million active users, if Facebook were a country it would be the third largest in the world.

Technology can overcome the barrier of distance and remoteness, making many places such as farms and small rural towns more attractive to live in. In developing countries, access to health care, education, markets and online banking is improving development and reducing poverty. Over five years, developing countries increased their share of the world's internet users from 44% in 2006 to 62% in 2011.

Geoactivities 8.5

Knowledge and understanding

1. What is the difference between primary data and secondary sources?
2. What is an infographic?
3. What are the five requirements of a good graph?
4. Explain how geographers can use maps and infographics as a part of geographical inquiry.

Inquiry and skills

5. Refer to 8.5.3.
   a. Name the three types of graphs used on this infographic.
   b. Which two countries have the largest numbers of internet users? Suggest a reason for this.
   c. What proportion of the world's internet users come from Africa? Suggest a reason for this statistic.
   d. Describe the distribution of places where 90% of the population is using the internet.
   e. Would you live in a place with limited or no internet access?
6. Use the internet to locate an infographic about Australia. It must reflect the liveability of the country.
   a. Write a brief report summarising the information presented in the infographic.
   b. See which class member can find the most interesting infographic.
7. Create an infographic about a place you consider highly liveable, using an online tool such as Piktochart or Canva.

ISBN 978 1 4586 6277 4

8.6 An urban world

All over the world people are choosing to live in large towns and cities. Cities are centres of wealth and economic opportunities. For the first time in history over 50% of the world's population is urban. It is estimated by 2030 this figure will reach 60% and total 5 billion people. The number of cities and towns is increasing at the expense of rural places in a process known as **rural–urban migration**.

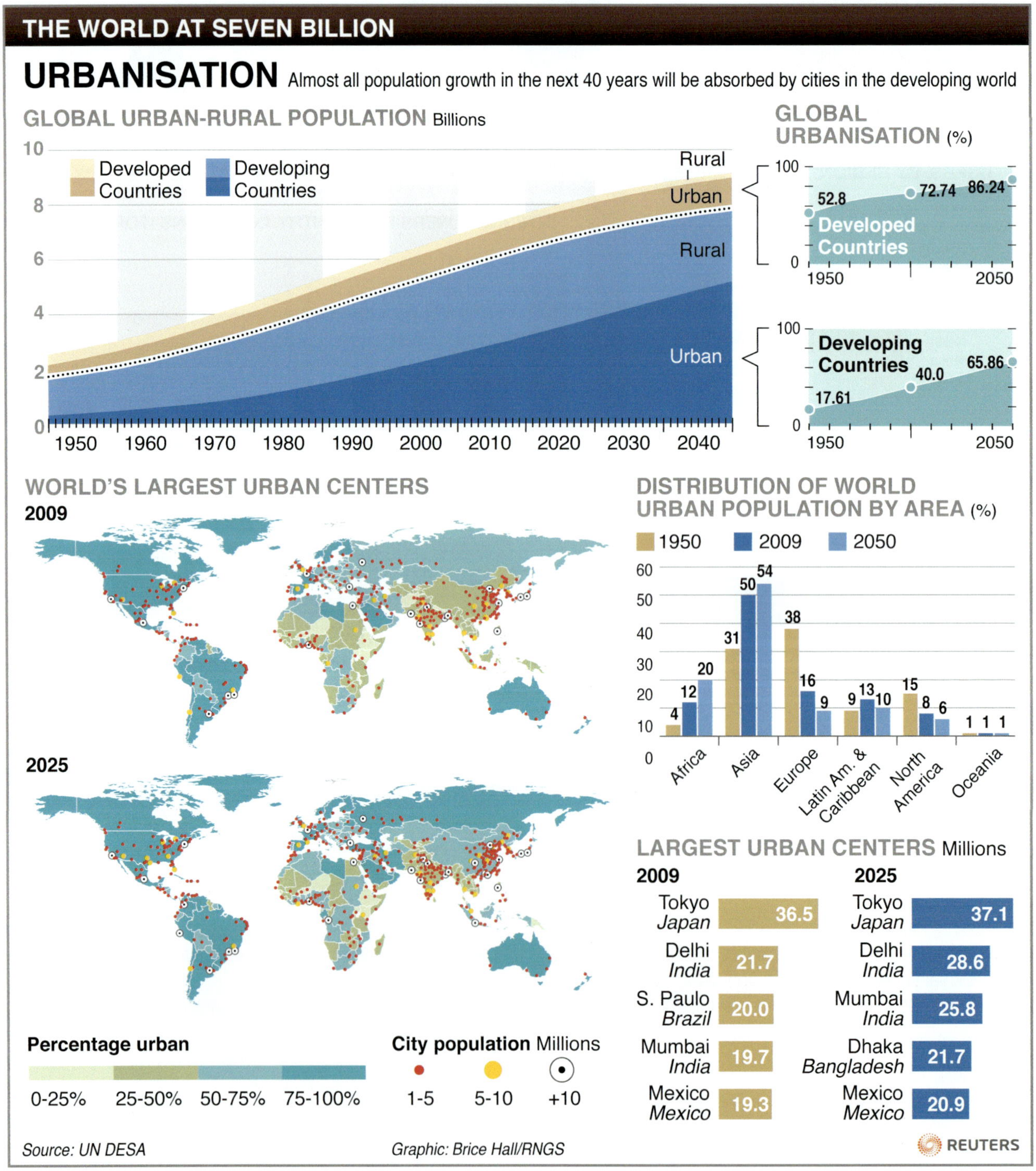

THE WORLD AT SEVEN BILLION

URBANISATION

Almost all population growth in the next 40 years will be absorbed by cities in the developing world

GLOBAL URBAN-RURAL POPULATION Billions

Developed Countries
Developing Countries
Rural
Urban
Rural
Urban
10
8
6
4
2
0
1950 1960 1970 1980 1990 2000 2010 2020 2030 2040

GLOBAL URBANISATION (%)

Developed Countries
100
0
52.8
72.74
86.24
1950 2050

Developing Countries
100
0
17.61
40.0
65.86
1950 2050

WORLD'S LARGEST URBAN CENTERS

2009

2025

Percentage urban
0-25% 25-50% 50-75% 75-100%

City population Millions
1-5 5-10 +10

DISTRIBUTION OF WORLD URBAN POPULATION BY AREA (%)

	1950	2009	2050
Africa	4	12	20
Asia	31	50	54
Europe	38	16	9
Latin Am. & Caribbean	9	13	10
North America	15	8	6
Oceania	1	1	1

LARGEST URBAN CENTERS Millions

2009		2025	
Tokyo *Japan*	36.5	Tokyo *Japan*	37.1
Delhi *India*	21.7	Delhi *India*	28.6
S. Paulo *Brazil*	20.0	Mumbai *India*	25.8
Mumbai *India*	19.7	Dhaka *Bangladesh*	21.7
Mexico *Mexico*	19.3	Mexico *Mexico*	20.9

Source: UN DESA

Graphic: Brice Hall/RNGS

REUTERS

8.6.1 Global urbanisation

ISBN 978 1 4586 6277 4

The urban revolution

The most rapid urban growth is occurring in developing countries within Asia and Africa. Africa's urban population is predicted to increase from 414 million to over 1.2 billion by 2050. Nigeria alone will add 200 million urban residents in the next 40 years. It is estimated:

- globally, 180000 people move into urban areas daily
- 60 million people move into cities each year in the developing world.

8.6.2 The city of Cairo sprawling towards the pyramids

Consequences of the urban revolution

Urbanisation in the developed world occurred with growth of industries, jobs and wealth. Cities had time to develop infrastructure and employment opportunities, and although these cities face issues such as congestion and pollution, the standard of living of most residents is high.

Cities in developing countries struggle to provide housing, employment, sanitation, education and basic healthcare facilities. Urbanisation has occurred without economic growth and planning. Millions live in poverty in unplanned slums. Cairo has 8 million slum dwellers in a population of 18 million, and has spread into the surrounding desert.

Despite the issues created by rapid urbanisation, millions of people choose to live in towns and cities. What is the attraction of cities? What makes cities liveable for millions of people?

Geoinfo

More than 70% of Africa's urban population lives in slums.

Geoactivities 8.6

Knowledge and understanding

1 Where is the fastest urban growth occurring?

2 What are the consequences of rapid urbanisation in the developing world?

Inquiry and skills

3 Refer to 8.6.1.

a Using the line graphs:

i How many people lived in urban areas in developed and developing countries in 2010? What are the projected figures for 2050?

ii How will the percentage of people living in urban places of the developing world have changed between 1950 and 2050?

b Using the column graph: Which continent (area) will have the greatest share of the world's urban population in 2050? What share will it have?

c Using the bar graph (bottom right):

i Which of the five largest cities in 2009 will increase its population the most by 2025?

ii How many of the five largest cities are in Asia in 2009 and 2025?

d Identify five changes in the number, size and location of large cities between 2009 and 2025.

4 Refer to 8.6.2.

a What type of photograph is this? What might explain the faded appearance of the houses in the background?

b Predict what might happen in the future. Suggest how this could be prevented.

c Investigate the growth of Cairo over time. Present as a timeline or graph.

ISBN 978 1 4586 6277 4

8.7 Why go to town?

There are stunning satellite images of Earth at night that reveal the extent of our urbanised world. While millions of people are born into cities, others migrate to them for the real or perceived opportunities they offer.

Internal migration to escape poverty

Internal migration is the largest global movement of people. Each year 740 million people relocate within nations, creating a rural to urban transition. Most migrants are seeking economic opportunities and a better quality of life.

In 2011, 377 million of India's 1.21 billion people were urban dwellers—a number projected to increase to 600 million by 2030. Under Indian inheritance laws, farms are divided for each new generation, over time making them too small to support large families. Erosion, deforestation and drought have reduced environmental quality and food production, leading to increased rural poverty. A lack of educational and health facilities, electricity and other infrastructure further contributes to poor liveability in rural India. These factors mean rural Indians are moving to large towns and cities.

However, the move to cities does not always mean a better quality of life, and many migrants join millions of earlier migrants living in slums.

Migrating to the 'bright lights'

In many developed countries (e.g. Australia and Canada), the 'bright lights'—including employment, lifestyle, educational, recreational, medical, sporting and cultural opportunities—attract rural migrants, especially young people. Increasing farm sizes, natural hazards (e.g. floods and droughts), mechanisation and lower incomes make rural life difficult and reduce employment opportunities. When young people leave a place, the age profile changes, leaving a large proportion of aged residents. Population loss leads to business closure and a cycle of rural decline—further reducing the liveability of rural places.

Destination cities

Cities are magnets for 214 million international migrants each year. Employment, business opportunities, family reunion, amenities, lifestyle,

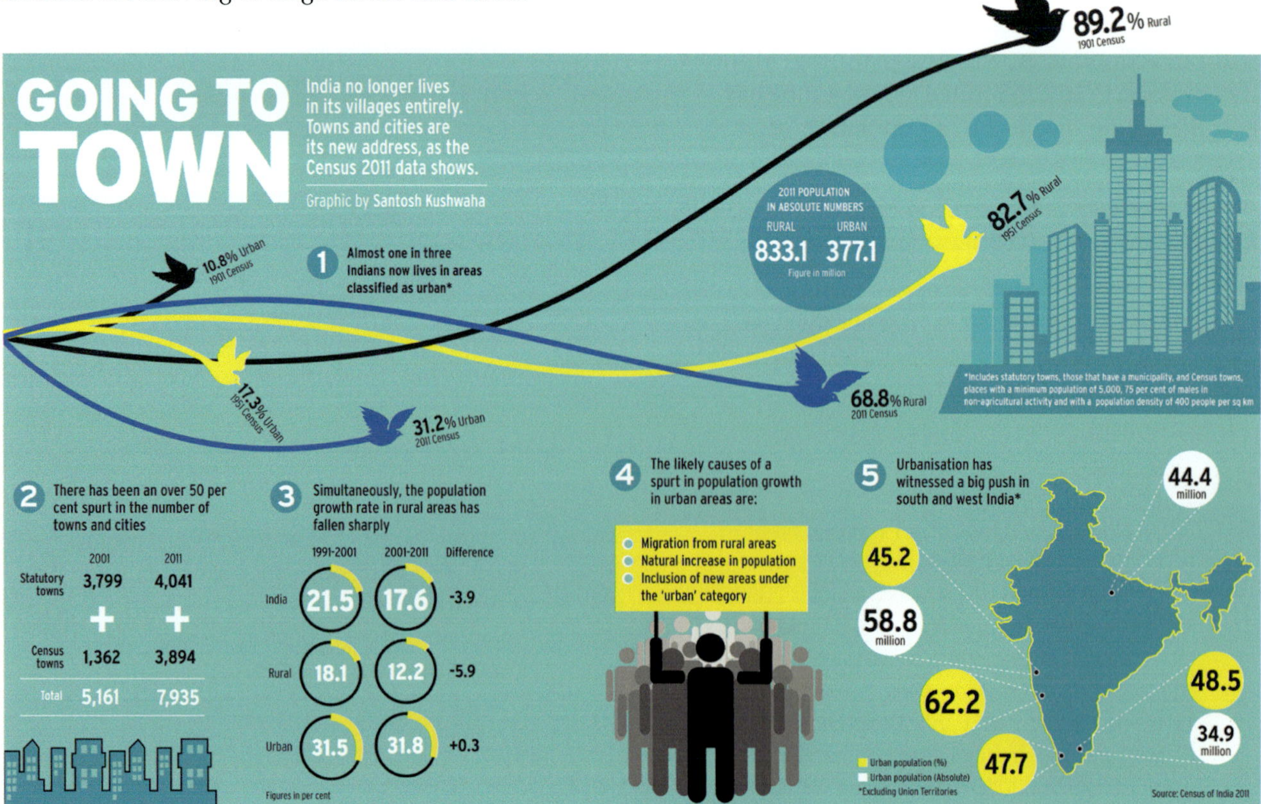

8.7.1 Urbanisation in India

ISBN 978 1 4586 6277 4

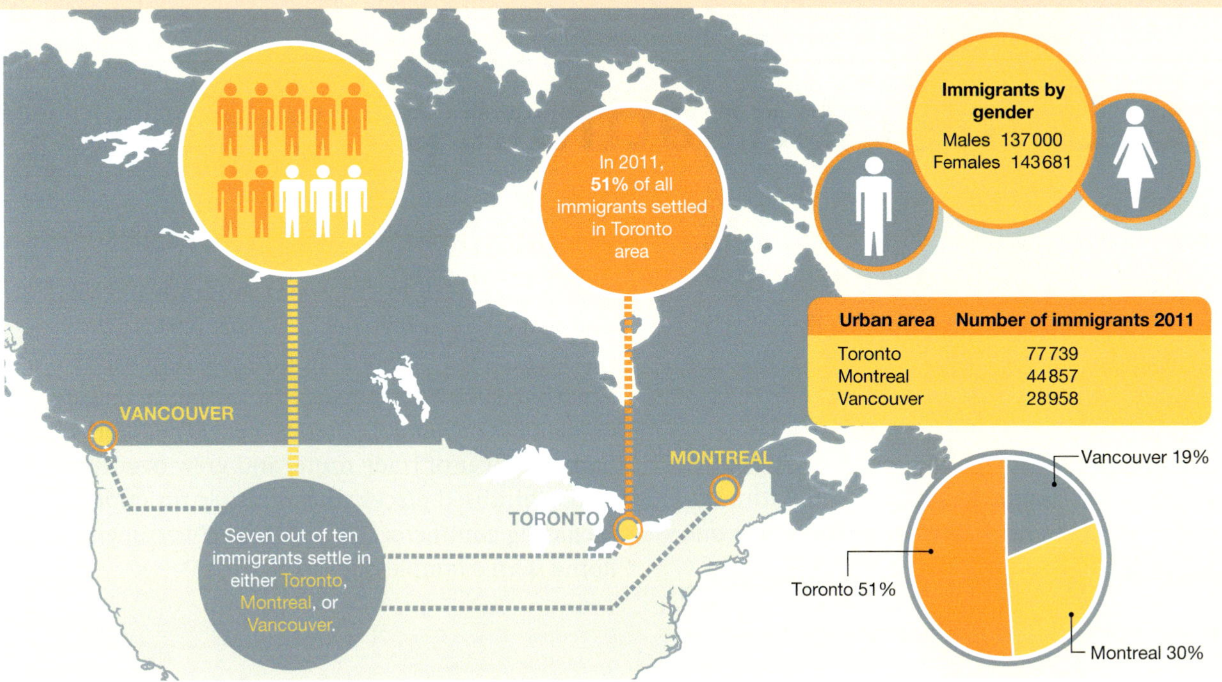

Urban area	Number of immigrants 2011
Toronto	77739
Montreal	44857
Vancouver	28958

8.7.2 Canadian immigration and urbanisation, 2011

safety and freedom attract migrants. Most arrive through the airports and ports of large cities in the developed world. Canada receives immigrants from over 200 countries with 7 out of 10 settling into three major cities. Most migrants arriving in Australia move to Sydney or Melbourne.

Migrants are attracted to the liveability of cities but the subsequent rapid urban growth leads to issues such as congestion and affordability. Making rural areas more liveable is one strategy to slow urban growth, but will it work?

Geoinfo

- Youth are 40% more likely than older generations to move from rural to urban areas.
- Globally, almost 90% of new migrants settle in cities.

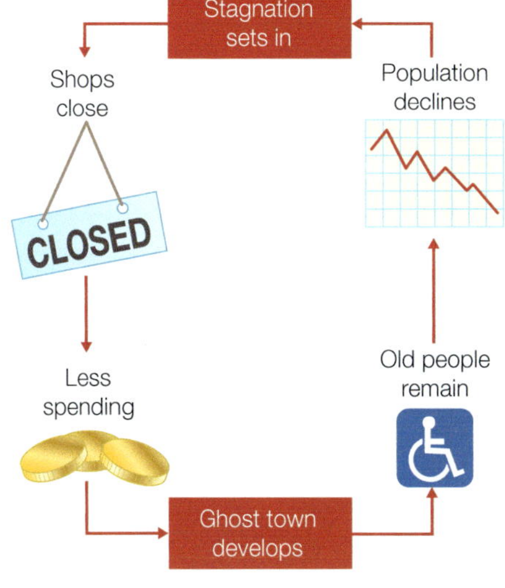

8.7.3 A cycle of rural decline and reduced liveability

Geoactivities 8.7

Knowledge and understanding

1. Explain the difference between international and internal migration.
2. Describe the urban transition occurring in India.
3. Explain why large cities become places of destination for international migrants.

Inquiry and skills

4. Refer to 8.7.1.
   a. Calculate the change in India's urbanisation between 1901 and 2011.
   b. How has the number of urban places changed?
   c. Where is the greatest urbanisation occurring in India?
   d. How many Indians were living in urban areas in 2011?
   e. List three factors contributing to India's urbanisation.
5. Refer to 8.7.2.
   a. Which Canadian city receives the greatest proportion of migrants?
   b. Describe the location of Canada's destination cities.
6. Refer to 8.7.3.
   a. Why is rural decline illustrated as a cycle?
   b. Do you think it is possible to stop rural decline by making rural places more liveable for young people? Explain your answer.

ISBN 978 1 4586 6277 4

8.8 Large cities attract people

The 21st century is witnessing some of the greatest ever changes to the way people live. Twitter grew to 500 million users in six years after its creation in 2006. In a similar way, the growth of super-sized cities is changing where millions of people live, with China leading the way. By 2025 the world's most populated country could have eight megacities (over 10 million people), two with over 20 million people, and 221 million cities (over one million people).

Historical growth of cities

People have lived in cities for thousands of years. The first cities were located in Asia, Africa and Europe. They developed around one important function, such as trade, defence, religion, culture or political power. Most cities were located on a harbour, river or trade route, and grew over time into centralised places with multiple functions, including commerce, the arts and government; Rome is an example of such a city.

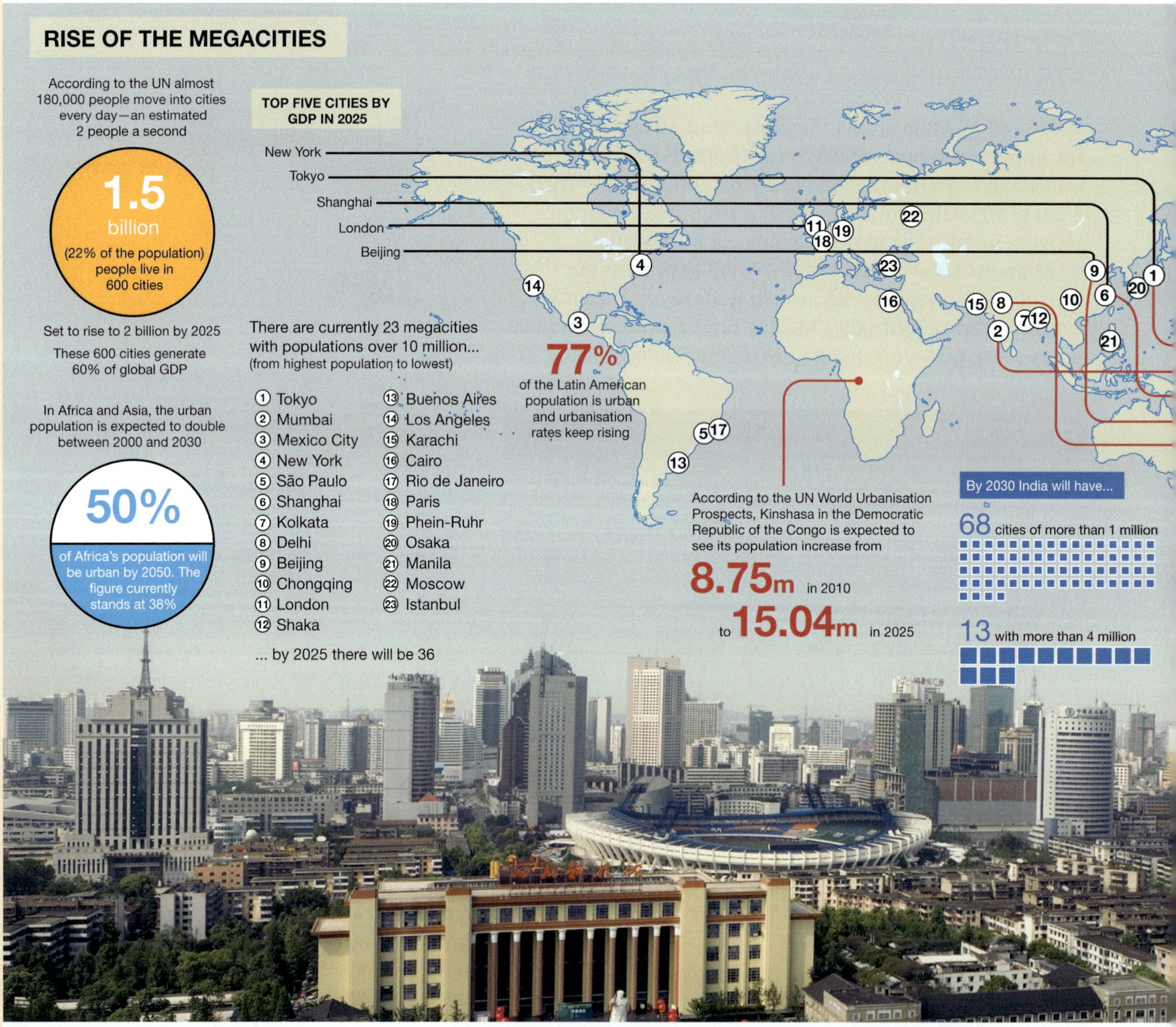

8.8.1 The rise of megacities

ISBN 978 1 4586 6277 4

Many of the early cities have disappeared (e.g. Babylon), while some have grown to become powerful world cities (e.g. London) and megacities (e.g. Mumbai).

Attraction of large cities

The Industrial Revolution, which resulted in the setting up of factories within cities, was a catalyst for urbanisation. However, **manufacturing** is no longer the only attraction for people living in large cities. Today's large cities are places where wealth is created. The top 25 cities in the world create over half of the world's wealth. Large cities are also places of opportunity and diversity, including employment, educational and health facilities, services, housing and entertainment.

Humans are social beings who choose to live where others do—the social attraction of cities creates liveability. Despite modern communication technologies, such as smartphones and the internet, people still like to be in close proximity with one another, centralising in towns and cities.

Urban population, 2030 forcasts

1 bn CHINA

590m INDIA

Global urban growth between 2005 and 2025 to come from China and India

40%

By 2030 China will have...

221 cities with 1 million or more inhabitants and 23 cities with five milliion or more

In 20 years, China's cities will have added 350 million people—more than the entire population of the United States today

TOP FIVE CITIES BY POPULATION IN 2025

Tokyo

Mumbai

Shanghai

Beijing

Delhi

75%

Percentage of Indian urban population earning an average of 80 rupees per day

6 mega cities with more than 10 million

These will need 2.5bn square metres of paved roads and 7400 kilometres of metros and subways—20 times what has been built in the last 20 years

In 2030 numerous Indian cities will have larger economies than many countries. Mumbai's GDP is projected to reach

$265bn

Geo**info**

- Population centres, built-up areas and urban centres are alternate names for cities and urban places across the world.
- The combined present-day populations of the USA, Brazil, Russia, Japan and Germany is the equivalent of the total projected population of Indian cities by 2050.

Geo**activities 8.8**

Knowledge and understanding

1. Name two changes to people's lives in the 21st century.
2. Why did early cities locate on harbours, rivers and trade routes?
3. What event provided the stimulus for rapid urbanisation?
4. Explain how industrialisation led to the growth of cities.
5. What does 'diversity' mean? How do large cities possess diversity?
6. If technology frees people from being close to each other, discuss why millions migrate to large towns and cities each year.

Inquiry and skills

7. Refer to 8.8.1.
   a. How many people move to cities every day?
   b. Describe the spatial distribution (location) of the world's megacities.
   c. Where is the greatest concentration of megacities? Name the cities in this area.
   d. Which continent has no megacities? Explain why.
   e. List the predicted top five cities in 2025 by population size and wealth (GDP).
   f. How many more megacities will there be in 2025 compared to now.
   g. Compare the predictions for urban growth in China and India.
8. Create a consequence diagram that shows how increased urban size can reduce liveability.

ISBN 978 1 4586 6277 4

8.9 Skyscrapers and sprawl

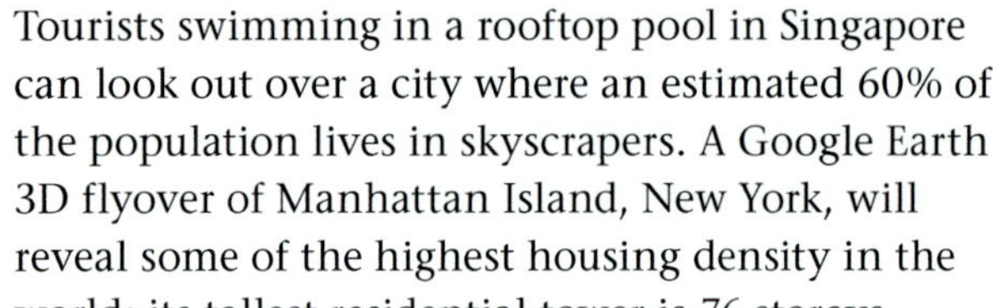

Tourists swimming in a rooftop pool in Singapore can look out over a city where an estimated 60% of the population lives in skyscrapers. A Google Earth 3D flyover of Manhattan Island, New York, will reveal some of the highest housing density in the world; its tallest residential tower is 76 storeys. **Housing density** is the number of dwellings in an area of land. Density reduces as you move from the centre of a city to the suburbs.

Housing density

The highest population and housing densities are in the largest megacities. Mumbai, India, houses an estimated 34 000 people per km^2 of land compared to 3800 per km^2 in Paris. High-density housing includes apartments and skyscrapers. In developing countries, where millions of people squeeze into slum dwellings on very small pieces of land, population and housing densities are also high. Low-density housing consists of single houses on individual blocks of land, as in the outer suburbs of Australian cities. Differences in housing density can be seen in photographs, maps and 3D models.

8.9.2 Models of housing density in London, Cairo and Mexico City were made from wood for a Global Cities exhibition at the Tate gallery, UK. The peaks show the highest residential densities.

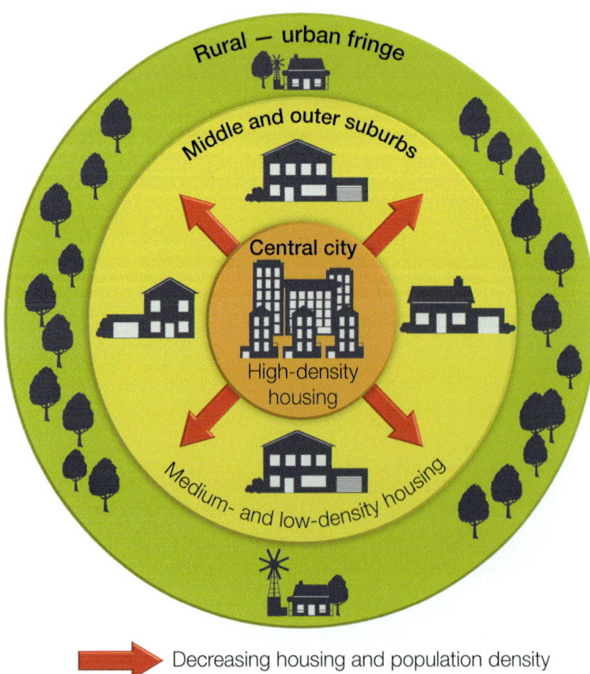

8.9.1 A simplified model of housing density in a city

CBD to the suburbs

The **central business district (CBD)** of a city offers a wide variety of employment opportunities. High- and medium-density housing is built close to the CBD or railway stations, bus routes and major roads so people can access jobs and the city lifestyle (e.g. restaurants and entertainment). Lower-density housing is found further from the CBD with townhouses, duplexes and separate houses on larger blocks of land; the main reason for this is the lower housing cost.

The suburbs provide a quieter, less hectic lifestyle, often preferred by young families. Income, stage of life (e.g. retirement), lifestyle choice, transport, employment, environment and safety are some factors that influence where people will live within a large city. Different places within cities—from the CBD through to outer suburbs—offer different conditions of liveability, determined by housing density, affordability, employment and the provision of services.

ISBN 978 1 4586 6277 4

	Advantages	Disadvantages
Low density	Quiet Cheaper land and housing—affordable Low-density housing Larger houses More room for families More privacy Natural environment	Lack of facilities (e.g. public transport, hospitals) Fewer employment opportunities Commuting to work wastes time and money Families need two cars Lack of entertainment and cultural facilities Isolation Loss of farmland and biodiversity Expensive to establish
High density	Walk to work Public transport Employment opportunities More facilities Close to everything Suits singles, couples and retirees Cultural and entertainment opportunities	Expensive housing—less affordable Smaller housing Less privacy Noisy Traffic congestion Poorer air quality Fewer open spaces Little or no backyard Higher crime rates

8.9.3 Low density versus high density

Sprawl versus tall

As cities grow, lower density housing spreads outwards from the city fringe into the surrounding farmland or natural environment. This is known as **urban sprawl**. When medium- and high-density housing is built on existing urban land it is called **urban consolidation**. In Australia, state governments and planners want to minimise urban sprawl. This has led to a growth in higher density housing.

8.9.4 High density residential living in Singapore

There is a lot of debate about which is better for the future liveability of cities.

In all large cities there is increasing pressure to house more people in a way that minimises the impact on the environment and reduces the urban footprint. This means energy- and water-efficient housing, low-carbon transport and even rooftop farming. Cities of the future need to be liveable and sustainable, and provide a variety of housing and transport to give people choices.

Geo**info**

A WWF campaign is targeting Perth's urban sprawl because of its impact on biodiversity.

Geo**activities 8.9**

Knowledge and understanding

1. What does the term 'housing density' mean?
2. Does high-density housing always mean high-rise buildings?
3. Why does high-density housing also mean high population density?
4. Where are the highest housing densities in a city? Explain why.
5. Describe how liveability changes with housing density across a city.

Inquiry and skills

6. Refer to 8.9.1.
   a. Why is this a simplified diagram?
   b. Explain how the diagram would change if the CBD was on one side of a river or close to the coast?
7. Refer to 8.9.2.
   a. Which city has the highest housing density peak? What does this mean?
   b. How does Mexico City's housing density differ from that of Cairo?
   c. What are the benefits of using 3D models to compare cities?
8. Refer to 8.9.3. The class will be divided into three teams discussing the topic: 'Tall is better than sprawl for the future liveability of cities'.
   - Group 1 will argue in favour of more high-density housing within a city.
   - Group 2 will argue in favour of more low-density housing on city fringes.
   - Group 3 will offer a compromise solution for future cities.

   Each team will investigate the role they are taking and present their arguments as an oral presentation.

Chapter 8

8.10 Skyscrapers and slums

An estimated 500000 residents of Cairo, Egypt, live in cemeteries (necropolises) because of a chronic shortage of housing. Known as the 'City of the Dead', the poor have turned the tombs into homes. While poor rural migrants try to create liveable places in cities, wealthy residents reside in luxury skyscrapers right next to them. This highlights inequalities in liveability within cities.

Worst conditions

Slums are poor, overcrowded, unhygienic places without adequate access to quality housing. These places also lack safe water, sanitation, garbage collection, employment, medical facilities, public transport and electricity.

Slums are usually located in unsafe locations, for example adjacent to railway lines, on low-lying flood-prone land, or on steep slopes susceptible to mudslides. Dwellings consist of one or two rooms, which barely qualify as 'adequate', and are constructed of locally available materials, such as plastic sheets, mud, wood, cardboard and metal.

Slums: the five deprivations

The United Nations Human Settlements Programme (UN-Habitat) defines a slum household as one that lacks one or more of the following:

- **Access to improved water**
  An adequate quantity of water that is affordable and available without excessive physical effort and time
- **Access to improved sanitation**
  Access to an excreta disposal system, either in the form of a private toilet or a public toilet shared with a reasonable number of people
- **Security of tenure**
  Evidence or documentation that can be used as proof of secure tenure status or for protection from forced evictions
- **Durability of housing**
  Permanent and adequate structure in a non-hazardous location, protecting its inhabitants from the extremes of climatic conditions such as rain, heat, cold or humidity
- **Sufficient living area**
  Not more than three people sharing the same room

UNICEF, The State of the World's Children, 2012

8.10.1 Lack of sanitation

Numbers

An estimated 1 billion people live in slums, a figure expected to double by 2030. According to the UNICEF State of the World's Children 2012 report, every third city child today lives in a slum. Dharavi in Mumbai (India), Kibera in Nairobi (Kenya) and Paraisopolis in São Paulo (Brazil) are famous slums among the largest in the world.

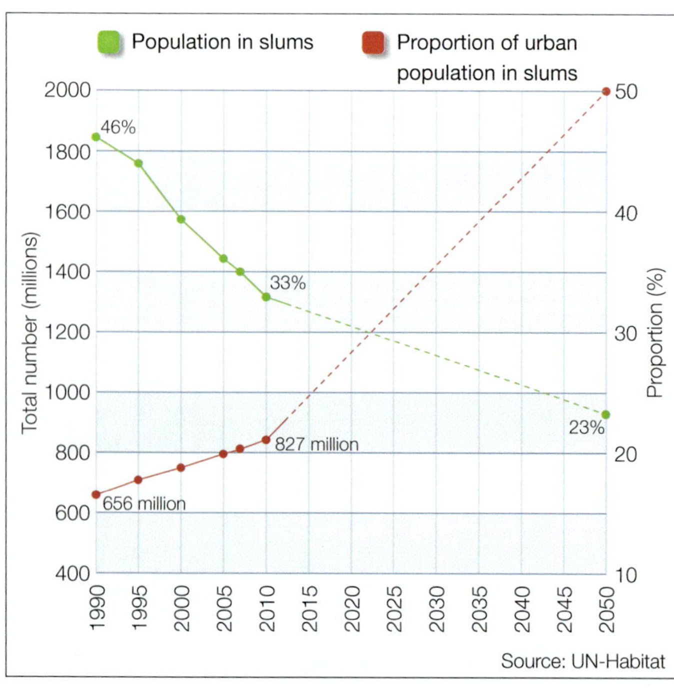

8.10.2 Numbers of people living in slums across the world

ISBN 978 1 4586 6277 4

Slum dwellers act to improve the liveability of their slums. People create their own jobs to improve their quality of life. Local industries develop, producing handicrafts like clay pots and recycling waste metals and plastics. Long-time slums develop a strong community identity, as people rely on each other to survive.

Trends

The proportion of people living in slums has fallen to meet Millennium Development Goal targets, but the number of people living in slums has increased.

The liveability of many slums is improving through the actions of community groups, governments and NGOs who are building public toilet blocks, installing solar power and replacing slums with apartment blocks serviced with plumbing and electricity.

The fascination of people from wealthier countries with slum dwellers after the movie *Slumdog Millionaire* has resulted in increased slum tourism, where tourists visit slums as a part of their travel itinerary. Opinions are divided as to whether this is a way to increase global understanding of slum life or an unfortunate invasion of privacy and a source of humiliation for slum dwellers. What do you think? Does slum tourism make slums more or less liveable?

8.10.3 Skyscrapers and slums. The Paraisopolis favela in São Paulo, Brazil, is located between several affluent communities made up of high-rise buildings and large single homes.

Geo**activities 8.10**

Knowledge and understanding

1 What are the five deprivations that can result in a place being called a slum?
2 What is the City of the Dead? Why do you think no-one forces people to leave these slums?
3 Why are slums often located on the worst land within a city?
4 Explain how slums are made more liveable.
5 Would you participate in a slum tour? Justify your answer.

Inquiry and skills

6 Refer to 8.10.1.
   a Construct a line drawing of the image.
   b How do such toilets both improve and reduce liveability?
   c Create a flow diagram to show the impact of this 'sanitation system' on the health of slum dwellers.
7 Refer to 8.10.2.
   a Describe the changing proportion of people living in slums between 1990 and 2010.
   b Compare the number of people living in slums in 2010 with the population of Australia (24 million).
   c Are these trends expected to continue?
8 Refer to 8.10.3.
   a If you lived in this luxury high-rise building in São Paulo how would you feel about the slums you could see? How could these slums impact on your life?
   b Under what conditions would you consider a slum to be a liveable place?

ISBN 978 1 4586 6277 4

8.11 Kibera slums and flying toilets

Each year World Toilet Day draws attention to the 2.6 million people without access to clean toilets. 'Flying toilets' are a hazard in Kibera, one of Africa's largest slums. A shortage of toilets forces people to use plastic bags, which are thrown onto rooftops or into drains. There is no waste collection service. Disease, unemployment, violence, water supplies, lack of government services (e.g. schools) and overcrowded housing are all serious issues in Kibera.

Kibera residents have little choice about where and how they live. A lack of services and facilities, high crime rates and poor environmental quality create low liveability. However, organisations and governments are working to improve the situation.

Mapmakers

Kibera is located in Nairobi, Kenya's capital. Most maps of Nairobi do not show the 2.5 km^2 with its mud houses and corrugated iron roofs, where population densities reach an estimated 90000 per km^2. Kibera is clearly visible on satellite photographs

8.11.2 Satellite image of Nairobi showing Kibera's slums

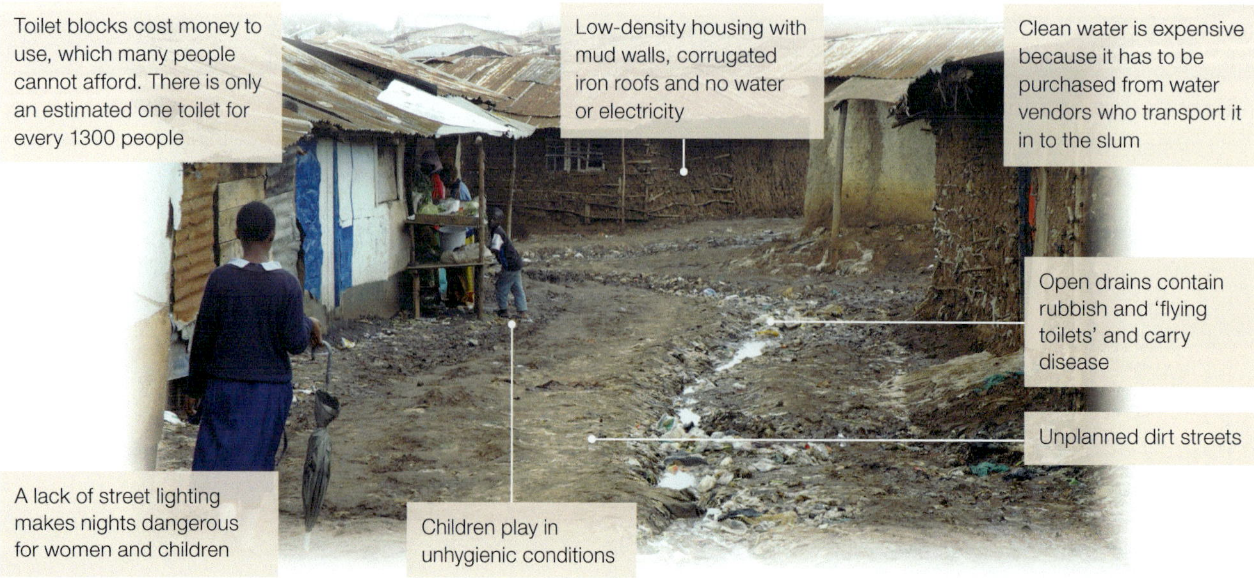

8.11.1 Living conditions in Kibera

ISBN 978 1 4586 6277 4

beside the planned streets and larger houses of official suburbs.

The Map Kibera project trained young community members to use GPS, cameras and computers. They collected information about services in Kibera (such as water, toilets and clinics) to create a digital community map. It is now an interactive community project.

Local people place comments on a blog and a television network uploads videos of news and upcoming events in the slum, and these are linked to the map. Aid organisations use the map to plan services, such as schools. The map has improved safety by warning people about violence or crime. Businesses are also shown on the map.

Women are heroes

Women in Kibera have a difficult life, and many are victims of violence. They work to support their families, as well as orphans whose parents have died of AIDS, disease or violence. Many women are involved in community projects, such as 'bag farming' to produce fresh vegetables.

The 'Women are Heroes' project placed images of women living in Kibera onto the roofs of houses to draw attention to the role they play in making Kibera a more liveable place.

Slum clearance and resettlement

The Kenyan government has started bulldozing the Kibera slums and building new housing. Most residents are happy to have clean, modern housing but others are worried that rents will be too high or they will lose their businesses when slums are cleared. Until Nairobi's population growth slows, many Kenyans will live in slums. What features would need to change for you to consider living in Kibera?

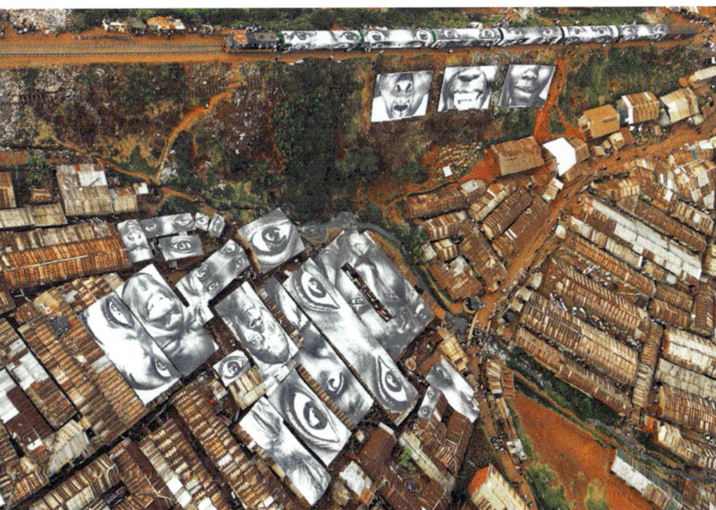

8.11.3 As part of the Women are Heroes project in Kibera, artist JR covered rooftops with blown-up photographs on canvas to celebrate the women who lived there. The covers keep houses waterproof in the wet season and send the message that women are the strength of the community.

Geo**info**

- HIV is estimated to affect over one-fifth of Kibera's population.
- Life expectancy in Kibera is around 50 years of age.

Geo**activities 8.11**

Knowledge and understanding

1. Where is Kibera?
2. What are flying toilets? Why are they used?
3. What would be the environmental impact of flying toilets?
4. Explain why Kibera is an unofficial settlement?
5. Why do residents of Kibera have little choice about where they live?
6. List three reasons why Kibera is a dangerous place.
7. Why is Map Kibera important for the community?
8. What is the aim of the Women are Heroes project?
9. Discuss the advantages and disadvantages of slum resettlement projects in Kibera.
10. List the features of Kibera that would make it unliveable for most Australians.
11. List the features of Kibera that make it liveable for Kenyans.

Inquiry and skills

12. Refer to 8.11.1.
    a. Draw a line drawing to show the main characteristics of the Kibera slum.
    b. Who might have built the toilet block? Give a reason for your answer.
13. Refer to 8.11.2.
    a. Explain why it is easy to identify the location of Kibera on the satellite photograph.
    b. State one advantage and one disadvantage of satellite photographs.
14. Go to the Map Kibera website (see Geolinks). Study the map of the entire Kibera slum.
    a. Calculate the distance of the slum from the centre of Nairobi.
    b. Compare the road pattern of Kibera with that of the suburbs around it.
    c. Identify facilities and services that have been mapped.
    d. Explain how the map has made the invisible visible and improved liveability.

ISBN 978 1 4586 6277 4

8.12 Future liveability: urban footprints

Cities consume more of the world's resources and produce more waste than any other human activity. It is estimated that London uses over a billion tonnes of resources and produces 18 million tonnes of waste each year. Urban air contains on average 10 times more dust particles, 5 times more sulphur dioxide and 10 times more carbon dioxide than country air. Environmental contamination, climate change, depletion of natural resources, loss of biodiversity and loss of farmland are consequences of urbanisation that reduce the liveability of all places on Earth.

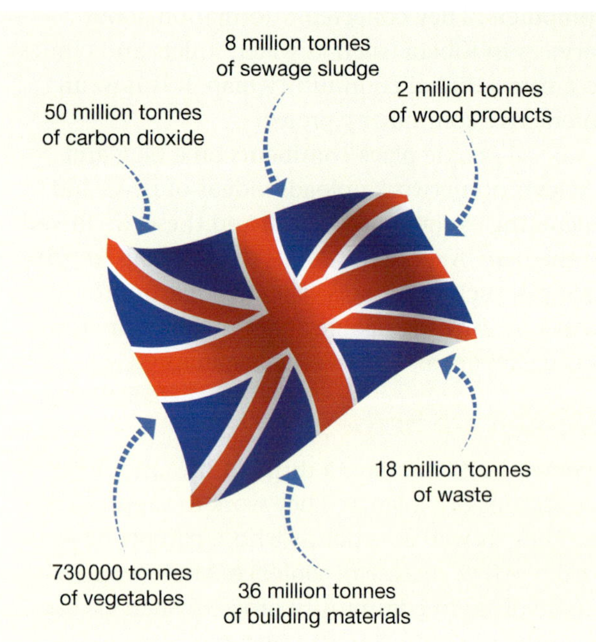

8.12.2 Ecological footprint of London

The ecological footprint of cities

A city's ecological footprint is the area of land and water needed to provide goods and services for its residents and manage its waste. London's ecological footprint is 125 times larger than its urban footprint, the area it occupies. Vancouver and Sydney, voted among the world's most liveable cities, have ecological footprints of 175 and 200 times their urban footprint. If everyone on Earth lived like residents of these cities the resources of three to four Earths would be required.

The footprint of cities is measured in global hectares. Biocapacity is the ability of Earth to

8.12.1 Ecological footprint of cities

ISBN 978 1 4586 6277 4

supply resources. The available biocapacity for the world's 7 billion people is less than 2 ha per person.

The term 'overshoot' is used when the ecological footprint exceeds biocapacity. Cities such as Adelaide have been called 'black holes' because they overshoot biocapacity.

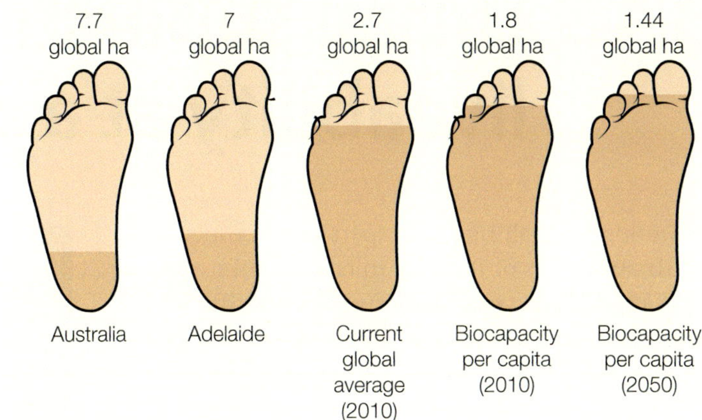

8.12.3 Ecological footprint of Australia's, Adelaide's and the Earth's biocapacity

Spatial variations

Cities in the developed world consume the most resources per person. An average American city with 650 000 people draws resources from 30 000 km^2, while an Indian city of similar size requires only 2800 km^2. Tokyo consumes 6.96 megatonnes of water daily compared to 0.9 in Dhaka.

Energy and food

The largest components of urban ecological footprints are energy and food. Food production requires water, land, energy and nutrients. Agriculture adds unwanted chemicals to soil and waterways, and takes valuable nutrients from rural environments. The production and transportation of food consumes energy, relying heavily on fossil fuels, generating CO_2 and contributing to climate change. Buildings are also high-energy consumers.

If cities do not change the way they function, an area the size of France, Germany and Spain (1.5 million km^2) will be added to Earth's urban ecological footprint by 2030. The quality of life in cities and on Earth itself is threatened by unsustainable development. This means our planet will become less liveable, unless ecological footprints are reduced.

Geoinfo

Cities consume 80% of Earth's resources while occupying less than 5% of the land.

Geoactivities 8.12

Knowledge and understanding

1 List the environmental consequences of urban growth.
2 Why does the ecological footprint of cities go beyond city boundaries?
3 Cities are said to be in 'overshoot'. What does this mean?
4 Use an example to illustrate the uneven impact of cities on Earth's resources.
5 Suggest why food and energy contribute to the large ecological footprint of cities.
6 What will happen to liveability if cities continue to develop in the same way as in the past?

Inquiry and skills

7 Refer to 8.12.1.
   a What resources are consumed by cities?
   b Where do these resources come from?
   c How do most cities deal with waste?
8 Refer to 8.12.2. In 2012 the population of metropolitan London (city plus suburbs) was 12 million.
   a Calculate the following components of London's ecological footprint:
      i CO_2 emissions per person
      ii wood products consumed per person
      iii waste generated per person
      iv vegetables consumed per person.
   b Suggest how each London resident could reduce these statistics.
   c If London does not reduce its footprint, how will its liveability change in the future?
9 Refer to 8.12.3.
   a What is happening to the world's biocapacity?
   b What was the average ecological footprint for each person on Earth in 2010?
   c What is the biocapacity available for each person on Earth in 2010?
   d What is the main message of this diagram?
   e Explain the connection between ecological footprints, biocapacity and liveability.

Chapter 8

ISBN 978 1 4586 6277 4

8.13 Urban superstructures: liveability at a cost

Technology is making skyscrapers one of the most sustainable types of housing in cities. China aims to build a new skyscraper every five days for the next three years. To provide heat, light and electricity to these buildings, however, China opens a new coal-fired power station every 10 days. At the same time, land reclamation projects are creating new urban islands, but not without impacting on marine ecosystems.

Dubai: desert transformed

You are sitting in your 92nd floor luxury apartment in the world's tallest building, overlooking the world's largest human-made islands and a spectacular sail-shaped skyscraper. You've visited the world's largest shopping mall, had an indoor ski and seen the world's largest marina. You are in Dubai, one of seven emirates that are the oil-rich United Arab Emirates (UAE).

Dubai transformed from a desert port to a **cosmopolitan** city of over 2 million people in two decades. A city irresistible to businesspeople, diplomats, workers, sports lovers and tourists was the vision of Dubai's royal family. Engineers overcame extreme environmental challenges including intense heat, violent storms, strong winds and wave erosion, to make the vision a reality. Now Dubai claims to be one of the world's most liveable cities. It boasts record-breaking superstructures, including the Burj Khalifa, Palm Islands, World Islands and Dubai Mall (complete with ski run).

Very liveable city

Most people living in Dubai are expatriates—citizens of other countries. They make up 90% of Dubai's retail, financial, trade, construction and tourism workforce. Only 17% of Dubai's population is Emirate.

Dubai is an economic and lifestyle choice. In addition to employment, luxury accommodation, shopping, entertainment, and recreational and sporting facilities, Dubai offers tax-free incomes, a high standard of living, a safe environment, a low crime rate and good educational facilities. Located between Europe, Asia and the Americas, Dubai is an ideal base for overseas travel.

8.13.1 The Burj Dubai (now called Burj Khalifa) is the world's tallest building

ISBN 978 1 4586 6277 4

8.13.2 A ski slope in the desert: an indoor ski slope in a Dubai shopping mall

Persian Gulf
Grid-patterned roads
Urban settlement
Artificial island
Burj al-Arab
Artificial harbour for recreational boats
Imported sand for surfers

8.13.3 The Burj al-Arab is a sail-shaped luxury hotel

Environmental costs

Behind Dubai's wealth, however, are thousands of poor construction workers from India, Pakistan, Bangladesh and the Philippines, making up 85% of Dubai's expatriates. Wages are low and housing is in crowded labour camps.

The UAE has one of the highest **ecological footprints (EF) per capita**. High CO_2 levels, traffic congestion, water use and waste disposal are problems. Dubai is very liveable but has yet to prove that it can also be ecologically sustainable.

Across the world, eco towers, green buildings and alternative energy sources are helping these new urban places become more sustainable.

Geo**info**

- 15–20% of the world's cranes are operating in Dubai.
- The Dubai World Cup is the world's richest horse race.

Geo**activities 8.13**

Knowledge and understanding

1 Why is Dubai a cosmopolitan city?
2 What was the vision of Dubai's royal family?
3 Discuss the factors that make Dubai a liveable city for rich and poor, young and old.
4 Suggest why skyscrapers are energy- and water-hungry buildings.
5 Explain why the UAE is not ecologically sustainable.

Inquiry and skills

6 Refer to 8.13.1.
  a How does temperature differ between the top and the bottom of the tower? Explain.
  b Why would the temperature difference create a demand for electricity?
  c What other features of the building would create a high demand for energy?
  d Dubai is in a desert. Where would the building materials come from?
7 Refer to 8.13.3. Draw and label a line drawing of features in the photo that contribute to Dubai's liveability.
8 Investigate how skyscrapers can be built to conserve environmental resources (see Geolinks).
9 Watch the documentary called *Strip the City: Dubai*. Explain how technology was used to create a liveable, cosmopolitan city in a desert.

Chapter 8

ISBN 978 1 4586 6277 4

Geothink

Interpreting images

1 Refer to 8.14.1a and 8.14.1b. The hourglass is used to represent the passing of time. These hourglass images represent changes to the places where people live and the impact of these changes on liveability.
   a What makes the hourglass an appropriate symbol for the transition of people from rural to urban places?
   b How does the shape of the hourglass represent the future distribution of world populations globally and within nations?
   c List ways that rural to urban migration impacts on the liveability of rural and urban places.
   d Explain the link between the second image and sustainability.
   e What is the relationship between the two images?

Connectedness

2 Refer to 8.14.2.
   a What do the blue lines on the map represent? Why is this important in calculating remoteness?
   b Use an atlas to name six of the most connected (least remote) countries and six of the least connected (most remote) countries on Earth. Justify your choices.
   c Colour and name these countries on a world map using an appropriate legend.
   d Find a map of world population density on the internet. What is the relationship between remoteness and population density? How does this reflect liveability?
   e Explain how technology is making remote places more liveable.

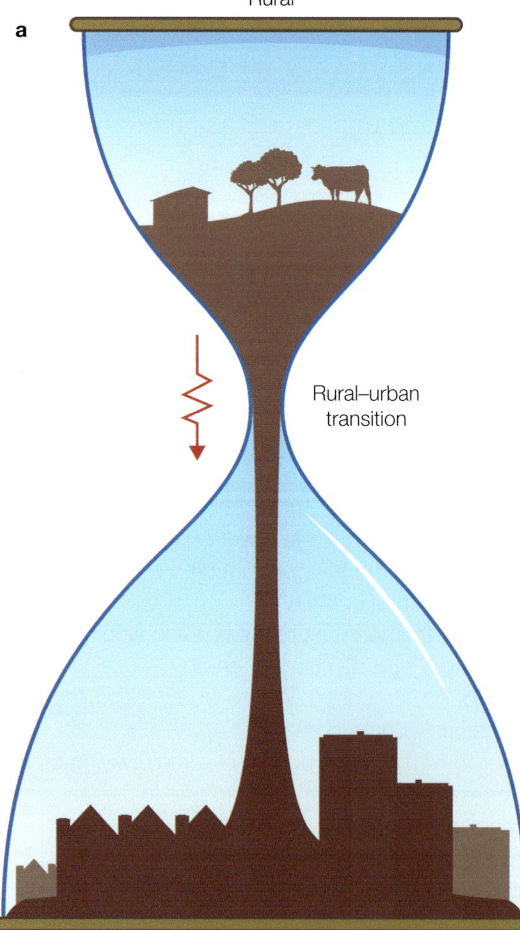

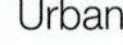

8.14.1 The hourglass represents changes to the places people live

ISBN 978 1 4586 6277 4

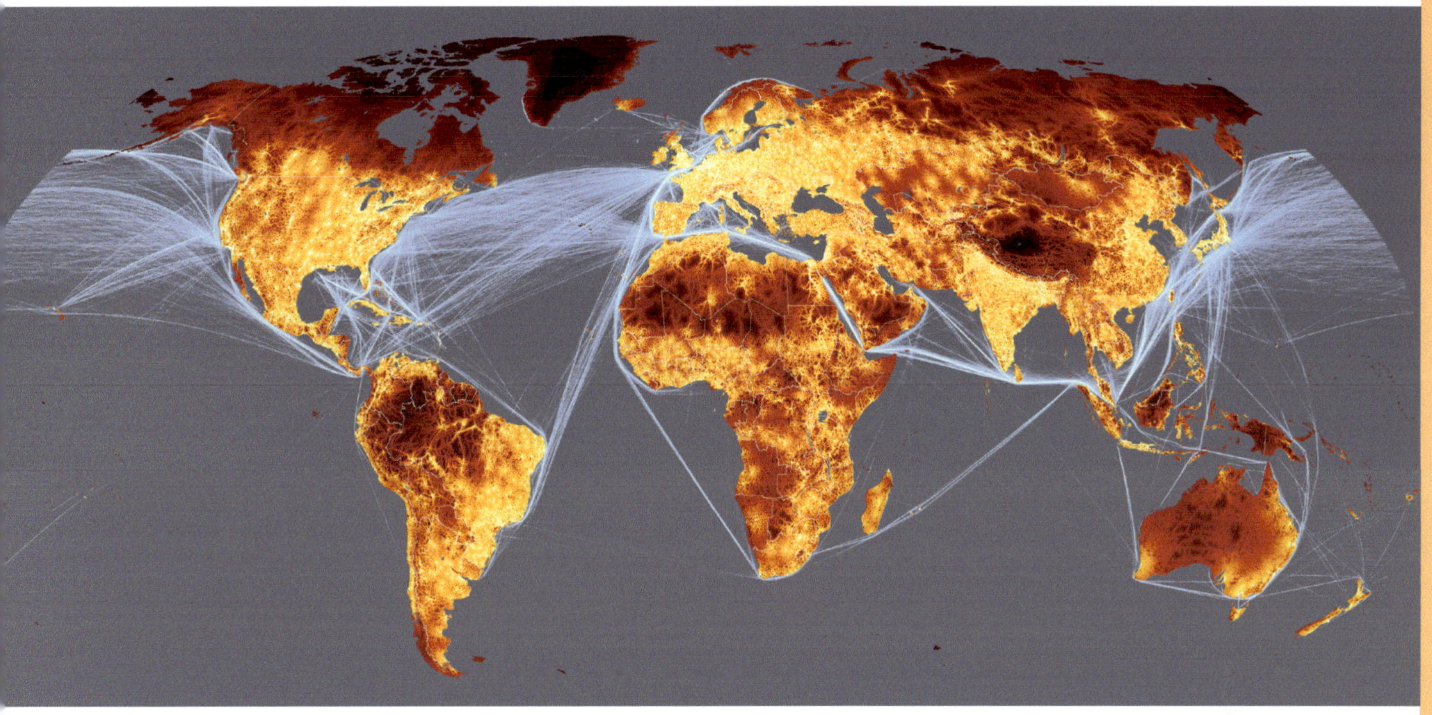

8.14.2 Global map of connectedness shown by travel times to cities. The darkest areas are least accessible, and therefore most remote. The blue lines show major shipping routes connecting the world's major port cities

People live in cities

Residents of Dharavi slum in Mumbai, India, recycle oilcans and resell them. Cottage recycling industries such as this provide employment for millions of urban slum dwellers globally.

3 Refer to 8.14.3.
   a Decide which statements around the photo are true, and which are false.
   b Turn the false statements into true ones.
   c Find a photo of a rural area, and annotate it with statements about the liveability of rural places.

8.14.3 What have I learned? True and false statements about the liveability of urban places

Chapter 8

ISBN 978 1 4586 6277 4

chapter 9

Population and development: challenges to liveability

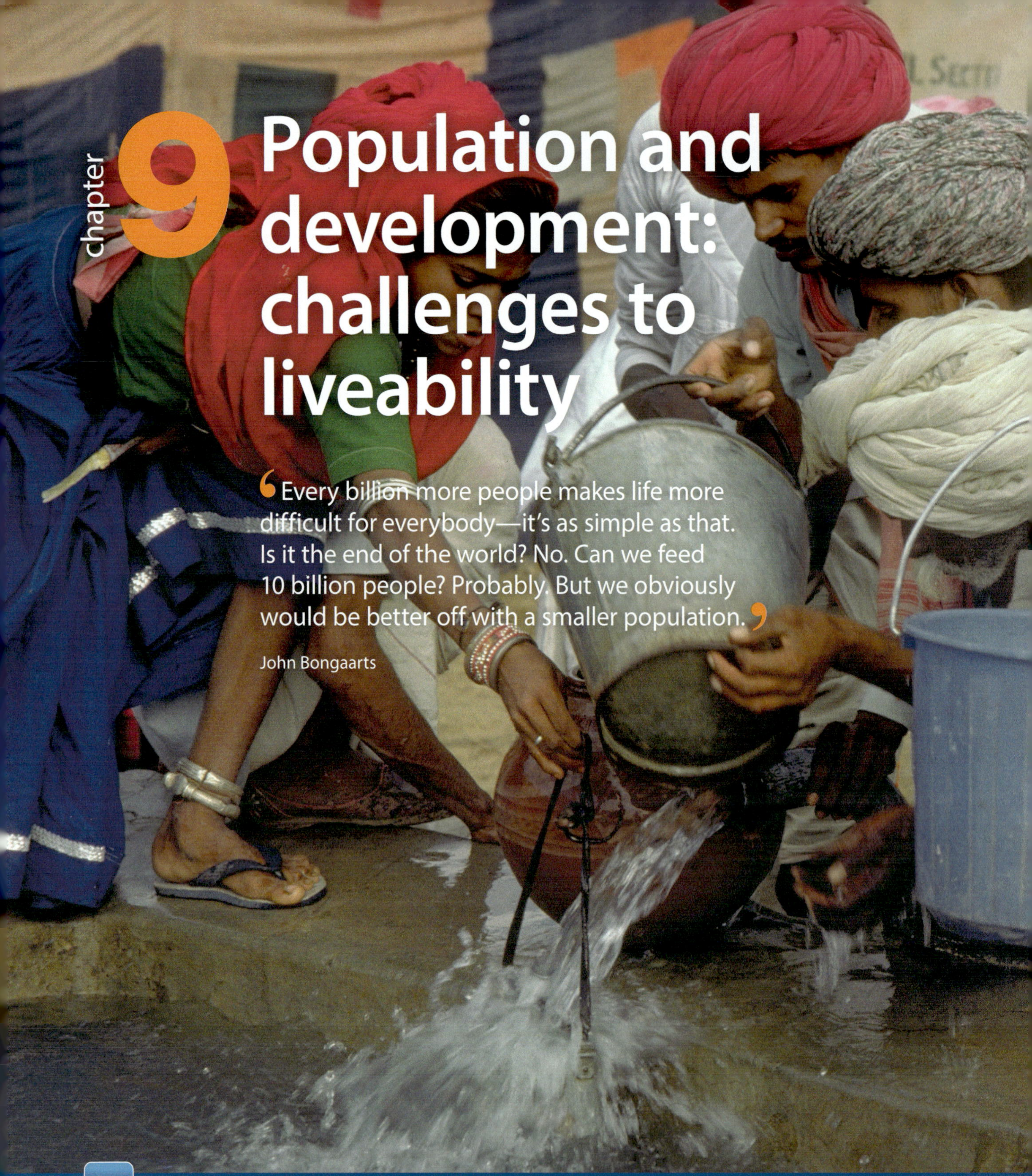

'Every billion more people makes life more difficult for everybody—it's as simple as that. Is it the end of the world? No. Can we feed 10 billion people? Probably. But we obviously would be better off with a smaller population.'

John Bongaarts

Aa Geovocab

dependent population: age groups in a population that are not a part of the workforce (e.g. too young or too old)

fertility rate: number of births per female

gender imbalance: when the number of males or females outnumbers the opposite sex

girl effect: when women are educated and empowered the fertility rate drops and incomes rise

Human Development Index (HDI): measurement of human development

population density: number of people per square kilometre

population distribution: spatial distribution of population across a place

remittances: payments sent by migrants to their families back home

ISBN 978 1 4586 6277 4

This chapter examines the influence of demographic characteristics and trends on the nature and liveability of places. Population size, growth, structure and distribution as well as migration have implications for how people live their lives and the locational choices they make. Levels of economic and human development in different places are influenced by the characteristics of their populations. For instance, the places with the fastest growing populations are some of the world's poorest, least developed nations. A person's age and gender has an impact on where they choose to live.

Collecting water from a well in the Indian state of Gujarat

Think, puzzle, explore

- **Place** How does immigration and emigration affect the character of places?
- **Space** How does the population vary over space?
- **Environment** How does population affect development?
- **Interconnection** What is the connection between development and liveability?
- **Sustainability** What is sustainable population growth?
- **Scale** How big is the world's population?
- **Change** How do populations grow and change over time in different places?

Geo**skills** in focus

- **Planning** a simple geographical inquiry
- **Evaluating** population profiles
- **Analysing** different types of graphs and maps, and changes over time
- **Communicating** statistical data using interactive web 2.0 tools

social status: position in society

sub-Saharan Africa: part of the African continent below the Sahara desert

United Nations High Commissioner for Refugees (UNHCR): UN agency concerned with protecting and enforcing the rights of refugees

ISBN 978 1 4586 6277 4

9.1 Planet under pressure

You can watch the world's population change every second on the internet, and discover the population size on the day you were born. Population size and growth influence what places are like to live in. Many people claim that the world is facing a population 'explosion' and the liveability of places and the planet itself will decline if growth continues. Approximately 140 people every minute and 70 million every year are being added to the planet. The current population is estimated to be over 7 billion, with more than 2 billion of these living in China and India. Demography is the study of population, including its size, structure and distribution.

Spiralling upwards

The last 1 billion people were added to the planet in the shortest time ever. Each decade the population has grown by different amounts ranging from 100 million to 900 million greater than the decade before. Population growth is slowing, but by 2050 the population could reach 9 billion. The most rapid growth is occurring in the developing world where large populations already live in places where their basic needs cannot be met. Of the world's 7 billion people, over 3 billion live on less than $2.50 a day and lack access to adequate shelter, safe water and health services.

9.1.1 Population milestones

Population	Year	Years to add 1 billion
1 billion	1804	From the first humans
2 billion	1927	123
3 billion	1960	33
4 billion	1974	14
5 billion	1987	13
6 billion	1999	12
7 billion	2011	12
8 billion	2025	14 (estimated)
Approx. 9 billion	2050	25 (estimated)

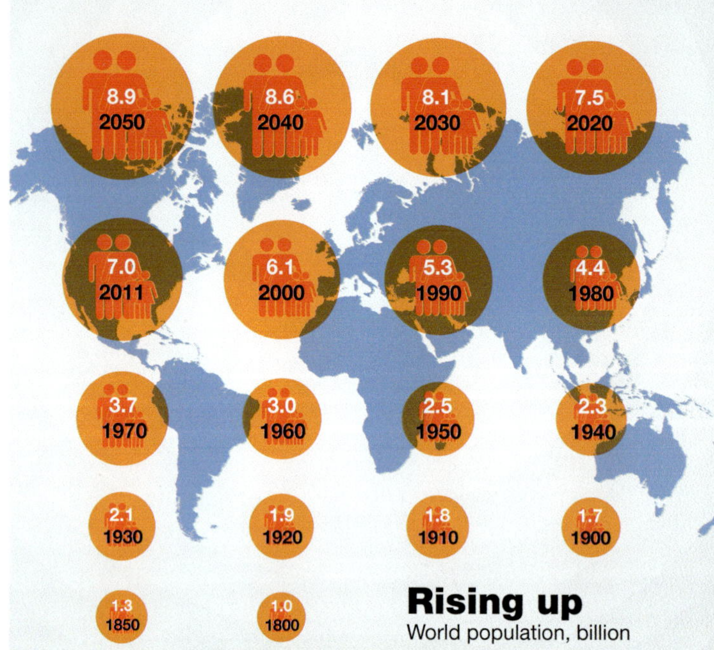

9.1.2 Changing world population by decade since 1800

Population growth

The global equation is simple. The natural increase in population is births minus deaths. The rate of natural increase is the birth rate minus the death rate. Currently there are five births for every two deaths.

For different places on Earth the population equation must take into account migration. The number of migrants that enter (immigration) and the number that leave (emigration) will affect the population of continents, countries and cities. In many developed countries, such as Germany and Australia, migration is an important part of population growth.

Change over time

Over time the populations of countries go through a *demographic transition*. When birth and death rates are both high or low population growth will be small. When there is a large gap between the birth rate and the death rate population growth is rapid. This is the situation in many developing countries, such as Nigeria. In some places, such as countries in Europe, populations are decreasing because birth rates have fallen below death rates.

ISBN 978 1 4586 6277 4

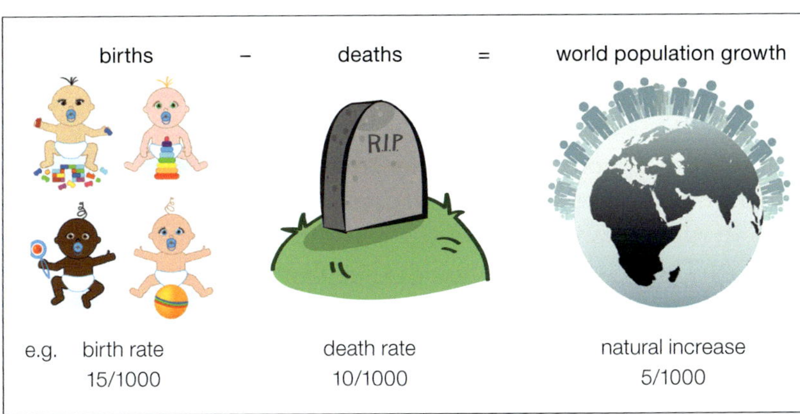

9.1.3 The global population equation

Geoinfo

- In 31 countries (e.g. Burkina Faso and Uganda) the population is likely to double by 2050.
- In 45 countries (e.g. Bulgaria and Japan) the population is likely to fall by 2050.

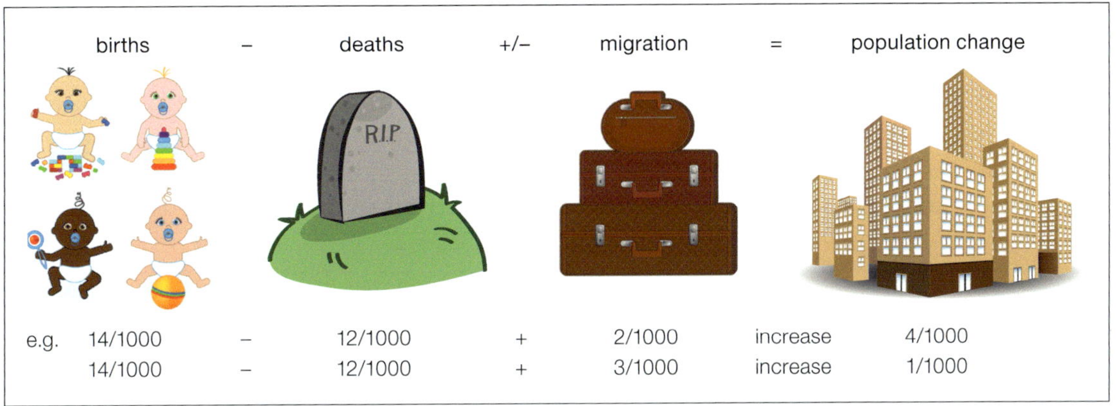

9.1.4 The country population equation

The size and growth of population influences liveability. When populations grow rapidly overcrowding puts pressure on resources. For instance, farms become too small to support large families so poverty increases. Controlling population growth is seen as a solution to ending poverty and improving the quality of life of people in the poorest countries of the world.

Geo**activities 9.1**

Knowledge and understanding

1 Identify the two factors affecting world population growth.

2 Explain why migration affects the population of countries but not the world.

3 What would cause the world's population growth to slow down?

4 What conditions would be needed for the world's population to stop growing?

5 Suggest a reason why the world's population might fall sometime in the future.

6 Examine the connection between population and liveability,

Inquiry and skills

7 Refer to 9.1.1 and 9.1.2.

   a Draw a line graph to show changes in the world's population between 1800 and 2050.

   b Referring to 9.1.1, add arrows to your graph to show when each billion milestone was reached.

   c Use the internet to find out the population of the world on the day you were born (see Geolinks). Add yourself to the graph.

   d Calculate the number of people added to the population since you were born.

   e Calculate the estimated population for the year you will be 50 years old.

8 View an online visualisation of population growth (see Geolinks).

   a Explain the technique used to demonstrate the growth of the world's population.

   b What might happen if the world glass overflows?

   c How could a situation of declining populations be added to this demonstration?

ISBN 978 1 4586 6277 4

9.2 Where do we all live?

Do you live in a place with lots of people? If you live in a city you may share the space with millions of people, like 4.1 million others in Melbourne. In rural places neighbours can be many kilometres away. The spread of people is known as **population distribution**, and it influences what places are like. Asia houses 60% of Earth's population while Antarctica has only a temporary population of about 1200 international scientists and other staff. China and India are the countries with the largest populations.

Trends

In the future Africa will have a greater share of the world's population because many countries (e.g. Chad) have high rates of natural increase. Nigeria—the seventh most populous country—could rank third by 2050. The 10 poorest countries in the world in 2012 were in Africa. In Liberia, an estimated 85% of the population already live on under $1 a day. Making places in Africa more liveable is difficult when population growth is rapid. Poverty is causing increased migration from Asia and Africa to places offering a better quality of life.

Places with more people

Population density is the number of people for every square kilometre of land. The highest densities occur in large Asian cities (e.g. Manila in the Philippines, with 34 000 per km^2). Populations are also denser along coastlines where climate and resources (e.g. water) can support more people. In Asia, the fertile floodplains of rivers support millions of rice farmers;

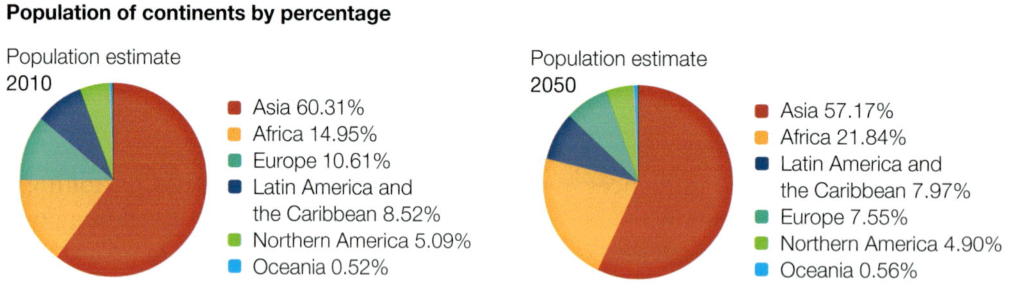

Population of continents by percentage

Population estimate 2010

- Asia 60.31%
- Africa 14.95%
- Europe 10.61%
- Latin America and the Caribbean 8.52%
- Northern America 5.09%
- Oceania 0.52%

Population estimate 2050

- Asia 57.17%
- Africa 21.84%
- Latin America and the Caribbean 7.97%
- Europe 7.55%
- Northern America 4.90%
- Oceania 0.56%

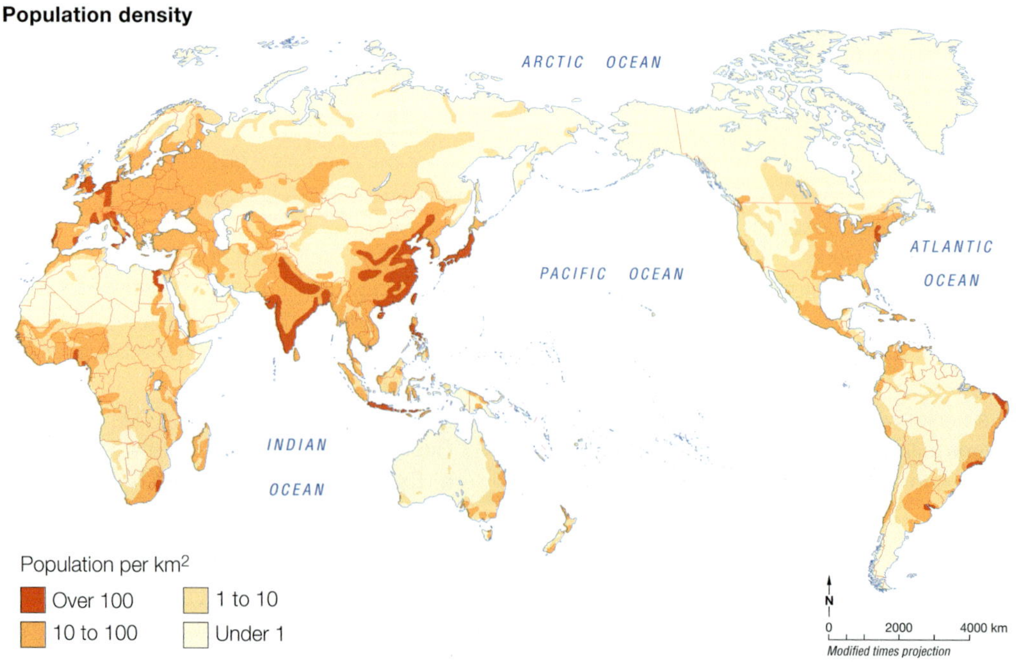

Population density

ARCTIC OCEAN
PACIFIC OCEAN
ATLANTIC OCEAN
INDIAN OCEAN

Population per km^2

- Over 100
- 10 to 100
- 1 to 10
- Under 1

N
0 2000 4000 km
Modified times projection

9.2.1 Population density and distribution

ISBN 978 1 4586 6277 4

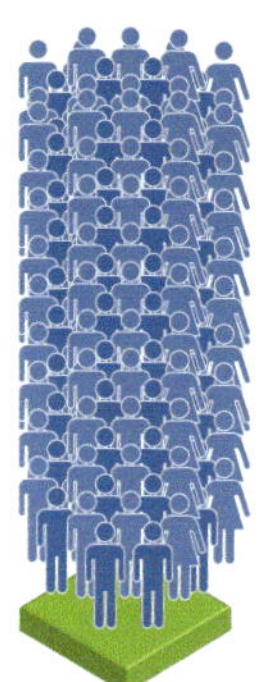

Monaco has 16 754 people per square kilometre

Singapore has 6336 people per square kilometre

Australia has 2.9 people per square kilometre

Mongolia has 1.7 people per square kilometre

9.2.2 What population density looks like

Shanghai, China

Sun City, Arizona, USA

Selenge province, Mongolia

9.2.3 Housing for different population densities

Geo**info**

- 99% of the world's population increases until 2050 will occur in developing countries.
- Every year 20 million infants are born in China.

for instance, 300 million people along India's Ganges River. Densities are lowest where environments are extreme, such as Central Australia and the Arctic. The character and liveability of places like Shanghai in China and remote Ulan Bator in Mongolia is influenced by population density.

Australia's population density

Australia has a small population (23 million) in a large country (7.5 million km^2) making the average population density only 2.9 per km^2. The highest densities occur in the eastern suburbs of Sydney where 8800 people live in each square kilometre. While this is lower than the densities in Asian cities, it is higher than in European cities such as Paris (3550).

At what point does population density start to reduce liveability? Do you think places with high or low population densities are more liveable?

Geo**activities 9.2**

Knowledge and understanding

1 Explain the difference between population distribution and population density.

2 Why will Africa's share of the world population increase in the future?

3 What is a floodplain? How can floodplains support high population densities?

4 Make a list of factors that would affect population distribution and density.

5 Describe two ways population density can influence the character of a place.

Inquiry and skills

6 Refer to 9.2.1.
- a Rank the continents in descending order according to their share of the world population.
- b Which continents will change their ranking by 2050?
- c Describe the pattern of population density in Australia and India.

7 Refer to 9.2.3.
- a List the advantages and disadvantages of living in Shanghai and Mongolia.
- b Briefly explain how population density can impact on the liveability of a place.

8 Research world population (see Geolinks).
- a Name four countries with a similar population to Australia.
- b How many of the top 10 most populous countries are in Asia? Africa? Europe?

Chapter 9

ISBN 978 1 4586 6277 4

9.3 The youngest places on earth

Why do photographs of African countries often show children living in poverty? Children make up 27% of the world's population, but the distribution is uneven. India is home to over 350 million children under 15 years of age—this is more than the population of the USA. In parts of Africa the proportion of children in a population can be as high as 50%. In Chad, one of Africa's poorest countries, the proportion is 45%.

Childhood poverty and hunger

Children have no choice about where they live, but the places they live influence their quality of life.

9.3.1 Countries with the youngest populations, 2014

Country	% aged <15
Niger	50%
Uganda, Chad	48%
Mali, Angola, Somalia	47%
Afghanistan, Gambia, Zambia	46%
Burkino Faso, Mozambique, Timor Leste, Malawi, Congo, Tanzania, Burundi	45%

Source: The World Bank

The places least able to provide the shelter, food, water, health care and education because of poverty have the highest birth rates and largest percentage of children. As a result, millions of children in the poorest countries suffer hunger on a daily basis, many are homeless and child labour is common. The character of these places is different from that of the places where most Australian children live.

Too many babies

Birth rates are determined by the **fertility rate**—the number of children born per woman. Why is this rate so high in the poorest countries of the developing world? In the poorest developing countries:

- children are needed to work on farms and to care for ageing parents
- children die from disease and malnutrition so more children are born to compensate
- low levels of female education, employment and status in society plus a lack of contraception give women little choice in the number of children they have.

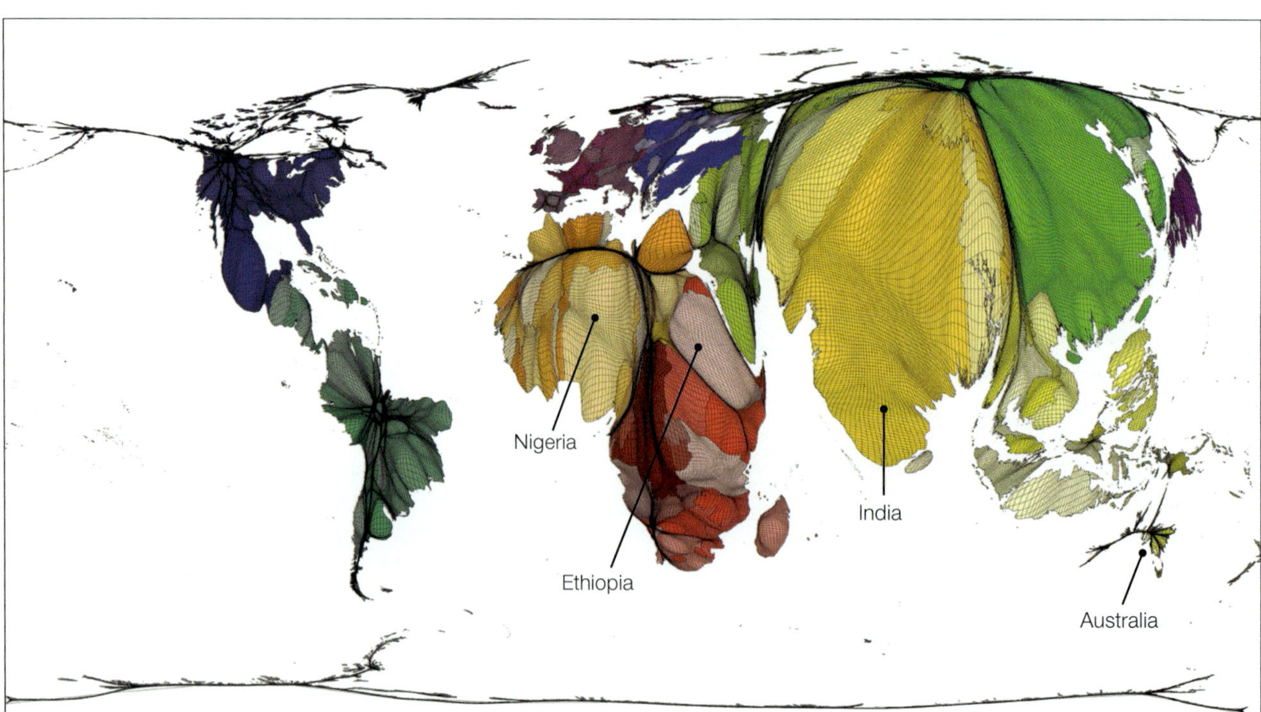

9.3.2 The world of children: the world remapped to show where most children live. In this map each child is given the equivalent amount of land. As a result countries with a large percentage of children appear bigger than they would on a map showing the size of the land.

ISBN 978 1 4586 6277 4

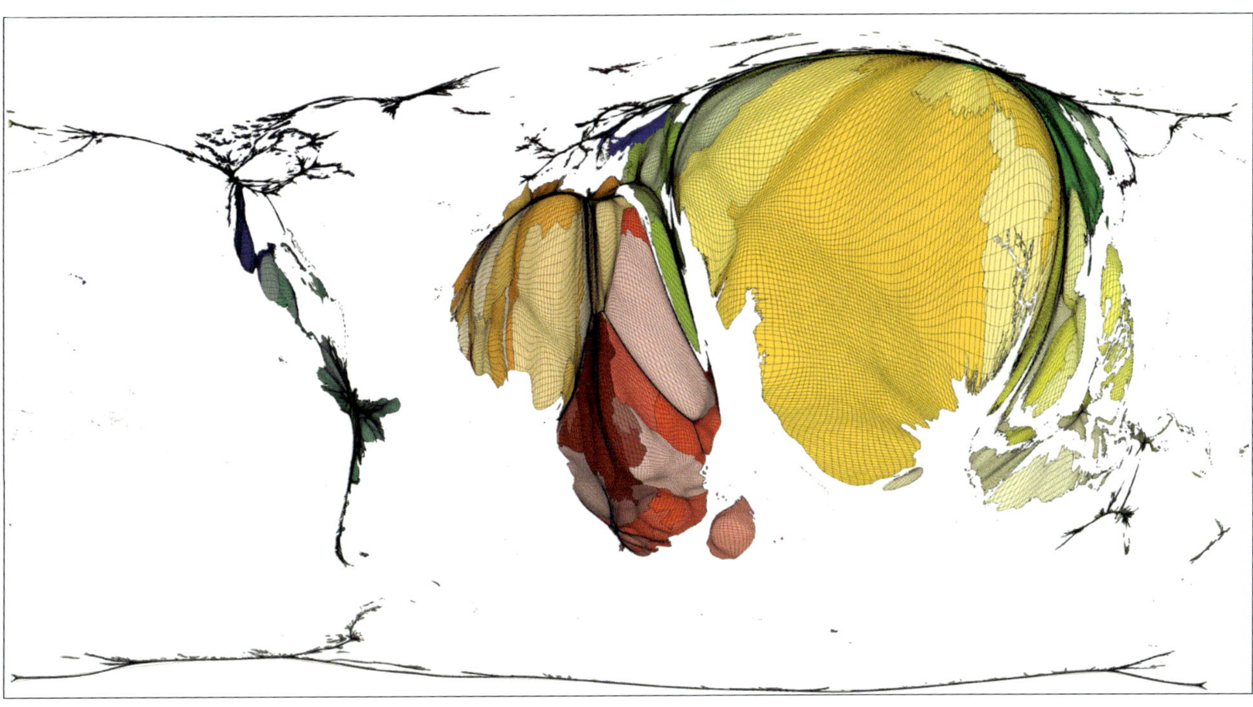

9.3.3 The world of childhood hunger: this map shows the distribution of hungry children. India, Nigeria and Ethiopia are even more out of proportion than in 9.3.2.

To understand the decisions made by poor people about the number of children they have, it is necessary to imagine yourself in their situation.

> Imagine yourself a dirt-poor (male) peasant 50 years ago. Your fields are in the middle of nowhere. Your village has no school, hospital or government services, certainly no pensions. Disease is rampant and security fragile. Ploughing and reaping are done by hand. If the harvest is normal, you usually have enough to go round. In these circumstances, the benefit of an extra pair of hands to gather the harvest outweighs the cost of feeding an extra mouth (which falls on your wife more than you). And when you can no longer work in the fields, your children will be the only ones to look after you. In such a society, all the incentives point to having large families.

Source: www.economist.com

When fertility rates decline, fewer children are born, wealth increases and quality of life improves, making places more liveable for children.

Geoinfo

- Six million children under five die every year as a result of hunger.
- Educated mothers tend to send their children to school, helping to break the cycle of poverty.

Geoactivities 9.3

Knowledge and understanding

1. Describe the uneven distribution of the world's children.
2. Which regions have the highest fertility rates?
3. Explain the link between fertility rates and poverty.
4. Why do women in poor countries have so many children?
5. Suggest changes that might help women have smaller families.
6. Explain how the quality of life for children depends on the place they are born.

Inquiry and skills

7. Refer to 9.3.2 and 9.3.3.
   a. What techniques are used in these maps to show the uneven distribution of the world's children and child hunger?
   b. Why do developed countries appear so small on each of these maps? Give an example.
   c. Are these types of maps an effective way to show comparisons between countries?
   d. How could these maps be improved to increase your understanding of population?
   e. Which map shows the greatest inequalities in liveability for children?
8. Refer to the quoted text and suggest one action a government could take to reduce fertility rates in poor countries. Explain how it would work.

Chapter 9

ISBN 978 1 4586 6277 4

9.4 Babies: the changing character of places

The characteristics of places and the quality of life they offer in the future will depend on the number of babies born in the next 50 years. Global fertility rates have fallen rapidly to an average of 2.5 births per woman. In some countries (e.g. South Korea) the rate is well below this average, while in others (e.g. Mali) the rate remains high. In Germany the number of births has dropped so much that the population is shrinking. German women are being encouraged to have more children. Can declining fertility have both positive and negative impacts on places?

Declining fertility

Factors responsible for the decline in fertility rates include:

- government programs (e.g. China's One Child Policy)
- contraception and family planning
- better health care
- later ages of marriage
- education and employment of women
- changing **social status** of women.

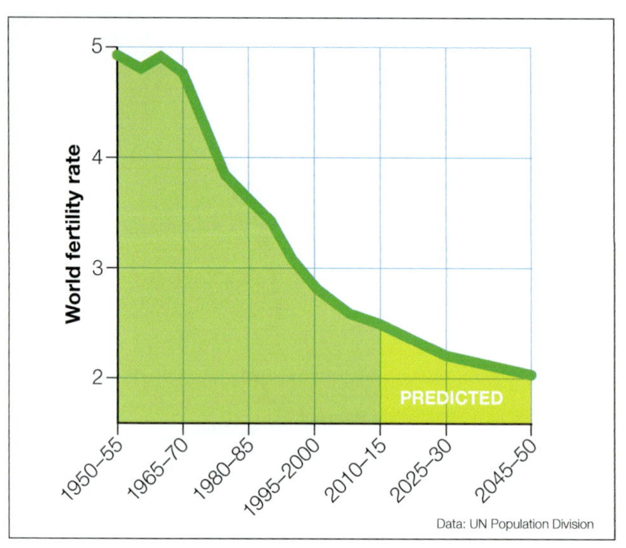

9.4.2 The falling global fertility rate

In South Korea government campaigns used slogans such as 'Give birth without thought and keep living like a beggar' to reduce fertility and improve quality of life. Now the fertility rate is so low the latest campaigns encourage women to have more babies, because there are too many older people in the population.

1970s: Singapore's 'Please stop at two' campaign

Indonesia's 'Two is enough' campaign

9.4.1 Campaigns to promote reduced fertility rates

Brazil's soap opera effect

Brazil is the largest country in South America, with a population of 191 million. The fertility rate has fallen dramatically from 6.3 to well below 2. Evening soap operas, or telenovelas, are believed to be partly responsible. Television followed the spread of electricity across Brazil, interrupting traditional ways of life and introducing families to soap operas. The small, affluent families shown in these soapies

ISBN 978 1 4586 6277 4

9.4.3 The soap opera effect: 90% of female characters in the average novela have just one child or none, which may have influenced Brazilian women to desire smaller families

9.4.4 Brazil's shrinking families

became the dream of many Brazilians. Now many Brazilian communities have smaller, wealthier families enjoying a better quality of life.

Needing more babies

In France and South Korea governments have introduced policies including paid maternity leave, shorter working days and improved access to childcare to boost fertility. A shortage of children is now considered undesirable in many places.

If the world is better off with less people why are countries concerned about declining populations? Does a shrinking population make a place more liveable?

Geo**info**

In 2012 Singapore introduced National Night to encourage couples to have children because the fertility rate dropped to 0.78.

Geo**activities 9.4**

Knowledge and understanding

1 What has happened to global fertility rates?
2 Give three reasons for the decline in global fertility rates.
3 How has South Korea's population policy changed over time?
4 What unusual explanation is given for Brazil's changing population characteristics? How has the life of Brazilian families changed?
5 Use examples to explain how declining fertility can both improve and reduce the liveability of places.

Inquiry and skills

6 Refer to 9.4.1. Explain how these population campaigns might change the character of these communities.
7 Refer to 9.4.2.
   a How did the global fertility rate change between 1950 and 2010?
   b What is the global fertility rate predicted for 2050?
   c Will all countries be at that level in 2050? Explain why.
8 If the world is better off with fewer people why are countries concerned about declining populations? What is the link to liveability? Present group ideas in a class discussion.

Chapter 9

ISBN 978 1 4586 6277 4

9.5 The grey tsunami: benefit or burden?

As a young person who will enter the workforce in a few years time, it is possible that you will pay higher taxes and retire later than your parents. This is because the world's population is ageing. Every month, across the world, over a million people turn 60, while in the USA someone reaches their sixtieth birthday every 7 seconds. Places with higher proportions of older residents will need buildings, facilities and services to make them more liveable for that group, such as ramps, retirement homes and mobility aids.

Children born today have a greater chance of reaching 100 years than any previous generation. The oldest person who ever lived, a French woman named Jeanne Calment, lived to be 122 years and 164 days old. She was a 'super-centenarian'—one of a small group who live longer than 110 years of age. Centenarians are the fastest growing older age group and most centenarians live in wealthier, healthier countries. Okinawa in Japan has the largest number of centenarians in the world.

Places with older people

Better health care means that people are living longer. The average global life expectancy is now 67.2 years, although there are huge global variations. Two-thirds of the 750 million people aged 60 and over live in poorer or developing countries—over half are in Asia. China has approximately 170 million older persons, more than the population of Russia. The smallest numbers of older people are in the poorest African nations where conflict, famine and AIDS keep life expectancies low.

Typical Japanese tofu and fish meal

Older Okinawan woman collects fresh seaweed

Longevity and Okinawa

Okinawa—a group of Japanese islands—is famous for the longevity of its population. Japanese people live longer than anyone else in the world, and Okinawans live longer than their fellow Japanese. Okinawa has the highest percentage of centenarians in the world.

What they eat and how much they eat contributes to their longevity. Common meals include stir-fried vegetables, sweet potatoes and tofu, which are high in nutrients and low in kilojoules. Their philosophy of 'Hare hachi bu' (eat until your stomach is 80% full) helps explain the absence of obesity.

Their low-stress and active lifestyle also explains their longevity. Their slower pace of life, known as 'Okinawa time' is one reason the islanders remain stress free.

When Okinawans migrate to other developed countries they lose their longevity. For instance, the life expectancy of Okinawans living in Brazil is 17 years lower than that of Okinawans in Japan.

ISBN 978 1 4586 6277 4

The 'oldest' places

When families have fewer children, the elderly portion of the population increases. Developed countries such as Japan and Italy have the highest proportions of 60-plus people in their populations. These are also the places where there are smaller numbers of young people. In 2011, 31% of Japan's population were 60 and over. The character of a community changes as it ages.

	15–64	Over 65	Ratio of the non-working and working
2005			1:8
2020			1:4
2050			1:1

9.5.1 Age dependency ratios: 38% of South Korea's population is predicted to be over 65 years of age by 2050

9.5.2 New Zealand's ageing population

Geoinfo

The Japanese government sends greeting cards to people aged over 65 once a month to tackle the social issue of aged loneliness.

Avoiding the burden

When the proportion of older people grows each working person needs to contribute more to support them; for instance, by paying higher taxes. This is known as age dependency. To avoid the burden on working people governments are increasing their intake of young immigrant families and encouraging more babies.

Would you prefer to live in a community with a large number of older or younger persons? What will your community be like when you are 100 years old?

Geoactivities 9.5

Knowledge and understanding

1 What is the main reason the 60-plus age group is increasing?
2 Where are the largest numbers of older people?
3 Where is the largest proportion of older people?
4 What is meant by the term 'grey tsunami'?
5 Why are Okinawans legendary?
6 Why is an ageing population a concern for governments and working people?
7 Explain the connection between the age structure of populations and the character and liveability of places.

Inquiry and skills

8 Refer to the box on Okinawa. Explain the longevity of the Okinawan people.
9 Refer to 9.5.1. Explain how the ageing population in South Korea will impact on the working population in 2020 and 2050.
10 Refer to 9.5.2.
  a How is the character of New Zealand communities portrayed in this cartoon?
  b How can the community character be changed for the future?
11 Use the web 2.0 Gapminder program to track changes in life expectancy over time (see Geolinks).
  a Name four countries with a life expectancy below 50 years. In which region(s) are they located?
  b Name four countries with a life expectancy over 80 years. In which region(s) are they located?
  c Explain the link between income and life expectancy.
  d Use the interactive maps to track the changing life expectancy of one country. Describe the changes and suggest reasons. Identify possible issues for liveability.
12 Investigate the age structure of your community and discuss consequences for the liveability of different age groups.

Chapter 9

ISBN 978 1 4586 6277 4

9.6 Skill: population profiles

A population profile is a visual image of population in the form of a graph. Geographers use these images to compare the population characteristics of places and study how communities change over time. Population profiles are a valuable tool for interpreting population trends.

A population profile

Population profiles are specialised graphs showing the age and sex structure of a population. They incorporate:

- a double bar graph with a central axis showing age groups in five-year cohorts (age groups)
- female population (shown on the right) and male population (on the left)
- the number or percentage of people for each age category.

The use of percentages allows the comparison of countries with different sized populations. The age groups below 15 and above 60 are generally known as the **dependent population**. These groups are often not working and depend more on governments for the provision of income support and facilities (e.g. aged pensions, retirement homes and childcare centres). The characteristics of places and the facilities they provide will influence their liveability for different age groups.

Population pyramid

Population profiles were previously known as population pyramids. The pyramid shape has a wide base and a narrow top representing a large percentage or number of people in the younger age groups, decreasing to a small percentage or number in the oldest age groups. Places where birth rates are high and average life expectancy is low have pyramid-shaped population profiles. These places are in the developing world (e.g. Kenya).

Shapes and imbalances

Population profiles change over time. Most of the developed world (e.g. Japan) and some developing countries (e.g. China) no longer have a pyramid population structure.

Gender and age imbalances are clearly visible on population profiles. An imbalance between males and females is a growing issue in China and India where there is a preference for boys over girls. **Gender imbalance** can result from war, social preferences or government policies. A shortage of girls leads to female kidnapping and makes some places more dangerous for women.

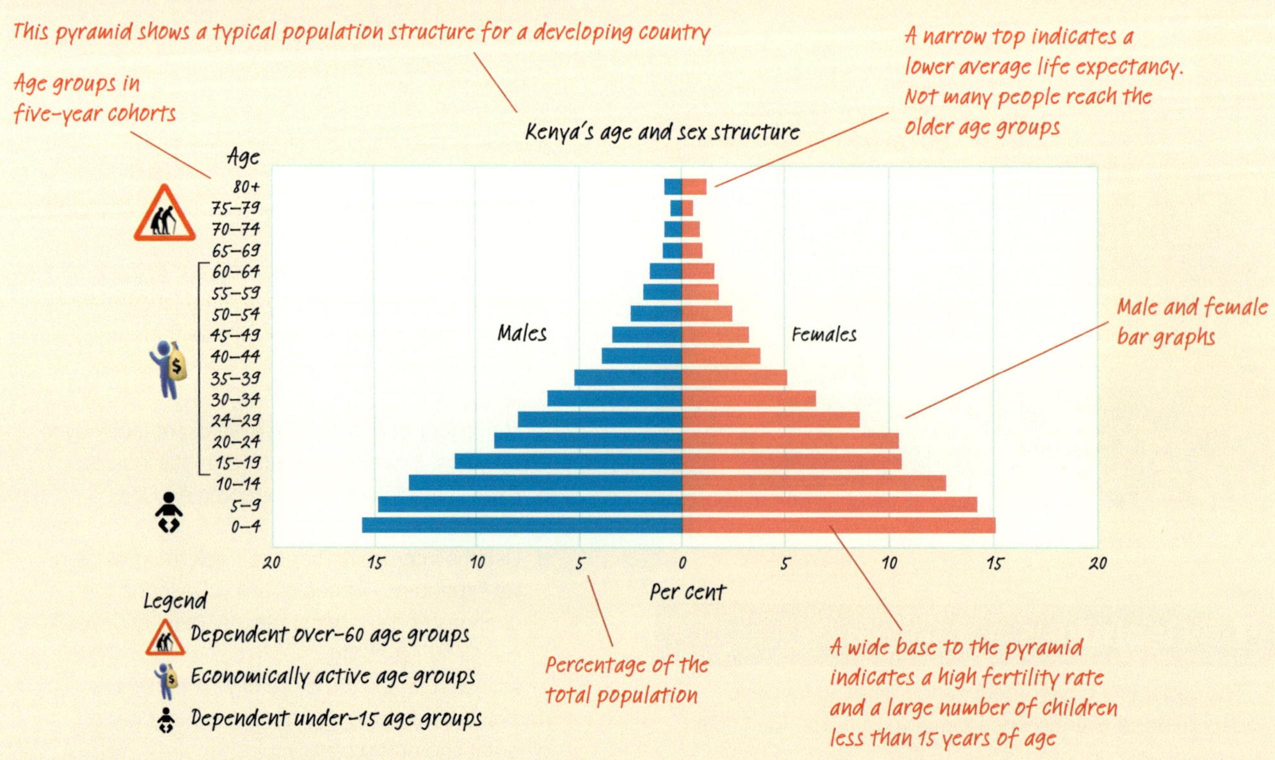

9.6.1 Features of a population profile

ISBN 978 1 4586 6277 4

9.6.2 Gender imbalance: in India there are far fewer girls born each year than boys. Some families use ultrasound technology to determine the gender of fetuses and then abort the females.

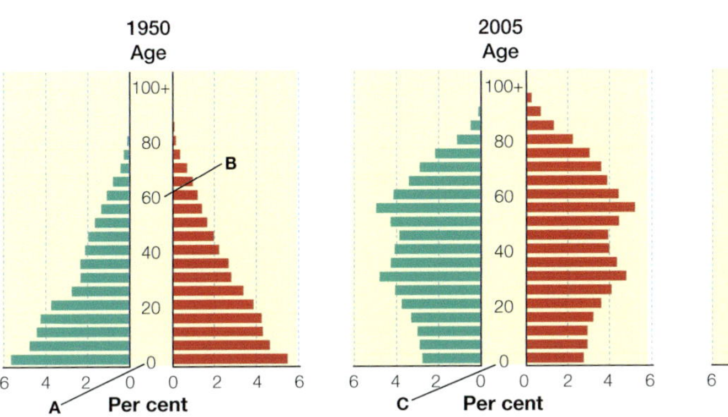

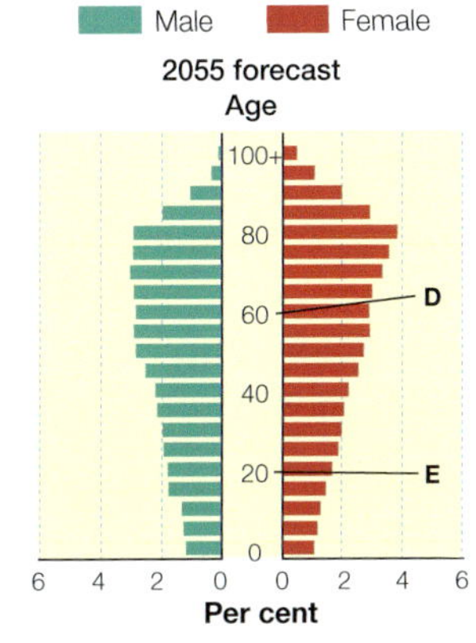

9.6.3 Japan's changing population profile

Geoinfo

A natural gender ratio is around 105 boys per 100 girls. In 2011, China had 118 newborn boys for every 100 girls.

Geoactivities 9.6

Knowledge and understanding

1. What does the lowest bar of a population profile represent?
2. What is the difference between the bars on the left and those on the right?
3. How would you figure out the percentage of people in Kenya in 2010 aged 10 to 14 in 9.6.1?
4. Which age group in 9.6.1 has the fewest people?
5. What does the population profile of a developing country look like?
6. Why do population profiles change over time?
7. What is meant by the term 'dependent population' when referring to population profiles?
8. List the information that can be interpreted from population profiles.
9. Why would population profiles be a useful tool for governments?

Inquiry and skills

10. Refer to 9.6.1.
    a. What percentage of Kenya's population is under 5 years of age?
    b. Kenya's population is approximately 41 million. How many people are aged 35 to 39?
    c. What are the three largest dependent age groups in Kenya? Why does this matter?
11. Refer to 9.6.2. How would the gender imbalance in India look on a population profile?
12. Refer to 9.6.3.
    a. Describe how the shape of Japan's population profile has changed over time.
    b. What is happening at A, B, C, D and E on the Japanese population profiles?
    c. What population issues will Japan face in 2055?
13. Compare the challenges facing the governments of Niger and Germany in relation to their population structures.
    a. Locate population profiles for Niger and Germany.
    b. Identify population issues for each country.
    c. Research data about those issues.
    d. Analyse the data to make a link to liveability.
    e. Report your findings as a multimedia presentation.
14. View an animated version of Australia's population profile (see Geolinks).
    a. Describe the projected change in Australia's population over time (1971–2056).
    b. Compare the population structure of your state or territory with one other. List the main differences.

ISBN 978 1 4586 6277 4

9.7 Migration, liveability and change

Will you always live in Australia, your current state or territory, your town? What would make you move and where would you go? Migration changes the distribution of the world's population and changes the character of places. Most people move to places they believe will be more liveable.

Today, over 215 million people do not live in the country of their birth—more Chinese live outside of China than French people live in France. It is predicted that by 2050 the number of international migrants could reach 405 million or 7% of the world's population.

Assisted by modern communications and transport, distance is no barrier to moving across the world.

Why migrate?

Migration can be voluntary or forced, legal or illegal, international or internal, rural or urban. If places become unliveable or other places offer a better quality of life, people consider moving.

Conditions that force people from a location are known as *push factors*. Factors that attract people to another location are called *pull factors*.

9.7.1 Push and pull migration factors

Push factors	Pull factors
Lack of jobs and resources, poverty	Employment opportunities
Lack of facilities (e.g. schools, hospitals and shops)	Facilities for health care, education and entertainment
War and natural disasters	Political stability, safe environments and aid
Lack of freedom	Freedom of speech, movement and religion
Human rights abuses	Human rights protections
Boring lifestyle	Exciting lifestyle and opportunities
High cost of living	Affordable living
Inhospitable climate	Better climate
Climate change effects	Higher land, new farmland

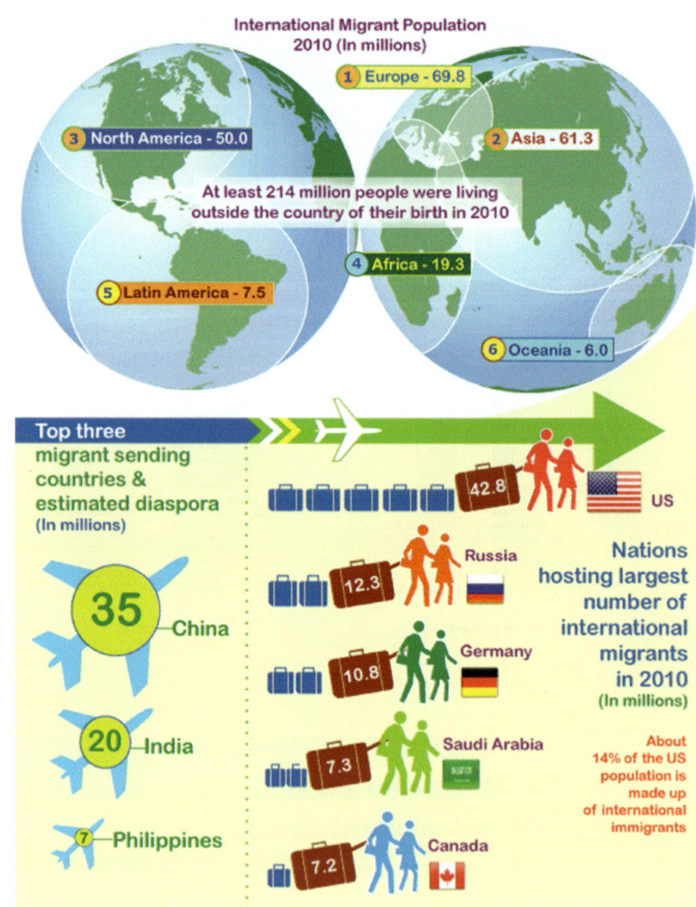

9.7.2 Global migration

Spatial patterns

Each year thousands of migrants cross international borders in search of work and a better life (e.g. from Mexico to the USA). Economic migrants from countries such as the Philippines work to send money (**remittances**) home to their families.

Environmental migration is predicted to increase with climate change and the depletion of resources, such as water. The liveability of low-lying coastal places and those short of water is under threat.

9.7.3 Top 5 remittance receiving countries, 2014

Country	Remittances received, USD billion
India	70
China	64
Philippines	28
Mexico	25
Nigeria	21

Source: World Bank

ISBN 978 1 4586 6277 4

Migration changes places

Most international migrants move to large cities in developed countries. Internal migration is mainly rural to urban. Developed countries and cities offer opportunities for work and a better quality of life. There are positive and negative impacts on the destination and source of migrants because migration changes the character of places.

9.7.4 Potential impacts of migration

	Source country	Destination country
Positive impacts	• Reduced population • Reduced birth rates • Income from remittances reduces poverty and improves standard of living • New skills when migrants return	• Reduces declining and ageing populations • Average age of population decreases • New foods and cultures • Solves labour shortages • Increased racial tolerance
Negative impacts	• Loss of workers for farms • Brain drain • Average age of populations increases • Parents separated from children and relatives	• Racial tension • Exploitation of cheap labour • Pressure on housing and infrastructure • Overpopulation of cities

Migrating from Puebla to New York City

The New York City area is now home to over half a million *poblanos*, natives of the Mexican state of Puebla. Ricardo and Aldea got to New York by crossing the 49 °C Sonoran Desert. They work 70-hour weeks for less than the minimum wage.

Migration has had a profound impact on villages in Puebla, such as Piaxia. Most of the 1600 current residents of Piaxia are either children or elderly. The mayor claims that 'maybe three out of four of my constituents live in New York'. The hundreds of millions of dollars sent back each year are having a dramatic effect on rural communities in Puebla. Forty years ago virtually all of the houses were made of palm thatches. Now they are mostly brick and concrete. The towns also have new restaurants, taxis, video arcades, cybercafes and newly paved streets—all made possible by remittances.

Most youth consider the prospect of migration—migrating to jobs in New York has become the norm.

Source: http://geo-mexico.com

Geoinfo

Countries with a high percentage of migrants are Qatar (87%) and the UAE (70%). Low-percentage countries include Nigeria (0.7 %) and Indonesia (0.1%).

Geoactivities 9.7

Knowledge and understanding

1 Define each of these types of migration in your own words: international, internal, voluntary, forced, illegal, legal, rural and urban.

2 Explain the difference between push and pull factors. Refer to liveability and use examples.

3 What are remittances and why are they important?

4 How are refugees different from economic migrants?

Inquiry and skills

5 Refer to 9.7.1.
Which push factors would be most relevant to the following situations?
 a a farming family in Somalia
 b a teenager living in rural Australia
 c an unemployed father of eight living in the Philippines.

6 Refer to 9.7.2.
 a Describe the global distribution of migrants in 2010. Suggest reasons for this pattern.
 b List the three main source and destination countries for migrants in 2010.
 c Research current statistics in international migration. Create your own infographic using a web 2.0 tool.

7 Refer to the article on migrating from Puebla. Work in groups to investigate current trends in the migration of one of these groups:
- Mexicans into the USA
- Africans into Europe.

Present a digital report. Include statistics, maps, graphs, cartoons, images and key facts. In your answer refer to places and liveability.

8 Refer to 9.7.3, 9.7.4 and the article.
 a List the positive impacts of Mexican migrants to New York on rural towns in Mexico.
 b Suggest positive and negative impacts of Mexican migrants on New York City.
 c 'The positive impacts of migration and remittances on the livability of places outweighs the negative impacts.' Discuss with reference to source and destination countries.
 d How has the character of Piaxia changed?

ISBN 978 1 4586 6277 4

9.8 Forced to flee: unliveable places

For millions of people, the world is an unsafe place. Over 43 million people live where they do because they were forced to leave their homes and seek refuge in a safer location. They are displaced people, of which only 15 million are refugees.

The recent crisis in North Africa has seen thousands of people lose their lives while crossing the Mediterranean Sea seeking refuge in Europe.

Seeking refuge

Although all displaced people seek safety and protection, they are legally divided into three groups:

- *Refugees*—seek refuge in another country due to persecution or a genuine fear for their life and freedom. The largest refugee populations come from developing countries (e.g. Afghanistan) and seek refuge in neighbouring countries. Under the UN Refugee Convention the **United Nations High Commissioner for Refugees (UNHCR)** has a legal obligation to protect refugees.
- *Asylum seekers*—seek protection in another country but their claim for refugee status is incomplete. The UN Refugee Convention and UN Declaration of Human Rights state that people have a right to seek asylum. It is not illegal. In 2010 people from 200 countries lodged over 350000 applications for asylum in 44 developed countries. Over 90% of asylum seekers arriving by boat in Australia prove to be genuine refugees.
- *Internally Displaced Persons (IDPs)*—are forced to seek refuge but within their own country. Their own government has a responsibility to protect and assist them but often fails to do so. Globally, the number of IDPs is about twice the number of refugees. Armed conflict, human rights abuses, development projects and natural disasters can cause internal displacement. Haiti's 2010 earthquake made over a million people homeless. Too poor to move or pay rent, 500000 Haitians remain in camps set up after the disaster.

Most conflict and disaster zones (e.g. Iraq and Somalia) produce a mix of refugees, IDPs and asylum seekers.

9.8.1 The requirements for refugee status under the UN Refugee Convention

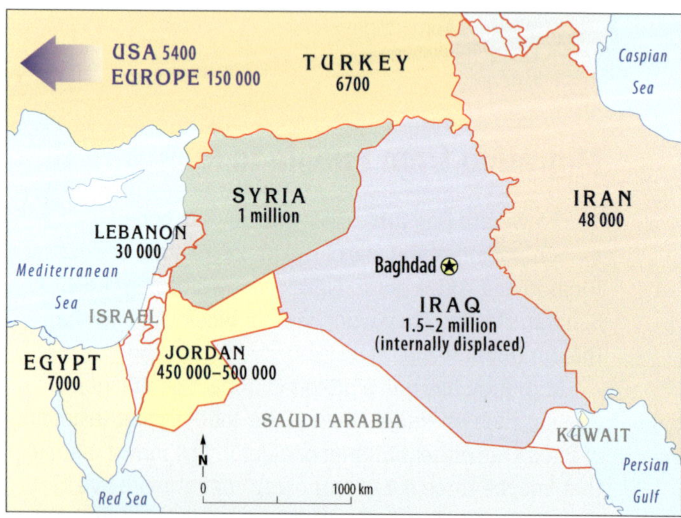

9.8.2 Conflicts can create both refugees and IDPs

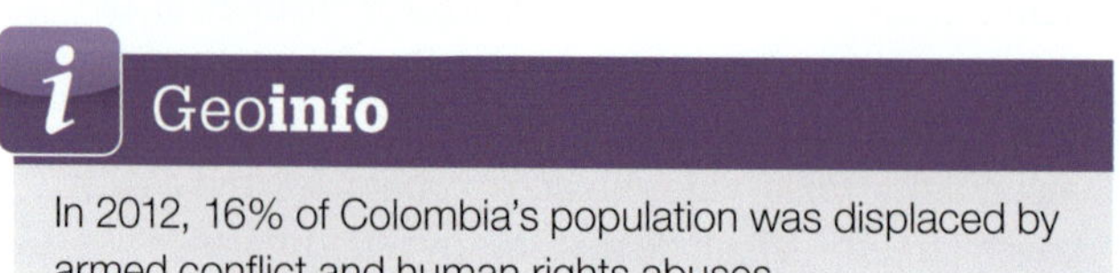

Geoinfo

In 2012, 16% of Colombia's population was displaced by armed conflict and human rights abuses.

ISBN 978 1 4586 6277 4

9.8.3 Refugees head for different places for a variety of different reasons

Cities are the new refugee camps

Despite popular belief, most refugees no longer live in refugee camps. Instead:

- eighty per cent of refugees live in their *neighbouring countries*.
- fifty per cent are living in *urban areas* (e.g. Nairobi and Kenya).
- less than 30% live in *refugee camps* (e.g. Dadaab—see 7.6).
- most IDPs live with *family or friends* in urban and rural areas or in special camps.

Going home

Most displaced people hope to eventually return to their homes, but for many it will never be safe to return. They may face years of uncertainty before finding a new permanent home—somewhere more liveable where their quality of life will be improved. The arrival of large refugee populations impacts on the economic, social and environmental characteristics of places. What do you think those impacts might be?

Geo**activities 9.8**

Knowledge and understanding

1 Explain the legal differences between refugees and IDPs.
2 Is it legal to seek asylum in another country?
3 Why do you think most asylum seekers apply to settle into richer countries?
4 Explain the connections between liveability and refugees.
5 In groups discuss why most refugees come from developing countries.
6 Draw a mind map to summarise the key points about people who have fled their homes to seek refuge in another place OR use a web 2.0 mind mapping tool such as bubbl.us.

Inquiry and skills

7 Refer to 9.8.1. Which of the requirements for refugee status do you think might be hardest to prove? Explain why.
8 Refer to 9.8.2.
  a Use statistics and examples to describe the distribution of Iraqis since 2003.
  b Calculate the percentage of Iraqi people living away from their homes who are IDPs.
  c Suggest reasons for the large number of refugees in Iraq's neighbouring countries.
9 Search the internet for an up-to-date infographic on global refugee patterns. In small groups develop a set of six questions to be answered by your classmates. At least two questions must relate to liveability.

ISBN 978 1 4586 6277 4

9.9 Population and human development

Just 20% of the world's population consumes 80% of the world's resources. This inequality means millions of people are living in poor, unhygienic, dangerous, overcrowded and barely liveable communities. The terms 'developing' and 'developed' are used when referring to countries and regions of the world. GDP per capita is used to compare levels of development. For instance, according to World Bank statistics in 2014, Australia's GDP per capita was $US61 887, compared to Ethiopia's $US567.

Measuring human development

Human development is more than a measure of wealth. It is about people's quality of life, including health, literacy, education level, employment, life expectancy, freedom and equality. The United Nations **Human Development Index (HDI)** ranks countries using measurements of life expectancy, literacy, education and GDP per capita. Countries are ranked from 0 to 1. The 10 countries with the lowest HDI in 2014 were all in Africa. People living in poverty often migrate to developed countries seeking opportunities to improve their lives.

9.9.2 The Human Development Index 2014

Top 10 ranked countries in 2014	HDI	Bottom 10 ranked countries 2014	HDI
1 Norway	0.944	**178** Mozambique	0.393
2 Australia	0.933	**179** Guinea	0.392
3 Switzerland	0.917	**180** Burundi	0.389
4 Netherlands	0.915	**181** Burkino Faso	0.388
5 United States	0.914	**182** Eritrea	0.381
6 Germany	0.911	**183** Sierra Leone	0.374
7 New Zealand	0.910	**184** Chad	0.341
8 Canada	0.902	**185** Central African Republic	0.341
9 Singapore	0.901	**186** Democratic Republic of the Congo	0.338
10 Denmark	0.900	**187** Niger	0.337

Source: http://hdr.undp.org/en/

Geo**info**

By 2015, another 200 million people will be added to the populations of the world's least developed countries, making future development difficult.

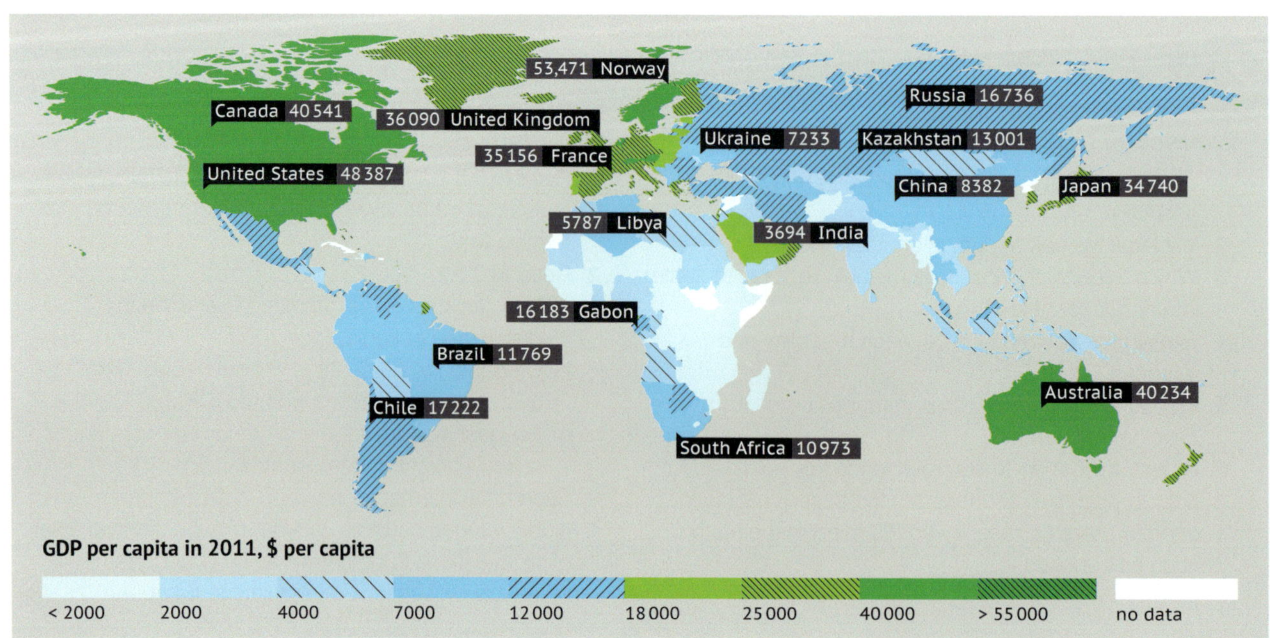

9.9.1 Countries mapped by GDP per capita in 2011

ISBN 978 1 4586 6277 4

BUT STILL TODAY

1 in 4 women around the world cannot read this sentence.

Girls make up 53% of the children out of school.

98% of people who can't read live in developing countries.

9.9.3 The girl effect

Improving development and liveability

The Millennium Development Goals focused on reducing global poverty by improving the health of the poorest people by 2015. From when they were implemented until 2014, global poverty had been halved and safe drinking water provided to 90% of the world's population. Each improvement has made communities healthier and safer to live in. In **sub-Saharan Africa** extreme poverty dropped 28% between 1990 and 2015; however, those living on less than $1.25 a day remains high at 41% with many still without access to electricity and limited access to education and health facilities. This part of Africa ranks poorly on liveability.

One strategy: the girl effect

Women are central to improving levels of human development in poor countries. When women are given equality, power and education there are improvements in family education, health, income and nutrition. This results in falling fertility rates, slower population growth and reduced poverty. This is known as the **girl effect**—educated women have fewer children because they provide better health care for each child. Given assistance with agriculture, women will increase food production in rural areas and invest profits in the education and health of their children.

Geo**activities 9.9**

Knowledge and understanding

1. Describe the uneven use of the world's resources.
2. How does human development affect people's quality of life and the liveability of places?
3. Explain the 'girl effect'. Why is it seen as a solution to poverty?
4. What will happen if future human development is not sustainable development?

Inquiry and skills

5. Refer to 9.9.1 and 9.9.2.
   a. What GDP per capita amount divides the world into developed and developing countries?
   b. Which two continents contain the least developed countries?
   c. What different information does the GDP per capita map and the HDI table give us?
6. Refer to 9.9.3.
   a. Why do women in Africa have a lower quality of life than men?
   b. Draw a mind map to summarise the effects of educating women.
   c. What is the impact of empowering women in agriculture?
   d. Using the internet, investigate the girl effect (see Geolinks). Contribute findings to a class discussion and create a mind map.

Chapter 9

ISBN 978 1 4586 6277 4

9.10 Interactive tools for geographers: development

Geographers investigate the characteristics of places and make comparisons and judgements, such as which place offers a better quality of life. Sorting through statistics, selecting relevant data and processing it can be difficult and time consuming.

Interactive tools such as Gapminder are web 2.0 tools that turn statistics into visual images including maps and graphs. Data comes from reliable sources, such as the World Bank and the United Nations. In some interactive tools, animations are used to show change over time. Visual images make comparing places easy and interesting. The value of interactive tools comes from:

- choosing variables (e.g. population size and life expectancy)
- comparing the characteristics of places (e.g. Australia and Nigeria)
- visualising change over time (e.g. HDI for India)
- understanding relationships (e.g. income and birth rates)
- testing ideas (e.g. infant mortality will fall when education improves).

Gapminder World

Gapminder World creates scatter graphs using factors that affect human development, such as infant mortality and CO_2 emissions. You can mix and match the data, change the year, and watch the graphs change over time. Passing the cursor over each bubble will show the name of a country. The visuals can be included in student presentations.

StatPlanet

StatPlanet creates graphs and choropleth maps, which show values using colours or shading. Countries and regions can be compared for quality of life and liveability.

Diamond Trees and Worldshapin

Two new tools use statistics to create interesting shapes:

- The *HDI Diamond Tree* uses diamond shapes to represent the HDI. The trunk of the tree represents the total HDI. The branches (diamonds) show the value of the individual HDI components.

When bubbles group along a line there is a clear relationship between the two features

Circles represent population size

Choose the feature you want on each axis

Interactive component shows each country's location change over time (press play)

Use colour to represent different global regions

Choose countries you want to see or track

9.10.1 A Gapminder graph showing life expectancy and income

ISBN 978 1 4586 6277 4

- *Worldshapin* creates a unique shape for each continent or country based around six measures of development—education, population, living standards, health, workplace equality and CO_2 emissions.

Both of these interactive tools allow comparisons to be made between places, over time and for individual indicators of development.

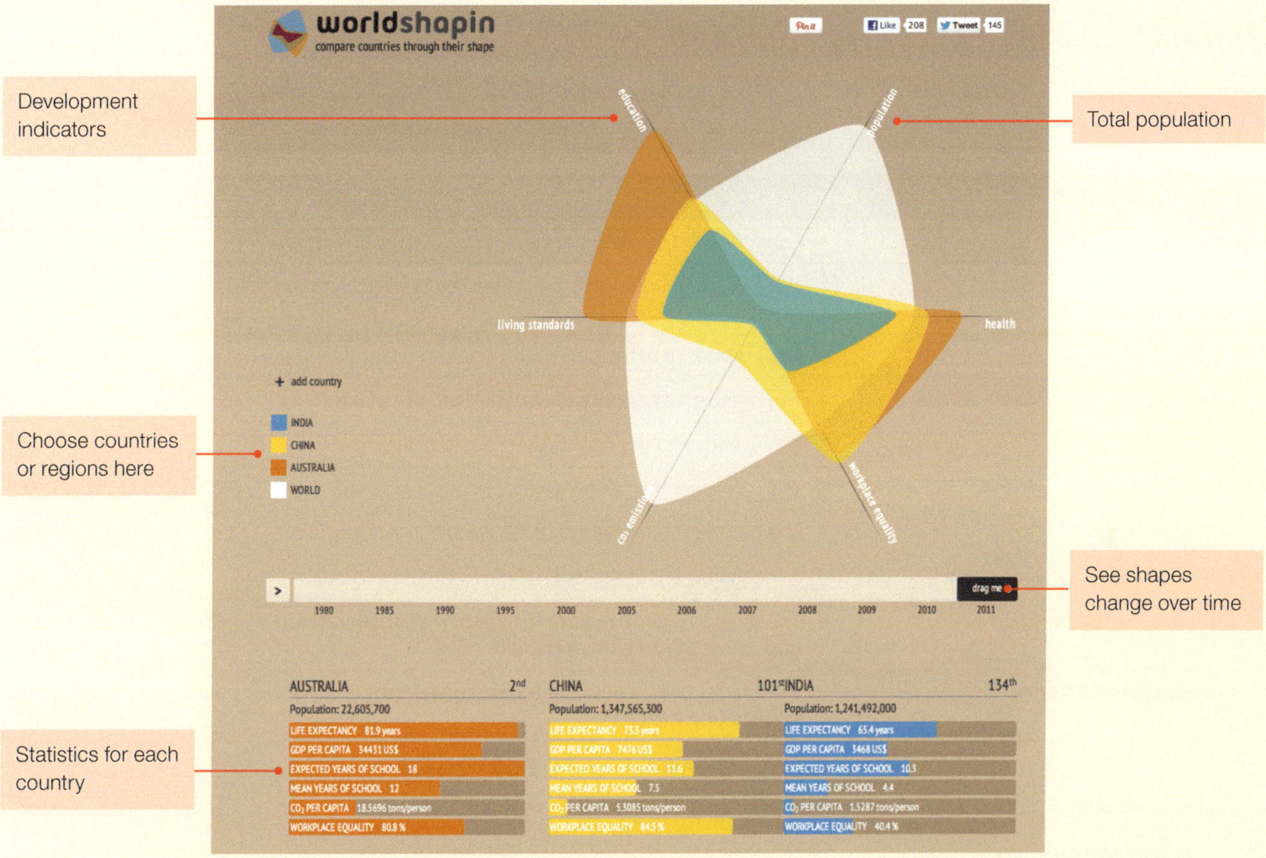

Development indicators

Total population

Choose countries or regions here

See shapes change over time

Statistics for each country

9.10.2 Worldshapin visual comparing Australia, China and India to the world

Geo**activities 9.10**

Knowledge and understanding

1. Name two sources of statistics used by tools such as Gapminder.
2. What features do Gapminder, Statplanet, HDI Diamond Tree and Worldshapin have in common?
3. What is the main aim of both Worldshapin and HDI Diamond Trees?
4. Discuss the advantages of interactive tools for geographers.
5. Discuss any disadvantages of these tools.
6. Explain how these tools could be used to compare the liveability of different places.

Inquiry and skills

7. Refer to 9.10.1.
   a. Name the two variables in the graph.
   b. What type of graph is this?
   c. Does life expectancy increase as income increases? Support your answer.
   d. What is the life expectancy and income per person in Swaziland?
   e. Is Swaziland a developed or developing country?
8. Refer to 9.10.2.
   a. How does Australia compare to India and China in the areas of health and education?
   b. How does India compare to the world in terms of workplace equality and living standards?
   c. Is this type of visual representation of statistical data useful? Explain.
9. Refer to the Gapminder World or worldshapin websites to develop your skills in the use of interactive tools, and to compare the development of different places.
   a. Create one image (e.g. map, graph or shape) using your choice of statistics relevant to liveability.
   b. Import your image into a Word document. Describe what your imported item illustrates about liveability. Create a suitable title for your document.

ISBN 978 1 4586 6277 4

Geo**think**

Population and liveability roundup

1 Refer to 9.11.1. Match the letter for each population concept terms with the matching definition.

9.11.1 Population concepts and definitions

Concept	Definition
A Birth rate	**1** A person who moves to another place for work
B Death rate	**2** The growth of population calculated by subtracting the death rate from the birth rate
C Natural increase	**3** The number of births per 1000 people in a population
D Demographic transition	**4** The number of people per square kilometre of land
E Population growth	**5** The average number of children per female
F Fertility rate	**6** The movement of people to a new location to live
G Migration	**7** The number of deaths per 1000 people in a population
H Human Development Index (HDI)	**8** A measurement of the level of development in a nation
I Population density	**9** The change over time from high birth and death rates in a population to low birth and death rates
J Population distribution	**10** The spatial distribution of a population across an area such as a country or a continent
K Economic migrant	**11** The average length of a person's life at birth
L Life expectancy	**12** An increase in the number of people

9.11.2 Population growth has an impact on liveability and quality of life for people in developing countries

ISBN 978 1 4586 6277 4

2 Refer to 9.11.1 and the text in this chapter.
  a List the factors that cause populations to decline.
  b What factors lead to population growth?
  c What is the impact of population growth on population density?
  d What are the impacts of population growth on the natural environment?
  e What do you think the term 'commons' refers to in 'tragedy of the commons'?
  f What would you add to the mind map to link population and liveability?

People forced to move

3 Refer to 9.11.3 and answer the questions on the photograph.
4 Use the theme 'People move' to create a mind map about migration, using words, images and colours. Include references to the liveability and levels of development in your map.

9.11.3 People forced to move

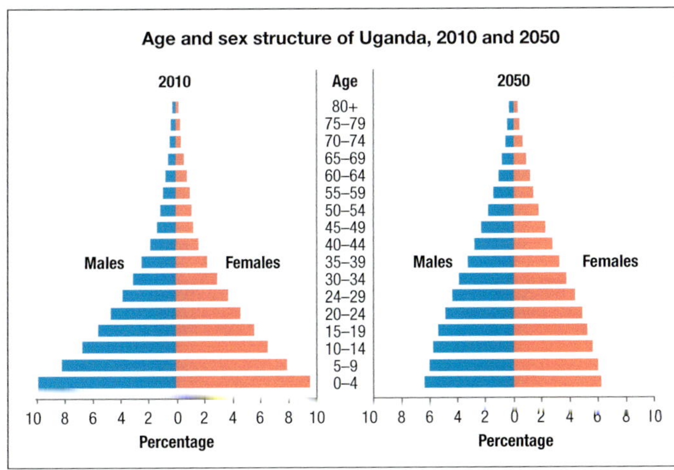

9.11.4 Population profile for Uganda

Population profiles

5 Refer to 9.11.4.
  a What percentage of Uganda's population was aged less than 15 years in 2010?
  b What percentage of Uganda's population is predicted to be less than 15 years by 2050?
  c Suggest reasons for the change in population of those less than 15 years of age.
  d Give evidence to explain how the life expectancy of people in Uganda is increasing.
  e What is happening to the size of the working age population in Uganda?
  f What evidence from the graphs suggest that liveability in Uganda is improving?

Chapter 9

ISBN 978 1 4586 6277 4

chapter 10

Strategies to enhance liveability

'The future is not somewhere we are going, but something we are creating.'

Professor Ian Lowe

Geovocab

eco-city: sustainable city

food security: availability and access to sufficient nutritious food now and in the future

gentrification: when decayed areas are improved and higher-income people move in

green infrastructure: green spaces (e.g. parks and gardens)

inclusivity: including people of all ages and backgrounds

integrated transport: types of transport linked together as a system

mass transit: transport that moves large numbers of people

microcredit: small loan given to poor people to help them move out of poverty

resilient: able to recover from changes

retrofitting: adding new technology to something old

ISBN 978 1 4586 6277 4

The liveability of a place can be enhanced when strategies are implemented to address contemporary issues affecting people's wellbeing, such as their health and happiness. Strategies vary from small to large scale, simple to complex, local to global, and from simple technology to sophisticated new technology. Strategies need to be relevant and appropriate to individual situations, and they need to be sustainable. Successful strategies originating in Europe have become models for developing liveable, sustainable communities.

Surrounded by medium-density housing, this plaza market place in Paris is a happy meeting place for local families. With no cars and high walkability, it is enjoyed by people of all ages

sustainability: using Earth's resources to ensure their availability in the future
urban decay: when part of a city falls into disrepair or misuse
urban renewal: developing an urban area that has deteriorated over time
zero carbon: does not produce carbon emissions from human activities

Think, puzzle, explore

- **Place** What does it mean to enhance liveability in different places?
- **Space** How can your street, neighbourhood and community be made more liveable?
- **Interconnection** How is sustainability and liveability interconnected?
- **Sustainability** Why do European countries lead the world in liveability and sustainability?
- **Scale** Why are strategies on a local scale most effective?
- **Change** What strategies have been implemented in different places across Australia to enhance liveability and sustainability?
- **Environment** What changes can be made to improve the environmental quality of places?

Geo**skills** in focus

- **Planning** strategies that enhance liveability in Australia and Europe
- **Evaluating** primary and secondary data for accuracy and reliability
- **Researching** using web 2.0 tools and geographical information systems
- **Communicating** using maps, graphs, statistics, photographs, interviews, surveys and ICT in geographical presentations
- **Reflecting** on ideas that might enhance liveability and sustainability in different places

ISBN 978 1 4586 6277 4

10.1 Enhancing liveability and sustainability

During summer in Paris, the Georges Pompidou expressway, which runs through the city, is closed for a month. It is converted into Paris Plages (Paris Beach), with palm trees, sand, umbrellas, a climbing wall, swimming pools, bicycle rentals, food stalls and entertainment. Paris Plages became an annual event after attracting 2 million people in the first year. This success led to:

- the permanent narrowing of the expressway on the right bank of the Seine River to make way for pedestrian walkways, cafes and bars
- a car-free zone on the left bank where parks, floating gardens, sports courts, restaurants and flower markets replaced the freeway.

The French Prime Minister stated, 'It's about reducing pollution and automobile traffic, and giving Parisians more opportunities for happiness.' Giving the river back to the Parisians is an example of strategies implemented to make the city a better place to live.

Appropriate strategies

Improving the quality of people's lives requires actions that are appropriate and relevant. There is no single strategy to create liveable, sustainable places.

Most strategies have multiple purposes and benefits. Cycling, car-free zones, walkable streets, zero-carbon programs, recycling and mass transit transport systems are improving cities across Europe.

At the same time, slum clearance and sanitation projects are making a difference in some of the

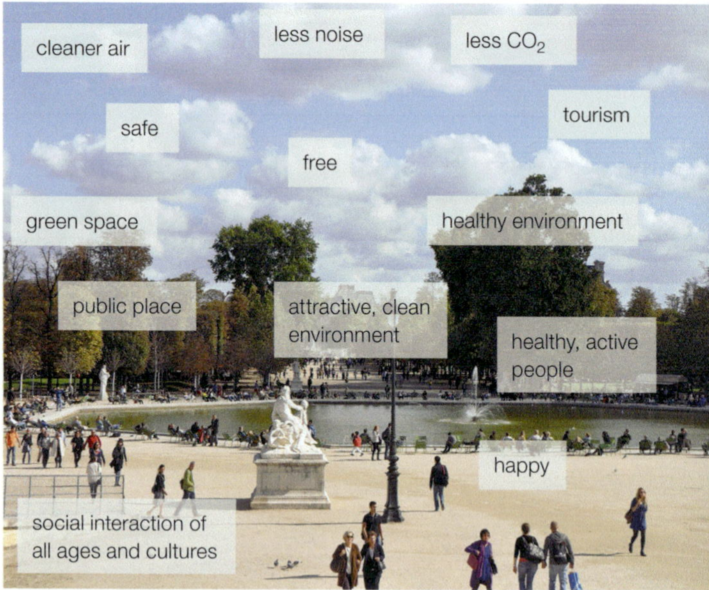

10.1.2 Places for people

10.1.1 Paris: people not traffic

ISBN 978 1 4586 6277 4

world's poorest rural and urban communities. In Kenya, PeePoo disposable toilet bags have improved sanitation, reduced disease, provided fertiliser for farming, increased **food security** and made places safer for females—all for a few cents per bag.

Old and new

New communities are being created to be liveable and sustainable. They use the latest technology and often allocate millions of dollars to investment, such as the sustainable Masdar City in the UAE.

Improvements are being made to existing places, such as:

- *introducing new practices and behaviours* (e.g. cycling to work in Copenhagen)
- *building facilities and infrastructure* (e.g. Australia's National Broadband Network, or NBN)
- *adapting what is already there* (e.g. retrofitting homes with solar panels in Seattle, USA)
- *replacing old with new* (e.g. urban renewal projects, such as Docklands in Melbourne).

10.1.3 Futuristic transport for carbon-neutral Masdar City

The language of liveability

Governments, organisations, businesses and individuals use a variety of terms when referring to strategies to improve liveability and sustainability. This is the new language of planning for a better future.

Walkability, connectedness, **inclusivity**, **retrofitting**, **integrated transport**, energy efficiency, low carbon, zero carbon, ecological footprint, **eco-city**, 20-minute neighbourhood, placemaking, places for people, **mass transit**, mobility, accessibility, public places, green spaces, green living, pocket parks, social interaction, One Planet Living, alternative energy, resource conservation, pedestrian-friendly, **resilient:** this is just some of the language of liveability.

Geo**info**

The NBN will provide greater access to specialist health care for people in rural Australia, and access to Virtual English Tuition for newly arrived migrants and refugees.

Geo**activities 10.1**

Knowledge and understanding

1. Identify one aspect of the Paris Plages project that improves the liveability of Paris.
2. Explain one aspect of the Paris Plages project that promotes sustainability.
3. Why do you think there is no single strategy for creating liveable and sustainable places?
4. Develop a class dictionary from the liveability and sustainability language list. Work in pairs to develop definitions. Reach a consensus through class discussion. Produce a classroom display of terms and definitions.
5. Draw a mind map to show the ways places can be made more liveable and sustainable.

Inquiry and skills

6. Refer to 10.1.1.
   a. Name two places where Paris Plages are located.
   b. Use an online French to English dictionary to redraw a version of the advertisement for Paris Plages in English.
   c. What evidence suggests that the Paris Plages were safe places for people to enjoy their summer?
7. Refer to 10.1.2.
   a. Which benefits of 'places for people' in Paris will reduce the city's ecological footprint?
   b. Rank the top five benefits according to your own ideas about liveability. Compare your rankings with those of another student and discuss your choices. Modify your answer if needed.
8. Refer to 10.1.3 and an atlas.
   a. Where is Masdar City located?
   b. Explain how new technology aims to improve the sustainability of a place.
   c. In groups design a liveable eco-city. Present your design as a Poster or Prezi.

ISBN 978 1 4586 6277 4

10.2 Strategies: a matter of scale

A strategy is a method or plan used to bring about a desired future. To improve liveability for the future, strategies have been implemented at a variety of scales from the global and national to local neighbourhoods, streets and buildings.

Global cooperation

The commitment of countries to make the world a more liveable place was demonstrated at the Rio+20 Conference on Sustainable Development in 2012. UN organisations such as UNICEF and UNEP provide guidance to countries to develop strategies tailored to their particular requirements. The Kyoto Protocol set targets for reducing greenhouse emissions and the Millennium Development Goals (MDGs) set targets to reduce poverty. Global organisations are unified in the goal to improve the liveability of places and the wellbeing of all people.

10.2.1 Strategies range from global to local

Scale					
Global	National	State	City	Neighbourhood	Building
Examples of government strategies					
• Kyoto Protocol • MDGs • RIO+20 • UNEP • UNICEF	• Liveable Cities Program • Creating Places for People: The Australian Urban Design Protocol • National Broadband Network (NBN)	• South Australia: Greenhouse strategy • Tasmania: Liveable Places Development Program • Queensland: Blueprint for the Bush	• Sustainable living strategy for South Perth • Brisbane City Council Sustainability Agency • EzyGreen Energy Reduction Program	• New suburbs created or older suburbs renewed • Green Square, Sydney • Southbank, Brisbane • Docklands, Melbourne	• Water tanks • Solar panels • Recycling • Insulation

10.2.2 Rio+20 Conference: a fish sculpture made of plastic bottles and a sit-in by indigenous tribes send a message to the conference delegates and the world about the unsustainable use of world resources.

ISBN 978 1 4586 6277 4

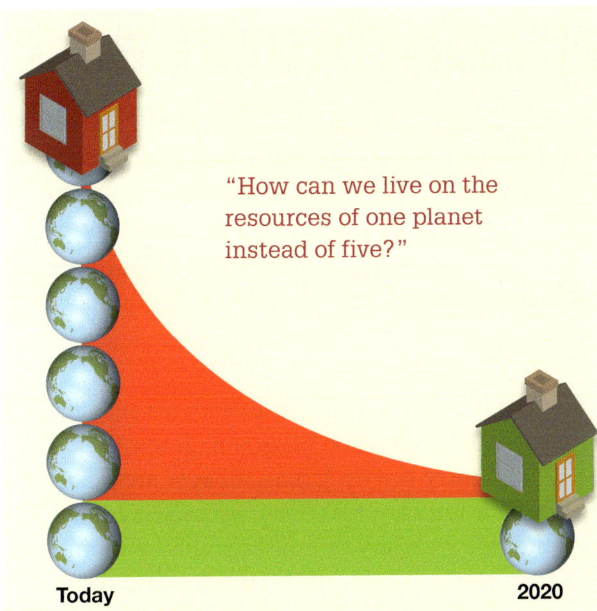

10.2.3 The One Planet Living goal is to reduce ecological footprints

Geoinfo

UNICEF works in more than 90 countries to improve water supplies and sanitation, improving the liveability of those places.

Global frameworks

One Planet Living (OPL) is a global initiative based on 10 principles of sustainability developed by BioRegional and WWF, a non-government organisation (NGO). OPL is about enjoying a high quality of life within our fair share of Earth's resources. If everyone lived like the average resident in the USA we would need five planets to live on.

The developers of Barangaroo in Sydney and Masdar City, UAE, created One Planet Action Plans prior to development. The aim of OPL is to reduce ecological footprints by implementing all or some of the 10 principles, such as **zero carbon** and sustainable water use.

Local action

Sonoma Mountain Village in California, USA, is a One Planet Community—a title given to developments that use all 10 principles, conduct ecological footprinting and encourage healthy, sustainable lifestyles. Sonoma Mountain Village aims for One Planet targets, including:

- 100% power from renewable energy
- reducing waste going to landfill by 98%
- reducing CO_2 emissions by 82%
- producing 65% of food from local farms and gardens
- increasing parks and open spaces and restoring native habitats
- creating a friendlier, safer community where health, happiness and wellbeing are priorities.

Sonoma Mountain Village targets go beyond those set by the state of California in areas such as green building, renewable energy and transportation. It is action at the local scale that brings strategies to reality.

Geoactivities 10.2

Knowledge and understanding

1 Why is action needed at a variety of scales to make places liveable and sustainable?
2 Give an example of a global strategy to improve liveability.
3 What role do government and non-government organisations play in improving the places we live in?
4 Briefly explain the concept of One Planet Living.

Inquiry and skills

5 Refer to 10.2.1.
   a At which scale do most practical strategies take place? Give an example.
   b At what levels would strategies be implemented before changes could be made in neighbourhoods and streets?
6 Refer to 10.2.2.
   a What do the two images represent?
   b Suggest how these images could be used to influence the behaviours of people.
7 Refer to 10.2.3.
   a Explain what the diagram illustrates.
   b Discuss the impact on liveability if there is no change in the future.
8 Investigate Somana Mountain Village to learn more about One Planet Living. Think up questions you want answers to. As a class discuss how the principles could be used in your community and how it would change the way you live and the liveability of your community.

ISBN 978 1 4586 6277 4

10.3 Creating better communities

Imagine a one-stop bus shelter where you can buy coffee and food, charge your phone, borrow books and listen to music. Do facilities such as this make places better to live in?

Across the world the addition of cycle paths, cafes, playgrounds, skate-parks, wheelchair ramps, affordable housing and efficient transport are making cities better places for living. People are encouraged to become involved in creating better cities through forums, surveys, interactive websites, blogs and competitions. In Britain school students were challenged to design their 'Dream Street'.

Experimental bus shelter in Paris

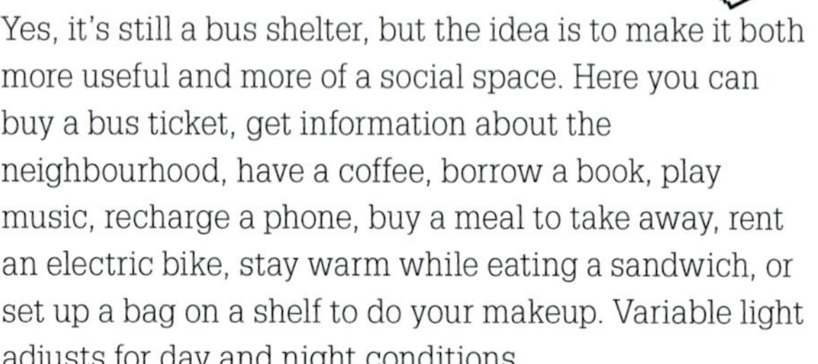

Yes, it's still a bus shelter, but the idea is to make it both more useful and more of a social space. Here you can buy a bus ticket, get information about the neighbourhood, have a coffee, borrow a book, play music, recharge a phone, buy a meal to take away, rent an electric bike, stay warm while eating a sandwich, or set up a bag on a shelf to do your makeup. Variable light adjusts for day and night conditions.

Source: www.humantransit.org

A liveable city

A liveable city is one that people want to live in now and in the future. It provides choices in housing and transport, and provides the services needed for a good quality of life (e.g. sanitation). The most liveable cities have attractive, clean environments and residents who feel included, safe and connected to each other.

Buildings, streetscapes, suburbs and **urban renewal** projects are designed to improve liveability. Elizabeth Quay, Perth, features a riverfront promenade, apartments, shops, cafes, restaurants and entertainment.

Urban villages

In Australian cities, 'village' developments encourage people to connect to each other, much like they do in small country towns. Urban villages are places to work, live and play, like Green Square, Sydney. They provide a variety of housing, employment opportunities and community services around a town centre, where everything is in walking distance. In the USA similar developments are called 20-minute neighbourhoods.

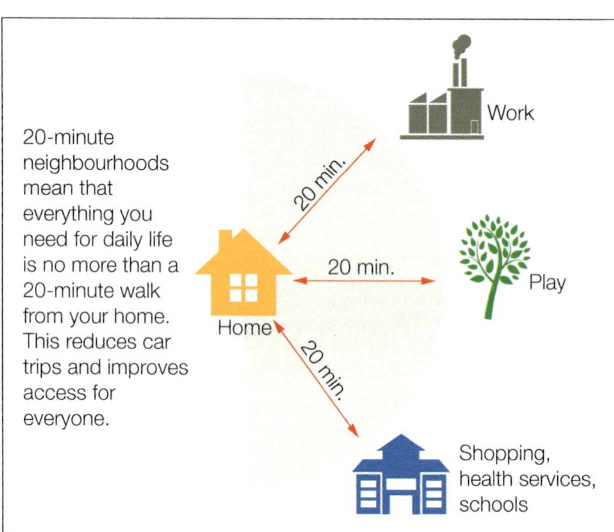

10.3.1 Twenty-minute neighbourhoods

Geoinfo

The Barangaroo redevelopment in Sydney is based on the concept of an urban village.

In Europe the town square or piazza has been a focus for economic, religious, social and cultural life for thousands of years. Here people come together for markets, festivals and celebrations, and social interaction takes place on a daily basis (e.g. Pamplona town square, Spain).

ISBN 978 1 4586 6277 4

10.3.2 Pamplona town square, Spain. Revellers at the opening of the San Fermin festival, which features the 'Running of the Bulls' and other traditional celebrations.

Living hell to living well

In Bogota, Colombia, a dangerous polluted city was transformed by the introduction of car-free days, a bus transit system, bicycle paths, footpaths, pedestrian-only streets and attractive open spaces. Governments and planners are working to make streets, neighbourhoods and cities places that provide a good quality of life for all. But if cities are to become more liveable in the future their ecological footprints must also be reduced and they must become more sustainable.

10.3.3 Bogota, Colombia, with its separate walking and cycle paths has become a model of liveability for cities in other developing countries

A failure to reduce CO_2 emissions, conserve water, minimise waste and use fewer natural resources will impact on environmental quality and future liveability.

Geoactivities 10.3

Knowledge and understanding

1 Make a list of improvements being made to cities. Name a group of people who would benefit from each change you have listed (e.g. families, children and the elderly).

2 Why are people becoming involved in city planning and change?

3 How are 'urban villages' in Australia similar to town squares in Europe?

4 Explain the meaning of the heading 'From living hell to living well'.

5 How does Bogota promote healthy living for everyone in the community?

Inquiry and skills

6 Refer to the bus shelter article.
- a If this was your school bus stop, what services should it provide?
- b How does this bus shelter make the location safer?

7 Refer to 10.3.1.
- a What makes 20-minute neighbourhoods liveable?
- b Does the place where you live qualify as a 20-minute neighbourhood? Explain.
- c Would you like to live in a 20-minute neighbourhood or urban village?

8 Refer to 10.3.2.
- a What are the advantages of a town centre or town square?
- b Describe the housing density around Pamplona's town square.
- c How would housing density affect the use of the square?

9 Find images of town squares in five different European cities. Bookmark the sites. Using Google Earth complete the following activities:
- a Locate your five cities and use the key to place a marker on the map.
- b When the dropdown box appears name each city and upload your photos.
- c Link your cities together to create a tour.
- d Save your tour and share it with another student in your class.

This activity can also be completed using Google Tour Builder.

ISBN 978 1 4586 6277 4

10.4 Places for people

The pedestrian-friendly, attractive Seven Dials neighbourhood in London is an iconic example of a liveable community. What is it that makes communities such as Seven Dials more attractive to live in than others?

Communities can vary in size from a city to a neighbourhood. For most people, their local neighbourhood is their community. *Placemaking* is a strategy that focuses on improving people's quality of life and creating places where every person can enjoy every day.

Great places to live

There are different perspectives on what features make a great place. Generally, the best communities are healthy, safe and attractive. They provide basic needs and services, and create opportunities for people to play, relax, enjoy and interact. Some traits important for good communities, such as 'charm', are intangible or qualitative—they are hard to measure and are influenced by personal perspectives. Quantitative traits (e.g. crime rates) are easily measured. Governments and planners focus on quantitative traits in their placemaking plans for new or improved communities.

Placemaking and walkability

The car has been responsible for the declining quality of life in many communities worldwide.

10.4.2 London's Seven Dials community, with narrow streets and alleyways that cater to pedestrians

Car-dominated communities suffer congestion, noise, pollution, danger and a lack of social interaction, which all impact on the wellbeing of people and the environment.

Places that rank highly on liveability (e.g. Copenhagen and Melbourne) are pedestrian-friendly. When people walk they interact, are more relaxed and healthy. Add cycling and an affordable mass transit system (e.g. buses and trams) and the need for cars diminishes. Fewer cars mean cleaner, safer and healthier communities. Walkability is a key trait for liveability. To get people walking, a walk must be useful, safe, comfortable and interesting. Walkability and transport enhance people's social connections and improve wellbeing.

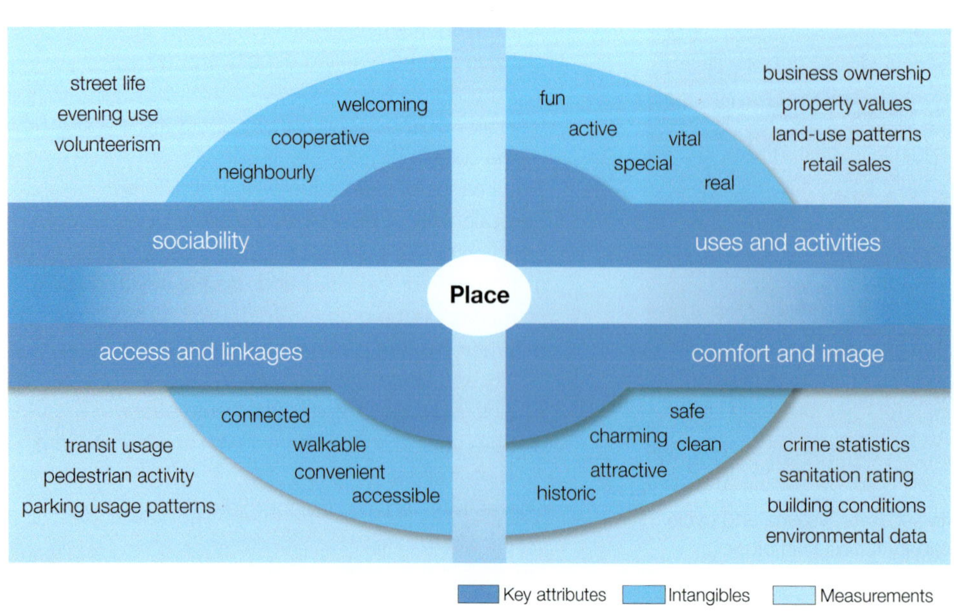

10.4.1 The traits that make places great to live in

ISBN 978 1 4586 6277 4

Placemaking: Melbourne

In the 1980s Melbourne's inner city was described as 'an empty useless city centre' in the *Age* newspaper. A famous Danish architect, Jan Gehl, advised planners that people are attracted to vibrant, bustling places. Melbourne needed to attract people back to the city. Strategies used to achieve this included:

- wider footpaths for pedestrians
- street furniture, trees, lamps and outdoor cafes
- an integrated mass transit system—trams, buses and trains
- a focus on walkability and cycling
- opening laneways to connect streets and give people interesting places to meet.

10.4.3 Melbourne's makeover: Centre Place bustles with shoppers and traders, and is famous for its tiny cafes and vibrant street art

Today Melbourne has a vibrant inner-city community with a rapidly growing population. People want to be in the city to live, work and play. Places such as Seven Dials and the city of Melbourne must continually adapt to new challenges to remain vibrant, liveable communities. Placemaking is a continuous process of change that also strengthens community identity.

Geo**info**

- Road vehicles produce 25% of harmful fuel emissions.
- Walking adds to a sense of community because people come into greater contact.

Geo**activities 10.4**

Knowledge and understanding

1. What do most people consider to be their community?
2. What does 'placemaking' mean?
3. Explain how placemaking is achieved.
4. What role does walkability play in creating great places to live?
5. Describe strategies to increase people's walking.
6. 'Walking must be useful, safe, comfortable and interesting.' Explain in your own words what this statement means.
7. Discuss the key to improving the liveability of the city of Melbourne and how it was achieved.

Inquiry and skills

8. Refer to 10.4.1.
   a. List four intangible features you consider important in a great place.
   b. Suggest ways a Year 7 student could measure the street life, land use, pedestrian activity and building conditions in a place.
   c. Measure the liveability of a place near your school or home.
   d. Write a statement about your community for each of the four measures of liveability, e.g. 'My community is safe (or unsafe) because …'
   e. Suggest four ways your community could be made more liveable.
9. Refer to 10.4.2.
   a. Identify four features of Seven Dials that make it a desirable place to live.
   b. Why do you think businesses have been attracted to this neighbourhood?
   c. Use the principles of walkability to explain why Seven Dials attracts walkers.

ISBN 978 1 4586 6277 4

10.5 Liveable streets

The nature of the street we live on can influence our quality of life. Over time, many streets have become car-oriented, noisy and dangerous places. Every year thousands of pedestrians are killed on multi-lane streets, the focus of which is moving vehicles. People living near these streets are exposed to higher levels of pollution and noise, and spend more time indoors. Liveable streets focus on people. They are comfortable, welcoming and safe places where people can live, play, socialise, travel and shop. These streets bring people together and foster a strong sense of community.

Traffic and safety

Making streets safer, cleaner and quieter usually involves eliminating or reducing and slowing down vehicle traffic, in favour of pedestrians. Strategies to achieve this include:

- *pedestrian-only shopping streets* (e.g. Times Square, New York, and Queen Street Mall, Brisbane)
- *mass transit systems* (e.g. buses and trams)
- *reduced number of car lanes*
- *traffic-calming devices* (e.g. speed humps)
- *designated bike lanes* and *timed traffic lights* for cyclists
- *widened pedestrian paths* that encourage walking
- *clean energy use* (e.g. electric cars and transit vehicles).

Attractive, sustainable and safe

The liveability of streets improves when they become more attractive for people to spend time in. This will happen when car traffic is reduced in conjunction with changes to infrastructure.

10.5.1 Creating a liveable street: Amsterdam Avenue and West 76th Street, New York, is re-created as a liveable, complete street

ISBN 978 1 4586 6277 4

10.5.2 Ten strategies to create liveable streets

A Vendors Help make streets into destinations rather than places to be driven through	**B Dedicated bus lanes** Get buses out of traffic and make trips faster and predictable	**C Raised, textured crosswalks** Create natural speed bumps and make pedestrians more visible	**D Pedestrian street lamps** Lighting shouldn't just be for cars	**E Curb extensions** Safety features reducing crossing distances; by narrowing the street they reduce drivers' speeds
F Separate bike lanes Dedicated lanes provide physical protection and encourage bike use	**G Speed bumps** Slow traffic	**H Street trees and plantings** Provide shade and oxygen, make the street look nicer, increase traffic safety, improve business	**I Bollards** Prevent motorists parking on footpaths and can be used to stop cars entering streets	**J Pedestrian-friendly lights** Traffic lights with a leading pedestrian interval gives pedestrians time to cross before cars turn corners

Infrastructure introduced could include:

- *greening* with trees and gardens
- *street furniture* (e.g. seats and shelters)
- *street lighting*
- *water-absorbing street surfaces* (e.g. paving, lawns and gardens)
- *activities and services* (e.g. cafes, markets, shops and entertainment)
- *well-maintained infrastructure and buildings*.

Streets with more pedestrians feel safer. This is known as the 'eyes on the street' principle. Where infrastructure and buildings become rundown or vandalised people feel less safe. This is known as the 'broken window theory'. People who feel safe are more likely to leave home and mix with their neighbours.

Complete streets

> A liveable street is a roadway designed to accommodate the needs of all users—drivers, transit vehicles, bicyclists, and pedestrians of all kinds (disabled, elderly, children, and lingerers).

When streets cater for everyone they are described as a 'complete streets'. There are social, economic and environmental benefits of liveable streets. A small number of changes can make any street more liveable.

European cities lead the world in creating complete and liveable streets, such as Las Ramblas in Barcelona, Spain. Creating liveable streets is easier where population and housing densities are higher, as in inner-city suburbs. In low-density suburbs associated with urban sprawl or large towns, liveable streets are usually linked to shops, transport facilities and public spaces.

Geo**info**

An estimated one-quarter of New York's land area is taken up by streets used mainly by private car owners.

Geo**activities 10.5**

Knowledge and understanding

1. List the benefits of removing cars from streets.
2. What non-car strategies can be used to make streets safer and healthier places?
3. Identify two pieces of infrastructure that can make streets more liveable. Explain.
4. 'This is something everyone knows: A well-used city street is apt to be a safe street. A deserted city street is apt to be unsafe.' Do you agree with this quote? Refer to the 'eyes on the street' principle in your answer.
5. Explain the 'broken window theory'.
6. What makes a complete street a liveable street?

Inquiry and skills

7. Refer to 10.5.1 and 10.5.2.
   a. Match the numbers on the image with the strategies in 10.5.2.
   b. Rank the strategies from 1 to 10 according to the impact you think they would have on the liveability of the street for teenagers your age.
   c. Choose an image of a street from the internet or photograph your own street. Use ICT drawing tools to add five changes that will make the street more liveable for students your age and younger children. Discuss with another member of your class your suggestions and the impact they will have on the quality of life.

ISBN 978 1 4586 6277 4

10.6 Green places and open spaces

Singapore's Gardens by the Bay project contains 18 steel supertrees, up to 50 m tall and covered with bromeliads, orchids, ferns and climbers. Two giant conservatories house plants from different climate zones. Chinese, Malay and Indian themed gardens cover the site. The $1 billion plus project covers 101 ha of Mariner Bay and contain 700 000 plants. Singapore plans to add another 900 ha of parks and gardens within 10–15 years. Why? In a high-density, high-rise city, nature and public spaces are essential for improving liveability. Gardens and public spaces make better places to live, work, play and raise families.

Liveability, parks and open spaces

Generally, the amount of green and public space per person declines as cities grow. Strategies are now being introduced to reverse this trend. The economic, social, and environmental benefit of trees, parks, gardens and open spaces is recognised worldwide. Children's physical and mental health improves where they have access to playgrounds and parks. Adults with access to parks are healthier, less stressed and more caring of others. Neighbourhoods are stronger and safer where shady streets, parks and public spaces bring people together. What's more, plants benefit the environment by removing pollutants and reducing temperatures.

The COMMUNITY BENEFITS

NATURE MAKES YOU NICER: COMMUNITIES WITH MORE GREENSPACE HAVE LOWER RATES OF CRIME & VIOLENCE

ROOM WITH A VIEW

GET A GLIMPSE OF GREEN

HOSPITAL PATIENTS WHO CAN SEE TREES & OTHER GREENERY FROM THEIR HOSPITAL ROOMS RECOVER FASTER AND REQUIRE LESS PAIN MEDICATION.

NATURAL VIEWS AT WORK LEAD TO INCREASED JOB SATISFACTION, BETTER CONCENTRATION, DECREASED MENTAL FATIGUE AND LOWER STRESS LEVELS.

"HUMANS ARE DISAPPEARING FROM THE OUTDOORS AT A RATE THAT WOULD MAKE THEM TOP ANY CONSERVATIONIST'S LIST OF ENDANGERED SPECIES."

TIM GILL, THE ECOLOGIST

The COGNITIVE BENEFITS

SPENDING TIME IN NATURE INCREASES CREATIVITY, CURIOSITY, AND PROBLEM SOLVING ABILITY

HEALTH BENEFITS

NATURE IS THE BEST NURTURE

MANY STUDIES SHOW SIGNIFICANT HEALTH GAINS FOR THOSE IN CONTACT WITH NATURE

REDUCED ANXIETY & DEPRESSION

DECREASED STRESS

INCREASED ENERGY

INCREASED IMMUNITY

50% LOWER RISK OF DIABETES

INCREASED VITAMIN D PRODUCTION

INCREASED WEIGHT LOSS & FITNESS

REDUCED SYMPTOMS OF A.D.D.

★ 50% LOWER RISK OF HEART ATTACK ★ 30% LOWER RISK OF COLON CANCER ★

SUGGESTED DOSAGE

CONTACT WITH NATURE IS AN AFFORDABLE, ACCESSIBLE AND EQUITABLE FORM OF PREVENTATIVE AND RESTORATIVE MEDICINE.

2 MINUTES

STRESS IS RELIEVED WITHIN MINUTES OF EXPOSURE TO NATURE (AS MEASURED BY MUSCLE TENSION, BLOOD PRESSURE AND BRAIN ACTIVITY)

2 HOURS

MEMORY PERFORMANCE AND ATTENTION SPAN IMPROVES 20% AFTER SPENDING AN HOUR INTERACTING WITH NATURE

2 DAYS

LEVELS OF CANCER FIGHTING WHITE BLOOD CELLS INCREASE 50% AFTER SPENDING TWO OR MORE CONSECUTIVE DAYS IN NATURE

10.6.1 Green spaces have health benefits

ISBN 978 1 4586 6277 4

10.6.2 Gardens by the Bay, Singapore

Green infrastructure is now considered as important for people's quality of life as hard infrastructure like transport, water and electricity. Community leaders, city planners and architects now include these features into urban development projects.

Creative solutions for limited space

Where space is limited, 'pocket parks', 'parklets' or 'people spots' are turning the smallest spaces (e.g. single car parking spots) into places to relax and socialise. San Francisco's Pavement to Parks strategy introduced 30 parklets throughout the city. Other cities, such as Chicago, are now following the trend.

Neighbourhoods with parklets have:

- more pedestrians
- an increased sense of community
- an increase in positive perceptions of the neighborhood
- an economic benefit for nearby businesses because people stop, stay around longer and shop.

10.6.3 There are many parklets along New York's High Line

In the most densely settled cities (e.g. Hong Kong) architects are proposing skyscrapers with vertical gardens to renaturalise the city, reduce smog and pollution, and bring the benefits of greenery to residents.

New York City

In 2007, New York City had less green space per person than any other US city. The city's mayor pledged that all New Yorkers would live within a 10-minute walk of a park and every neighbourhood would have a plaza by 2030. Public spaces like Times Square were converted to plazas, schoolyards were turned into attractive playgrounds, abandoned sites were made into parks, and natural places like waterfronts were reclaimed for public use. By 2012, over 75% of New Yorkers could easily access a park or public space, such as the High Line.

Geo**info**

One hundred trees can remove five tonnes of CO_2 from the atmosphere each year.

Geo**activities 10.6**

Knowledge and understanding

1. What is meant by 'green infrastructure'?
2. How does a park differ from a public or open space?
3. How can parks and public spaces create stronger communities?
4. How is a shortage of space for parks and public spaces overcome?
5. Explain how New York City fulfilled its promise to residents.

Inquiry and skills

6. Refer to 10.6.1. Create a mind map to show the benefits of nature and trees. On your mind map have separate branches for social, economic, environmental and health benefits.
7. Refer to 10.6.2 and the internet. Design a new green space for your community. Locate the site on a sketch map. Label your design to show features that will make your community more liveable and sustainable.
8. Refer to 10.6.3.
   a. Investigate New York city's famous High Line. Present findings as a digital presentation.
   b. Identify places in your school grounds suitable for a parklet. Create a design.

ISBN 978 1 4586 6277 4

10.7 Sustainable and liveable cities

Sustainability is about living within Earth's biocapacity. A sustainable eco-city or green city minimises resource consumption and waste production while creating economic growth and employment for its people and improving their quality of life.

Becoming sustainable

While there is no agreed definition of a sustainable city or a list of targets to be met, there are certain initiatives that can be used to make cities sustainable and therefore more liveable. These include:

- renewable energy programs (e.g. wind and solar) to reduce CO_2 emissions
- water- and energy-efficient buildings and suburbs
- efficient and affordable public transport, electric car stations and bike corridors
- walkable neighbourhoods
- urban agriculture and tree planting
- recycling and waste composting
- higher density housing.

Citizenship for sustainability

Governments, non-government organisations, businesses and individuals are becoming sustainability-focused at local, national and international levels. The WWF's 'One Planet' campaign, Sustainable Cities International, the Australian Conservation Foundation, the United Nations Environment Program and the Global Footprint Network all promote sustainable urban development.

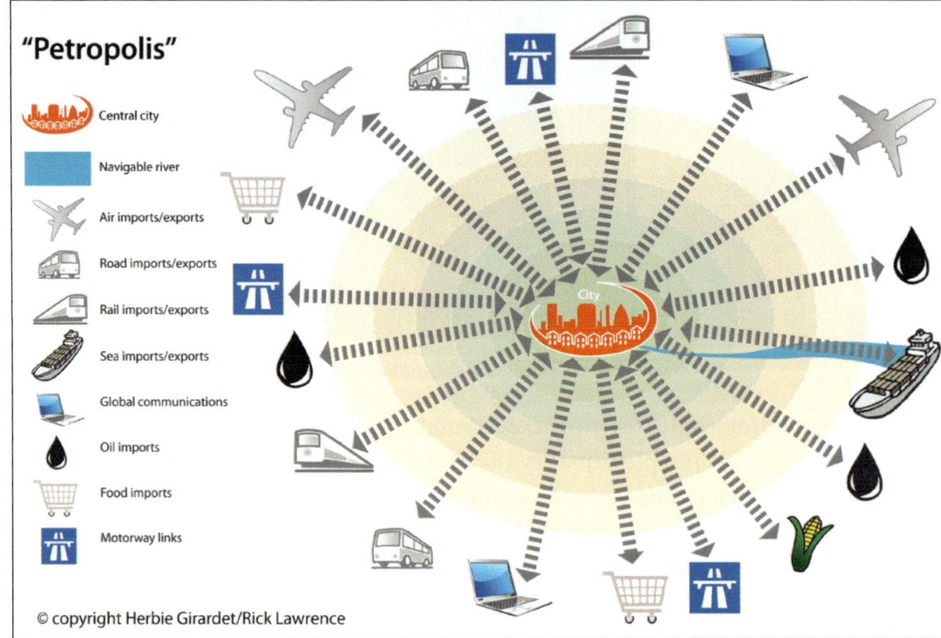

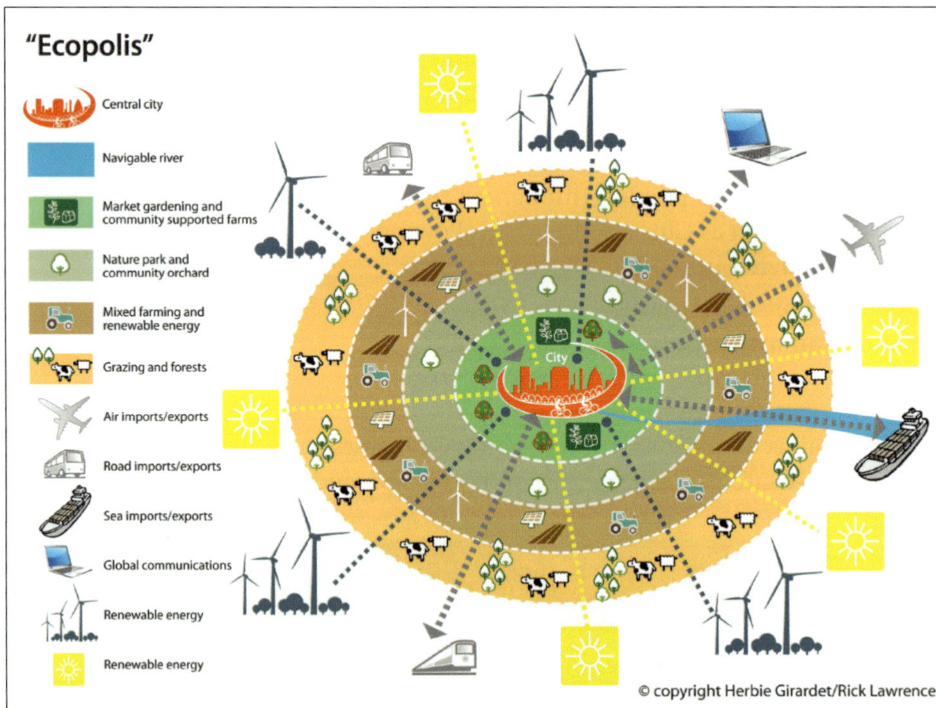

10.7.1 Models of a petropolis and ecopolis

Sustainable cities should regenerate the natural ecosystems on which they depend, as well as minimise future impacts. This concept of a regenerative city—an *ecopolis*—contrasts with a *petropolis*, or resource-hungry city, with little concern for its environmental impact.

ISBN 978 1 4586 6277 4

Small and large-scale initiatives

Effective sustainability programs have been implemented at a variety of scales from individual buildings to neighbourhoods, suburbs and whole cities:

- Melbourne's 1200 Buildings Program encourages retrofitting older commercial buildings to reduce energy usage and create green jobs. For example, the energy rating of 530 Collins Street was upgraded from 2 to 5 stars and the energy used for lighting was halved.
- Brisbane's Santos Place is a 6-star green building with double-glazing, external sun shading, sensored light switches and rooftop rainwater

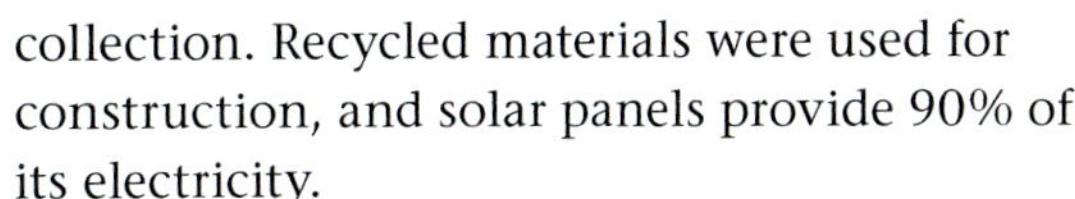

collection. Recycled materials were used for construction, and solar panels provide 90% of its electricity.
- Masdar City, UAE, will rely entirely on solar and other renewable energy sources to become the first zero-carbon, zero-waste city in the world.
- In New York City increasing greenhouses and rooftop gardens are supplying local supermarkets.
- Copenhagen has a cycling superhighway network where timed traffic lights, tilted garbage cans and conversation lanes encourage people to ride to work.

What is happening in your community to enhance sustainability and liveability?

10.7.2 Artist's impression of a planned eco-friendly New York

Geo**info**

'Design will save the world' is the motto of Inhabitat, one of many blogs devoted to sustainable design for the future.

Geo**activities 10.7**

Knowledge and understanding

1. What is the main aim of creating sustainable cities?
2. Sustainability strategies are implemented at different scales. What does this mean?
3. What groups are involved in planning and implementing strategies for sustainable cities?
4. Explain the motto 'design will save the world'.

Inquiry and skills

5. Refer to 10.7.1.
   - a Why is there a dominance of transport in the petropolis city model?
   - b What is the source of energy in a petropolis?
   - c What is missing in the ecopolis city model?
   - d Why are the images for transport drawn smaller in the ecopolis model?
   - e Describe the main source of food for each city model.
   - f Which model best fits your capital city?
   - g List the obstacles that might prevent cities from becoming more sustainable.
6. Refer to 10.7.2.
   - a Create a line drawing of each image. Label the sustainability initiatives you can identify.
   - b Discuss why sustainable neighbourhoods are more liveable.

ISBN 978 1 4586 6277 4

10.8 From 'take, make, waste' to 'reduce, reuse, recycle'

The future liveability of places depends on how Earth's resources are used. Sustainable living is about minimising human impact on natural resources while providing a satisfactory quality of life. A liveable and sustainable future involves a change from the 'take, make, waste' attitude to resources of the past to a 'reduce, reuse, recycle' approach. Places where people can live sustainably require three fundamental features:

- lifestyle choices that have a small footprint (e.g. recycling and walking)
- places designed to conserve resources (e.g. green buildings and eco-cities)
- infrastructure that supports sustainable living (e.g. solar power and cycle paths).

Buildings

Buildings consume 25% of Earth's resources and produce large volumes of waste. To support sustainable living, existing buildings can be **retrofitted** with solar panels and water tanks. New buildings are designed to 'green' standards such as BASIX, the Building Sustainability Index of NSW. Strategies to achieve green targets include installing rainwater tanks and water-saving fixtures, insulation, passive solar design, natural lighting and native gardens.

10.8.1 Buildings use resources and produce waste

Resources and waste	Estimated contribution from buildings
Energy	40% of global energy
Water	25% of available water
Timber	55% of non-fuel-use timber
Solid waste	40% of global waste is from construction or demolition
CO_2 emissions	40% of global production

Rooftop gardens and green walls reduce energy consumption and minimise rainwater runoff. The ecological footprints of buildings can be minimised when rooftops are used to grow food and produce wind and solar energy. Australia's greenest office buildings—1 Bligh Street, Sydney, and Council House 2 (CH2), Melbourne—have 6-star sustainability ratings and are considered office blocks of the future.

10.8.2 Sustainable 6-star office building at 1 Bligh Street, Sydney

Neighbourhoods and suburbs

Malmö in Sweden is recognised globally for its sustainable, liveable neighbourhoods created in old industrial districts. It uses 100% renewable energy from solar, wind and geothermal sources. Bike paths and energy-efficient gas-powered buses also enhance sustainable living in Malmö. In a similar way, the Carlton Eco-Neighbourhood in Melbourne uses natural lighting and ventilation, solar hot water, water recycling and rainwater capture systems. Public parks and landscaping with footpaths, cycle-ways and playgrounds enhance liveability.

Eco-cities

In places with rapid population growth and urbanisation, new cities have been designed with reduced energy consumption, waste production targets and features that enhance liveability. Tianjin Eco-city in China will be a model for new cities in the developing world, where increasing populations need to be housed in a liveable yet sustainable way.

When the places people live in support a sustainable lifestyle and have the features that are important for liveability, the result is healthier, safer and happier communities.

ISBN 978 1 4586 6277 4

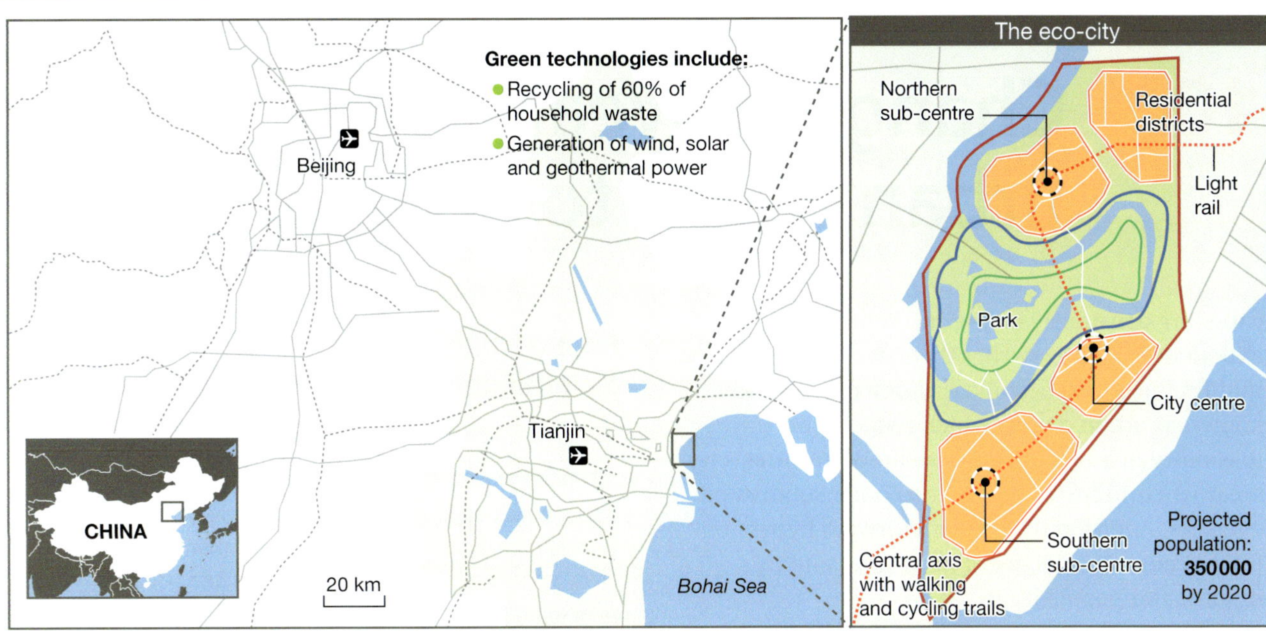

10.8.3 Tianjin Eco-city

10.8.4 Clean energy (wind and solar) is being used to power eco-cities like Tianjin

Geo**activities 10.8**

Knowledge and understanding

1 What do you understand by the 'take, make and waste' attitude to resources? Why does this attitude need to change?
2 Describe the essential features for sustainable living.
3 Why do buildings need to be made more sustainable?
4 Explain how rooftops can be used to make buildings more sustainable.
5 Sustainability is more than just energy efficiency. Explain what this means using examples.
6 Discuss why eco-cities are important for places with rapid population growth.

Inquiry and skills

7 Refer to 10.8.1.
   a Suggest one strategy to reduce the use of energy, water or timber in a school building.
   b List (from largest to smallest) at least six sources of waste generated in buildings.
8 Refer to 10.8.3 and 10.8.4. Write a report explaining why Tianjin will be a sustainable and liveable city when completed.
9 Choose one of the following places to investigate sustainable living: Malmö, Sweden; Freiberg, Germany; Sydney Central; Barrangaroo; or a sustainable building, neighbourhood or city approved by your teacher. For your chosen place complete the following activities in a digital format.
   a Locate the place on a map.
   b Collect at least four images illustrating the sustainable features of your chosen place.
   c Annotate each image to identify features relevant to sustainability and liveability.
   d Present your findings using a web 2.0 tool.

ISBN 978 1 4586 6277 4

10.9 Changing liveability: urban blight to renewal

Cities are constantly changing. While some experience rapid growth, others decline. Over time buildings age and become run-down in a process known as **urban decay**, which reduces the liveability of a place. This is common in inner-city areas when industry relocates to outer suburbs or overseas. Additionally, new motorways bypassing towns cause the closure of businesses and loss of jobs, and people leave.

Cities can become more liveable if planners refurbish decaying areas into exciting new precincts, such as Salamanca Place in Hobart, Melbourne Docklands, South Bank in Brisbane, and Barrangaroo and the 'Block' in Sydney. Newcastle, New South Wales, experienced urban decline with the closure of the steelworks in 1999. A major renewal project in Newcastle, called the Honeysuckle Development, converted old buildings into museums, restaurants and art galleries, and constructed boardwalks and marinas. This was part of a process called **gentrification**.

Fixing the Block: Pemulwuy Project $70 million redevelopment

In 1973 a grant of 8000 m^2 to the 'Block' in Redfern was given to the Aboriginal Housing Company (AHC) to house Indigenous people. Containing 102 houses, the Block became the birthplace of Aboriginal urban land rights. Over a period of time most were burnt and became dilapidated. By 2011 all buildings had been demolished.

A plan to reinvent the Block includes flats for students, a childcare centre, new residential blocks, retailers and businesses. Future tenants will include those who were evicted from the Block to allow for its demolition.

Joyce Ingram, Aboriginal elder, leaves the Block

From this...

to this

10.9.1 Honeysuckle Development in Newcastle: from urban blight to urban renewal

10.9.2 Advantages and disadvantages of urban renewal

Advantages	Disadvantages
• Use of existing infrastructure • Cost of upgrading roads, water supply, sewerage system and communications less costly than new infrastructure • Close to work—reduce greenhouse gas emissions • Renovation of historic terrace homes—recycling	• Lack of space—no backyards • Expensive shops • Noise from neighbours and transport • Difficulties parking • Transport bottlenecks • Higher house prices forcing less well-off people into outer suburbs

ISBN 978 1 4586 6277 4

10.9.3 The transformation of Barrangaroo will reconnect the Sydney CBD with the harbour, provide open spaces on the waterfront for all to enjoy, and include residential and commercial tower buildings

Global overview

Urban renewal replenishes housing stock, facilities, services and the environment, making more liveable places. Urban redevelopment in parts of New York City increased liveability when new housing projects were built and areas allocated to parks. Rio de Janeiro (Brazil) is planning for the Olympic Games in 2016, so the government is renovating 700 km of infrastructure, including water supplies, sanitation, electricity and telecommunications.

Around the world urban areas have been rebuilt to improve their liveability. Examples include:

- *Detroit, USA*—Three large car manufacturers—Ford, General Motors and Chrysler—were located in Detroit. Competition from car manufacturers in Japan and South Korea saw the American car industry decline. Unemployment increased and buildings became dilapidated. Over the past 10 years, 25% of Detroit's population have moved out of the city, leaving a third of the area vacant. At present, Detroit is undergoing a multimillion dollar facelift to improve liveability.
- *London's Docklands area*—In the 1960s, large container ships carrying cargo required deep water, causing the closure of the docks located in shallower water. Unemployed people abandoned their homes and crime increased. In response, the old port area was redeveloped into an upmarket commercial and residential area. London's Docklands now encompass a high-rise commercial zone containing Britain's tallest building.
- *Chinese cities*—Urban renewal as a result of the 2008 Beijing Olympic Games and the 2010 Shanghai World Expo improved the liveability of many areas. New buildings were constructed and transport infrastructure improved. However, many Chinese people fear that traditional buildings being replaced by skyscrapers could destroy their culture and social connections.

Geoactivities 10.9

Knowledge and understanding

1 Describe how urban decline impacts on liveability.
2 Discuss how urban renewal in parts of London and China has improved liveability.
3 Discuss the advantages and disadvantages of urban renewal for the liveability of places.

Inquiry and understanding

4 Refer to the text about the 'Block'.
   a Explain how urban renewal aims to make the Block a more liveable place.
   b Investigate the changes to one area in Australia experiencing decline and how it has been renewed for improved liveability.
5 Investigate any changes in your local area that have made it more liveable. Include photographs and maps.

ISBN 978 1 4586 6277 4

10.10 Farming our cities

Most of the world's population now live in urban places. Minimising the importation of resources such as food, and the production of waste products (e.g. food packaging) can reduce the ecological footprint of towns and cities, and contribute to sustainability. Urban agriculture can reduce food footprints by transforming cities from being food importers and consumers to food producers and waste recyclers. With New York City rooftop farms, community gardens in Melbourne, and 'sack gardens' in Sao Paulo's slums in Brazil, urban agriculture is creating stronger, healthier and happier communities.

Benefits

Farming in towns and cities can have health, social, economic and ecological benefits, including:

- energy savings from reduced transportation or 'food miles'
- a cleaner environment—reduced CO_2 production, pollution and waste
- more attractive green communities with lower temperatures
- cheaper food due to lower transportation, distribution and packing costs
- stronger, safer communities where people know and care for each other
- reduced tension between neighbours and different ethnic groups
- healthier people through reduced hunger and malnutrition
- reduced poverty through job creation and the sale of produce
- important community events, such as farmers markets.

Rooftops and vertical sack farms

Urban agriculture can reduce hunger and poverty. Corporations and individuals donate land, or governments release unused public land. In Mexico City, a new 'Green Roof Program' is creating 10 000 m^2 of rooftop agriculture a year. In Sao Paulo, the non-government organisation Cities Without Hunger uses farming to reduce health and social issues such as crime, hunger, poor sanitation and unemployment. Self-help projects pop up wherever there is a space.

Where land is limited (e.g. slums and high-rise apartments), sacks and growbags are used. In Kenya, vertical sack gardens grow plants at different levels. French NGO Solidarités provides the training, seedlings and sacks. In developed countries like Japan vertical greenhouses and office farms produce high volumes of food.

Livestock, including chickens, pigs and even bees, are raised on urban farms. New York's InterContinental Barclay Hotel has a rooftop apiary to produce honey, and the bees also pollinate plants throughout the city. Europe's largest urban farm—Mudchute in London—has 200 animals, 70 community allotments for crops, and an educational program.

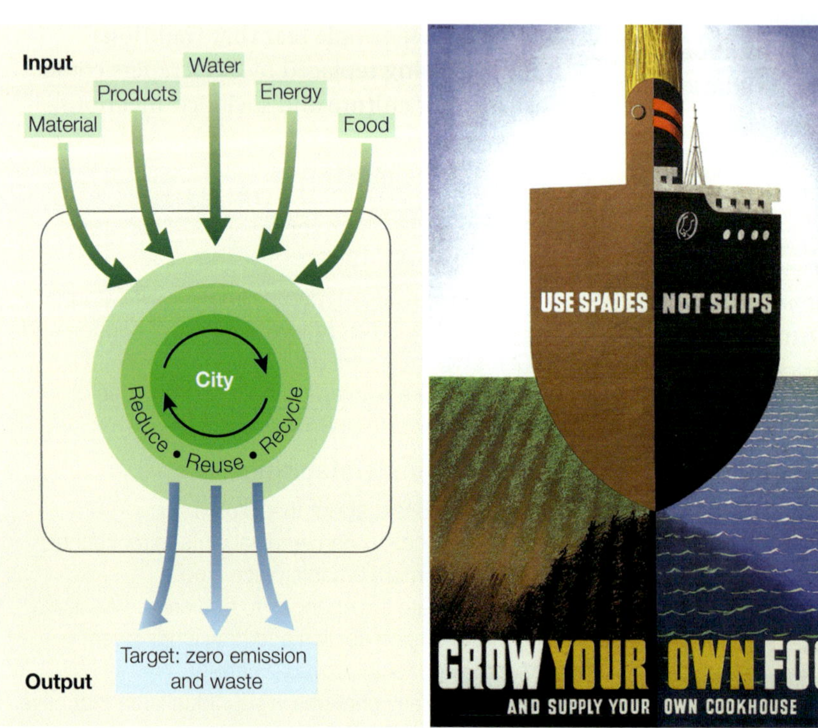

10.10.1 Urban sustainability: a sustainable city model and a poster urging people in post-war USA to get gardening

ISBN 978 1 4586 6277 4

Urban agriculture is an important strategy for achieving sustainable urban development and addressing the social, economic and health issues that affect liveability.

10.10.2 Urban agriculture looks different across the planet

Geo**info**

- Urban farmers do not usually own the land.
- 'Guerrilla' farmers use land without permission.

Geo**activities 10.10**

Knowledge and understanding

1 How can urban agriculture reduce the footprint of cities?

2 Draw a square, then divide it into four equal quadrants. In each square list the health, social, economic and ecological benefits of urban agriculture.

3 Identify four sources of land for urban agriculture.

4 Assess the importance of urban agriculture as a strategy for improving the liveability and sustainability of cities.

Inquiry and skills

5 Refer to 10.10.1.
   - a What is the aim of a sustainable city?
   - b What is the message in the poster?
   - c Explain the link between the poster, urban agriculture and sustainability.

6 Refer to 10.10.2.
   - a Match each image with the correct label.
     - office farm in Tokyo
     - apiary project in New York
     - rooftop farm in New York
     - vertical bag farming in Nairobi
   - b Which image is most likely to match the following description?
     - a community garden
     - to reduce poverty
     - green lifestyle
     - commercial farming

7 Visit the website of the Five Borough Farm in New York (see Geolinks). Use the infographic to answer the following questions.
   - a Briefly explain the difference between the four types of urban agriculture.
   - b List the essential ingredients for urban agriculture.
   - c Explain how urban agriculture can improve the liveability of city communities.

8 Collect images of future urban agriculture proposals. Present them using web 2.0 tools such as Prezi and Glogster. Explain how one of the proposals, e.g. vertical greenhouses, would work to enhance liveability.

ISBN 978 1 4586 6277 4

10.11 Strategies reducing poverty

Orissa state in India is one of the poorest places in the world, with high levels of illiteracy and child mortality. When a smartphone was given to each family in the village of Juanga, school attendance improved by 20% and reported diseases dropped by over 50%. Children and mothers earned points by scanning their phones when attending school and healthcare classes. The points could be redeemed for food, clothing and medicine.

In 2000 the Millennium Development Goals focused global attention on places with low liveability like Juanga. Poverty and other development issues can be reduced with creative and affordable strategies, such as mobile phones, **microcredit** and sanitation projects.

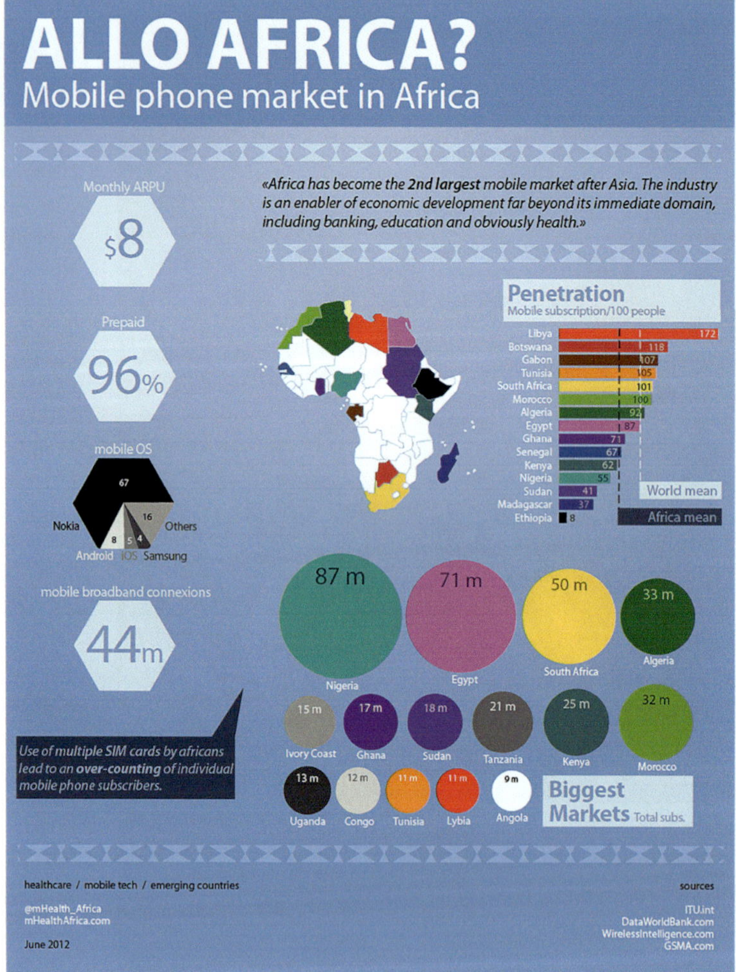

10.11.1 The use of mobile phones in Africa

Mobile phones

The extension of submarine cable networks (e.g. from Europe to Africa) has enabled the rapid expansion of mobile phone services into the world's poorest countries. Farm incomes, health, employment, safety and the status of women have improved where mobile phones and appropriate apps have been introduced.

Farmers can access weather forecasts and advice from governments and NGOs. M-farm provides up-to-date market information and links farmers to buyers. Dairy farmers use iCow to monitor each cow, receive SMS reminders about inoculation dates and obtain advice about breeding. Farmers are producing more food and earning higher incomes. Women feel safer and better able to provide health care for their families using mHealth. Mobile banking enables people without bank accounts to receive and send money. Education is also being delivered by phone.

Geoinfo

- There are currently five major international cables in place providing bandwidth to the African continent.
- One in five Africans had a smartphone in 2012.

Microcredit

Microcredit is a small loan with a fee. The loans allow the poorest people to buy things they need to increase their income. This can mean starting a small business or improving farm output with the addition of a piece of equipment, seeds, fertilisers or animals. Many microcredit providers are joint ventures between governments, banks and businesses (e.g. Benetton). In Bangladesh, women purchasing mobile phone kits become village phone operators charging a small fee for people without phones to make and receive calls.

Sanitation and waste removal

Approximately 59% of Indian households have a mobile phone, yet only 47% have a toilet. Open defecation spreads disease and significantly reduces

ISBN 978 1 4586 6277 4

10.11.2 Microcredit can alleviate poverty and improve liveability

10.11.3 Sanergy provides a sustainable solution to the sanitation crisis in Kenya

the liveability of urban slums in developing countries. An integrated sanitation and waste system in Kenya called Sanergy provides clean, affordable toilets to slum communities, collects the waste for conversion into energy and fertiliser, and provides sustainable employment for local people.

Innovative small-scale projects are successfully improving the liveability of the world's poorest places in conjunction with larger-scale government, non-government and international aid programs.

Geo**activities 10.11**

Knowledge and understanding

1 How and why has the liveability of Juanga village improved?
2 What development enabled the spread of mobile phones into developing countries?
3 Construct a mind map to illustrate improvements associated with mobile phones in the world's poorest countries.
4 What is the purpose of microcredit?
5 Explain why improved sanitation is essential to improving liveability in urban slums.

Inquiry and skills

6 Refer to 10.11.1.
   a Which country has the highest number of mobile phone subscriptions?
   b Which country has the lowest number?
   c Which global mobile phone market is larger than Africa's?
   d Explain how liveability in the countries shown in the bar graph might be improved by mobile phone technology.
7 Refer to 10.11.2.
   a Write a paragraph explaining how microcredit can improve the quality of life of people in Africa's poorest countries
   b What is self-sufficiency? Suggest why this is important.
   c There are differences of opinion about the success of microcredit as a strategy to reduce poverty. Investigate the arguments for and against microcredit. Present your findings in a table.
8 Refer to 10.11.3. Investigate the integrated Sanergy sanitation system. Explain how the system can improve the liveability of places where poverty means lack of access to services and facilities.
9 'It is easier to improve the liveability of places in poor countries suffering from poverty than it is in developed countries.' Discuss the statement and justify your opinions. Present a short written report summarising your arguments.

ISBN 978 1 4586 6277 4

10.12 Learning from Europe

In 2012, European cities filled eight of the top 10 rankings in the Mercer Quality of Living Survey. Governments, architects and town planners worldwide look to Europe for inspiration when planning for liveability and sustainability.

European buildings

Most of Europe's older buildings were built at a time when 'sustainability' was not a buzz-word—they depended upon natural ventilation and natural daylight, shading from the sun, eaves, shutters, balconies on which to grow plants, dry washing and sit outside, and thick walls and insulated roofs to keep the buildings cool in summer and warm in winter. Many of these older buildings, therefore, have good opportunities for retrofitting, now that we can combine good passive design with good technologies.

Source: www.thefifthestate.co.au

Towns and cities in Europe have a long history compared to settlements in Australia and USA, and developed features like public squares and high-density housing now considered to be desirable for liveability. Narrow streets were made for people, not cars, and buildings contained many passive energy efficient features. Many European cities, such as Oslo and Copenhagen, now lead the world in sustainable living with their new technologies and strategies such as zero carbon targets.

Features of European cities

Features shared by many European cities include:

- compact size—smaller in area than places of similar population in Australia
- greater percentage of people living in medium-density apartments
- public places (e.g. parks, gardens, piazzas and green roofs)
- pedestrian-only or shared streets
- many people who walk or cycle
- well-developed public transport systems used by many
- use of smaller motor vehicles and smart cars
- strong support for local agriculture and farmers markets
- buildings retrofitted for energy and water efficiency
- adoption of low carbon energy solutions (e.g. Copenhagen Zero Carbon Target for 2025).

10.12.1 A liveable community in Paris

ISBN 978 1 4586 6277 4

Higher-density living

Many features of European towns and cities stem from higher density living. When more people live in an area they are less likely to need cars and more likely to walk or cycle. Narrow roads and apartment living necessitate smaller cars and public spaces for relaxing, socialising and play. An Ecodensity strategy in Vancouver, Canada, aims to create walkable, affordable and energy-efficient communities based on European models.

10.12.2 Electric smart cars are common in European cities (e.g. Florence, Italy)

Sustainability

Amsterdam in the Netherlands is one of Europe's greenest cities—from 2015 all new buildings are required to be energy-neutral. Many buildings already add more energy to the power grid than they use. New offices along the city's IJ waterfront use water to heat and cool the buildings, and they produce solar energy and recycle water.

Copenhagen in Denmark has over 400 km of cycling tracks, and cyclists make up about 35% of daily commuters. Green bike routes improve access to recreational facilities while 'pocket parks' will give 90% of residents walkable access to a green space by 2025. The city has a mandatory green roof policy and the city plans to be carbon neutral by 2025.

In Oslo in Norway, intelligent lighting adjusts intensity to suit traffic and weather conditions while bio-methane from waste powers mass transit and heating. Car- and bike-sharing programs are in place along with 400 charging stations for an increasing number of electric vehicles, which receive free parking and toll-free access to all roads.

10.12.3 Copenhagen's vision of the future

Feature	2010	2025
Carbon emissions in tonnes	2 500 000	0
Electric capacity from windmills MW	46	360
Km of bike lanes	369	482
% of citizens travelling to work on bikes	33	50
% of citizens living within minutes walk of a public park	63	90
1000s of people living in the city	535	640

Source: www.sustainia.me

Geoinfo

Biomass used for heating saves carbon emissions equivalent to 60 000 cars a year.

Geoactivities 10.12

Knowledge and understanding

1 Why are governments and planners looking to Europe for inspiration?

2 Explain the following concepts: medium-density housing, green roofs, shared streets and retro-fitting.

3 What can we learn from the construction of old European buildings?

4 List four features of European cities that contribute to liveability.

5 Why are public spaces and public transport important in areas of high housing density?

6 Explain what Amsterdam, Copenhagen and Oslo have in common.

Inquiry and skills

7 Refer to 10.12.1.
   - a Propose three additions to this streetscape to improve its liveability for people your age.
   - b Explain what makes this place a walkable neighbourhood.

8 Refer to 10.12.2 and research smart cars.
   - a What are the advantages and disadvantages of electric vehicles?
   - b Explain how one of the disadvantages you listed could be overcome.

9 Refer to 10.12.3. Discuss how Copenhagen in 2025 will be a more liveable and sustainable city.

ISBN 978 1 4586 6277 4

10.13 Enhancing liveability in Australia

Australia is one of the world's most developed nations. Its resources, environments and lifestyles are envied worldwide and many cities rank highly in global liveability and quality of life rankings. Despite these attributes, Australians believe their communities can be improved to address issues such as traffic congestion, public transport, affordability and sustainability.

Strategies

To enhance liveability the Australian Government has introduced:

- the *Liveable Cities Program* to fund better urban design, reduce car dependency, create open spaces and address climate change
- the *Urban Design Protocol* for Australian cities, which is a guideline for creating liveable and sustainable communities
- *Clean Energy Future*, which introduced a price on carbon and promotes clean energy
- the *Renewable Energy Target*, which aims to achieve 20% of electricity from renewable sources by 2020
- the *Low Carbon Communities* program to fund energy efficiency in public buildings and amenities (e.g. LED street lighting)
- the *Green Vehicle Guide* to help motorists choose fuel-efficient low-carbon vehicles.

North, south, west and east

State, territory and local governments are committed to enhancing places for living in. Brisbane's 'Growing a Green Heart Together' is about communities taking action for a clean, green city. Conserving resources, minimising waste and improving transport options are pieces of the heart.

In the remote Aboriginal community of Warburton, WA, the provision of a community college and meeting place, an arts precinct, an orange farming initiative and culturally appropriate housing is creating a greener, healthier and happier community.

Adelaide most liveable city: new poll

Adelaide is Australia's most liveable city according to a landmark new survey of city dwellers' views of their own city. The *My City: The People's Verdict* survey of 5231 people in all capital cities, and Newcastle and Wollongong, was undertaken by the Property Council. Respondents ranked the importance of 17 key attributes of cities and assessed their own city against these attributes.

The study crowns Adelaide as Australia's most liveable city, closely followed by Canberra, with Melbourne and Perth tying off for third place.

Australians scored their cities highly on the following attributes:

- recreational outdoor environments
- natural environments
- school and educational facilities
- good climate.

But Australians were scathing about their city's performance in the following areas:

- public transport services (42% approval)
- roads and traffic congestion (41% approval)
- environmental sustainability and climate change (37% approval)
- providing quality affordable housing (34% approval).

Source: www.propertyoz.com.au/Article/Resource.aspx?media=1978

ISBN 978 1 4586 6277 4

PIECES OF THE GREEN HEART

Actions fostered by *Growing a Green Heart Together* will contribute to Council's 2026 Vision of a clean, green city. *Growing a Green Heart Together* links several environmental citywide outcomes. Achieving this vision will strengthen Brisbane's community and economy.

Towards Zero Waste:
- no such thing as waste
- we purchase the right amount of the right things
- litter prevention, not cleaning up afterwards.

Green and Active Transport:
- high-quality public transport
- connected bikeways and walkways
- good alternatives to the car for getting around.

Water Smart:
- healthy river, creeks, bays and waterways
- using appropriate alternative water sources.

Food in the City:
- established community gardens
- local food production
- shared access to gardening space.

Cleaner and Sustainable Energy Use:
- carbon neutral
- affordable, available clean energy
- efficient energy use: buildings, spaces and vehicles.

Green and Biodiverse City:
- nature conservation
- diverse and protected habitat, flora and fauna
- 40% of city as natural habitat
- wildlife corridors.

Clean Air:
- pollution free.

10.13.1 Brisbane City Council's sustainability initiative

10.13.2 Enhancing liveability for young people with the Interactive Fountain, Perth

An interactive water feature in Forrest Place, Perth, recognises the importance of children in liveability planning. 'Building Places for Young People' is an initiative of the WA Commissioner for Children and Young People. Water-sensitive urban design (WSUD) is used across the country to manage water, create green spaces and enhance biodiversity. For instance, new suburbs in Canberra are created around WSUD principles and Adelaide uses recycled water on its green belt.

As in Europe, a key feature of liveability is successful public places. In Darwin the waterfront precinct offers an attractive public place to be enjoyed by young and old, a mix of cultures, locals and tourists. Hobart's Salamanca precinct is located by the waterfront and features beautiful historic sandstone buildings. It is home to a famous weekly market as well as numerous cultural events throughout the year. Hobart is also one of Australia's most affordable cities.

ISBN 978 1 4586 6277 4

10.13.3 A highlight of Hobart is the historic Salamanca precinct, where people can live, work, shop, dine and relax

10.13.5 A renewed waterfront precinct offers high liveability to the people of Darwin, with beaches, a wave pool, apartment accommodation, shops and restaurants, playgrounds, open spaces and transport connections

Green places, energy and transport

Melbourne's cyclist-friendly strategies gained it the title 'Bike City' from the Union Cycliste Internationale (UCI), and the provision of new bike lanes in Sydney has increased the number of commuter cyclists. In Adelaide, an extension of the tramline improved public transport, created green spaces and reduced car usage. A 100% solar-powered electric bus named Tindo ('sun' in the Indigenous Kaurna language) has been added to Adelaide's public transport system.

The City of Sydney is promoting green roofs and walls for new developments, such as One Central

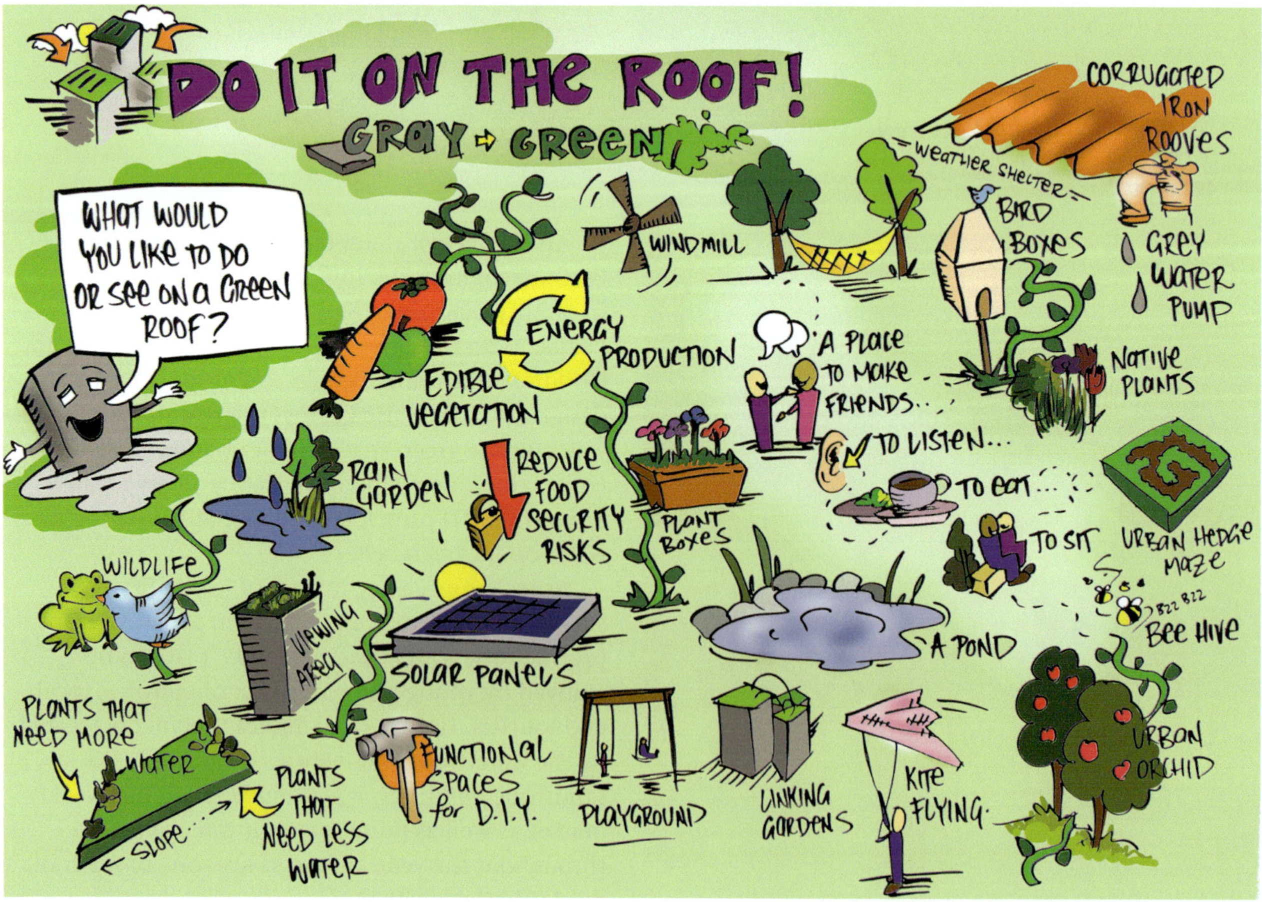

10.13.4 Melbourne's Do it on the Roof initiative

ISBN 978 1 4586 6277 4

Park, Ultimo, which has a 130-metre vertical garden. In 2014, Sydney City had 83 green roofs in place and over 50 approved for development. Melbourne's 'Do it on the Roof' campaign seeks better use of wasted roof spaces for greening the city and providing renewable energy, while on remote King Island, Tasmania, wind and solar energy means lower energy and living costs plus lower fossil fuel use.

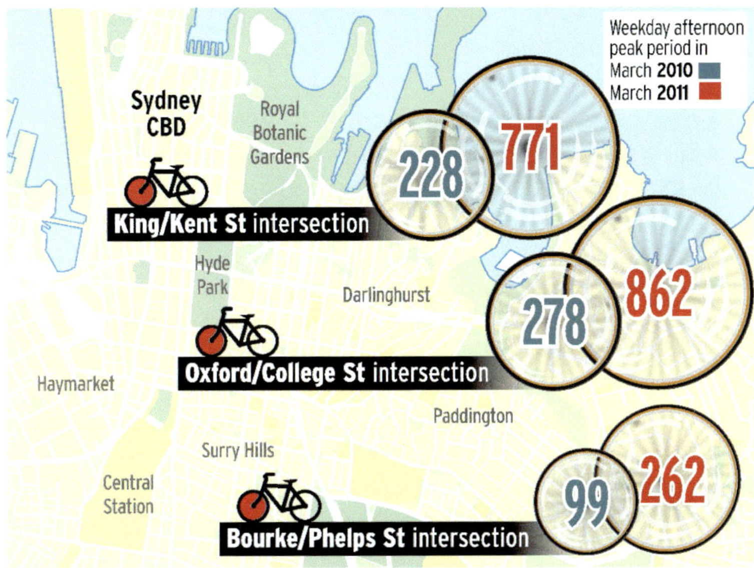

10.13.6 Number of bikes on a weekday afternoon (peak) in Sydney

Geoinfo

- The average new home in Australia is 215 m^2 compared to 76 m^2 in London.
- Mass transit, cycling and walking are considered the best transport options for cities.

Geoactivities 10.13

Knowledge and understanding

1 What aspects of liveability concern Australian citizens?

2 Identify one Australian government strategy to improve liveability.

3 Describe one strategy to improve liveability in a place you are familiar with.

4 What is WSUD? Why do you think WSUD is important in Australia?

5 Outline two strategies improving liveability in remote places.

Inquiry and skills

6 Refer to the article on Adelaide.

a How is the People's Verdict Survey different to other measures of liveability?

b In what areas did Australian cities perform highly and poorly in this survey?

c Suggest one way the performance of Australian cities could be improved.

7 Refer to 10.13.1.

a List the seven targets for a green, clean, Brisbane.

b Why is 'Growing a Green Heart Together' a community plan?

8 Refer to 10.13.2.

a Explain how an interactive fountain could improve the liveability of a place.

b What other facilities could create better communities for children?

9 Refer to 10.13.3 and 10.13.5. Identify features of Darwin's waterfront and Salamanca in Hobart that create high liveability.

10 Refer to 10.13.4.

a 'If we steal the ground for a building we can give it back to nature on the roof'. Discuss the meaning of this quote.

b Work in pairs to design a green roof that will improve the liveability and sustainability of your school. You have a budget of $10 000. Choose a school building. Imagine the building has a flat rooftop. Estimate the dimensions and area of the rooftop by measuring the building at ground level. Draw a plan of the roof with a scale. Add the features you would like to see on the green roof. When your plan is completed draw a picture to show what the roof will look like in 2 years time.

c Have a class vote to determine the most popular design. Create a SWOT analysis for the most popular design. Use the SWOT analysis to justify turning the plan into reality.

11 Refer to 10.13.6.

a Calculate the total increase in weekday afternoon cyclists at three Sydney intersections between March 2010 March and March 2011.

b Which intersection had the greatest increase in cyclists?

c How does cycling improve liveability for individuals and the city as a whole?

ISBN 978 1 4586 6277 4

Geothink

Attributes of a liveable place

1 Refer to 10.14.1.
   a Identify the attributes of a liveable place in the image.
   b List features that enhance liveability for people of all ages not shown in the image.
   c Create a portfolio of annotated photographs that illustrate the liveability of places in Australia. Present this as an e-portfolio.

2 Describe one benefit for each of the following liveability strategies.
   a Complete streets
   b Walkability
   c Green infrastructure
   d Urban agriculture
   e One Planet Living
   f Eyes on the street
   g Microcredit
   h Water-sensitive urban design
   i Renewable energy

3 Refer to 10.14.2. In the *Australian Urban Design Protocol*, a liveable community is comfortable, vibrant, safe and walkable. Use the list of attributes in the protocol to evaluate the liveability of your community.
   a Write a statement about your community for each of the four measures of liveability in the *Australian Urban Design Protocol*; for example, My community is safe (or unsafe) because …
   b Suggest four ways your community could be made more liveable (in your opinion).

New transport hierarchy

4 Refer to 10.14.3.
   a What is the most important consideration in this planning hierarchy?
   b How is this a change from the past?
   c Why is change considered important for enhancing liveability?
   d Why are cars at the bottom of the hierarchy? Is this fair for all types of communities? Explain your answer.

10.14.1 Attributes of a liveable place

Principles	Outcomes	Attributes—how it helps to achieve world-class urban design
Comfortable	Comfortable and welcoming	• It feels comfortable to walk through, sit, stand, play, talk, read or just relax and contemplate • It is not too exposed to unpleasant noise, wind, heat, rain, traffic or pollution • You can freely use the place, or at least part of it, without having to pay • You can be yourself and feel included as part of the community • It caters for people with various physical capabilities, the old and the young
Vibrant	Vibrant, with people around	• You can see that there are other people around • People are enjoying themselves and each other's company • There are places to meet and interact, play, explore, recreate and unwind • It is a place you want to visit, experience or live in
Safe	Feels safe	• It feels safe and secure, even at night or on your own • There aren't signs of decay such as graffiti, rubbish, weeds or derelict buildings and places • Roads and paths are safe for adults and children to walk or ride their bikes
Walkable	Enjoyable and easy to walk and bicycle around	• It prioritises people walking or riding before vehicles • It is easy to get around on foot, bike, wheelchair, pushing a pram or wheeling luggage • Buildings and streets feel like they're the right size and type for that place • It encourages physical activity and social interaction, and promotes a healthy lifestyle

10.14.2 Evaluating liveability

ISBN 978 1 4586 6277 4

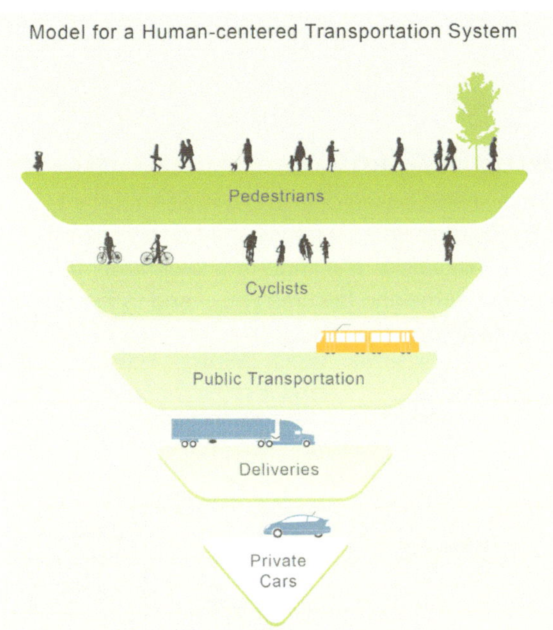

10.14.3 Transport planning hierarchy

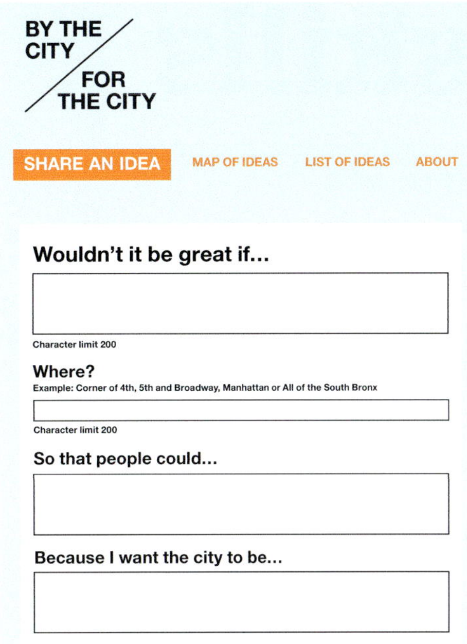
BY THE CITY / FOR THE CITY

SHARE AN IDEA | MAP OF IDEAS | LIST OF IDEAS | ABOUT

Wouldn't it be great if...

Character limit 200

Where?

Example: Corner of 4th, 5th and Broadway, Manhattan or All of the South Bronx

Character limit 200

So that people could...

Because I want the city to be...

10.14.4 This was an online survey for New York citizens in 2011. It can be applied to any place seeking input from citizens on their ideas to enhance liveability.

Planning liveable places

5 Refer to 10.14.4. Develop a simple survey form like this one. Complete three forms suggesting strategies you would like to see used in your street, your community and your town or rural district to enhance liveability. Discuss with your class.

6 Refer to 10.14.5.
   a List goods and services provided by local farms.
   b What would be the role of community gardens and urban orchards?
   c How could a person living at point A travel to work in the city centre?
   d Is the transport system integrated?
   e What would be the purpose of each interchange?
   f How are water and waste used sustainably?
   g What is the purpose of the farmers markets?
   h Name one aspect of liveability that is difficult to show on a map.

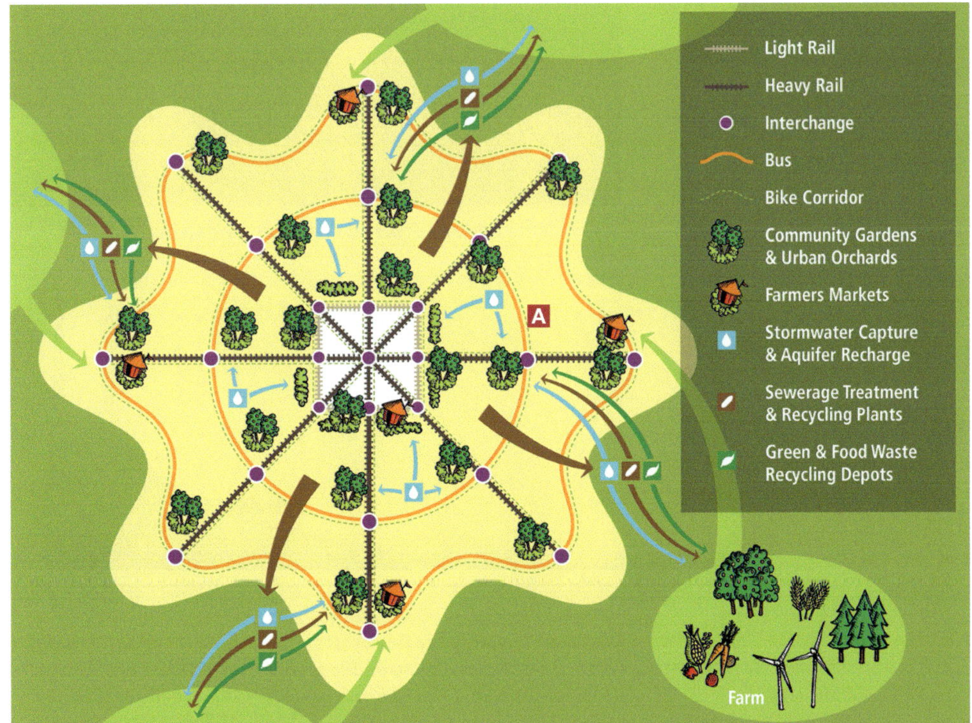

10.14.5 Planning liveable places

Chapter 10

Geoskills

Geographical skills are used to make a range of decisions and source information.

- **Primary sources** of information are original materials collected by the researcher. They include interviews, surveys, questionnaires, measurements and photographs, along with field sketches, diagrams, maps and statistics.
- **Secondary sources** include a wide range of print and online publications and resources, such as published newspapers, journals, magazines, photos and images.

To create presentations of your research, you can use a wide range of web 2.0 presentation tools. The following geographical tools are to be integrated in geographical knowledge.

Selecting photographs

There are four types of photograph commonly used by geographers to show different points of view. These are: **ground**, **oblique**, **aerial** or **satellite views**. They can be used to study landforms and undertake impact assessments. Be sure to carefully consider the captions you use (and include source lines where relevant).

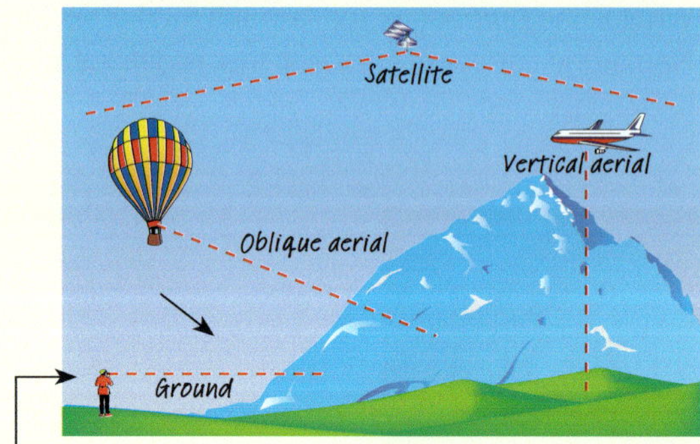

Ground photos are taken at ground level. Scale is larger closer to the camera

Drawing sketches

Creating a sketch based on a photo is an important geographical skill. Using a pencil and tracing paper makes the task easier. It is important to consider the types of landforms and features you intend to focus on, and to write appropriate labels.

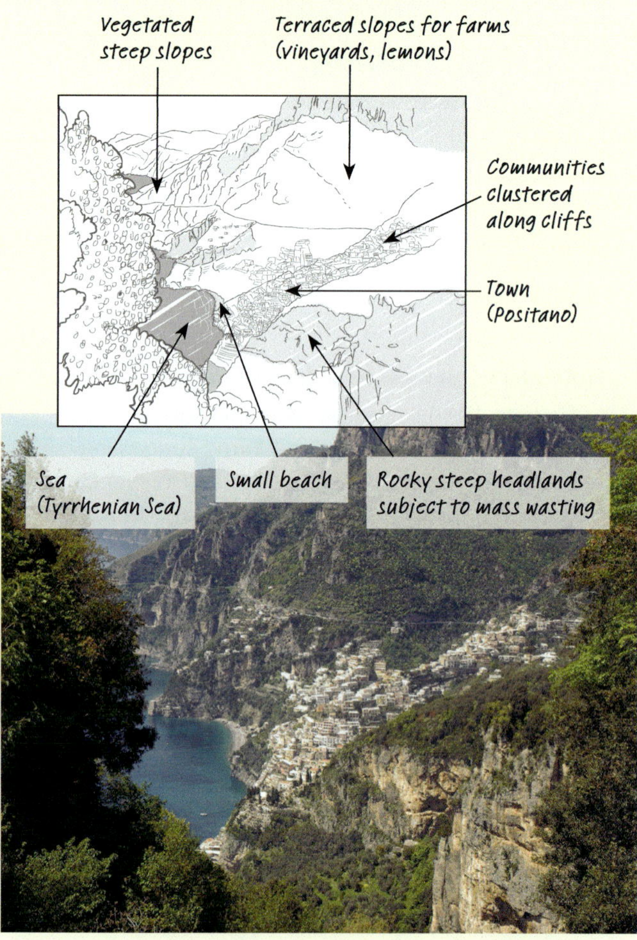

In 1997, Italy's Amalfi Coast was listed as a World Heritage Site

Drawing a transect

A transect is a simplified diagram similar to a cross-section, but represented as an external view of the landscape viewed from the ground level. A transect can show natural and human features, and illustrates changes in the terrain and built environment.

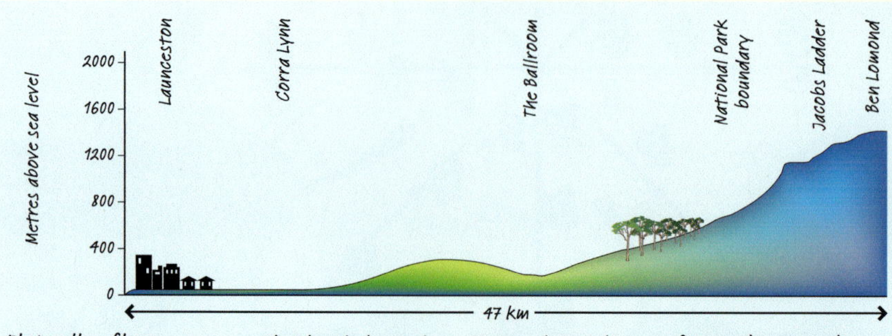

Flat valley floor, vegetation cleared for settlements—location of the city of Launceston in Tasmania

Flat land cleared for farms along fertile river valley. Farm houses and roads

Rising slope where some forests remain. Sheep and cattle grazing

Steep slope, cooler climate, no settlement, mostly National Park with ski village near summit of Ben Lomond

ISBN 978 1 4586 6277 4

Map essentials

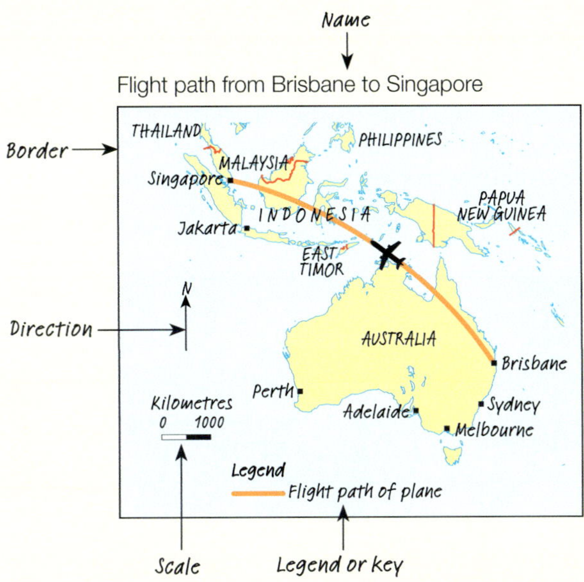

To help us understand them, maps are often labelled with BOLTSS:

- ***b****order*—a line around the map
- ***o****rientation*—the compass direction, indicated by arrows showing North
- ***l****egend* (or Key)—explanation of symbols, such as green for trees and blue for rivers. Note: a key is more specific to an individual map.
- ***t****itle*—the name of the map indicating its area, region and focus
- ***s****cale*—the measurement converting the real distance on the ground to the map dimensions. Scales are often represented as a fraction, e.g. 1:100 000.
- ***s****ource*—the source line for reproduction, map projection or data.

Latitude and longitude

The imaginary lines running parallel to the equator are called **lines of latitude**, and are measured in degrees. The equator is at 0°, and divides the Earth into the northern and southern hemispheres. For example:

- North Pole 90°N
- South Pole 90°S
- Tropic of Cancer 23½°N
- Tropic of Capricorn 23½°S.

The imaginary lines running between the North and South Pole are called **meridians**, or **lines of longitude**. They are also measured in degrees. For example:

- the meridian of Greenwich (UK) is 0°
- the lines to the right of the meridian of Greenwich are east, and the lines to the left are west, running from 0° to 180°
- the International Date Line is 180° and runs through the Pacific Ocean.

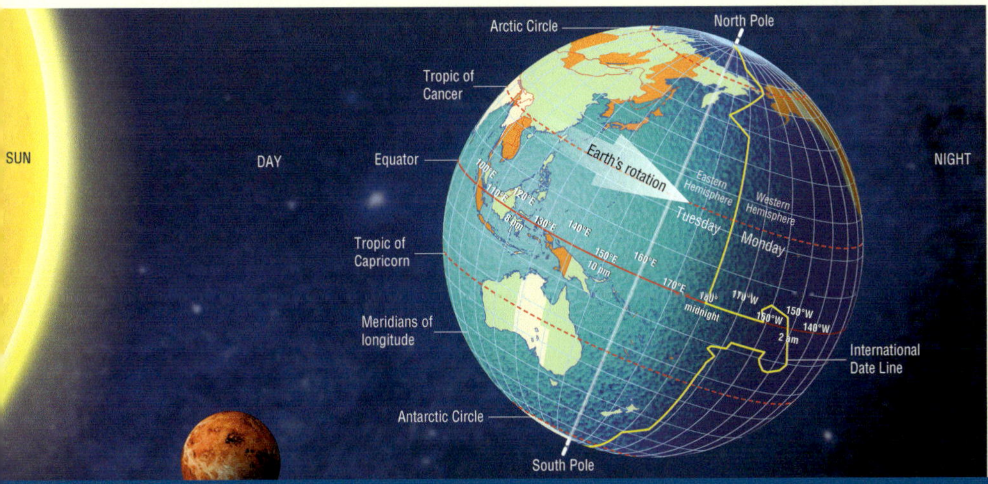

Time zones are based on meridians of longitude. There are 24 hours in the day and 24 time zones. Each meridian zone is 15° wide, which is the same distance the sun appears to travel in the sky every hour.

Atlas indexes give degrees (°) and minutes ('). Latitude is always given before longitude. In the example, the reference for Wellington in New Zealand is latitude 41°17'S and longitude 174°47'E.

ISBN 978 1 4586 6277 4

Scale, direction and bearing

Scale

A map scale refers to the relationship between distance on the ground and the corresponding distance on a map. Scale can be represented as a line, ratio, fraction or a statement. For example on a 1:100000 scale map, 1 centimetre on the map represents 1 kilometre on the ground.

Scale can be represented as a:

- line (e.g. 1 centimetre represents 1 kilometre)

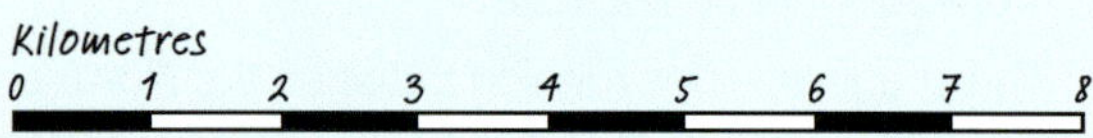

- fraction: 1/100 000
- ratio 1:100 000
- statement: 1 centimetre on the map represents 1000 metres or 100000 centimetres on the ground.

Maps can be small or large scale. Examples of small scale maps are wall world maps or a map of Australia. Examples of large scale maps are maps of a local area or a bushwalking track. This type of map includes more details.

Map direction

A map and a compass are important to determine direction. You might use a map and compass when participating in the sport of orienteering or navigating across a mountain range.

Direction has four cardinal points (N, S, E, W). These are further divided into eight points (e.g. NE, SW) and 16 points (e.g. ENE, WSW).

Topographic maps usually contain three types of north:

- true north (TN), which is the true direction of the North Pole
- grid north (GN), which follows the north–south grid lines on a topographic map and is used to read direction, because bearings are plotted from GN
- magnetic north (MN), which is the direction that the north point of a compass needle points; MN alters continually with the world's magnetic poles.

The difference between TN and GN is called the *grid convergence*. The difference between MN and GN is known as the *grid–magnetic angle*. The diagram below is a direction indicator from a topographic map of the Alligator Rivers in the Northern Territory.

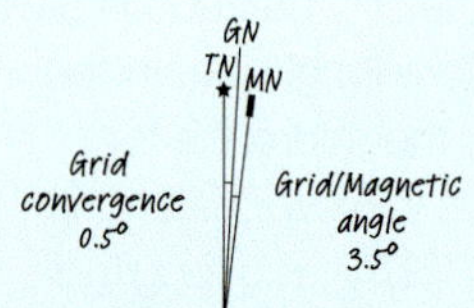

True North (TN), Grid North (GN) and Magnetic North (MN) are shown diagrammatically for the centre of the map

MN is correct for 2000 and moves easterly by less than 0.1° in 10 years

Direction and bearing

Direction can also be given as a bearing. Bearing is divided into 360 degrees (as in a circle) and is usually measured on a map using a protractor. On the diagram below, the direction from O to A is NE and the bearing is 66° 30'.

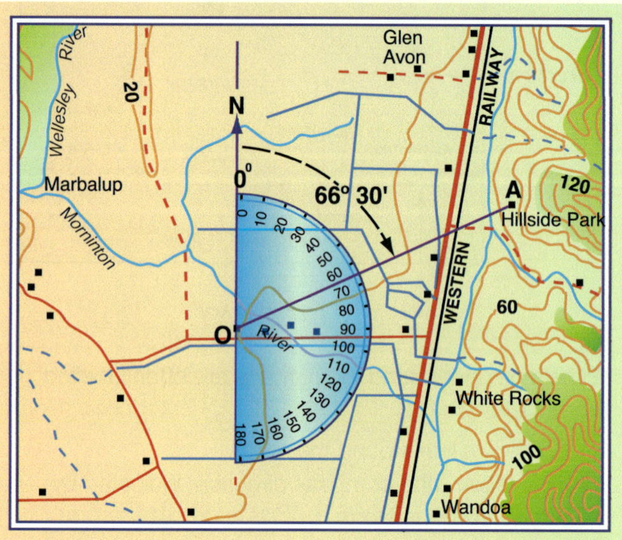

Wind direction

If a cyclone or fire is approaching your home you will need to understand the location of the impending disaster. What does it mean when the television announcement declares that wind is coming from the NE (direction) or from 45 degrees (bearing)?

You name a wind according to the direction it comes *from*. For example, a southerly comes from the south and generally brings cooler weather to Australia's southern states.

An arrow symbol is used to show wind direction. Wind speed is shown by the feathers on the arrow. Each feather represents a unit of wind speed measured in knots or kilometres per hour.

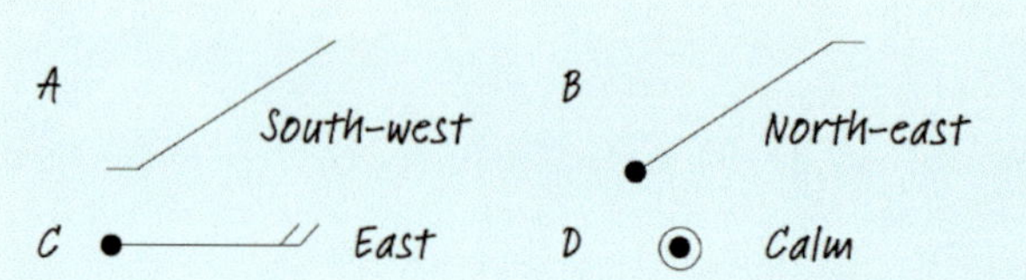

See Geolinks for more information about scale, map direction and bearing.

ISBN 978 1 4586 6277 4

Using grid coordinates

Alphanumerical systems

In an **alphanumerical system**, map grids are marked by the alphabet along one axis and numbers along the other, in order to provide a coordinate. For example, on this map you can see the windmill is located in D4.

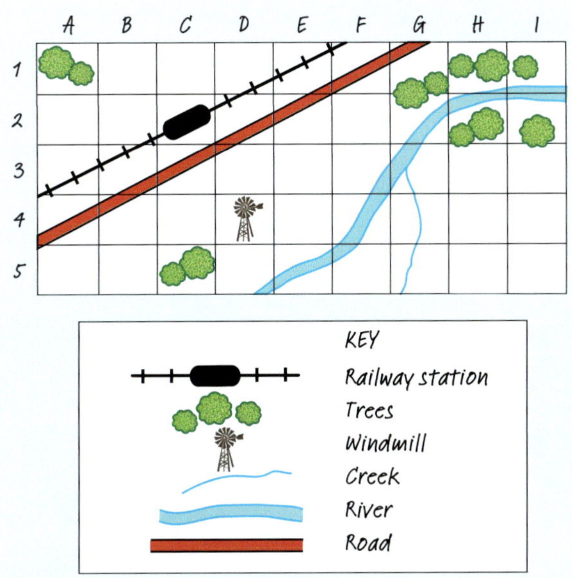

Area and grid references

An **area reference (AR)** is a four-figure map coordinate formed by the intersection of eastings and northings (e.g. AR 1943 in the diagram below). Area references are useful for locating large features, such as towns or lakes. A **grid reference (GR)** is a six figure map coordinate used by geographers to locate smaller features, such as a building. To find these extra numbers, each easting and northing is subdivided into 10 equal parts (e.g. GR 196437 in the diagram below).

In the topographic map at the bottom of this page, grid references can be used to locate a campsite in a nature reserve.

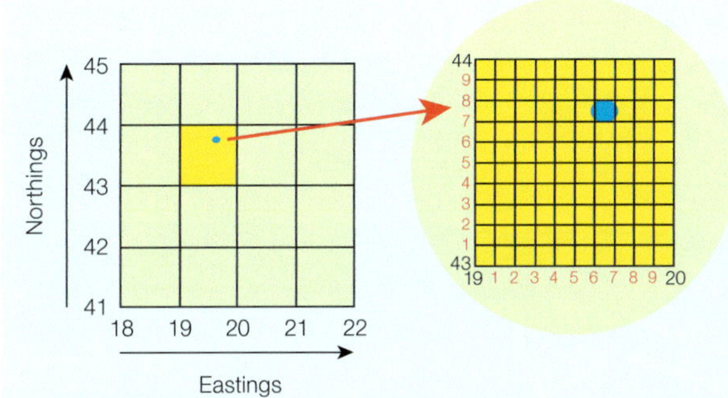

Sample area reference (AR 1943) and grid reference (GR 196437).

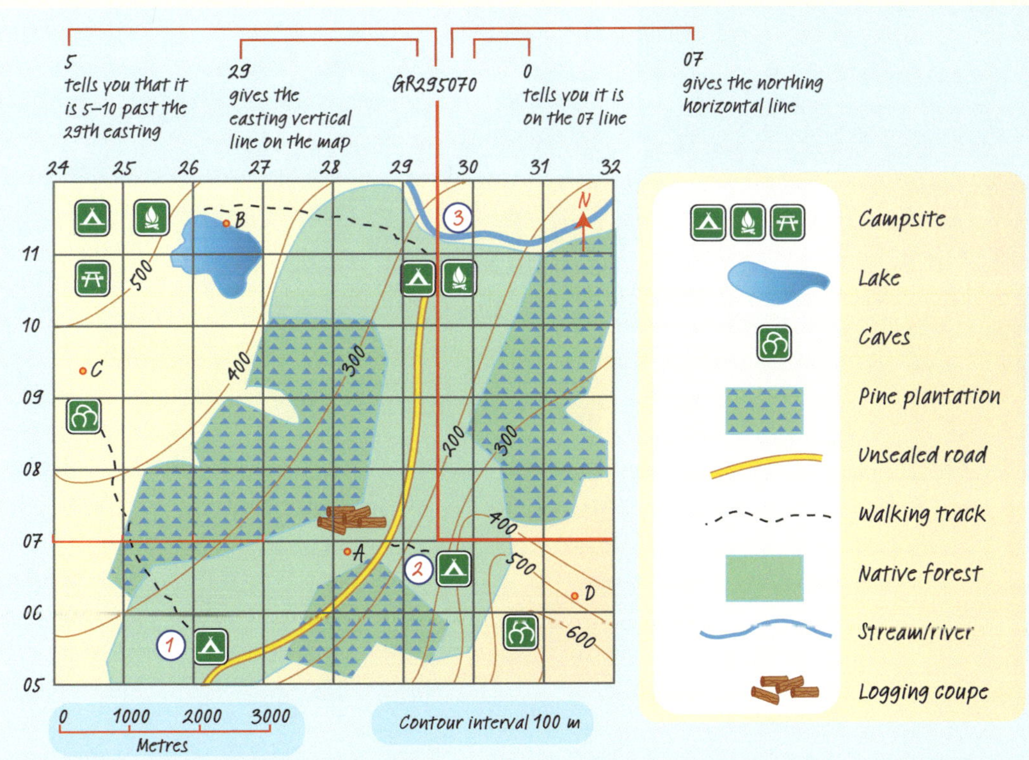

A topographic map of a nature reserve, showing the location of a camp site at GR 295070.

Topographic maps

The word 'topography' comes from the Greek words 'topos' (place) and 'graphos' (write). Topography is the study of the surface features of Earth and other astronomical bodies. A topographic map (also called a physical map) is a detailed and accurate representation of these surface features developed through measurements from satellite imagery, vertical aerial photography and ground research. It is necessary to regularly update topographic maps to reflect changes to the Earth's surface.

Topographic maps are very detailed, representing both natural and human elements. They show:

- *relief*—to represent elevation and terrain (e.g. mountains, valleys and plains)
- *hydrography*—waterways (including oceans, seas, rivers, lakes, wetlands and coast)
- *vegetation*—natural habitats, wooded and cleared areas, plantations and agricultural land use
- *built features*—human settlement and infrastructure (e.g. roads, airports, industrial areas, suburban sprawl).
- *place names*—landforms, towns and cities, regions and countries (state and national borders).

Aerial photograph of Trinity Inlet, Cairns

Topographic map of Trinity Inlet, Cairns

ISBN 978 1 4586 6277 4

Relief maps

Relief is represented on topographic (physical) maps using contour lines, colours, spot heights and symbols.

Contour lines

Contour lines are lines that join places of equal height above sea level. The contour lines always increase or decrease by the same amount. The difference between the spacing of each contour is known is the contour interval. Contours may indicate the presence of different features in a landscape, as outlined in more detail in the accompanying annotated diagrams on this page.

Gradient

The gradient of an area is the angle of the slope between two points. It is calculated by dividing the vertical difference in height (rise) by the horizontal distance (run).

$$\text{Gradient} = \frac{\text{Vertical distance in height (rise)}}{\text{Horizontal distance (run)}}$$

Gradient can be represented in four ways. The following example has a distance between two points of 1000 m and a rise of 100 m.

1 Fraction: $\frac{1}{5}$
2 Ratio: 1:5
3 Words: For every 5 m there is a rise of 1 m. Or there is a 1 m rise for every 5 m.
4 Percentage: 20%

Contour lines give a good indication of the gradient of an area. The closer the lines on a topographic map the steeper the gradient and vice versa. Surveyors and engineers need to accurately measure gradients when building dams, roads, railways and construction sites. Gradients are also important in activities like rock climbing, bushwalking and skiing.

Steeper gradients have greater runoff and potential erosion (removal of sediment). They also pose potential hazards in terms of landslides, rock falls, mudslides and avalanches. Authorities use topographic maps to identify and manage these areas of risk.

Landslides may be caused by any combination of water saturation and steep slopes. Most slides are triggered after an extended period of heavy rainfall when the ground is wet for a prolonged period of time.

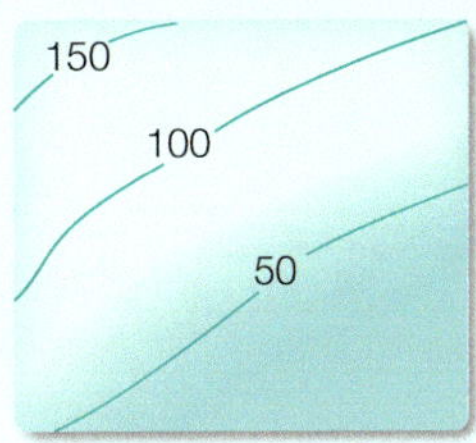

Flat land with gentle slopes has widely spaced contours.

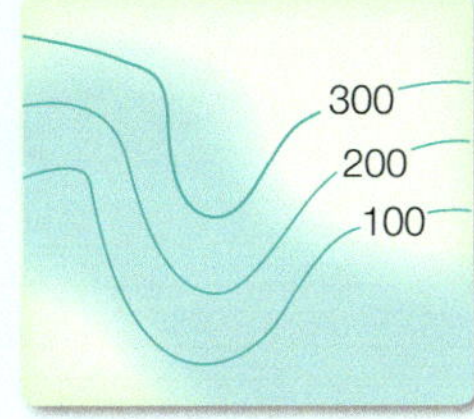

Spurs are extensions of hills or mountains with contours that point towards lowest land.

Knolls or round hills have evenly spaced circular contours.

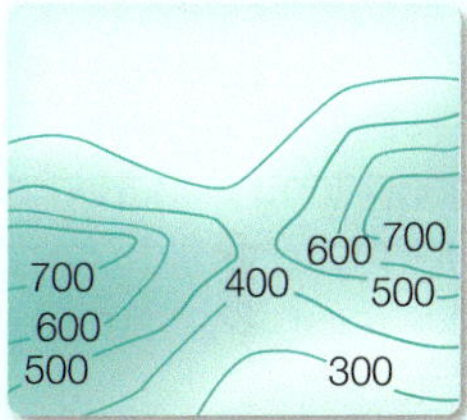

Saddles have two areas of higher land with a dip between them.

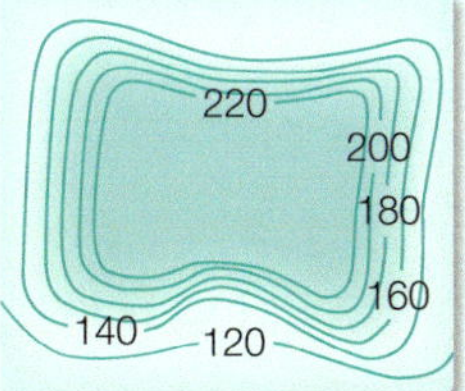

Plateaus are large areas of flat land with closely spaced contours at the steep edges.

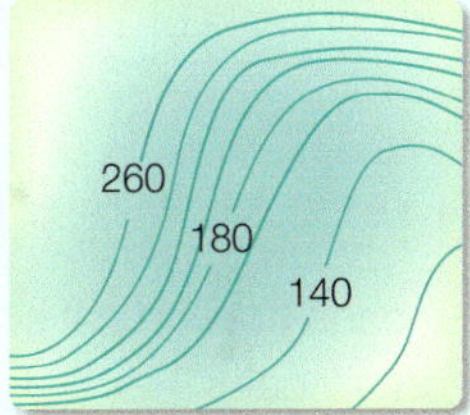

Steep slopes or cliffs have very closely spaced contours.

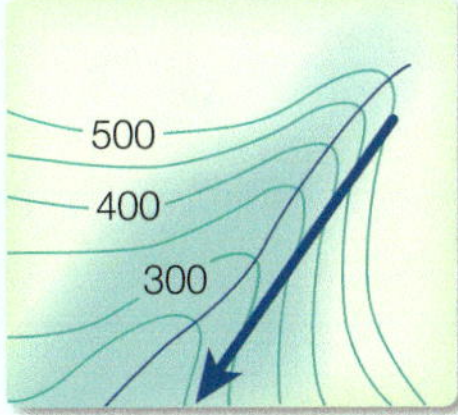

River Valleys flow from the highest to lowest point.

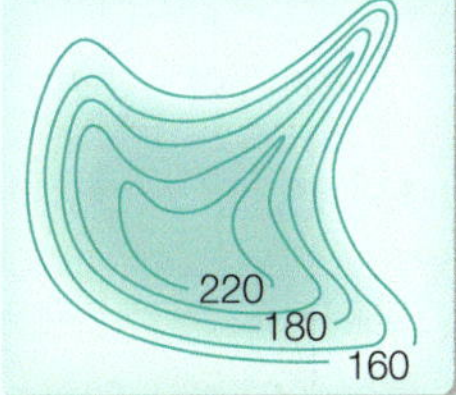

Ridges are the long, narrow tops of hills or mountains with contours that point away from the high land.

Contour patterns of some common landforms

ISBN 978 1 4586 6277 4

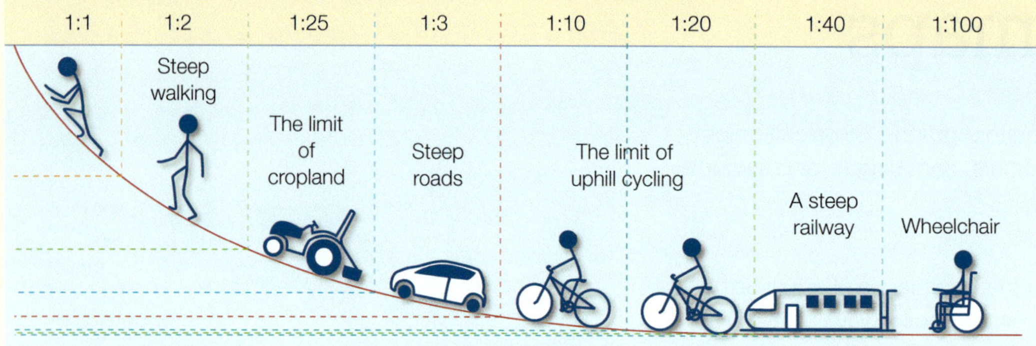

The gradient of slope determines the limits of human activities

Vertical exaggeration

For a cross-section to fit on a page and be visually useful it has to be distorted vertically. Each cross-section has its own vertical exaggeration (VE) that is a measure of this distortion of shape. The VE is written on the cross-section, along with the horizontal and vertical scales, to ensure that users do not make false interpretations of the relative relief. The formula for calculating VE is:

$$VE = \frac{\text{vertical scale}}{\text{horizontal scale}}$$

If a cross-section is constructed with a:

- horizontal scale of 1 cm to 1 km (a ratio of 1:100 000)
- vertical scale of 1 cm to 200 m (a ratio of 1:20 000)

then

$$VE = \frac{1/100\,000}{1/20\,000} = 5$$

This means the vertical scale is five times that of the horizontal scale.

Steps in constructing a cross-section

1. Choose two points on a map for your transect. Label them A and B.
2. Place an edge of a sheet of paper along the transect between points A and B. Write the heights underneath these two marks.
3. Work carefully from left to right; mark where each contour line meets the edge of the paper.
4. Record the height of each contour under the mark.
5. Draw a graph to plot the contours. The horizontal axis should be the same length as the distance between points A to B on the map.
6. Choose a suitable scale for the vertical axis.
7. Use the contour heights marked to plot each point on the graph.
8. Join the points with a smooth line to show the cross-section (or profile) between points A and B.
9. Give the cross-section a border, title and labels.
10. Show the horizontal and vertical scales and the VE.

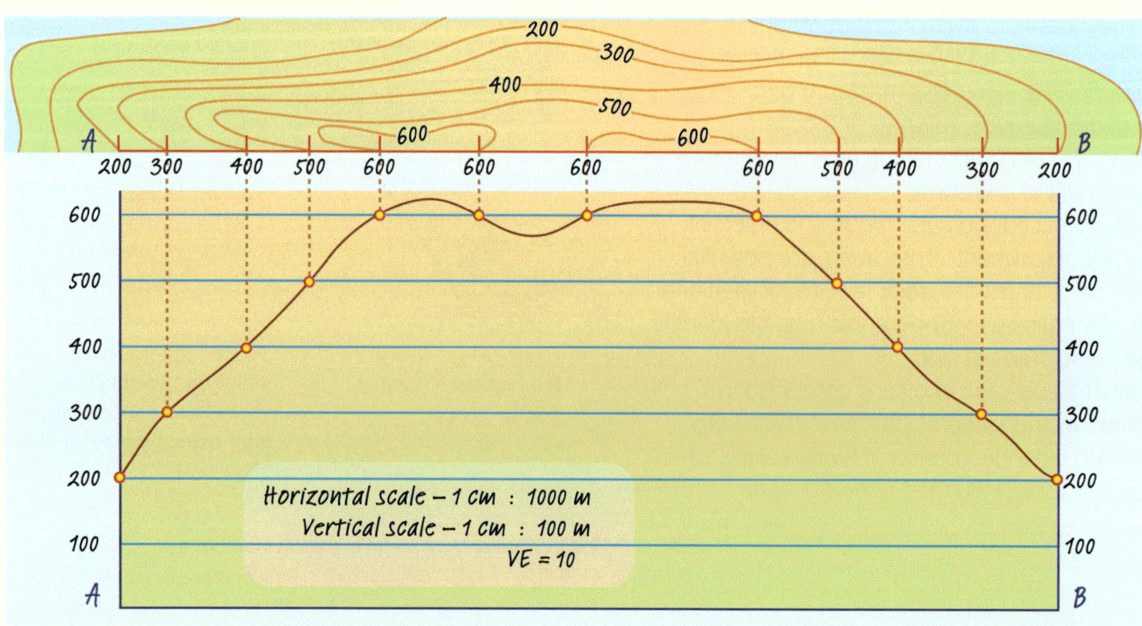

Constructing a cross-section from a topographic map

ISBN 978 1 4586 6277 4

Thematic maps

Maps are excellent tools to illustrate the spatial distribution of geographical data such as location of deserts, human wellbeing indicators, biodiversity hotspots and impacts of climate change on different places. In some cases, accuracy of borders and geographical features is sacrificed to the demonstration of key points or summary conclusions

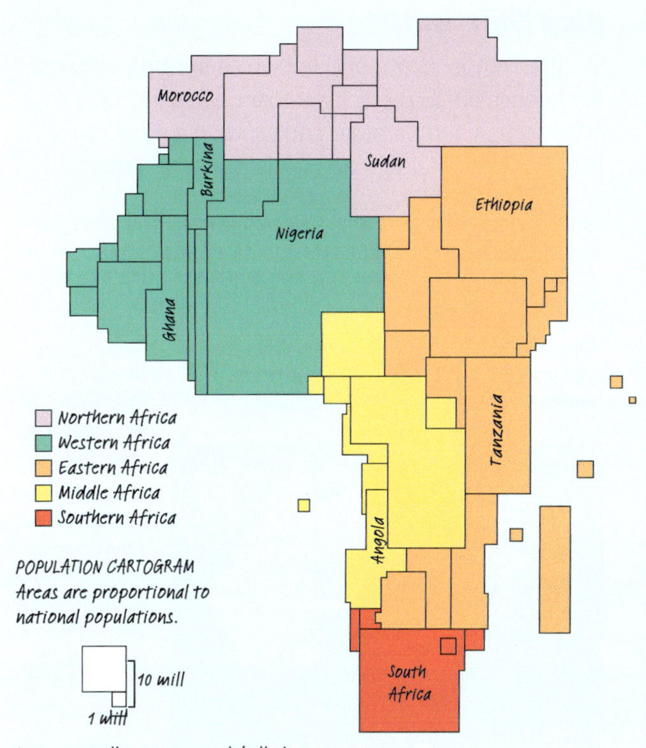

Population cartograms

A cartogram shows the size of a country or state in reference to a variable statistic, such as population or Gross Domestic Product (GDP). This creates (in some cases) a distortion of the actual physical size of the country or land area to map and compare the relevant data. For example, in this map of Africa, the comparative size of different countries is distorted to show that a proportionately larger number of people live in Nigeria while fewer people live in South Africa.

Choropleth maps

Choropleth maps show the variable density or measurable quantity of objects or population by using different colours, shades, patterns and symbols.

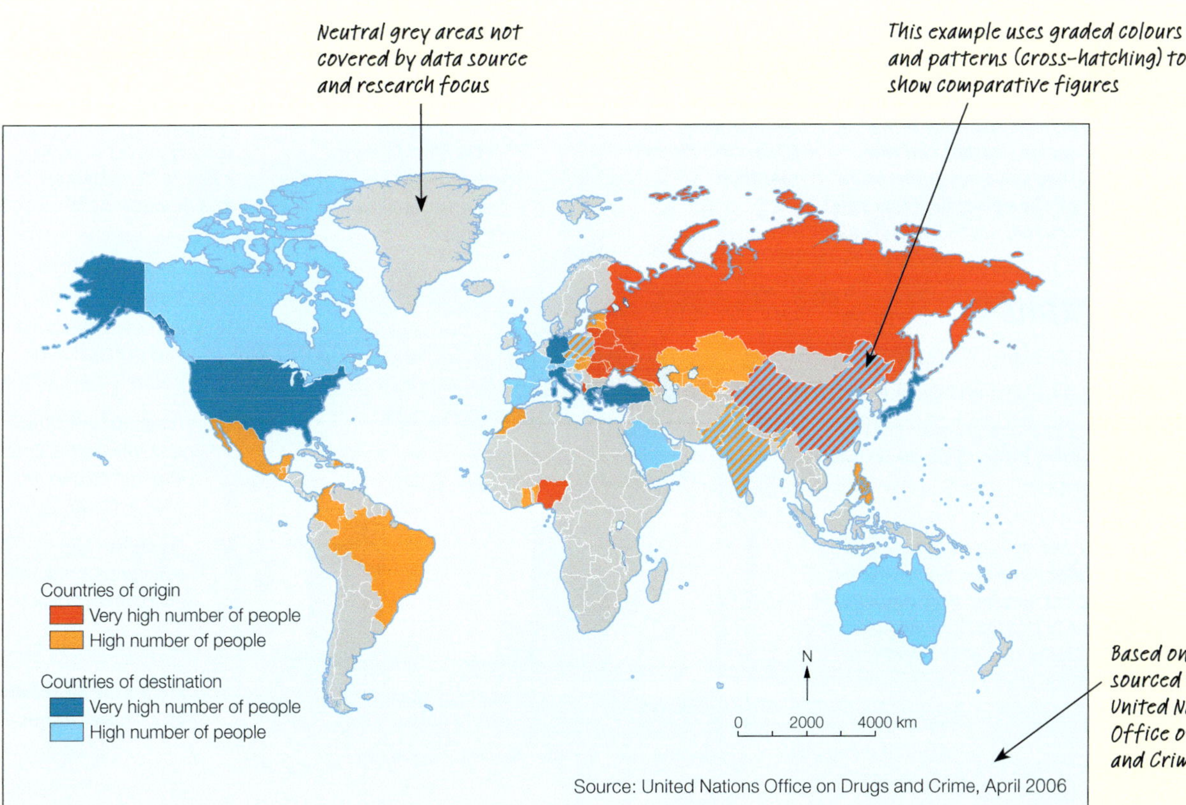

Choropleth map of key source and destination countries in human trafficking

ISBN 978 1 4586 6277 4

Weather and climate

Weather maps

Weather maps, commonly known as **synoptic charts**, aid forecasters to predict weather conditions including precipitation, temperature and wind on a particular day. Here is an annotated map of Australia's weather.

Isoline maps, often called **isopleths**, show how hot or cold it is in different parts of the world at a particular season or time of year. For example, in the maps shown on this page, you can see global average temperatures in January and July.

Winds blow in a clockwise direction around a low pressure system, and in an anticlockwise direction around a high pressure system in the southern hemisphere. It is the opposite direction in the northern hemisphere

High pressure system—high air pressure in the centre is 1026 hPa

Isobars—lines joining places with the same air pressure

Air pressure measured by a barometer in hectopascals, 1008 hPa

Wind direction and speed. You name a wind from where it comes from. The wind at Townsville comes from the NE. It is blowing between 5–13 km/h

Low pressure system—lowest air pressure in the centre is 1015 hPa

Rainfall or precipitation that has fallen in the last 30 hours

Cold fronts coming from Antarctica—looks like ice along a line. They bring cooler weather and rain. They move in the direction of the points i.e. from poles towards southern Australia

Wind—it is windier and seas are rougher when the isobars are closer together

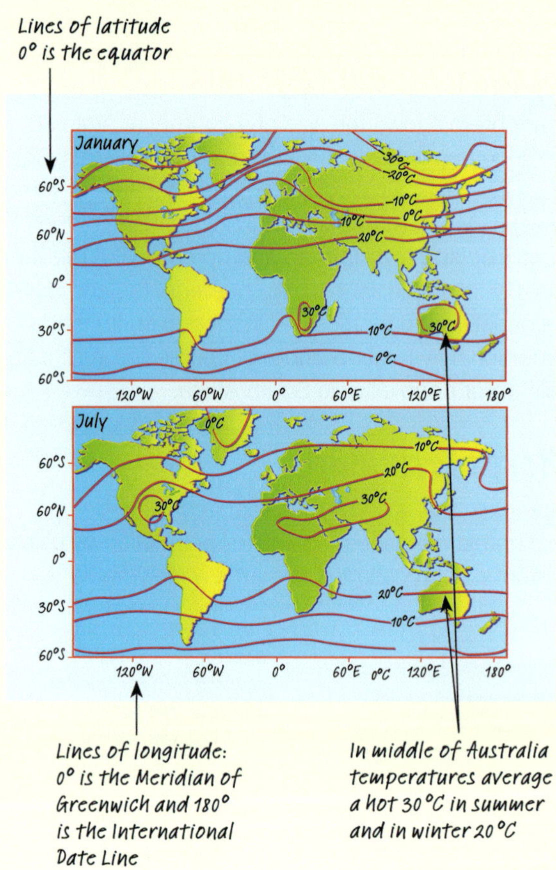

Climate graphs

Variations in temperature, precipitation, humidity and winds are referred to in the short term as weather (daily) and in the long term as climate. Here is an example of a climate graph:

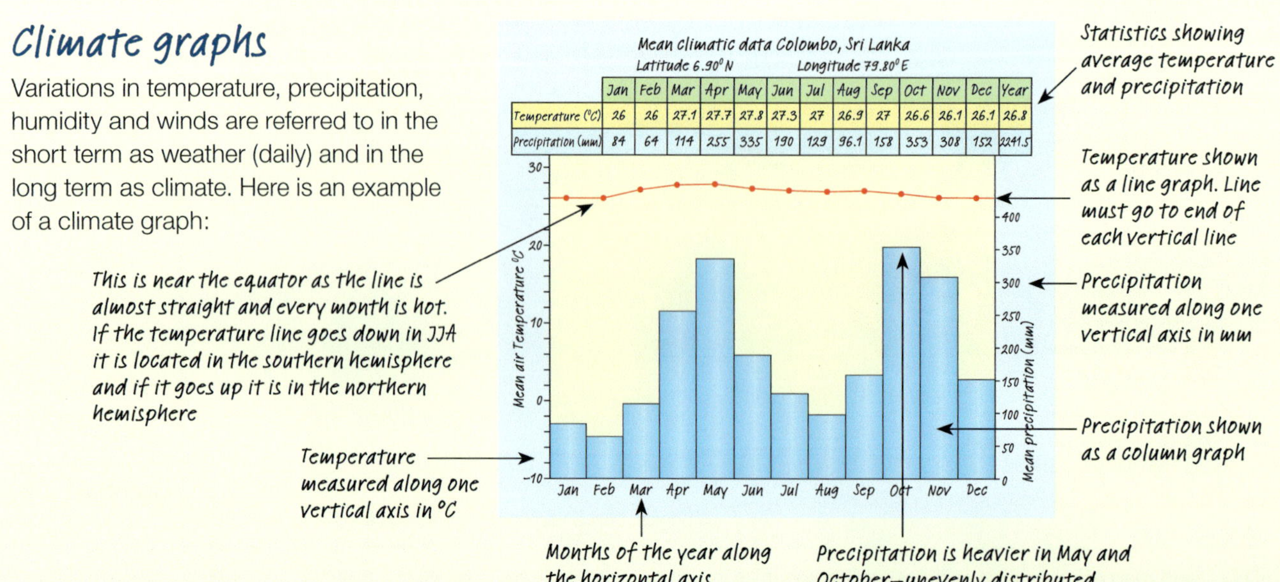

Mean climatic data Colombo, Sri Lanka
Latitude 6.90° N Longitude 79.80° E

	Jan	Feb	Mar	Apr	May	Jun	Jul	Aug	Sep	Oct	Nov	Dec	Year
Temperature (°C)	26	26	27.1	27.7	27.8	27.3	27	26.9	27	26.6	26.1	26.1	26.8
Precipitation (mm)	84	64	114	255	335	190	129	96.1	158	353	308	152	2241.5

ISBN 978 1 4586 6277 4

Data and graphs

Graphs are a form of data visualisation that helps make information more accessible. Graphs are used by geographers to show different types of data and comparative statistics across different regions and data subsets and also to show developments over time. Sometimes these figures contrast actual and projected (future) figures. Graphs can be produced manually or generated from an Excel spreadsheet using a graphics program. Following are a selection of the main graph types.

Data sources

Up-to-date data and comparative statistics are best sourced from official agencies such as the United Nations, state and federal government departments like the Australian Bureau of Statistics and the Bureau of Meteorology and the US Geological Survey (see Geolinks for more reference sources).

Bar and column graphs

Bar graphs are drawn horizontally and are used to compare sizes of data between different places at a single point in time. Column graphs are drawn vertically and are used for showing a single variable over different regions or periods of time. Multiple column graphs compare a number of sets of data with the same category, as in the following example.

Cumulative line graphs

A line or linear graph connects points as markers to show change. These can be used to feature actual and projected figures to indicate past, current and future trends. The cumulative line graph featured here, shows subsets of data from different regions over the time period in question for the production of global industrial roundwood.

Title or caption of graph

Shows that North America has the greatest access to communications

Access to communications by world regions

Number per 1000 people

0
200
400
600
800
1000

World
Europe
Middle East & North Africa
Sub-Saharan Africa
Asia (excluding Middle East)
North America
Central America & Carribean
South America
Oceania
Australia

Televisions
Internet users

Major world region

Scale gives measurement or volume (sometimes as a percentage)

Legend or key for data

Label axis title tells you what is being measured

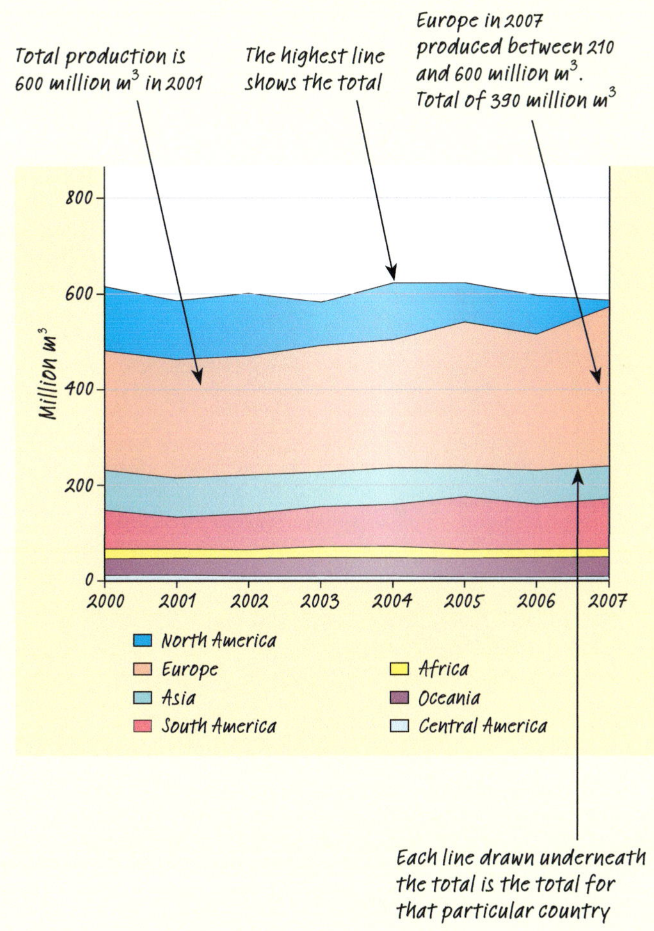

ISBN 978 1 4586 6277 4

Geographic information systems (GIS)

Geographic information systems bring together information from multiple data sources. Each set of data is created as a layer, culminating in a series of layers on top each other. To create these data sets, the data must be tied to a specific location on the Earth's surface. GIS applications are useful for the analysis of patterns and trends, and are often used in urban planning and cartography, environmental impact assessment reports and natural resource management.

A GIS is an information management system, incorporating computer hardware and software, that also enables users to access, create and share knowledge through mobile communications networks.

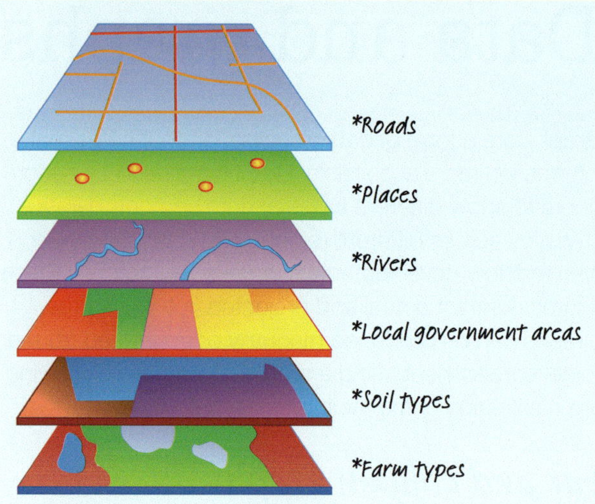

Pie graphs

Pie graphs are used to show proportions as a percentage. Sectors are labelled or coloured with a key to differentiate them. They are usually drawn clockwise from the 12 o'clock position, showing the largest segment first to the smallest segment last.

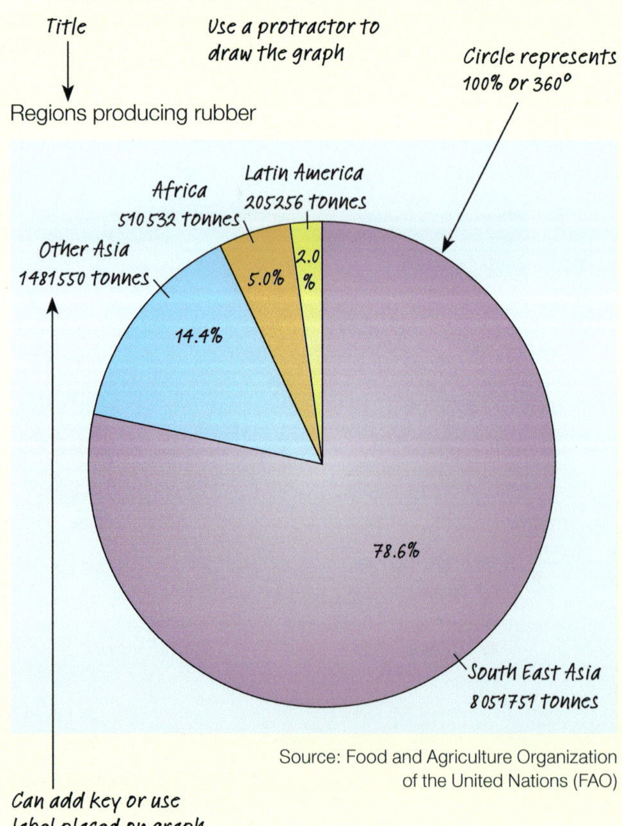

Source: Food and Agriculture Organization of the United Nations (FAO)

Population profiles

Despite the name, these graphs aren't always pyramid-shaped. The example below gives an overview of Australia's population profile by gender. From this graph, it is easy to see at a glance that there is a smaller percentage of young people than older people (aged 65 years or more).

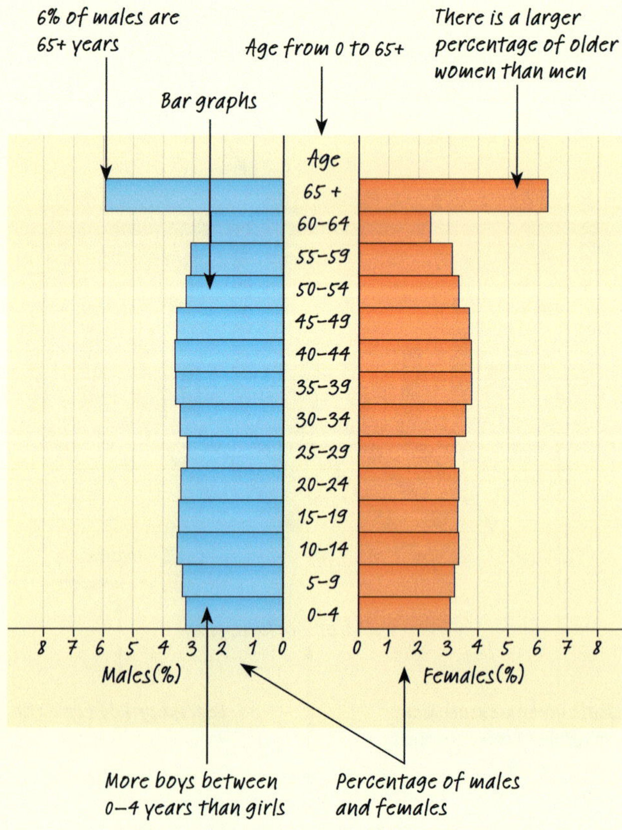

Source: FAO

ISBN 978 1 4586 6277 4

Ternary graphs

Ternary graphs have three related variables that, when combined, add up to 100%. An example is three components in a soil sample (clay, silt and sand).

When reading a ternary graph on soil composition, soil that is high in clay at point A would contain 80% clay, 10% silt and 10% sand, which adds up to 100%.

What are the soil components at B, C and D?

Here are the answers:

B – 10% clay, 80% silt, 10% sand
C – 10% clay, 40% silt, 50% sand
D – 30% clay, 0% silt, 70% sand

See Geolinks for more information about ternary graphs.

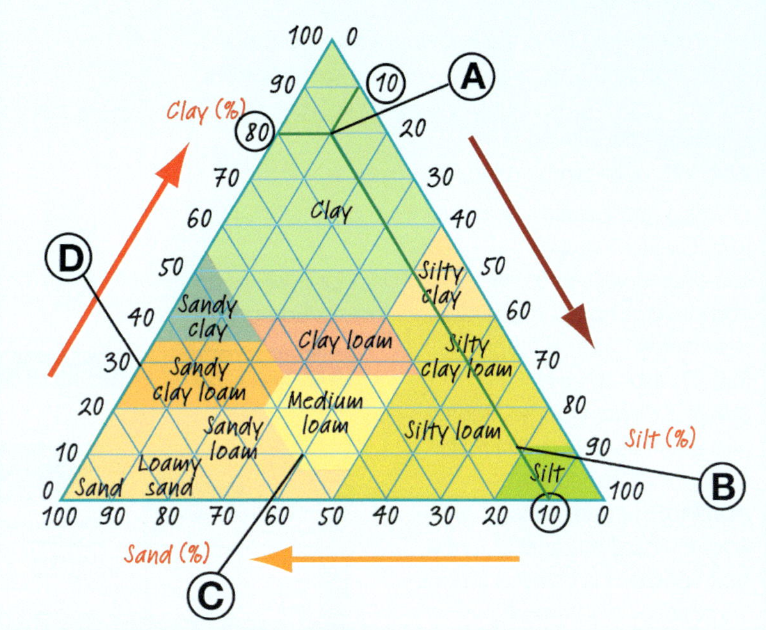

Communicating geographical information

Concept maps

Concept maps show the links between ideas. They often clarify our thinking and show relationships between the cause and the effect of a geographical topic. The following concept map shows world population growth and its links to poverty and affluence.

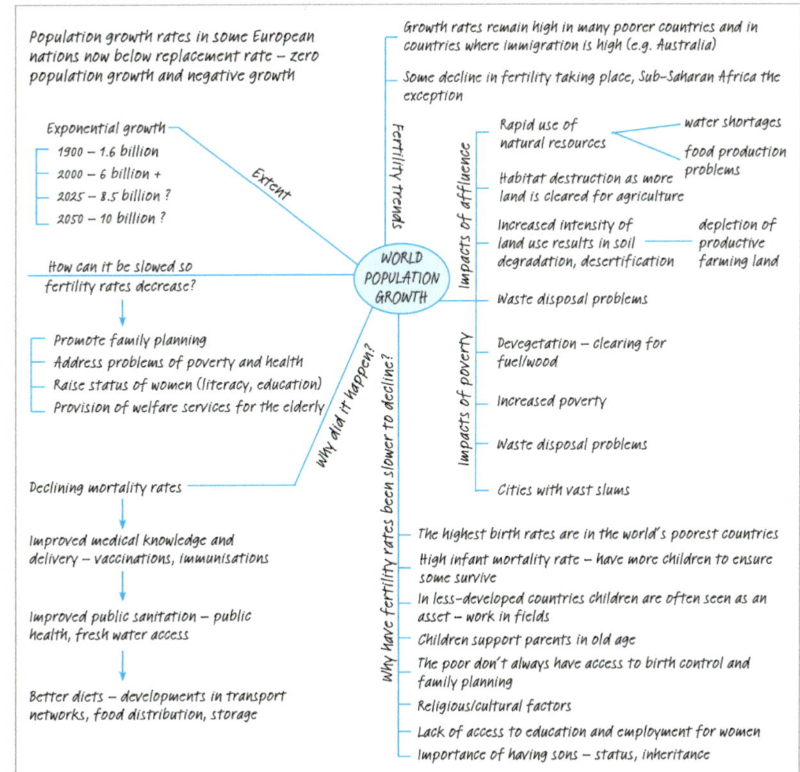

ISBN 978 1 4586 6277 4

Web 2.0 tools

Online presentation tools are known as web 2.0 tools. These tools provide exciting new ways for geographers to research geographical inquiry questions and communicate their findings.

Web 2.0 tools can be used for:

- drawing graphs using statistics from questionnaires (e.g. Create A graph)
- creating surveys and collating the data for analysis and conversion to graphs (e.g. SurveyMonkey, Edmodo)
- developing creative maps (e.g. Scribble Maps)
- creating movies, photo stories and animations (e.g. Animoto, Xtranormal, Tubechop)
- developing presentations with links to visual material that have been personally created (primary sources) or are from other sources such as the internet and magazines (secondary sources) (e.g. Prezi, Glogster EDU).

Creating surveys and collating data

SurveyMonkey

(www.surveymonkey.com):
use SurveyMonkey to create surveys and send the link to your geography class (or survey sample). You will receive the completed surveys online. Results can be presented as graphs or tables.

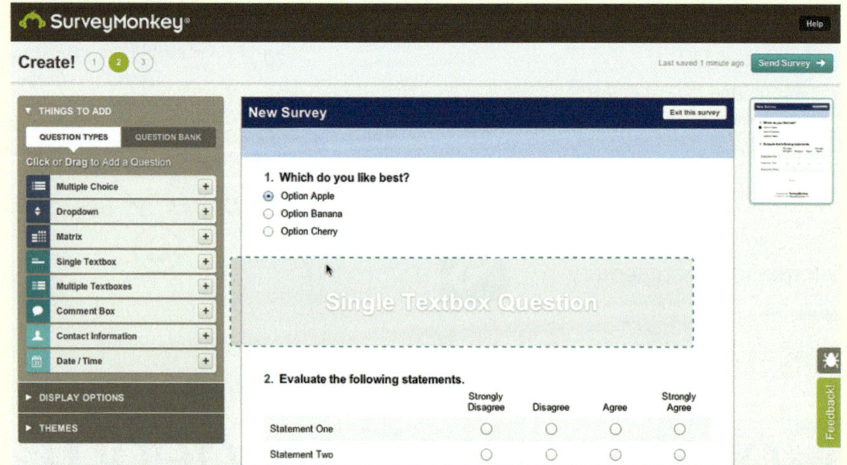

Creating maps

Scribble Maps

(http://scribblemaps.com):
using base maps, you create an outline of places with movable boundaries and labels. You can add travel paths, a legend and photographs of geographical places.

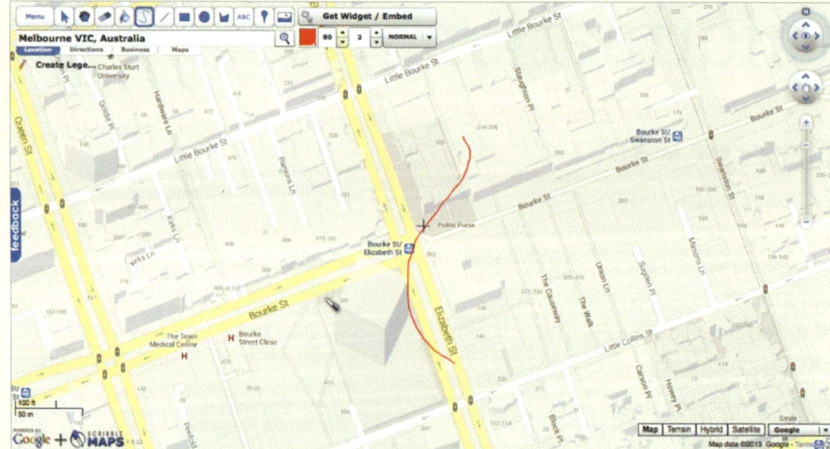

StatPlanet

(www.statsilk.com/software/statplanet):
use StatPlanet to create interactive maps, such as the population map shown below, using geographical data that you import into the program. You can also create interactive graphs and charts.

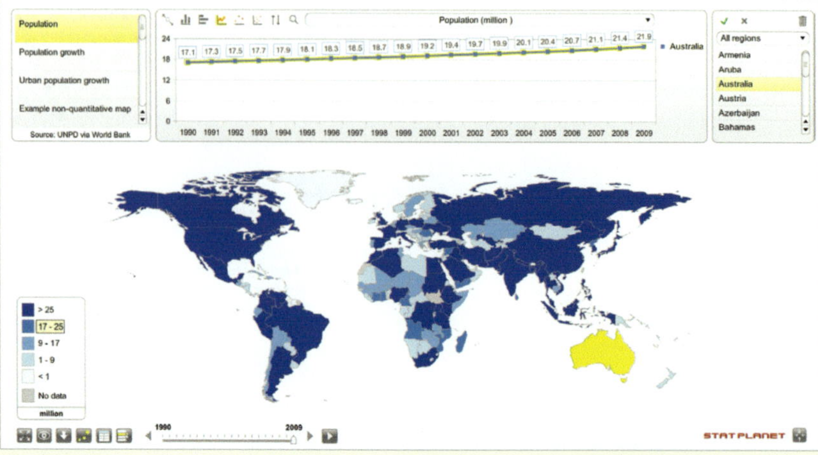

ISBN 978 1 4586 6277 4

Creating presentations

Glogster EDU
(www.glogster.com):
A GLOG is an interactive online poster used to present a variety of geographical information using maps, graphs and active links to information in video clips (e.g. from YouTube) and websites.

Prezi
(http://prezi.com/):
Prezi is a zooming presentation tool that allows you to integrate geographical photo-graphs, Word documents, PDF files, video clips, graphs, tables and other tools into one document.

Creating infographics

Sites such as Visual.ly (http://create.visual.ly/) and Infogr.am (www.infogr.am/) enable you to present information visually and graphically using geographical tools such as maps, graphs, diagrams as well as short pieces of text. Large quantities of information are presented in a small space and in a visually attractive format such as in 'Australia Defined'.

Cartoons

Cartoons often address current geographical issues such as climate change, environmental degradation, human rights abuses and asylum seekers. The power of the cartoon lies in its ability to present complex issues in a simplified form. Cartoons develop critical thinking and initiate classroom discussion and debate. They may inspire students to draw their own cartoons on a selected geographical issue.

See Geolinks for more information about using web 2.0 tools for geography.

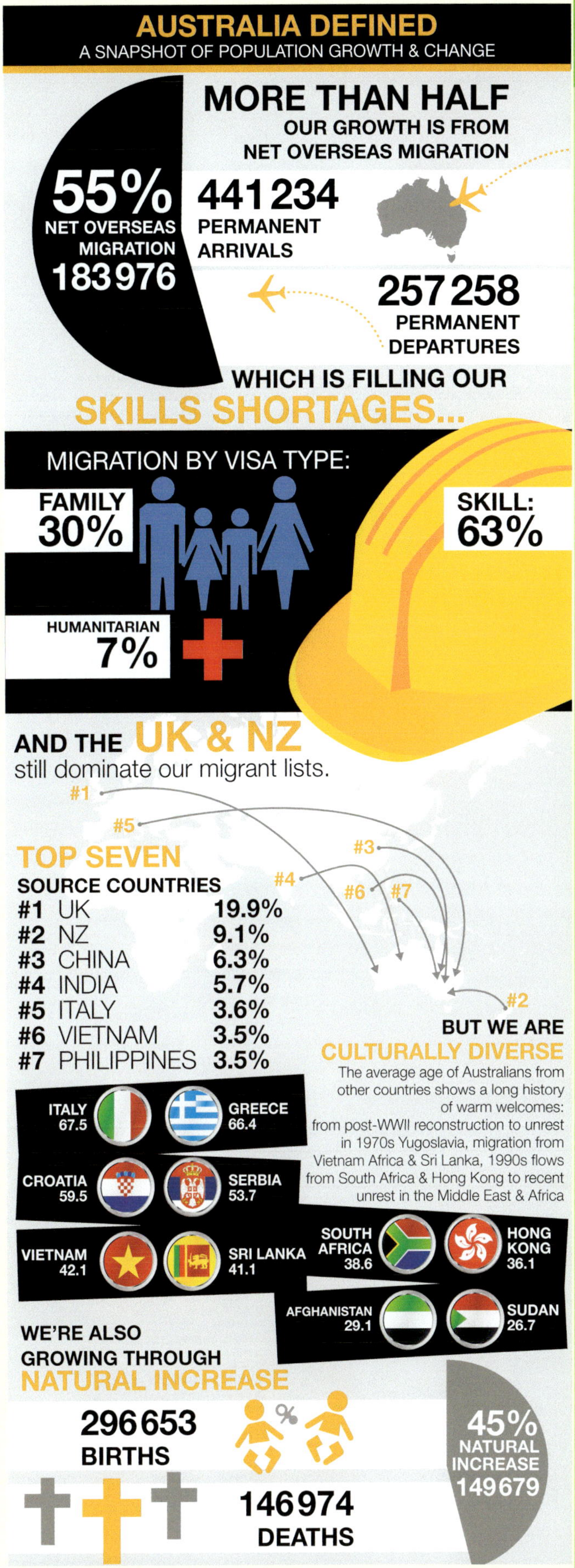

AUSTRALIA DEFINED
A SNAPSHOT OF POPULATION GROWTH & CHANGE

MORE THAN HALF
OUR GROWTH IS FROM NET OVERSEAS MIGRATION

55% NET OVERSEAS MIGRATION **183976**

441234 PERMANENT ARRIVALS

257258 PERMANENT DEPARTURES

WHICH IS FILLING OUR **SKILLS SHORTAGES...**

MIGRATION BY VISA TYPE:
FAMILY 30%
SKILL: 63%
HUMANITARIAN 7%

AND THE **UK & NZ** still dominate our migrant lists.

TOP SEVEN SOURCE COUNTRIES

#1	UK	19.9%
#2	NZ	9.1%
#3	CHINA	6.3%
#4	INDIA	5.7%
#5	ITALY	3.6%
#6	VIETNAM	3.5%
#7	PHILIPPINES	3.5%

BUT WE ARE **CULTURALLY DIVERSE**
The average age of Australians from other countries shows a long history of warm welcomes: from post-WWII reconstruction to unrest in 1970s Yugoslavia, migration from Vietnam Africa & Sri Lanka, 1990s flows from South Africa & Hong Kong to recent unrest in the Middle East & Africa

ITALY 67.5
GREECE 66.4
CROATIA 59.5
SERBIA 53.7
VIETNAM 42.1
SRI LANKA 41.1
SOUTH AFRICA 38.6
HONG KONG 36.1
AFGHANISTAN 29.1
SUDAN 26.7

WE'RE ALSO GROWING THROUGH **NATURAL INCREASE**

296653 BIRTHS

146974 DEATHS

45% NATURAL INCREASE **149679**

ISBN 978 1 4586 6277 4

ISBN 978 1 4586 6277 4

ISBN 978 1 4586 6277 4

Index

ISBN 978 1 4586 6277 4

ISBN 978 1 4586 6277 4

ISBN 978 1 4586 6277 4

ISBN 978 1 4586 6277 4

ISBN 978 1 4586 6277 4

ISBN 978 1 4586 6277 4

ISBN 978 1 4586 6277 4

ISBN 978 1 4586 6277 4

ISBN 978 1 4586 6277 4

ISBN 978 1 4586 6277 4

ISBN 978 1 4586 6277 4

ISBN 978 1 4586 6277 4

ISBN 978 1 4586 6277 4

ISBN 978 1 4586 6277 4

ISBN 978 1 4586 6277 4

ISBN 978 1 4586 6277 4

ISBN 978 1 4586 6277 4

ISBN 978 1 4586 6277 4

Acknowledgements

The author and publisher are grateful to the following for permission to reproduce copyright material:

PHOTOGRAPHS

AAP Images//Mark Thiessen, National Geographic, **23**, **155**; AFP, **182**, **188**, **353**; Alamy/Accent Alaska.com, **216**, /All Canada Photos, **217**, /Elisabeth Blanchet, **237**, /BrazilPhotos.com, **230**, /Dennis Chang – Singapore, **270**, /Directphoto.org, **270**, /DREAMSTOCK, **vi**, /Chris Howes/Wild Places Photography, **137**, /Simon Grosset, **286**, /Images & Stories, **335**, /Brian Jannsen, **270**, /frans lemmens, **83**, /Richard Levine, **357**, /Londonstills.com, **344**, /peter mcdonald, **288**, /Stefano Paterna, **250**, /PhotoStock-Israel, **295**, /Pictorial Press Ltd, **353**, /Valery Rizzo, **357**, /Sampete, **270**, / John Quixley, **161**, /Michael Schmeling, **78**, /Jan Sochor, **343**, /David South, **317**, /Jane Sweeney, **v**, **4**, Alamy/Tetra Images, **80**, /WorldTravel, **285**, /Raymond Warren, **345**; Cartoon courtesy of John Allison, **288**; Animaps, **245**;Associated Press/A.K. Hendratmo, **9**; Aurora Photos/David McLain, **322**; Australia Post, **125**; John Bliss, **53**, **88**, **89**, **93**, **127**, **132**, **133**, **162**; Susan Bliss, **13**, **129**, **255**, **272**; CLTS Knowledge Hub, photo by Petra Bongartz, 2010, **357**; Professor AP Bradbury, Channel Coastal Observatory National Oceanography Centre, **144**; VINCENT CALLEBAUT ARCHITECTURES - WWW.VINCENT.CALLEBAUT.ORG, **219**; Cartoonstock, **232**, **262**; Lorraine Chaffer, **vii**, **32**, **33**, **126**, **245**, **250**, **276**, **277**, **278**, **292**, **301**, **309**, **355**, **360**, **361**; Lee Chapman/Tokyo Times, **232**; Christian Peacemaker Teams, **253**, © Commonwealth of Australia (Geoscience Australia) 2015. This product is released under the Creative Commons Attribution 4.0 International Licence. http://creativecommons.org/licenses/by/4.0/legalcode, **280**; Corbis/Peter Adams/JAI, **69**, /Noah Addis, **303**, / Theo Allofs, **6**, **107**, /Robin Bartholick, **154**, /Marco Becher/dpa, **230**, / Yann Arthus-Bertrand, **24**, **160**, /Bettmann, **320**, /Amit Bhargava, **359**, / Boris Roessler/dpa, **223**, **224**, /Steve Bowman, **154**, /TOLGA BOZOGLU/epa, **285**, /Dordo Brnobic/National Geographic Society, **6**, /Rob Brown/Hedgehog House/Minden Pictures, **84**, /Angelo Cavalli/Purestock/SuperStock, **71**, **107**, /Lloyd Cluff, **114**, /Ashley Cooper, **154**, /Adela Nistora/Demotix, **154**, /Destinations, **71**, /Education Images/UIG, **118**, / Ric Ergenbright, **vii**, /Dave Fleetham/Design Pics, **67**, /Amanda Hall, **349**, / Markus Hanke, **317**, /Robert Francis/Robert Harding World Imagery, **163**, /Patricio Robles Gil/Sierra Madre/Minden Pictures, **119**, /Paul Giamou/Aurora Photos, **134**, Jens Goerlick/FRALMM/Galleries, **136**, /Grand Tour, **99**, /Blaine Harrington III, **vi**, /Gavin Hellier/JAI, **7**, **272**, /Hervé Hughes/Hemis, **154**, /Image Source, **70**, /Imaginechina, **6**, **270**, /Frank Krahmer, **69**, /Ivan Vdovin/JAI, **10**, /Antonio Lacerda/epa, **340**, /Frans Lanting, **67**, **107**, /Xiaoyang Liu, **285**, /Jacob Maentz , **176**,/Eddy Marissen/ Foto Natura/Minden Pictures, **6**, /NASA, **57**, /NASA/epa/, **75**, /Kazuyoshi Nomachi, **7**, **106**, /Ocean, **66**, **99**, /Owaki – Kulla, **134**, Thierry Orban/Sygma, **181**, / POOL, **104**, /Roger Ressmeyer, **75**, /Dr. Richard Roscoe/Visuals Unlimited, **iii**, /Tui De Roy/Minden Pictures, **107**, /Phil Schermeister, **6**, /Jochen Schlenker, **209**, George Steinmetz, **107**, /John Stanmeyer/VII, **321**, / Oliver Strewe, **173**, /Jim Sugar, **57**, /TH-Foto and Jamal Nasrallah, **180**, / Wes Thompson, **6**, /Top Photo Corporation/TopPhoto, **107**, /Christian Franz Tragni, **176**, /Tyrone Turner/National Geographic Society , b**176**, / Penny Tweedie, **77**, /Jim Wark/AgStock Images, **286**, /Micah Wright/First Light, **57**, /Norbert Wu/Science Faction, **6**, **16**, /YONHAP/epa, **138**, /HOW HWEE YOUNG/epa, **353**; CoreLogic, **111**; Crown copyright © Department of Conservation, Te Papa Atawhai, **85**, **211**; Place Laboratory, Curtin University, **366**; Didi Foundation, **205**, **206**, **207**, **208**; Dreamstime.com/Galyna Andrushko, **2**, /Bcbounders, **3**, /Egal, **76**, /Johncarnemolla, **168**, /Jorge Duarte Estevao, **152**, /Joseph Gough, **42**, /Guarino, **338**, / Timhesterphotography, **76**, /Intrepix, **76**, /Jackq, **270**, /Joyfull, b**317**, /Chris Van Lennep, **175**, /Temistocle Lucarelli, **150**, /Lunamarina, **iii**, /Xiye, **15**, / Lorpic99, **21**, /Michaeldb, **315**, /Peter Montgomery, **159**, /Nijethorpe, **260**, / Nomadsoul1, **78**, /Photographerlondon, **184**, /Pniesen, **170**, /Romeo1232, **77**, /Rorem, **270**, /Teenbull, **102**, /Tiplyashina, **315**, /Tottoro, **315**, /Luis2007, **77**, /Vhcreative, **78**, /Vladzetter, **263**, /Kun Yang, **147**; Dr Norman C Duke and MangroveWatch Ltd, **174**;Robert Dun, **37**;Fairfax Photos/Glenn Campbell, **169**, **275**, /SMH, **364**; Flickr: pizzodisevo, **42**; Gainesville Marina, **143**; Getty Images /Timothy Allen, **289**; /Artie Photography (Artie Ng), **iv**, / Ashley Cooper pics, **266**, /Auscape/UIG/, **46**, /Bill Bachman, **76**, /John W Banagan, **42**, Daniel Berehulak, **311**, /Arya Bima/AFP, **221**, /De Agostini/C. Sappa, **73**, /Carsten Peter, **158**, /JEAN CLAUDE-CHAPON/AFP, **229**, / DigitalGlobe/ScapeWare3d, **9**, **304**, /Grant Dixon, **120**, /Greg Elm, **285**, / Manfred Gottschalk, **52**, Manfred Gottschalk, Uluru Kata Tjuta National Park, **123**, /Martin Gray, **156**, /Charles Gullung, **288**, /IMAGEMORE Co, Ltd, **20**, /Steven Kazlowski, **114**, /Daniel Loiselle, **144**, /Jean-Pierre De Mann, **21**, /Kylie McLaughlin, **362**, /G. R. 'Dick' Roberts/NSIL, **17**, /NOAA, **23**, /Randy Olsen, **166**, /Jochen Schlenker, **10**, /Science Photo Library, **60**, /SPL, **76**, / katsumi.takahashi, **119**, /David Wall Photo, **265**, /Peter Walton Photography, **121**, /RAVEENDRAN/AFP, **325**; GlogsterEDU, **245**; Google Tour Builder, **279**; Jeppe Hein Appearing Rooms Forrest Place, 2012, Courtesy Johann König, Berlin and 303 Gallery, New York. Photo by Jeppe Hein., **363**; Benjamin D. Hennig, Worldmapper Project, www.viewsoftheworld.net, **318**, **319**; IDE.fr, **338**; iStockphoto /boggy22, **203**, /francois-rouxm **247**, / oversnap, **8**, /AlexSava, **9**, /gniedzieska, **10**, /RobertoGennaro, **236**; Jenolan Caves, **14**, **18**, **19**; London Pentagram, **300**; POLYGRAM/AUSTRALIAN FILM FINANCE/THE KOBAL COLLECTION/LOCKWOOD, ELISE, **125**; MapsForFree.com/Hans Braxmeier, **77**; Marine National Facility, **8**; David McCandless, **267**; Millenium House/Jannine Doyle, **80**; © Pascale Zintzen/MSF, **260**; National Geographic, **92**; National Museum of Australia, **172**; naturepl.com/Anup Shah, **82**; Newspix/Jason Busch, **71**, Brendan Esposito, **180**; Tom Pfeiffer/www.VolcanoDiscovery.com, **230**; Tropical Islands Hotel, Germany, **247**; Poster by Dany Pepin, **226**; Picture Media/Philimon Bulawayo, **252**, /Nacho Doce, **270**, /Jose Miguel Gomez, **343**, /Omar Ibrahim, **329**, /Reuters/MTSAT/JMA, **81**, /Yuriko Nakao, **203**, /Reuters, **209**; Cartoon by Punch Limited, **288**; Image courtesy of Satellite Imaging Corporation, **100**; Cartoon by Seppo, **228**, **229**; Shutterstock/aquapix, **144**, /A_Sh, **134**, Syaheir Azizan, **218**, /Pierre-Yves Babelon, **99**, /Artur Bogacki, **140**, /Dan Breckwoldt, **164**, /R.A.R. de Bruijn Holding BV, **iv**, /Costazzurra, **79**, /Emi Cristea, **135**, /Darrenp, **210**, /Distinctive Images, **322**, /Don Fink, **113**, /Fotos593, **202**, /GTS Production, **168**, /Josef Hanus, **264**, /Anton_Ivanov, **157**, /lev radin, **270**, /Iv Nikolny, **254**, GOOD, **254**, /KeikR, **176**, /fuyu liu, **20**, /Jose Marques, **315**, /meunierd, **302**, /Photopictures, **156**, /Razvy, **235**, /Jane Rix, **24**, /rvlsoft, **241**, /saiko3p, **251**, /Michael Schmeling, **238**, / Irina Schmidt, **309**, /Space Factory, **285**, /S-F, **308**, /John T Takai, **315**, / think4photop, **50**, /travelight, **124**, /Tupungato, **243**, /worldswildlifewonders, **167**, /Igor Zh, **v**; Cartoon Andy Singer, www.andysinger.com, **239**; SPIEGEL ONLINE International, **222**; Stairs Studio Inc., **244**; StatSilk (2015). StatPlanet: Interactive Data Visualization and Mapping Software. www.statsilk.com, **245**; Torfaen County Borough Council and Michael Blackmore, **137**; © UNICEF/NYHQ1997-0095/DONNA DECESARE, **226**; Andrew Watkins, **151**; Wikimedia Commons/Sémhur, **35**, /Strebe, **78**, /Strebe (using NASA imagery and Geocart map projection software), **79**; Wikipedia/Atmospheric Radiation Measurement Program, **153**, /Frost Design, **165**; © 2008 Kent Williams, **164**; Jim Woodmencey, **201**.

OTHER MATERIAL

Quote from *Mixed views aired over Great Ocean Road rezoning*, ABC News, 13 November 2012, http://www.abc.net.au/news/2012-11-13/mixed-views-aired-over-great-ocean-rd-rezoning/4368808, **42**; Extract from 'Bush burns ease global warming' by Cheryl Jones, *The Australian* 29 April 2009, **167**; Image Australian National Maritime Museum, **171**; Graphic based on Barrett DJ, Couch CA, Metcalfe DJ, Lytton L, Adhikary DP and Schmidt RK (2013) Methodology for bioregional assessments of the impacts of coal seam gas and coal mining development on water resources. A report prepared for the Independent Expert Scientific Committee on Coal Seam Gas and Large Coal Mining Development through the Department of the Environment, **154**; Quote from *Mountains from Space: Peaks and Ranges of the Seven Continents* by Stefan Dech et all, Harry N Abrams 2005, **4**; Infographic Graham Brookman 'Design for Life – Permaculture – The Food Forest Story' (Movie - The Food Forest 2011), **367**; Image Cairns Airport, **281**; Data from CanadaFAQ.ca, **297**, Infographic Centre for Indigenous Environmental Resources (CIER), **274**; Data © Commonwealth of Australia (Geoscience Australia) 2013; **194**; Commonwealth of Australia, Creating Places for People: an urban design protocol for Australian cities, http://www.urbandesign.gov.au, Department of Infrastructure and Regional Development, Creative Commons (CC) Attribution 3.0 Australia licence,

ISBN 978 1 4586 6277 4

368; Graph CopperBridge Media, **326**;Graphic Jessamy Gee - Think in Colour, **364**; *Keys to Geography*, 2nd edition, Australian Geography Teachers Association, Macmillan Education Australia, 2010, pp.50-1, images reproduced by permission of Australian Geography Teachers Association, **39**; Text Geo-Mexico.com, **327**; Graphic, **350**; copyright Herbie Girardet/ Rick Lawrence, Data courtesy CCICED, **52**; Data from DEEWR, Australian Jobs 2012, **262**; Infographic Guardian News & Media Ltd 2012, photo © Jose Fuste Raga/Corbis, **298**; Cartoon by Andrzej Krauze, **199**; Graphic mhealthafrica.com, **348**; NCHRP Web-Only Document 184: Going the Distance Together: A Citizen's Guide to Context Sensitive Solutions for Better Transportation, reproduced with permission, **258**, Data from Russell-Smith, Jeremy et al. (2007). "Bushfire 'down under': patterns and implications of contemporary Australian landscape burning." In: *International Journal of Wildland Fire* 16(4): 361-377. Copyright © International Association of Wildland Fire USA 2007. Published by CSIRO PUBLISHING, Collingwood, Victoria Australia. Reproduced with permission. http://www.publish.csiro.au/nid/114/paper/WF07018.htm, **97**; Morphy, H. and Morphy F., 'Tasting the Waters: Discriminating Identities in the Waters of Blue Mud Bay', *Journal of Material Culture*, Vol. 11, No. 1-2, 67-85 (2006), SAGE Publications, **171**; Smith, Ashley W, 1948-. We've got to either (A) Let more people into N.Z. or (B) Have more babies. MG business - mercantile gazette, 9 December 2002. Ref: DX-023-130. Alexander Turnbull Library, Wellington, New Zealand. http://natlib.govt.nz/records/22740493, **322**; Mara Sofferin, **348**; Michael Sorkin Studio,Terreform, **351**; Dorothy Tang, **151**, Diagram by Dennis Tasa, **64**; Infographic ©UNICEF/NYHQ2012/socialandcivicmedia/ oliviermarie, **268**; Infographic University of Adelaide, **289**; Infographic Ben VanderVeen, **234**.

The authors and publisher would like to acknowledge the following:

Infographic The Amish Village, Lancaster County, PA, www.theamishvillage.net, **235**; Architectus, **352**; Data Australian Bureau of Statistics, **279**; Australian Tourist Publications, **280**; Infographic, Borderzine, **257**; BOP Consulting, 2012, World Cities Culture Report, Greater London Authority, London, August 2012, **269**; Facebook/Paul Butler, **80**; Map by Matthew Coller, **173**; Extract from 'The waste in Tasmania's forests: most timber left to rot' by Andrew Macintosh and Richard Denniss, Crikey, Friday 14 December 2012, **129**; Image by Nicolle R. Fuller, Sayo-Art LLC with permission of W. H. Freeman and Company, New York, **306**; Quote from 'Go forth and multiply a lot less', The Economist, Oct 29th, 2009, **320**, Data from © 2013 Gallup, Inc., **289**; Get Up!, **104**, **155**; GOOD, **240**, **346**; Google Maps, **76**; GRID-Arendal/Lawrence Hislop, **286**; M. Norton-Griffiths, D Herlocker, L Pennycuick, African 'The patterns of rainfall in the Serengeti ecosystem, Tanzania', *Journal of Ecology*, vol. 13, issue 3–4, pp 347–74, Dec 1975 (pub'd online 29/4/08), **92**; Copyright © 2012-2013 Impercar, **103**; Images Hunter Development Corporation, **354**; Map based on Journal of Alpine Research, rga.revues.org, **159**; The Japan News, **222**; © European Union 2010, **311**; Diagram based on Haiti Humanitarian Response info graphic created by William Pitzer, Infoartz.com, for the Knight Foundation, released under Creative Commons License 2011, http://cdn.theatlantic.com/static/mt/assets/science/haititech.jpg, **192**; JPALMZ, **81**; Tobias Karlhuber, **139**; The timing of the seasons applies generally to the northeast Arnhem Land coastal region. This diagram has been compiled by Katie Hayne in consultation with Dr Raymattja Marika. The information was modified from data collected by Markus Barber through direct observation and conversations with Djambawa Marawili and Ngulpurr Marawili at Blue Mud Bay. The circular diagram is based on a diagram by Pat Faulkner. See: Barber, M. (2005) "Where the Clouds Stand: Australian Aboriginal Relationships to Water, Place, and the Marine Environment in Blue Mud Bay, Northern Territory" PhD Thesis, The Australian National University, p90., **265**; Image Masdar City, **339**, NASA GSFC Landsat/LDCM EPO Team, **35**; NASA, **49**, **185**, **191**; NASA Earth Observatory, **51**; M. E. J. Newman, **81**; The Philips Center for Health and Well-being, **243**; © Photonics Research Group -Ugent-IMEC 2009, **80**; Graph Population Reference Bureau and National Coordinating Agency for Population and Development, Kenya Population Data Sheet 2011 (Washington, DC: PRB, 2011), **323**; RIA Novosti, **103**, **141**,**200**, **219**, **256**, **330**; Image by Ferdi Rizkiyanto, **310**; UCAR/COMET, **216**; Infographic United Nations, **183**; USAID (50.usaid.gov/learning-out-of-poverty), **331**; WeatherOnline, **247**; Willard Worden circa 1910, Courtesy Greg Garr Collection, **136**; WorldClimate (www.worldclimate.com), **94**.